Design-Process-Technology Co-optimization for Manufacturability IX

John L. Sturtevant
Luigi Capodieci
Editors

25–26 February 2015
San Jose, California, United States

Sponsored by
SPIE

Cosponsored by
Hitachi High Technologies America, Inc. (United States)

Published by
SPIE

Volume 9427

Proceedings of SPIE 0277-786X, V. 9427

SPIE is an international society advancing an interdisciplinary approach to the science and application of light.

Please use the following format to cite material from this book:
Author(s), "Title of Paper," in *Design-Process-Technology Co-optimization for Manufacturability IX*, edited by John L. Sturtevant, Luigi Capodieci, Proceedings of SPIE Vol. 9427 (SPIE, Bellingham, WA, 2015) Article CID Number.

ISSN: 0277-786X
ISBN: 9781628415292

Published by
SPIE
P.O. Box 10, Bellingham, Washington 98227-0010 USA
Telephone +1 360 676 3290 (Pacific Time) · Fax +1 360 647 1445
SPIE.org

Printed in the United States of America.

Publication of record for individual papers is online in the SPIE Digital Library.

SPIEDigitalLibrary.org

Paper Numbering: Proceedings of SPIE follow an e-First publication model, with papers published first online and then in print. Papers are published as they are submitted and meet publication criteria. A unique citation identifier (CID) number is assigned to each article at the time of the first publication. Utilization of CIDs allows articles to be fully citable as soon as they are published online, and connects the same identifier to all online, print, and electronic versions of the publication. SPIE uses a six-digit CID article numbering system in which:

- The first four digits correspond to the SPIE volume number.
- The last two digits indicate publication order within the volume using a Base 36 numbering system employing both numerals and letters. These two-number sets start with 00, 01, 02, 03, 04, 05, 06, 07, 08, 09, 0A, 0B … 0Z, followed by 10-1Z, 20-2Z, etc.

The CID Number appears on each page of the manuscript. The complete citation is used on the first page, and an abbreviated version on subsequent pages.

Contents

POSTER SESSION

Authors

Numbers in the index correspond to the last two digits of the six-digit citation identifier (CID) article numbering system used in Proceedings of SPIE. The first four digits reflect the volume number. Base 36 numbering is employed for the last two digits and indicates the order of articles within the volume. Numbers start with 00, 01, 02, 03, 04, 05, 06, 07, 08, 09, 0A, 0B...0Z, followed by 10-1Z, 20-2Z, etc.

Conference Committee

Symposium Chair

Mircea V. Dusa, ASML US, Inc. (United States)

Symposium Co-chair

Bruce W. Smith, Rochester Institute of Technology (United States)

Conference Chair

John L. Sturtevant, Mentor Graphics Corporation (United States)

Conference Co-chair

Luigi Capodieci, GLOBALFOUNDRIES Inc. (United States)

Conference Program Committee

Robert Aitken, ARM, Inc. (United States)
Jason P. Cain, Advanced Micro Devices, Inc. (United States)
Fang-Cheng Chang, Cadence Design Systems, Inc. (United States)
Lars W. Liebmann, IBM Corporation (United States)
Ru-Gun Liu, Taiwan Semiconductor Manufacturing Company Ltd. (Taiwan)
Mark E. Mason, Texas Instruments Inc. (United States)
Andrew R. Neureuther, University of California, Berkeley (United States)
Shigeki Nojima, Toshiba Corporation (Japan)
David Z. Pan, The University of Texas at Austin (United States)
Chul-Hong Park, SAMSUNG Electronics Company, Ltd. (Korea, Republic of)
Michael L. Rieger, Synopsys, Inc. (United States)
Vivek K. Singh, Intel Corporation (United States)
Chi-Min Yuan, Freescale Semiconductor, Inc. (United States)

Session Chairs

1 Invited Session I
John L. Sturtevant, Mentor Graphics Corporation (United States)
Luigi Capodieci, GLOBALFOUNDRIES Inc. (United States)

2 Layout Patterns Applications
John L. Sturtevant, Mentor Graphics Corporation (United States)
Luigi Capodieci, GLOBALFOUNDRIES Inc. (United States)

3 Multipatterning
Lars W. Liebmann, IBM Corporation (United States)
Shigeki Nojima, Toshiba Corporation (Japan)

4 Invited Session II
Robert Aitken, ARM, Inc. (United States)
Michael L. Rieger, Synopsys, Inc. (United States)

5 Layout Optimization and Verification I
Robert Aitken, ARM, Inc. (United States)
Michael L. Rieger, Synopsys, Inc. (United States)

6 Design Interaction with Metrology: Joint Session with Conference 9424
Alexander Starikov, I&I Consulting (United States)
Jason P. Cain, Advanced Micro Devices, Inc. (United States)

7 DFM (Design and Litho Optimization): Joint Session with Conference 9426
Jongwook Kye, GLOBALFOUNDRIES Inc. (United States)
Andrew R. Neureuther, University of California, Berkeley (United States)

8 Invited Session III
Chi-Min Yuan, Freescale Semiconductor, Inc. (United States)
Ru-Gun Liu, Taiwan Semiconductor Manufacturing Company, Ltd. (Taiwan)

9 Circuit Variability
Chi-Min Yuan, Freescale Semiconductor, Inc. (United States)
Hsu-Ting Huang, Taiwan Semiconductor Manufacturing Company, Ltd. (United States)

10 DSA Design for Manufacturability: Joint Session with Conferences 9423 and 9426
Michael A. Guillorn, IBM Thomas J. Watson Research Center (United States)
Sachiko Kobayashi, Toshiba Corporation (Japan)
Vivek K. Singh, Intel Corporation (United States)

11 Layout and Optimization and Verification II
Luigi Capodieci, GLOBALFOUNDRIES Inc. (United States)
Chul-Hong Park, SAMSUNG Electronics Company, Ltd. (Korea, Republic of)
David Z. Pan, The University of Texas at Austin (United States)

The daunting complexity of scaling to 7NM without EUV: Pushing DTCO to the extreme

Lars Liebmann[a], Albert Chu[a], Paul Gutwin[b]
[a] IBM, Hopewell Junction, NY; [b] Cadence, San Jose, CA

ABSTRACT

This paper reviews the most critical components of a 'holistic' DTCO flow for an advanced technology node and in doing so quantifies the differences between 7nm technology node definitions implemented with extreme ultraviolet and 193nm immersion lithography. The DTCO topics covered include: setting scaling targets for critical pitches, gear-ratios, and cell height; defining a set of patterning solutions, required RET restrictions, and resulting patterning cost; compiling physical design objectives to achieve power, performance, and area scaling; developing a set of standard cell logic cell architectures; and finally assessing achievable cell-level as well as macro-level scaling.

Keywords: Design technology co-optimization, standard cell logic design, cell architecture optimization, gear-ratio definition, cell-abstractions

1. INTRODUCTION

The history of semiconductor lithography has shown repeatedly that, given a choice, most of the microelectronics industry will adopt shorter wavelength or higher numerical aperture (NA) patterning solutions over resolution enhancement technique (RET) extended incumbent solutions in an attempt to maintain a constant level of design and process complexity. Even though frequency doubling RET such as off-axis-illumination (OAI) and alternating-phase-shifted-mask (altPSM) were already demonstrated in 365nm wavelength lithography [1], the industry took the path of hardware enabled scaling until the lack of options beyond 193nm wavelength immersion lithography (193i) forced the adoption of increasingly complex and design-invasive RET as shown in figure 1. The 14nm technology node introduced the process and design communities to the complexity of double patterning. Then the 10nm node increased the level of pain further by forcing triple patterning and sidewall image transfer solutions. While the end of scaling has been predicted and proven wrong many times over the 65 year history of VLSI, there is now a heightened sense that the escalation in complexity and the associated cost, on both the design and process sides of the technology, can not continue at the rate observed in the last three technology nodes. Against this backdrop of staggering increases in patterning, process, and design complexity, prominent semiconductor manufacturers, on the IDM as well as the foundry side of the business, have proclaimed that at least one more cycle of scaling, the 7nm technology node, is feasible without reliance on wavelength-enabled scaling afforded by extreme ultraviolet (EUV) lithography.

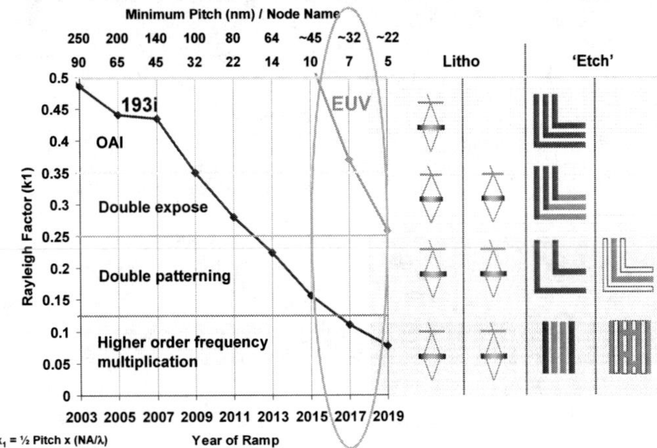

Figure 1, Scaling deeper into the 'sub-resolution' domain forces the introduction of increasingly complex and design-restrictive resolution enhancement techniques. In the 7nm node (circled), EUV has the potential to offer a major reset in patterning-induced design complexity.

Design-Process-Technology Co-optimization for Manufacturability IX, edited by John L. Sturtevant, Luigi Capodieci, Proc. of SPIE Vol. 9427, 942702 · © 2015 SPIE · CCC code: 0277-786X/15/$18 · doi: 10.1117/12.2175509

The rapidly expanding impact of patterning dominated design restrictions forced ever closer and earlier collaboration between designers, design tool providers, and process engineers in advanced technology nodes to achieve what is now known as design technology co-optimization (DTCO). While older technology nodes allowed process and design development to occur in parallel against design rules that were simply scaled by 70% from the previous node following a well established roadmap, advanced technology nodes require a more 'holistic' approach. To maintain acceptable power-performance-area-cost (PPAC) scaling, the technology architecture definition consisting of: primary feature pitches (e.g. wiring pitch, contacted poly pitch, and the fin pitch in finFET), connectivity information, cell architectures for logic and memory, and routing stacks for hierarchical wiring have to be carefully co-optimized alongside methodology rules and design tool specifications to enable critical elements of the physical design flow.

To illustrate key DTCO concepts, this paper focuses on the early definition of a potential 7nm technology node implementation. In the following chapters, it details five phases of a DTCO flow to demonstrate how this approach can be used to quantify the differences between 7nm node definitions based on either EUV or 193i patterning. While the phases of DTCO are presented sequentially here, it is important to stress the iterative nature of the DTCO process, every detail of the technology architecture definition is subject to continuous refinement throughout the optimization process leading to incremental or sometimes rather fundamental improvements in the technology definition over time. It is also important to mention that, while the work captured in this paper focuses on standard cell logic optimization, memory and analogue designs simultaneously go through the same DTCO process.

Specifically: after establishing pitch, gear-ratio, and cell height scaling targets for the sample 7nm node; the paper will compare level-specific RET-driven design restrictions and patterning cost; review design objectives for power, performance, and area scaling; contrast 193i and EUV compatible cell architectures; and finally assess achievable scaling at both the cell-level and the routed block-level.

2. SCALING TARGETS

The DTCO process starts by establishing dimensional targets for critical feature pitches such as the contacted poly pitch (i.e. the transistor gate pitch), wiring pitch, and the pitch of fins used in finFET. For the most part, these scaling targets are still set by basic $.5^{\frac{1}{2}}$ numerical scaling from the previous node, but DTCO has to weigh the competitive pressure to maintain an aggressive scaling roadmap against avoiding 'scaling cliffs' where small gains in pitch reduction come at a disproportionate increase in cost, complexity, or risk. The dimensions used herein for the four most important pitches in the technology definition are shown in table 1. Further, manufacturability, device performance, and patterning considerations in advanced technology nodes forces the use of restricted design rules [2] which limit the range of pitches available in each critical design level. The resulting limited set of available design grids necessitates that the numeric values chosen for the critical pitches form useable gear-ratios for the range of cell heights that are needed to design the products for which the technology definition is targeted. As illustrated in table 2, this phase of the DTCO process also defines an initial target for the smallest cell size that should be achievable with the technology definition. For the purpose of the 193i to EUV 7nm comparison we therefore set the scaling target as a 7.5T cell image with a metal pitch of 32nm, a contacted poly pitch of 40nm, and a set of devices using 3 active fins each.

Scaling Targets		Reasoning
Level	Pitch	
1x Metal (beol)	32nm	== 50% of beol pitch in 14/16NM node
Fins (fin)	24nm	== 3/4 of beol pitch 7.5T = 10 fins, 9T = 12 fins, 10T = 12 fins ... at uniform fin pitch
Contacted poly (cpp)	40nm	== 5/4 of beol pitch less critical gear ratio but good ballpark estimate
Vias, contacts (via)	45nm	== diagonal of beol Needed for wiring efficiency

Table 1, Hypothetical but realistic values for first level metal pitch, fin pitch, contacted poly pitch, and via pitch along with an explanation of the reasoning why the specific values where chosen for this work.

Cell Height Options based on Gear Ratio Compliance							
Cell Height	Metal Tracks	3	4.5	6	7.5	9	10.5
	Total Fins	4	6	8	10	12	14
Fin Efficiency	'Dummy' Fins	4	4	4	4	4	4
	Active Fins	0	1+1	2+2	3+3	4+4	5+5

Table 2, A metal to fin pitch gear ratio of 4:3 increases fin count by 2 for every 1.5 metal track increases in cell height. Based on the number of useable wiring tracks and number of active fins, the 7.5 track cell height is deemed the most aggressive scaling target that should be assessed for feasibility.

3. COMPARISON OF RET IMPLICATIONS

Based on an understanding of feature-specific resolution limits, RET-specific geometry constraints, and level-specific topology needs, patterning engineers work with designers to select appropriate patterning solutions for each critical design level. Working through multiple sets of patterning options to establish an outlook on design impact and patterning cost for various technology definitions is a necessary part of the DTCO process and not just an academic exercise done for this paper. For the purpose of this specific work, a sample 193i and a EUV solution have been defined in table 3.

			193i Litho		EUV Litho	
Level	Stack	Min. Pitch	RET	Restrict.	RET	Restrict.
Fin	1	24	SAQP	++++	SADP	++++
Poly	1	40	SADP	++++	SE	+
Tungsten Strap	1	40	LELE	+++	SE	+
Local Interconnect	1	40	LELELE	+++	SE	+
1x Metal	3	32	SAQP	++++	SE	++
1x Via	3	45	LELELE	+	SE	+
1.5x Metal	2	48	SADP	+++	SE	+
1.5x Via	2	68	LELE	++	SE	+
2x Metal	2	64	LELE	++	SE	+
2x Via	2	90	LELE	++	SE	+

Table 3, Critical design levels used in this sample technology definition, critical feature pitches, resolution enhancement techniques for each level for 193i and EUV patterning, and a qualitative assessment of patterning-driven design restrictions: + Forbidden pitch; ++ Preferred orientation (limited, restricted wrong way); +++ Unidirectional, limited pitches; ++++ Unidirectional grating

The RET information outlined in table 3 along with a relative cost multiplier for each RET, table 4a, can be used to gain first insights into how the two patterning approaches compare on cost as shown in table 4b. The similar cost of the two patterning solutions, certainly well within the uncertainty limits of this early estimation, highlights the challenge that an effective cost comparison requires a substantial amount of process detail and runs the risk of achieving useable levels of accuracy well after fundamental process choices (such as exposure tool purchases) had to be made. Significant also is the difference in design restrictions, qualitatively captured in table 3. Efficiently approximating the circuit-level design impact of these qualitative restrictions is essentially the goal of the remainder of the work presented here.

Normalized Wafer Cost	
193i SE	1
193i LELE	2.5
193i LELELE	3.5
193i SADP	3
193i SAQP	4.5
EUV SE	3
EUV SADP	8

			193i Litho		EUV Litho	
Level	#	Pitch	RET	Cost	RET	Cost
Fin	1	24	SAQP	4.5	SADP	8
Poly	1	40	SADP	3	SE	3
W Str.	1	40	LELE	2.5	SE	3
Loc. Int.	1	40	LELELE	3.5	SE	3
1x Metal	3	32	SAQP	13.5	SE	0
1x Via	3	45	LELELE	10.5	SE	9
1.5x Met.	2	48	SADP	6	SE	6
1.5x Via	2	68	LELE	5	SE	6
2x Metal	2	64	LELE	5	SE	6
2x Via	2	90	LELE	5	SE	6
			Total	58.5	Total	59

Table 4, a) (left) Estimate of relative patterning cost for various exposure solutions normalized to 193i single exposure lithography b) (right) cost comparison of the two hypothetical but realistic lithography stacks assumed in this work

4. OBJECTIVES FOR POWER, PERFORMANCE, AND AREA SCALING

While the lithography community often sees pitch reduction as the primary scaling objective, many elements affect to what degree actual designs can take advantage of the lithography enabled pitch scaling. A detailed assessment of node-to-node circuit performance uplift at constant power ventures into very sensitive details of the technology specification and process implementation and is outside the scope of this paper. However, general concepts that affect chip area, circuit performance, and overall technology competitiveness, as outlined in table 5, are well established and form the basis for a complex set of topological design objectives that guide the DTCO process.

Elements of power-performance-area (PPA) scaling:		
Area	Cell area	Width: number of poly tracks per logic function
	Placement	Allow arbitrary abutment
		Allow cells to be mirrored in X and Y
	Routing	Allow good pin access (≥3 hit-points per pin)
		Allow critical via constructs (diagonal of cpp and beol)
		Maintain good wiring efficiency (jogs, wire widths)
		Scale wire density of full stack (tracks per um)
Power Performance	Device width	Fin granularity (need multiple device widths abutted)
		Cell height granularity (need multiple cell heights)
	Wiring	Low resistance power rails
		Efficient transistor wiring (reduce via count, wire length)

Table 5, A compilation of topological layout goals related to power, performance, area, and cost scaling

5. CELL ARCHITECTURE COMPARISON

In addition to memory cells (primarily SRAM) the vehicle that churns the technology definition and drives the DTCO process in advanced technology nodes is a set of standard logic cells. Rather than starting with design rules that, after much work, designers ultimately determine to be inadequate against the scaling targets and design objectives, the DTCO process starts with sample logic cell architectures that help communicate geometric, structural, and connectivity needs as well as process constraints and yield limiters.

The iterative process of cell architecture optimization is illustrated here based on two standard logic cells, an and-or-invert (AOI) and a 2 finger inverter (INV). As illustrated in figure 2 on the left, the AOI is used for this exercise because it has high wiring density: 3 input pins, a bidirectional output pin, and an internal signal connection make this cell a good canary for efficient cell level wiring. The 2 finger inverter, figure 2 on the right, is used here because it is a very compact example of a cell in which neighboring poly tracks need to be strapped together to form a multi finger device. Input pins are labeled A-C and the output pins are labeled 'Out'. In addition to the power rails (labeled Vdd and Ground) at the horizontal cell boundaries, the M2 wiring tracks on which the router can connect signal wires to the cell are also shown. The intersection of the pins and the M2 tracks are potential pin access points. Even though only one connection has to be made to each pin, having multiple pin access points is very desirable to facilitate efficient routing as indicated in table 5.

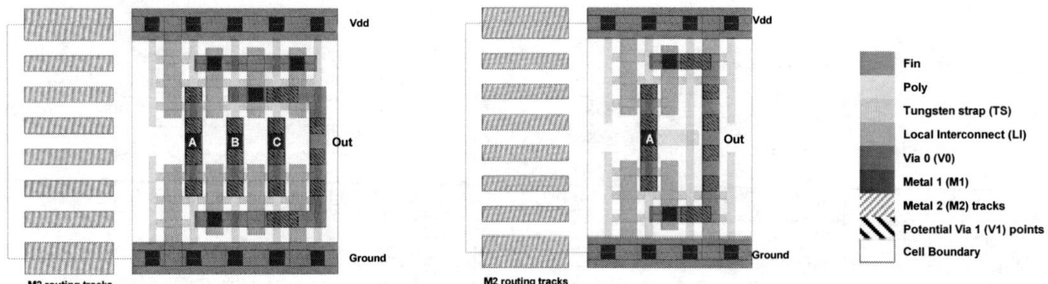

Figure 2, And-or-invert (left) and 2 finger inverter (right) are used herein to illustrate cell architecture optimization.

Figure 3 shows three iterations of cell architecture optimization for the AOI and INV. The top image shows a very conventional cell architecture as one might see in earlier finFET technology nodes. This cell architecture is popular due to several desirable attributes: a 2x wide M1 power rail shared with the neighboring cell provides lower resistance and higher resilience to electromigration failures, 6 M2 signal tracks help alleviate wire congestion in this compact 7.5T cell image, three or more access points per pin aid in efficient routing, an output pin in which the horizontal legs align with the M2 tracks provide potential to make this high current connection using a bar shaped via to lower resistance, and the local interconnect is used to efficiently strap poly fingers together without blocking any metal tracks. Unfortunately, as the design-rule-arc analysis in figure 4 illustrates, this cell architecture is not manufacturable at the 7.5T cell height. Design-rule-arc analysis is an important component of DTCO that helps communicate design pinch-points to the process engineers or, as in this case, identifies layout configurations that are simply not going to work. A design-rule-arc is formed by a string of shape-interactions that have to fit into a confined space of known size. The particular example shown in figure 4 describes a design-rule-arc that spans from the center of the cell to its horizontal border and is therefore 7.5 beol half-pitches long. Adding up all the design rules that contribute to this arc, one can clearly see that the required space far exceeds the available space. Furthermore, most of the components of this particular design-rule-arc are simple derivatives of the technology scaling targets (e.g. wire width, wire space, via size all equal ½ wire pitch) leaving only the metal past via extension and metal tip-to-side rules to absorb all the negative slack in this arc. Often a small amount of negative slack can be spread across several design rules, providing a very quantitative way to negotiate tighter process targets for high value design-rule-arcs. But in this case it is clear that this design-rule-arc is broken beyond negotiation and a new cell architecture needs to be explored, leading to the second cell image in Figure 3.

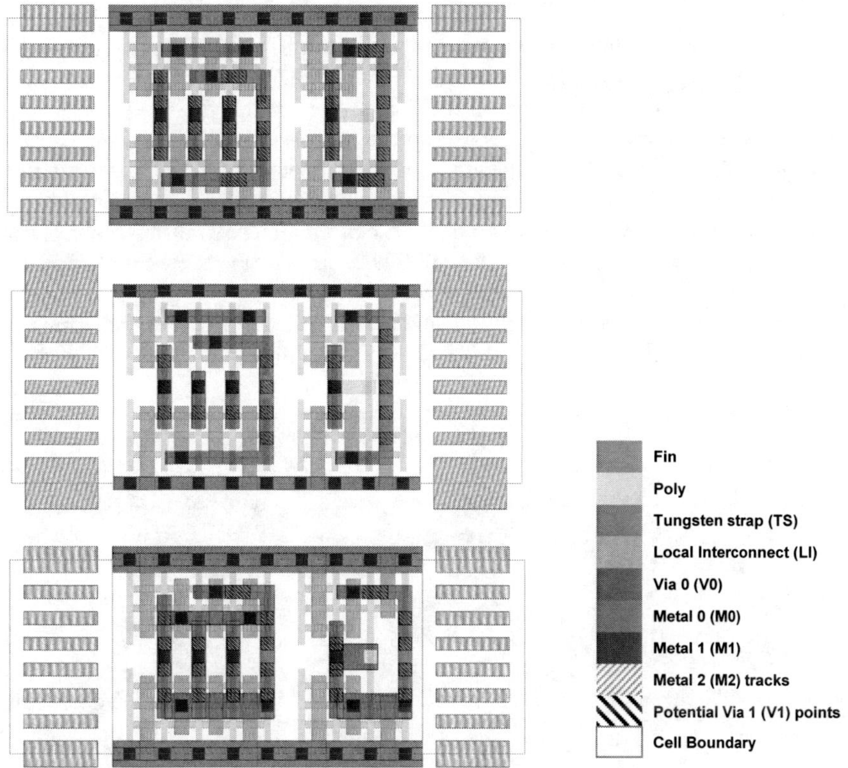

Figure 3, Three iterations of cell architecture optimization illustrating that, even with EUV, a second metal level is required to achieve a manufacturable 7.5T cell image in the 7nm node.

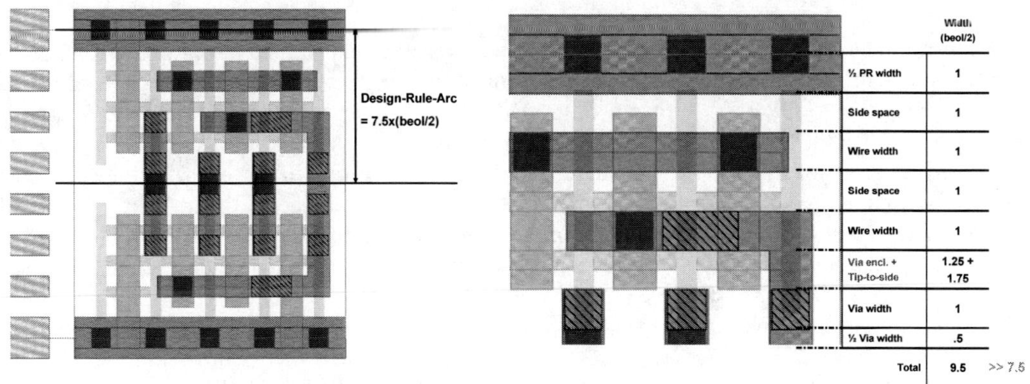

Figure 4, Sample design-rule-arc analysis as applied to the cell architecture optimization

To close the vertical design-rule-arc on M1, the second cell architecture in figure 3 sacrifices many of the desirable attributes outlined above: the M1 power rail is down to 1x width, there are only 5 M2 signal tracks, pin access is down to 2 points in several pins, and the output pin no longer aligns with the M2 tracks. On top of all these design trade-offs, further design-rule-arc analysis reveals more irresolvable constraints between the power taps formed by the tungsten strap and the signal connections on the first signal track next to the power rail. Once cells are stacked on top of each other there simply is not enough room to make a reliable connection to the power rail in one cell and land a via on the drain side of the transistor in the adjacent cell. To resolve this and other broken design-rule-arcs a second intra-cell wiring level has to be introduced as shown in the bottom cell architecture in figure 3. With this addition, all the positive design attributes of this cell image are restored, of course, at the cost of an additional metal level.

The fact that, even with EUV patterning, aggressive logic cell scaling requires an additional metal level raises hope that the more restrictive 193i patterning solution may compare quite favorably by providing a means of decomposing the bi-directional M1 design into two complimentary unidirectional metal layouts as shown in figure 5. The 'traditional' cell on the left with bidirectional metal is not compatible with higher-order frequency multiplication. A cell architecture using a single level of unidirectional metal and limited local interconnect wiring [3] (2nd from left) requires pins, and therefore poly connections, to be vertically staggered so that they do not interfere with each other. For small cell images these staggered pin configurations use up too much vertical space which is lost to device width and therefore the scaling objective of 3 active fins per device can not be met with this metal orientation. To overcome these constraints, the two cell images on the right of figure 5 use 'compound gratings', i.e. they decompose the bidirectional wiring level into two unidirectional levels. The two resulting cell architectures differ primarily in the sequence metal orientations: horizontal M0, vertical M1, and horizontal M2 (2nd from right) vs. vertical M0, horizontal M1, and vertical M2 (far right). Both cell images have their merits and would warrant further DTCO exploration, however, for the purpose of this paper, the cell image resulting in horizontal M2 wiring tracks was chosen to limit the number of variables that are altered in the comparison to the EUV-compatible cell image on the far left.

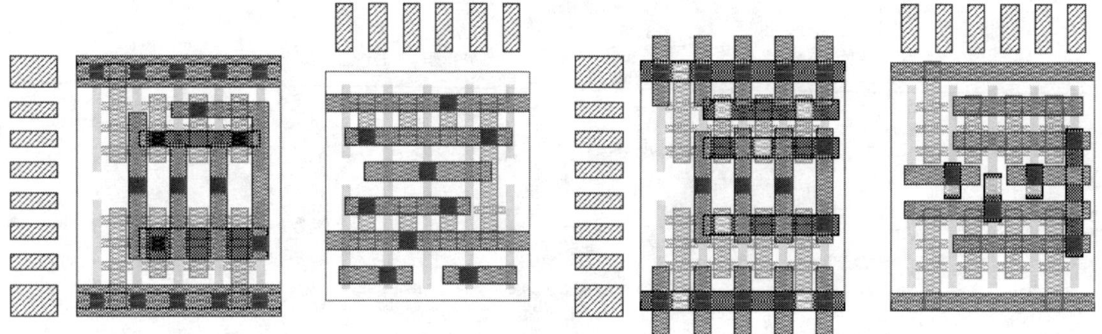

Figure 5, Sample cell images with bidirectional M1 (far left); unidirectional M1 perpendicular to poly with no additional metal in the cell (2nd from left); and 2 compound grating cell images differing in the sequence of metal orientations: horizontal M0, vertical M1, horizontal M2 (2nd from right) vs. vertical M0, horizontal M1, vertical M2 (far right)

As shown in table 3, the metal pitch chosen for this exercise can only be achieved with 193i patterning through higher order frequency multiplication such as self-aligned-quadruple-patterning. In addition to being practical only for unidirectional line patterns, SAQP has additional constrains that are a direct result of the complicated patterning scheme illustrated in figure 6. The top mandrel is the only feature in this grating that is lithographically formed. The bottom mandrel is formed by sidewall deposition onto the top mandrel and the final features in turn are formed by sidewall deposition onto the bottom mandrel. Since the bottom mandrel and the final sidewall features are deposited rather than printed, it is extremely costly to vary their dimension and fixed width should be assumed for each. Since the final sidewall pattern exists at a single width only, and it is more tolerable to designers to be constrained to a single wire space rather than a single wire width, in most cases SAQP for wires would be implemented in a 'sidewall is dielectric' tone. This in turn means that there are two ways to vary the wire width, e.g. to form wider power rails: changing the width of top mandrel shapes or changing the space between top mandrel shapes. Since every top mandrel edge results in 2 final sidewalls, the space between two wide wires (e.g. power rails) has to be filled with 2+2n sidewalls or (2+2n)-1 narrow wires. The net result of this lengthy discourse into SAQP fundamentals is that a SAQP-compatible cell image has to have 5 or 7 signal wires, as shown in Figure 7, meaning that the cell image in figure 5, while unidirectional, is not quite SAQP-compatible.

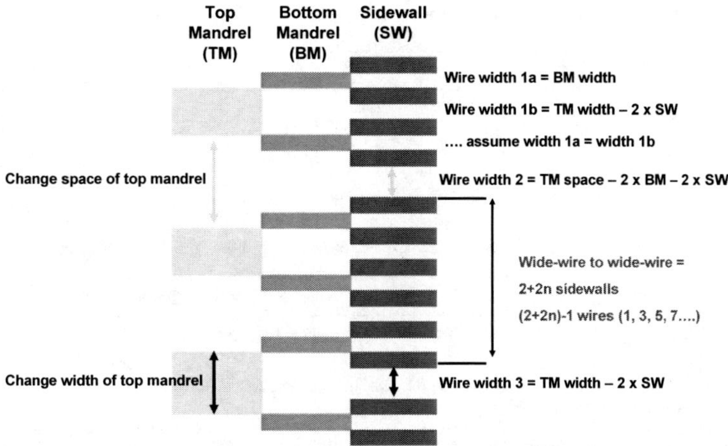

Figure 6, Schematic illustration of the SAQP process which requires 2+4n sidewalls between wider spaces.

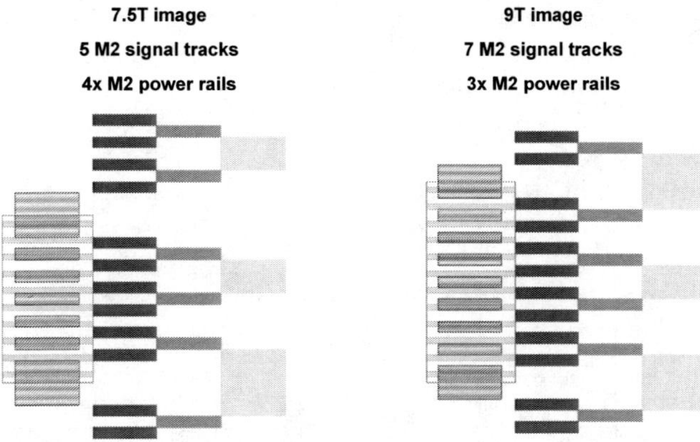

Figure 7, Unidirectional metal cell images with 5 or 7 signal tracks are SAQP-compatible.

The cell image shown in figure 8 addresses the odd signal wire requirement by widening the power-rails, and centering a group of 5 signal wires on the pins. The 4x wide power-rails are clearly overkill but are necessary to meet the fixed space requirements imposed by SAQP. This sample SAQP-compatible cell image runs M0 horizontally at 5/4 M2 pitch, M1 vertically at 5/4 M2 pitch, uses a stapled power rail that should provide reasonable electrical robustness, has 3 access points per pin, and has 5 M2 signal tracks.

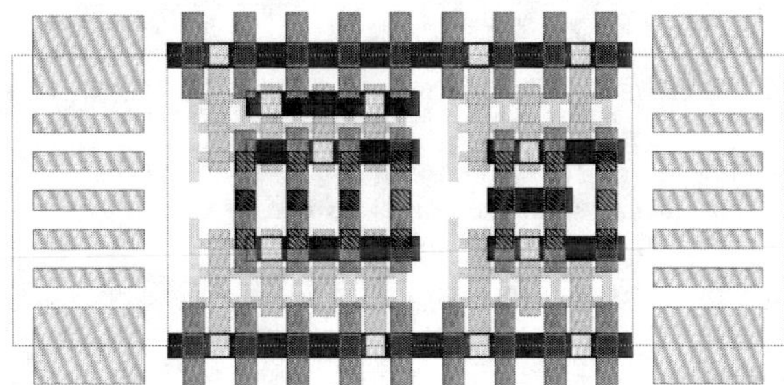

Figure 8, Potential SAQP-compatible cell architecture.

For the AOI and INV cells used in this baseline comparison, the EUV and 193i cell images compare quite favorably, as shown in table 6. While this detailed cell architecture work provided significantly more clarity on the scaling impact of the 193i patterning induced design restrictions outlined in chapter 3, it is still impossible to quantify the ultimate design impact of the cell image differences listed in table 6, for that further DTCO analysis is required.

	EUV	193i
	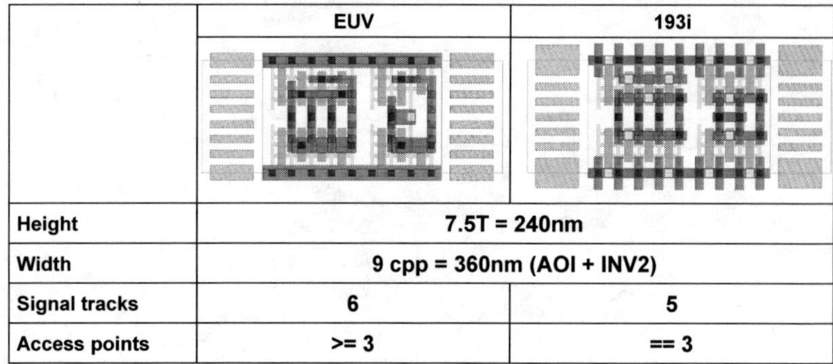	
Height	7.5T = 240nm	
Width	9 cpp = 360nm (AOI + INV2)	
Signal tracks	6	5
Access points	>= 3	== 3

Table 6, Quantitative design differences between the EUV and 193i cell architectures.

6. MACRO-LEVEL SCALING ASSESSMENT

Since the ultimately achievable technology scaling is very product and design specific, it is useful to separate the scaling estimate into scaling achievable for designs that are cell-area scaling limited, i.e. easy-to-route designs, and designs that are wire-density scaling limited, i.e. hard-to-route designs.

6.1 Cell-area limited scaling

To assess the technology definition for cell-area scaling limited designs, a good first estimate is achieved by summing the area scaling of a set of commonly used logic cells weighted by their relative utilization in a typical design as illustrated in table 7.

CELL GROUP	NAME	USAGE (%)	Cell Width (cpp)
INVERTER BUFFER	IV	4.5	3
	IVX2	0	4
	BF	3	4
	BFX2	7.5	5
NAND NOR	NAND2	3	4
	NAND2X2	6	6
	NAND3	3	5
	NOR2	3	4
	AND2	3	5
AOI OAI	AO22	3	9
	OAI22	3	6
	AOI21	4	5
	OAI211	2	6
MUX	MUX21	15	8
FLOP	SDFPQ	40	24
Area scaling estimate			Sum (usage*width*cpp*height*beol)

Table 7, Estimate of macro-level scaling for cell-area limited designs.

It is important to realize that this first estimate of macro-level scaling is based on cell-area scaling *targets* that are established based on the scaling achieved in the cell architecture definition using a smaller subset of logic cells (e.g. the AOI and INV used in this work). It is also important to note in table 7 that a large percentage of the macro area is occupied by more complex sequential logic cells like the MUX and FLOP. As figure 9 shows, the complex transistor connectivity necessary to design a MUX make compact design very challenging. To facilitate competitive scaling of these complex cell images, special constructs [4,5] are introduced to the technology definition. Two of these special constructs, a cross couple formed in bidirectional metal or local interconnect and an offset diffusion contact, are highlighted in the left image of figure 9. These special constructs solve very specific and high leverage connectivity problems without running the risk of allowing uncontrollable design freedom that would result from relaxing restrictions in the general purpose design rules. When rigid coarse grid design rules are no longer simply a design for manufacturability tool but a fundamental necessity for the chosen patterning solution, as is the case with SAQP, design rule waivers, even in the form of special constructs, are simply not possible. The right image in figure 9 is a rough estimate of what a MUX might look like when mapped into the SAQP-compatible cell architecture that was established in figure 8. Without the benefit of special constructs and aggressive use of bidirectional metal, the MUX grows in width by 4 poly tracks.

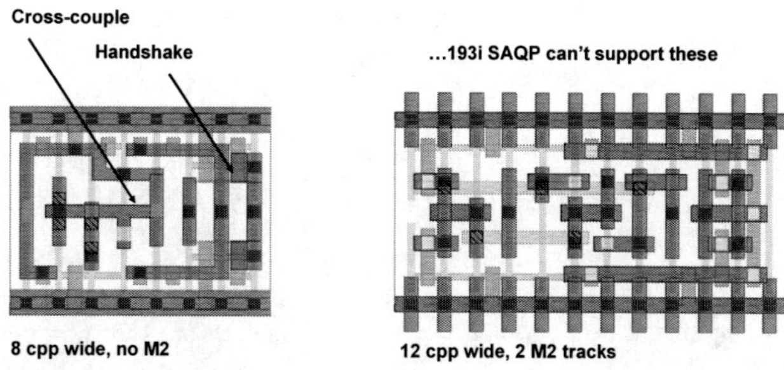

Figure 9, Estimate of MUX layout for EUV (left) and 193i (right).

Using the MUX as a guideline, the FLOP, which shares many of the same layout elements, is adjusted appropriately, leading to the scaling estimate comparison in table 8. This first-pass estimate shows 20% scaling degradation of the 193i compatible designs over the EUV compatible designs for logic blocks that are primarily cell-area scaling limited.

CELL GROUP	NAME	USAGE (%)	Cell Width (cpp) @ 7.5T	
			EUV	193
INVERTER BUFFER	IV	4.5	3	3
	IVX2	0	4	4
	BF	3	4	4
	BFX2	7.5	5	5
NAND NOR	NAND2	3	4	4
	NAND2X2	6	6	6
	NAND3	3	5	5
	NOR2	3	4	4
	AND2	3	5	5
AOI OAI	AO22	3	9	9
	OAI22	3	6	6
	AOI21	4	5	5
	OAI211	2	6	6
MUX	MUX21	15	8	12
FLOP	SDFPQ	40	24	29
Area scaling estimate			1.00	1.20

Table 8, Comparison of estimated block-level scaling for cell-area limited designs in EUV and 193i.

6.2 Router-limited scaling

For some designs the density with which logic cells can be wired together is a more important scaling limiter than the actual cell-area. For designs that require dense and complex wiring between logic cells, actual place-and-route experiments have to be conducted to assess the effectiveness of a given technology definition. The key parameter that has to be experimentally established is the 'utilization', i.e. the density with which a router can place logic cells before it can not find a routing solution with an acceptable number of design rule violations. In the past, this very important component of the scaling assessment could only be conducted once all the other phases of the DTCO process had been closed, a design rule manual had been finalized, a process design kit (PDK) had been published, and a sufficiently comprehensive standard cell library had been designed. At that point, a lot of time and resources had been invested in the technology definition under investigation and fundamental changes would potentially introduce significant schedule or technology risk. Here we introduce a patent pending approach to assess macro-level scaling in place-and-route experiments earlier in the DTCO process using *cell-abstractions*. These cell-abstractions, as shown in figure 10, accurately capture the key cell-architecture parameters such as function specific cell width, number of signal tracks, and number and placement of pin access points. Using these cell-abstractions, a library 'mock-up' can be generated that allows place-and-route experiments to contribute to the next cycle of DTCO refinement in parallel to detailed layout work that flushes out specific cell-level design pinch points as illustrated in figure 11.

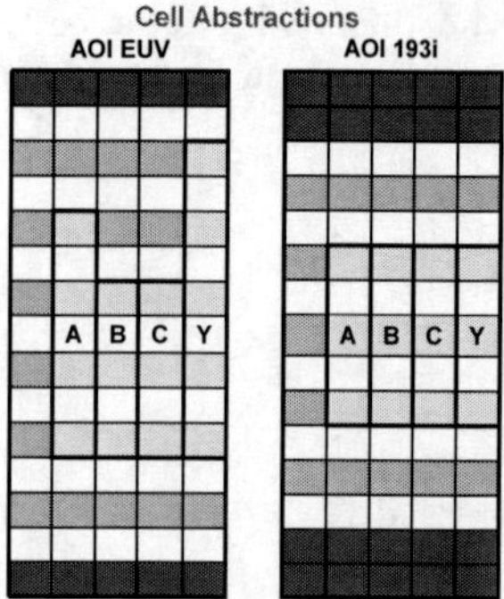

Figure 10, Cell-abstractions for the EUV (left) and 193i (right) versions of the AOI shown in table 6. Power rails are identified in dark blue, signal wiring tracks are light blue, and pin access points are shown in green.

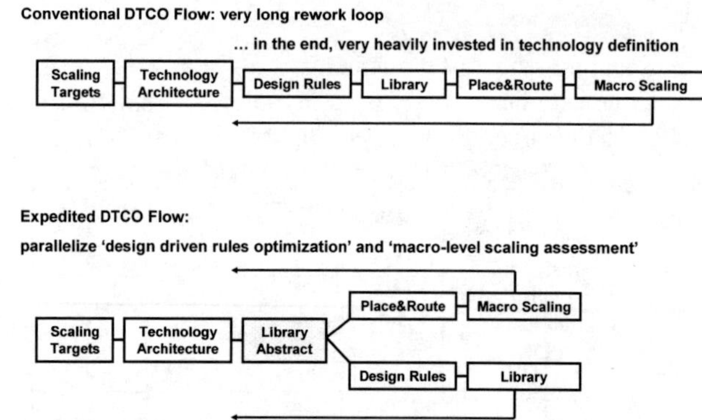

Figure 11, DTCO efficiency benefit afforded by the introduction of cell-abstracts in early routability assessments

The experiment used here to demonstrate the usefulness of cell-abstraction based place-and-route experiments and to further refine our estimate of the design impact of EUV vs 193i patterning, is based on a deliberately difficult to route design that was created by the IBM ASIC design team several technology nodes ago. This design, internally named 'Conehead', uses 11973 logic cells distributed as shown in table 9. It is worth noting that this design uses a different distribution of logic cells and therefore yields a different cell-area based scaling comparison than was shown for an 'average' design in table 8.

CELL GROUP	NAME	USAGE (%)	Cell Width (cpp) @ 7.5T	
			EUV	193i
INVERTER BUFFER	IV	7	3	3
	IVX2	1	4	4
	BF	0	4	4
	BFX2	0	5	5
NAND NOR	NAND2	7	4	4
	NAND2X2	0	6	6
	NAND3	3	5	5
	NOR2	9	4	4
	AND2	4	5	5
AOI OAI	AO22	9	9	9
	OAI22	9	6	6
	AOI21	17	5	5
	OAI211	6	6	6
MUX	MUX21	0	8	12
FLOP	SDFPQ	28	24	29
Area scaling estimate			1.00	1.13

Table 9, Cell usage distribution and cell-area based scaling comparison of the 'Conehead' design used for place-and-route experiments.

Using the metal stack shown in table 3 with unidirectional wiring for the 193i stack and preferred orientation wiring (i.e. wrong way wiring at relaxed pitch) for EUV, the place-and-route experiments involve: routing the design at decreasing total logic block area (i.e. increasing utilization or cell density), plotting the number of DRC violations remaining at the end of a finite number of detailed routing optimization cycles, and then recording the achieved block area based on the point where 300 DRC violations were exceeded (note: 300 is an arbitrary but realistic value). Figure 12 shows a 36% block-area difference between 193i and EUV designs using the 7.5T cell architectures and routing constraints established above.

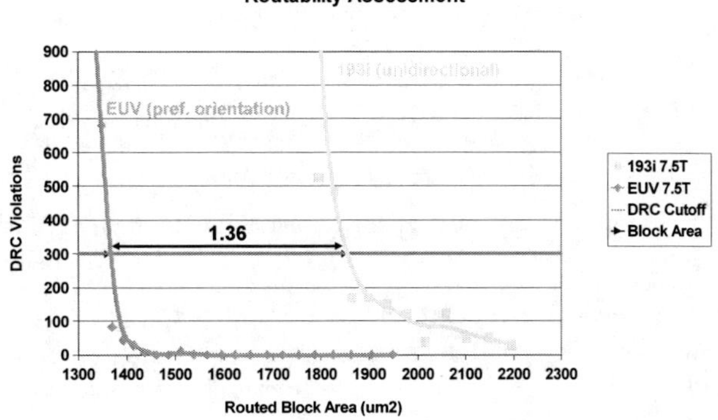

Figure 12, 36% difference in routed block area between 193i and EUV.

To further illustrate the efficiency of the cell-abstract based place-and-route experiments and to demonstrate the complex interdependence of all the parameters involved in the DTCO process, we repeated the above experiments for a 9T cell design. Using the same architecture as used for 7.5T the 9T cell-area limited scaling shown in table 10 is very predictable. The 9T block area is estimated to be 20% larger than the 7.5T area and the 193i designs remain 13% larger than the EUV designs. The results of the place-and-route experiments with 9T cell images are more surprising and are shown in figure 13. For this difficult to route design, the routed block area actually *decreases* by 20% in going from 7.5T to 9T for the 193i design while no change in block area is seen for 9T vs 7.5T in the EUV design. Finally, the routed block area difference between 193i and EUV designs reduces from 36% to 9% in going from 7.5T to 9T cells.

CELL GROUP	NAME	USAGE (%)	Cell Width (cpp) @ 7.5T		Cell Width (cpp) @ 9T	
			EUV	193i	EUV	193i
INVERTER BUFFER	IV	7	3	3	3	3
	IVX2	1	4	4	4	4
	BF	0	4	4	4	4
	BFX2	0	5	5	5	5
NAND NOR	NAND2	7	4	4	4	4
	NAND2X2	0	6	6	6	6
	NAND3	3	5	5	5	5
	NOR2	9	4	4	4	4
	AND2	4	5	5	5	5
AOI OAI	AO22	9	9	9	9	9
	OAI22	9	6	6	6	6
	AOI21	17	5	5	5	5
	OAI211	6	6	6	6	6
MUX	MUX21	0	8	12	8	12
FLOP	SDFPQ	28	24	29	24	29
Area scaling estimate			1.00	1.13	1.20	1.36

Table 10, Cell-area limited scaling estimate of the 'Conehead' design for 9T vs 7.5T cells.

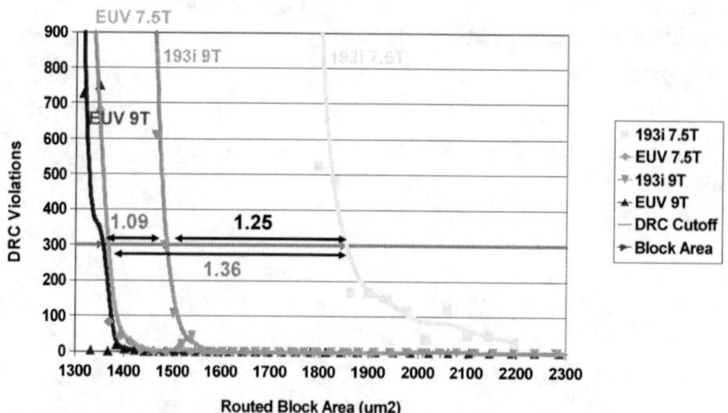

Figure 13, Larger cells improve routability for both EUV and 193i.

The results of these cell-abstraction based place and route experiments are summarized in table 11. These data clearly highlight the complexity of DTCO and demonstrate the need to include full macro level design assessment in the earliest phases of the technology optimization. These experiments further confirmed the not surprising but often ignored fact that the benefit of smaller cell size diminishes for hard-to-route layouts.

	Signal Tracks	Access Points	Wiring Constraint	Cell Area	Routed Area	Router Utilization
EUV 7.5T	6	>=3	Pref. Orient.	1.00	1.55	64%
193i 7.5T	5	==3	Unidirectional	1.13	2.11	53%
EUV 9T	7	>=4	Pref. Orient.	1.20	1.54	78%
193i 9T	6	==4	Unidirectional	1.35	1.69	80%

Table 11, Summary of cell-abstraction based place-and-route experiments.

7. CONCLUSION

In addition to illustrating several key concepts of DTCO in the various stages of a potential 7nm technology definition and highlighting the need for 'holistic' co-optimization of all design parameters in a comprehensive macro-level design flow, this work provided initial quantification of the differences between EUV and a 193i versions of the 7nm technology node. Patterning cost estimates for a purely EUV vs. a purely 193i based technology definition proved to be inconclusive. Macro-level design experiments using logic cell-abstractions showed that a conservative target for the area increase of 193i over EUV, for easy-to-route designs is in the ballpark of 10% while for hard-to-route designs it can be as high as 36%, depending on how aggressively the cell size target is set. While high performance designs that tend to use larger cell images may be able to scale to the 7nm node without the benefit of EUV, the bulk of the low-power, cost-oriented foundry business will have to do a lot more optimization to make a 193i-based 7nm node viable. A key element to this optimization is a DTCO flow that can efficiently explore the circuit level implications of a large variety of technology parameters.

REFERENCES

[1]Phase edge lithography for sub 0.1 mu m electrical channel length in a 200 MM full CMOS process, Agnello, P. Newman, T. Crabbe, E. Subbanna, S. Ganin, E. Liebmann, L. Comfort, J. Sunderland, Proceedings of the 1995 Symposium on VLSI Technology. Digest of Technical Papers - Symposium on VLSI Technology 1995. IEEE, Piscataway, NJ, USA,95CB35781. p 79-80

[2]Layout optimization at the pinnacle of optical lithography, Lars W. Liebmann, Greg A. Northrop, James Culp, Leon Sigal, Arnold Barish, Carlos A. Fonseca, Proceedings SPIE Vol. 5042, p. 1-14, Design and Process Integration for Microelectronic Manufacturing; Alexander Starikov; Ed., , Jul 2003

[3]Design implications of extremely restricted patterning, Kaushik Vaidyanathan, Renzhi Liu, Lars Liebmann, Kafai Lai, Andzrej J. Strojwas, Larry Pileggi, J. Micro/Nanolith. MEMS MOEMS 13(3), 031309 (Jul–Sep 2014)

[4]Bridging the resolution gap in 14 nm: designing for efficient transition to EUV (Invited Paper), Lars W. Liebmann, SPIE Design for Manufacturability through Design-Process Integration VI [8327 45], 2012

[5] M. Rashed, et al. , "Innovations in Special Constructs for Standard Cell Libraries in Sub 28nm Technologies", IEDM Technical Digest, 248-251 (2013).

High Coverage of Litho Hotspot Detection by Weak Pattern Scoring

Jinho Park[a], NamJae Kim[a], Jae-hyun Kang[a], Seung Weon Paek[a], Steve Kwon[a]
Marwah Shafee[b], Kareem Madkour[b], Wael ElManhawy[c]
Joe Kwan[d], and Jean-Marie Brunet[d],

[a] SAMSUNG Electronics Co., Ltd. Yongin, Gyeonggi-do 446-711 South Korea
[b] Mentor Graphics Corporation, Heliopolis, Cairo, Egypt, [c]Mentor Graphics Corporation, Wilsonville, OR USA, [d] Mentor Graphics Corporation, Fremont, CA USA

ABSTRACT

Achieving lithographic printability at advanced nodes can impose significant restrictions on physical design, including large numbers of complex design rule checks (DRC) and compute-intensive detailed process model checking. Early identifying of yield-limiter hotspots is essential for both foundries and designers to significantly improve process maturity. A real challenge is to scan the design space to identify hotspots, and decide the proper course of action regarding each hotspot.

Building a scored pattern library with real candidates for hotspots for both foundries and designers is of great value. Foundries are looking for the most used patterns to optimize their technology for and identify patterns that should be forbidden, while designers are looking for the patterns that are sensitive to their neighboring context to perform lithographic simulation with their context to decide if they are hotspots or not.[1]

In this paper we propose a framework to data mine designs to obtain set of representative patterns of each design, our aim is to sample the designs at locations that can be potential yield limiting. Though our aim is to keep the total number of patterns as small as possible to limit the complexity, still the designer is free to generate layouts results in several million of patterns that define the whole design space. In order to handle the large number of patterns that represent the design building block constructs, we need to prioritize the patterns according to their importance.

The proposed pattern classification methodology depends on giving scores to each pattern according to the severity of hotspots they cause, the probability of their presence in the design and the likelihood of causing a hotspot. The paper also shows how the scoring scheme helps foundries to optimize their master pattern libraries and priorities their efforts in advanced nodes.

Moreover, the paper demonstrates how the hotspot scoring helps in improving the runtime of lithographic simulation verification by identifying which patterns need to be optimized to correctly describe candidate hotspots, so that only potential problematic patterns are simulated.

Keywords: DFM Scoring, Pattern Library, Litho Hotspot Detection, Design Technology Co-Optimization (DTCO)

1. INTRODUCTION

In advanced technology nodes, there are more significant patterning issues due to RET limitations [2] [3]. These issues can happen in a DRC clean pattern and causes violations within a process window spanning ranges of defocus and energy dose. Regions in the design layout that have poor manufacturability characteristics even with the application of RET techniques are called Lithographic hotspots.

As lithography simulation runtimes continue to increase with advancing technology nodes, many DFM technology leaders are beginning to truly weigh the benefits of using pattern matching as a way to expedite some of their advanced verification processes. The benefit of improved runtimes through simulation can be obtained by reducing the amount of data being sent to simulation. By inserting a pattern matching operation, a system can be designed

Design-Process-Technology Co-optimization for Manufacturability IX, edited by John L. Sturtevant, Luigi Capodieci, Proc. of SPIE Vol. 9427, 942703 · © 2015 SPIE · CCC code: 0277-786X/15/$18 · doi: 10.1117/12.2087473

such that it only simulates in the vicinity of topologies that somewhat resemble hotspots while ignoring all other data [4].

However, for advanced nodes the manufacturing process window tightens and the number of patterns continues to rapidly increase. Comprehensive and compact test patterns are crucial to the development of new semiconductor technology [5,6].

Having a pattern library with real candidates for hotspots is of great value for both foundries and designers. Foundries need to know the most used patterns to optimize their technology for, and identify patterns that should be forbidden. On the other hand, designers are looking for the patterns that are sensitive to their neighboring context to perform lithographic simulation with their context to decide if they are hotspots or not.

In the proposed flow, a scoring technique is used to classify patterns according to the probability of occurrence of the patterns and their possibility of causing hotspots. Based on this, scoring a proper course of actions regarding each pattern is identified. The paper is organized as follows; the statistical framework "libScoring" proposed for patterns classification is described in section-2, and the methodology of the pattern library optimization is discussed in section-3. Section-4 shows the experiments and results of applying this framework to optimize the advanced node pattern-library. Finally, conclusion is drafted in section-5.

2. LIBRARY SCORING FRAME WORK

The PM library module that extracted simulated weak patterns requires the existence of a preformed pattern library. Each pattern existing within the library is described in an individualistic fashion with specific information pertaining to coordinate space of vertices, special pattern boundary considerations, per-edge allowable movement tolerances, or perhaps other pattern attributes. Assuming that a library has been developed and qualified for a given technology, the effects of applying it across an arbitrary design can result in identifying a set of locations which may, or may not, be hotspot locations. Such areas are referred to as candidate hotspot regions (shown in figure 1). These are the regions that will ultimately get passed to the litho checker for simulation and verification. The total area to be passed to simulation has been greatly reduced from the original amount of full chip data, thus we can expect a runtime performance benefit [4,7]. This is what we call the fast mode litho hotspot check flow.

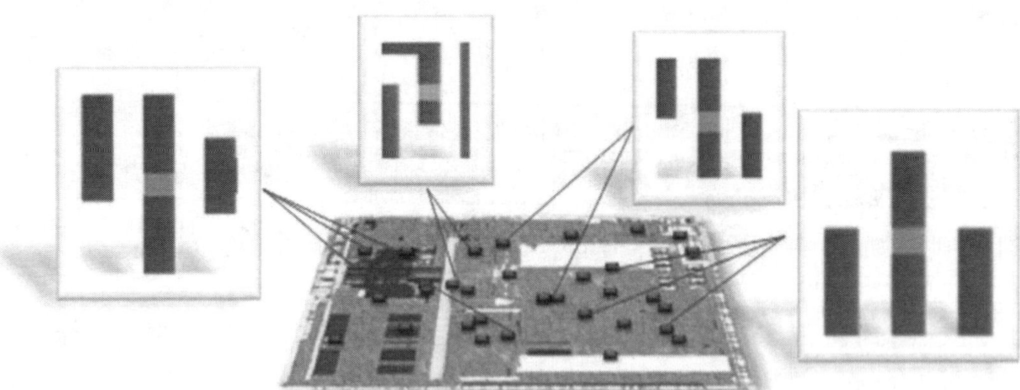

Figure 1. A pictorial representation of the candidate hotspot regions to be passed to simulation [4].

Fast-mode litho hotspot check flow using PM library should consider to cover full chip layout and to detect accurate results as same as full simulation, To extract PM library from litho simulation results without using logical filtering way causes overestimate run time or hotspot missing. Litho hotspot or weak patterns need to be optimized before being passed to PM library module.

Library-scoring "*libScoring*" framework starts by gathering information about each pattern in a library, and generates statistical scores for each pattern. The inputs to the framework are: a pattern matching library set of input layouts with their corresponding hotspots, and pattern matching results of this library on the input layouts. Figure 2 shows the inputs and outputs of the proposed framework.

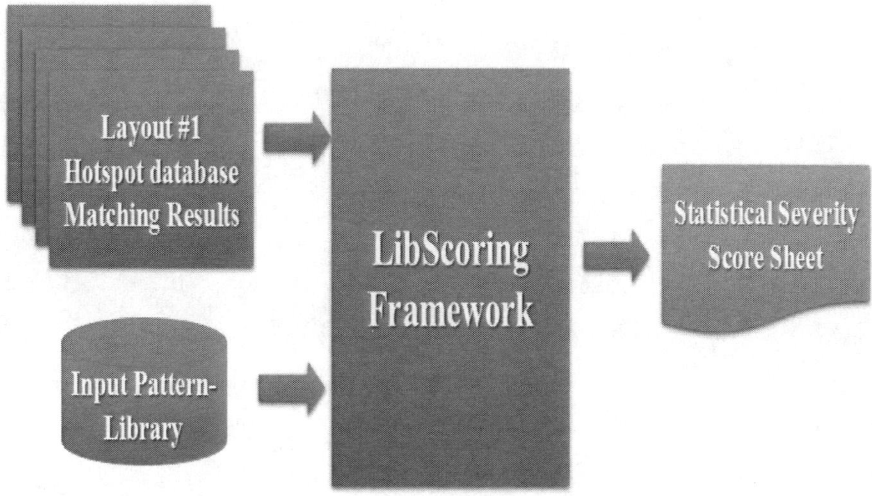

Figure 2: *LibScoring* frame work data flow

From the pattern-matching results, the number of occurrences of each pattern is calculated. While, the severity of the pattern comes from the litho-simulation information represented by the set of hotspots that are interacting with each match of each pattern. By processing the collected data, a score is calculated for each pattern that indicates its severity. Finally, a scoring sheet that summarizes all the statistical data and the calculated scores for all patterns is generated. An example for a score-sheet is shown in Table 1.

There are several scoring equations that can be used to do the scoring for patterns. One of these equations that is functional in statistical data is illustrated in Figure 3.

$$\mathbf{S(P}\boldsymbol{i}\mathbf{)} = \sum(\boldsymbol{a}.(\mathbf{NM}\boldsymbol{i})+ \boldsymbol{b}.(\mathbf{NMCH}\boldsymbol{i})+ \boldsymbol{c}.(\mathbf{NMWH}\boldsymbol{i}))$$

Where:

S(Pi) is the score of pattern number *i*

NMi is matching population of pattern *i*

NMCHi is matching counts of critical hotspot pattern *i*

NMWHi is matching counts of warning hotspot pattern *i*

a,b,c are weighting factors

Figure 3: Scoring Equation used to generate score for pattern

Pattern Name	Total Matches	Critical HS	Warning HS	Score
99	8445073	1508841	6184	94,681.10
50	4894482	958938	624440	70,361.78
67	3550591	554635	30	36,773.51
120	60636392	3728	7172	288.07
115	75563716	4898	2272	295.05
5	6318933	878	4932	119.06
118	8937997	496	1608	51.08
64	2322	224	19	29.39
52	1939548	137	0	9.46
112	2265705	96	2	6.59
46	5662	8	137	4.10
89	1409	0	10	0.28
76	1891	0	7	0.19
40	225	0	6	0.22
7	719	0	4	0.12
495	3222	1	0	0.12
79	1047	0	1	0.03
88	37	0	1	0.06
105	58692	0	0	0.0
259	2057	0	0	0.0

Table 1: Example for an output score-sheet

3. PATTERNS CLASSIFICATION AND OPTIMIZATION

Patterns Classification

The analysis of the generated statistical score sheet categorizes the patterns into four classes as shown in Figure 4. Patterns of the first class (high matches count and high severity) are of high importance and need to be fixed by reviewing and optimizing the pattern library. Patterns of the second class (high matches count and low severity) need to be optimized to define the problematic contexts that cause the hotspot to occur. Third class patterns (low matches count and high severity) need to be either fixed through the recipe or by forbidding them to occur in the design. Patterns of the fourth category (low matches count and low severity) can be easily removed from the library as they are not occurred frequently and also have low severity.

In the process of defining the pattern severity, a statistical study is done on the hotspot properties and check values through analyzing the histogram of the hotspot minCD and space distribution. The histogram defines the distributions of the space values of certain width checks or spacing checks which gives an indication on the degree of the hotspot severity and thus categorize it into a severe or marginal or weak hotspot category. Two examples of width and space histogram are shown in Figure 5.

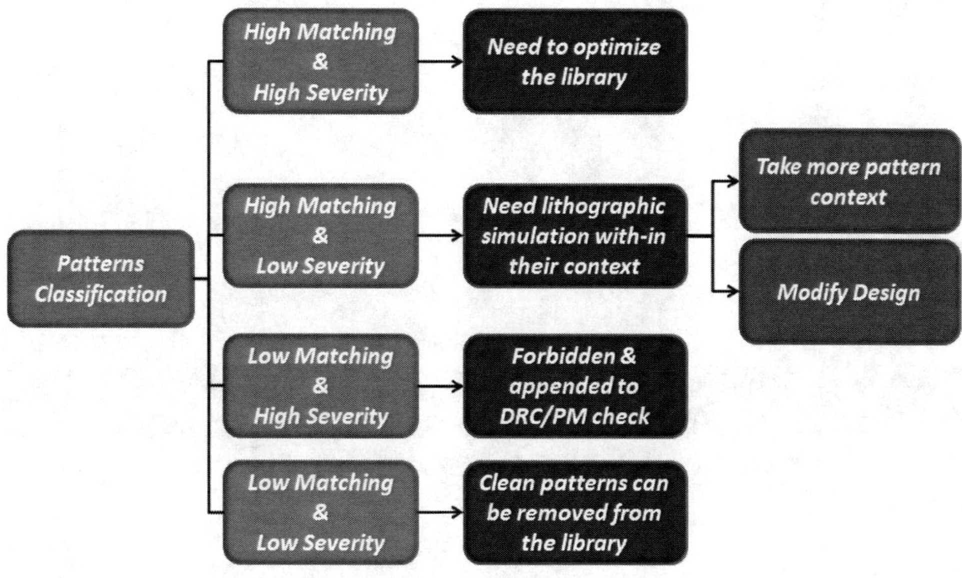

Figure 4: Patterns Classification

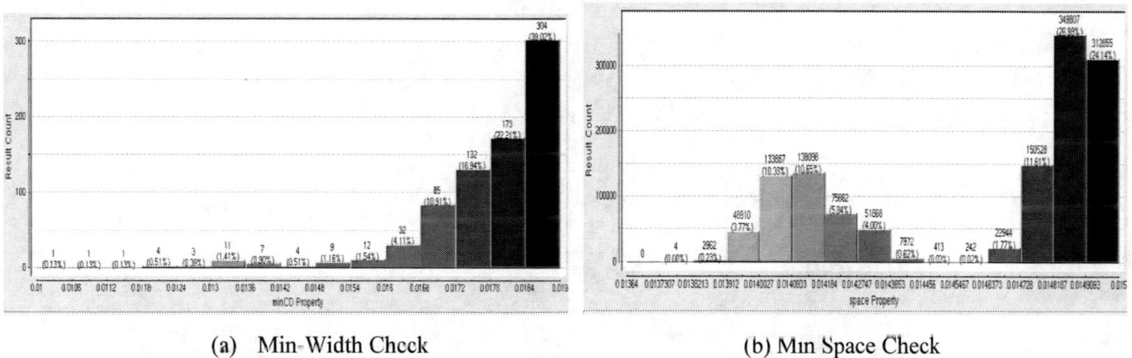

(a) Min-Width Check (b) Min Space Check

Figure 5: Histogram of CD Values of a Min-Width/Spacing Check

Patterns Optimization

LibScoring framework was used to evaluate a pattern library that consists of uniform capturing patterns from litho hotspots. All these uniform captured patterns are divided 4 categories as previously shown figure 4. An example of the first category patterns is shown in Figure 6(a). This pattern occurs extensively in the design and in the same time it mostly has severe hotspots inside the extent of the pattern. This case needs to be classified and separated so that the pattern marker is adjusted in center of the pattern. In other words, high severity with high frequency patterns is also optimized through pattern separation by focusing on real hotspot marker of points. Figure 6 shows the two adjusted patterns that represent the real hotspot where (a) is uniform pattern that clipped by regular value from hotspot marker. It shows extremely high existence in design where there is a high risk of litho hotspots among the found matches. After adjusting hotspot marker to reflect real hotspots, the uniform pattern is separated into two patterns as shown in (b) where each pattern shows reasonable matching counts.

(a) Uniform Pattern

(b) Separate "A" type

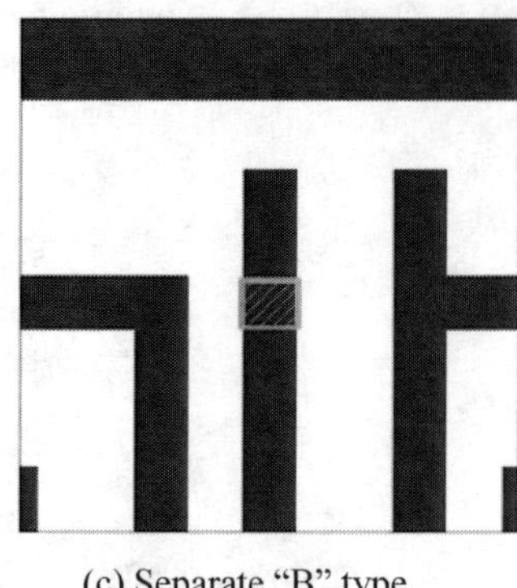

(c) Separate "B" type

Lib. Type		Matching Count	Hotspot Count	Score
Conventional	(a) Uniform extent	526,209	2,966	**343.74**
New	(b) Separate "A"	33,359	1,924	**265.40**
	(c) Separate "B"	12,512	1,042	**188.46**

Figure 6: Pattern optimization with pattern separation according to hotspot marker centering

Second category patterns are the patterns that have high matching count but low severity. When analyzing those patterns, it was found that those patterns are problematic only when put in certain context. Two actions can be made; either modifies the pattern such that the context polygons that cause the hotspots be in the pattern extent, or the designer should change their design to fix this hotspot. Figure 7 shows an example for a pattern that needs to be enlarged to take more contexts.

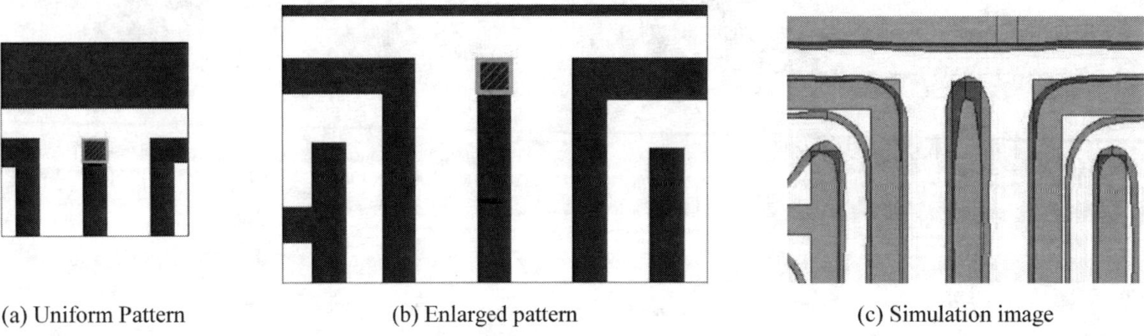

(a) Uniform Pattern (b) Enlarged pattern (c) Simulation image

Lib. Type		Matching Count	Hotspot Count	Score
Conventional	(a) Uniform Extent	84,000	520	**76.86**
New	(b) Enlarged Pattern	4,700	520	**123.99**

Figure 7: Litho hotspot results versus pattern clipping size

Figure 8 shows an example for 3rd category, low matching and high severity. The litho. hotspot is related with a standard cell that was DRC and Litho hotspot clean when checked with no context, but when placed abut to certain cells it causes severe width hotspot. To fix this hotspot, it is needed to change the design of the standard cell. When the litho hotspot is checked on library level, it is not possible to consider all of cell abutment combination. This third category case is able to send hotspot reports directly without any simulation or it is suitable for PM solution that is defined with certain key patterns that should be forbidden from the design.

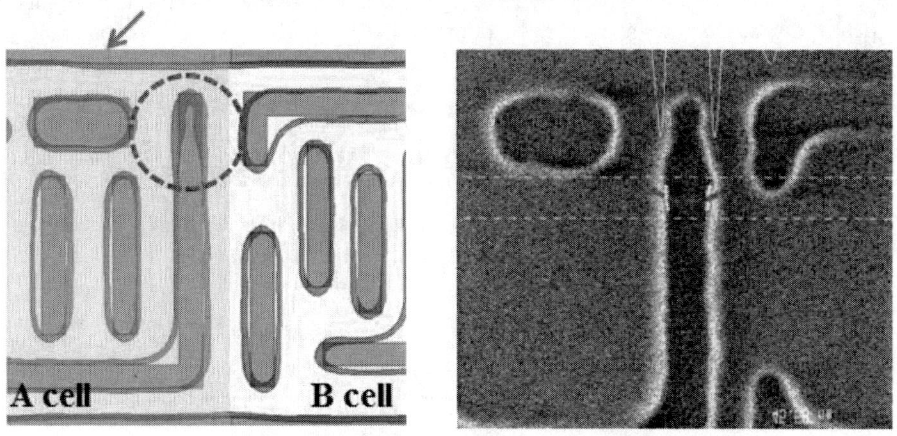

Figure 8: Problematic standard cell placement

Several patterns found that have low matching count but without showing any hotspots, those patterns were removed from the pattern-library.

4. EXPERIMENTS AND RESULTS

From the previous section, several patterns are changed and/or replaced by multiple patterns. Although the number of patterns in the library may be increased, but the resultant matched regions from this library is optimized, which consequently optimizes the litho-simulation that is needed after that. Several runtime and accuracy experiments are done using an optimized pattern library obtained from the proposed framework showed same level of accuracy like the original PM library and with better runtime performance. The runtime improvements in some cases were up to 6.7X as shown in Figure 9.

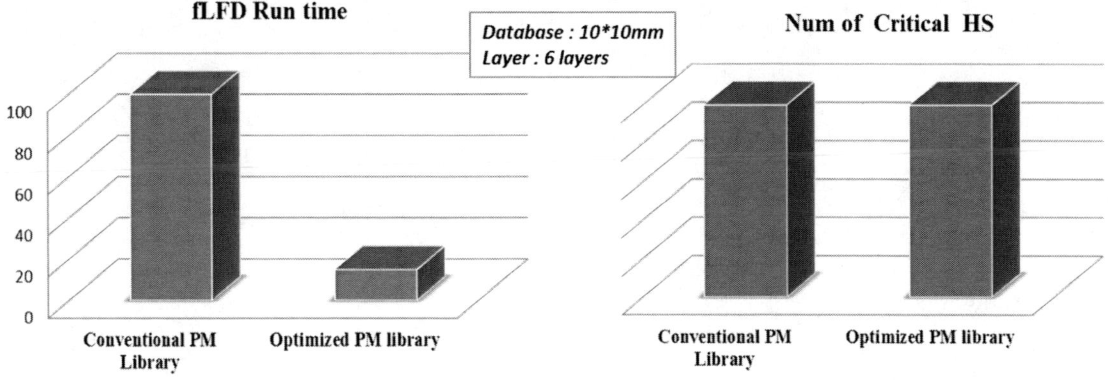

Figure 9: Runtime improvements and Hotspots results

5. CONCLUSION

A statistical framework to data mine designs to obtain classified patterns that represents the technology designs is presented. The proposed pattern classification methodology depends on giving scores to each pattern according to the severity of hotspots they cause, and the probability of their presence in the design. Based on this classification, a methodology to optimize a pattern library is developed and tested for both accuracy and performance. The analysis of the scored PM library helps both foundries and designers to either improve the technology recipe or optimize certain patterns

Moreover, the paper demonstrates how the hotspot scoring helps in improving the runtime of lithographic simulation verification by identifying which patterns need to be optimized to correctly describe candidate hotspots, so that only potential problematic patterns are simulated.

6. ACKNOWLEDGMENT

The authors would like to thank Samsung DFM team as well as Mentor Graphics DFM LFD team for their outstanding support and continuous drive towards creating an engineering solution that addresses the advanced nodes challenges.

7. REFERENCES

1) W. C. Tam and R. D. Blanton., "*To DFM or Not To DFM?*" Design Automation Conference (2011).
2) Wong, A., "*Resolution Enhancement Techniques in Optical Lithography*", SPIE Press (2001).
3) Schellenberg, F. M., "Resolution enhancement technology: The past, the present and extension for the feature", Proc SPIE 5377, (2004).
4) Mark C. Simmons, Jae-Hyun Kang, Youngkeun Kim, et al., "*A state-of-the-art hotspot recognition system for full chip verification with lithographic simulation*", Proc. SPIE (2011), Vol. 7974, 79740M
5) E. Teoh, V. Dai, L. Capodieci, Y.-C. Lai, and F. Gennari, "*Systematic Data Mining Using a Pattern Database to Accelerate Yield Ramp*", Proc SPIE, vol. 9053, Mar 2014.
6) Seung Weon Paek, Jae-Hyun Kang, Naya Ha, et al., "*Yield enhancement with DFM*", Proc. SPIE (2012), Vol. 8327
7) Jae-Hyun Kang, Byung-Moo Kim, Naya Ha, Sarah Mohamed, Kareem Madkour, Wael ElManhawy, et al., "*Model Based Hint for Litho Hotspot Fixing beyond 20nm node*", Proc. SPIE (2013), Vol. 8640

A pattern-based methodology for optimizing stitches in double-patterning technology

Lynn T.-N. Wang*[a], Sriram Madhavan[a], Vito Dai[a], and Luigi Capodieci[a]
[a]GLOBALFOUNDRIES, 2600 Great America Way, Santa Clara, CA, USA 95054

ABSTRACT

A pattern-based methodology for optimizing stitches is developed based on identifying stitch topologies and replacing them with pre-characterized fixing solutions in decomposed layouts. A topology-based library of stitches with predetermined fixing solutions is built. A pattern-based engine searches for matching topologies in the decomposed layouts. When a match is found, the engine opportunistically replaces the predetermined fixing solution: only a design rule check error-free replacement is preserved. The methodology is demonstrated on a 20nm layout design that contains over 67 million, first metal layer stitches. Results show that a small library containing 3 stitch topologies improves the stitch area regularity by 4x.

Keywords: DFM, design for manufacturability, double patterning, LELE, decomposition, stitching, pattern matching, topological patterns, layout regularization, optimization, lithography, misalignment, hotspots

1. INTRODUCTION

Resolution enhancement techniques (RET) such as optical proximity correction (OPC), attenuated phase-shift masks (att-PSM), and off-axis illumination have pushed 193nm (ArF) lithography to the fundamental resolution limits [1]. To enable further scaling, Litho-Etch-Litho-Etch (LELE) Double Patterning Technology (DPT) has become the key technology enabler for 20nm nodes and beyond [2]. This manufacturing process requires a change to the physical design flow, in which layout features on one mask layer are decomposed to complementary mask layers prior to manufacturing. This decomposition often causes conflicting mask layer assignments, so stitching has been introduced as a design technique to resolve these conflicts with minimal area penalties [3].

State-of-the-art DPT-compliant physical design methodologies use Design Rule Checks (DRCs) to guide layout design decisions for stitching. These DRCs ensure that sufficient overlay stitching areas are drawn to prevent electrical opens caused by process-induced variability, such as critical dimension (CD) variation, line-end pullback, and mask-to-mask misalignment. However, DRCs do not indicate where and how to stitch, which results in a large variety of stitches being drawn in custom, decomposed designs. As layout design regularity becomes increasingly more important for achieving high manufacturability and yield, the use of stitches needs to be better controlled [4]. Pattern matching methodologies have been demonstrated to simultaneously promote layout regularity and optimize stitches for manufacturability. These patterns may be model-based or empirical-based. Model-based patterns include the inverse Fourier transforms of Zernike polynomials [5]. Empirical-based patterns include exact layout clips containing lithography hotspots [6]. While both types of patterns are effective for guiding stitching decisions, the corresponding manual layout fixes are difficult to automate for production designs.

An automated topological pattern-based methodology for characterizing and fixing stitches is presented. This proposed methodology requires a topological pattern-based library of stitches to be built a priori, in which each topology is associated with a predetermined fixing solution. These empirically-based topologies are extracted from existing decomposed designs. A pattern matching engine then searches for matching topologies from the library in decomposed layout designs. If a match were found, the pattern matching engine would replace the existing stitch with the predetermined fixing solution. These modifications are opportunistic because only the DRC-clean replacements are preserved. By identifying stitch topologies in decomposed designs and opportunistically replacing them with consistent predetermined fixes, the design variability is reduced, and the design manufacturability is improved.

This paper is organized as follows: Section 2 contains the background on DPT-compliant stitches, state-of-the-art classification methods, and an automated topological pattern-based layout optimization flow. Section 3 discusses the

Design-Process-Technology Co-optimization for Manufacturability IX, edited by John L. Sturtevant, Luigi Capodieci, Proc. of SPIE Vol. 9427, 942704 · © 2015 SPIE · CCC code: 0277-786X/15/$18 · doi: 10.1117/12.2085955

experimental results generated from the classification methods and the layout optimization flow described in Section 2 for various 20nm designs. Section 4 concludes the paper.

2. BACKGROUND ON STITCHING, CLASSIFICATION METHODOLOGIES, AND TOPOLOGICAL PATTERN-BASED LAYOUT OPTIMIZATION FLOW

Stitching is a multi-patterning layout technique that enables the successful resolution of conflicts incurred during decomposition with minimal area penalties. This technique decomposes one feature from one mask layer to two complementary mask layers so that one larger feature can be manufactured from two smaller features. It involves two steps: splitting and stitching. See Fig. 1.

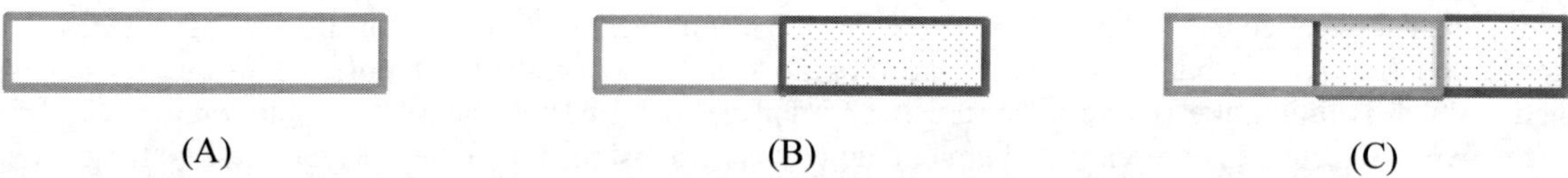

(A) (B) (C)

Fig. 1: Stitching is a multi-patterning layout technique that is commonly used to resolve conflicts. It takes the original feature in (A), splits the feature to two pieces in (B), and stitches it back together in (C).

An example of a DPT-compliant stitch is shown in Fig. 1. The orange (left) feature represents the first exposure layer, and the blue (right) feature represents the second exposure layer. The original feature in Fig. 1 (A) is split to two smaller features, with each sub-feature decomposed to a complementary mask layer as shown in (B). Then, at the location of the split, the line-ends are extended to create an overlap region to prevent electrical opens caused by process effects. This overlap region is termed a stitch (C).

Stitching usages vary widely across 20nm products. Table 1 shows the number of stitches used for three metal layers across five 20nm designs. The chip area for each design is also listed.

Design/ Layer	A ($93mm^2$)	B ($228mm^2$)	C ($54mm^2$)	D ($9mm^2$)	E ($36mm^2$)
M1	72	33	15	8	6
M2	5	31	15	8	6
M3	7	38	23	8	6
Total	84	112	53	24	18

* Counts in millions

Table 1: Stitch usages for three metal layers across five designs

Table 1 shows that the current production designs use millions of stitches. For example, Design A uses 72 million stitches for the M1 layer and 84 million stitches for the M1, M2, and M3 layers combined. The usage of stitches also varies significantly across the five designs, from the highest 112 million (i.e., Design B) to the lowest 18 million (i.e., Design E), which is a factor of approximately 10x difference. Thus, since there are millions of stitches that need to be characterized, automated design analysis tools are used. Also, differences in design styles lead to various styles of stitching. Using the analysis, automated DFM optimizations are defined and implemented to improve layout manufacturability.

2.1 Classification methodologies for layout characterization and analysis

This work uses an analysis flow that leverages three classification methodologies: rule-based, exact pattern-based, and topological pattern-based. They are executed in a sequential order. First, rule-based classification, i.e., design rule checks (DRC), are used to classify the stitches. Then, exact pattern-based classification is used to convert the DRCs to 200nm by 200nm patterns. Lastly, topological pattern-based classification is used to convert the exact patterns to topologies. See Fig. 2.

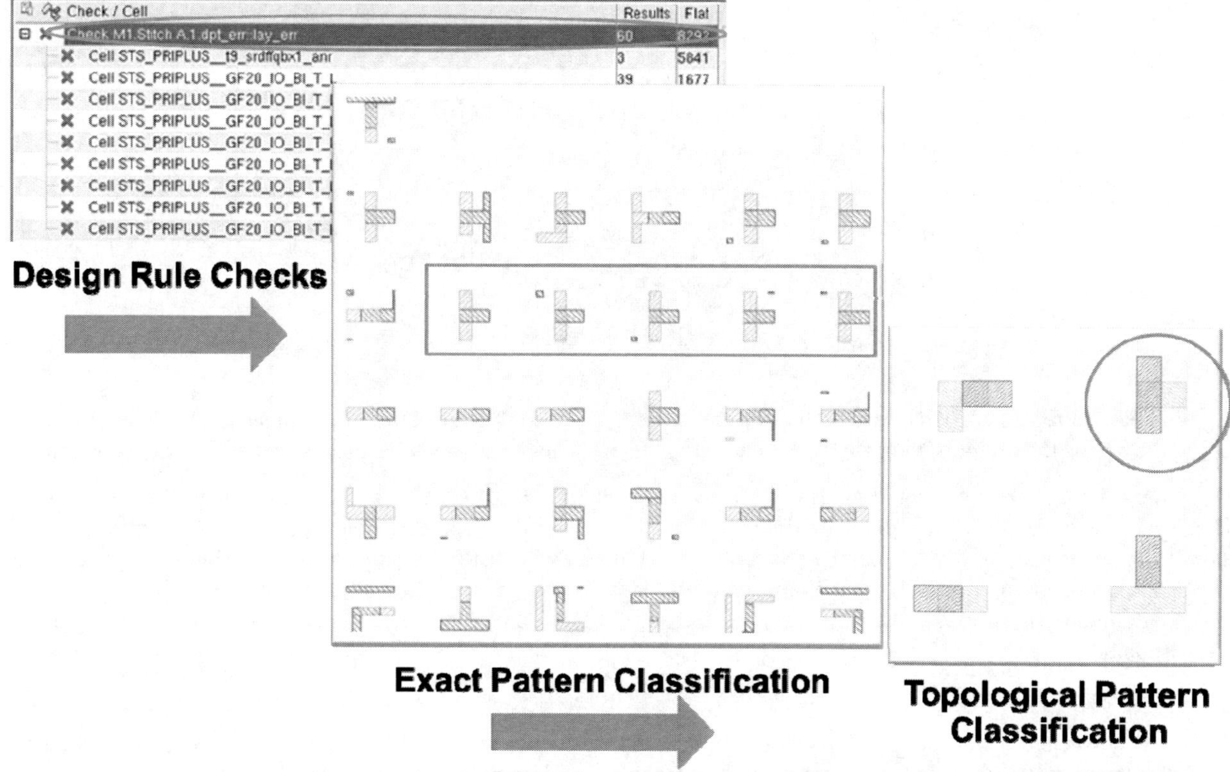

Fig. 2: Analysis flow using DRCs, exact pattern classification, and topological pattern classification

The sequential classification flow is described in Fig. 2. Rule-based classification using DRCs enables the examination of stitches based on cells. However, the rule-based classification methodology cannot automatically identify similar stitches within cells or among cells. To further eliminate the repetitions of similar stitches, exact pattern classification is used to reduce the stitches from DRC error markers to exact patterns. Further examination of the classified patterns indicates that there are many repetitions in the topologies, such as the cross-shaped stitch (shown by the rectangular box). Topological pattern classification is used to reduce the exact patterns to topologies. For example, the cross-shaped stitch patterns reduce to one topology (shown by the circle).

The following sub-sections describe each classification methodology in more detail:

2.1.1 Rule-based classification

Historically, the prevalent methodology for characterizing layout features, including stitches, is to employ the guidance from traditional DRCs. One typical DRC for stitches is a simple overlay check that ensures the intersections of the two complementary mask layers meet a minimum length requirement. See Fig. 3.

Fig.3: To guarantee the manufacturability of stitches, DRCs enforce that a minimum overlay length (dimension indicated by the arrow) is met for every stitch.

The DRC that checks the overlay length (dimension shown by the arrow in Fig. 3) ensures that there is enough overlap at the stitch to guarantee manufacturability. In addition to providing guidance for designing stitches, DRCs are also powerful analytical tools. Based on the error flags, they can quickly classify millions of stitches by checks or by cells.

2.1.2 Exact pattern-based classification

Another way to characterize layout designs is to use pattern classification [7]. This methodology provides a fast and accurate way for classifying a large number of layout features, including stitches, along with their neighboring features. By specifying four parameters: key, space, don't care, and a pattern radius, stitches along with their neighboring features can be quickly captured. See Fig. 4.

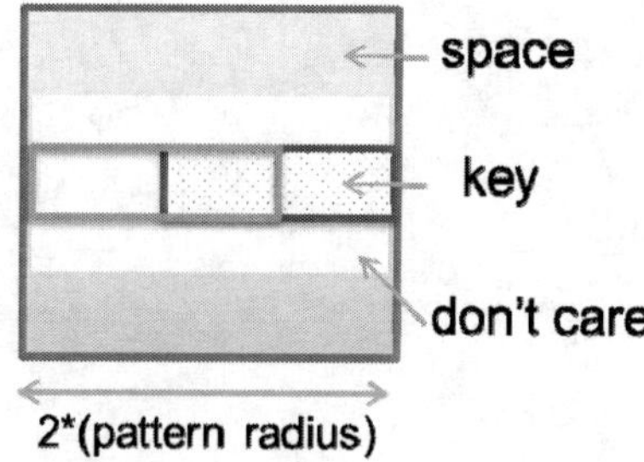

Fig 4: A pattern representation of a stitch is shown. The "key" is the pattern of interest. In this case, the stitch is the key. Using the key as the center of the pattern, the pattern radius provides the bounds for the automated layout extraction.

The same stitch shown in Section 2.1.1 is represented as a pattern in Fig.4. The "key" is the drawn feature, i.e., the stitch. The "space" is the empty space region surrounding the stitch. The "don't care" region is the tolerance band that enables the pattern classification to reduce similar "keys" with the dimensions that fall within the bands to the same pattern class. The size of the pattern clips is defined by the "pattern radius," in which each side is 2x the pattern radius. In this work, exact pattern classification is used. To do so, the tolerance of the "don't care" region is set to zero. The pattern radius is set to 100nm.

2.1.3 Topological pattern-based classification

A third way to classify layout designs is to use topological pattern-based classification [8]. This form of classification reduces patterns with the same root patterns, i.e., topologies, to the same topological class. In this work, the topological pattern classification is based on definitions of "scan lines" and "deltas." Each "scan line" corresponds to an edge in a pattern. Each "delta" corresponds to the range of dimensions between subsequent scan lines. See Fig. 5.

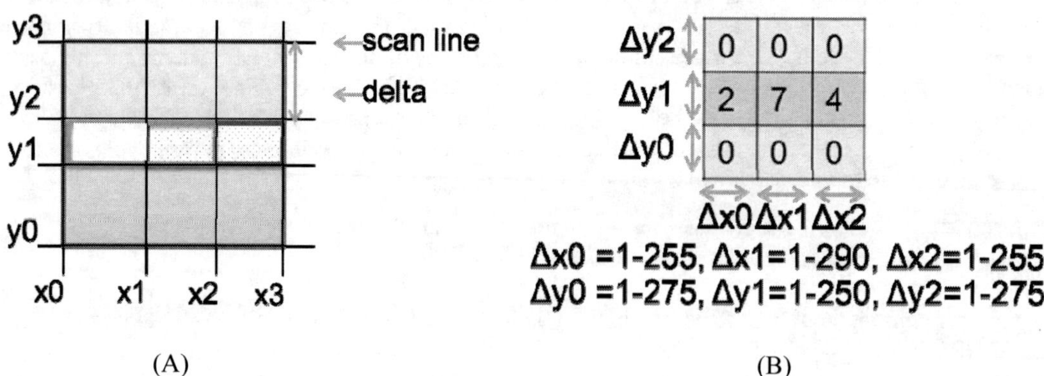

Fig. 5: A topological representation of a stitch is shown in (A), and the corresponding bitmap representation of (A) is shown in (B)

The same stitch shown in Section 2.1.1 and 2.1.2 is represented as a topological pattern in Fig. 5. Each edge is defined by a scan line. A 2D pattern has two sets of scan lines, one in the x-direction (i.e., x0 to x3) and another in the y-direction (i.e., y0 to y3). In between each subsequent scan line is a delta, a specified range of dimensions. A topological pattern shown in (A) is represented in bitmap notation shown in (B). Four layers are used: 0, 2, 4, 7. Layer 0 represents the empty space surrounding the feature, layer 2 represents the orange (left) feature, layer 4 represents the blue (right) feature, and layer 7 represents the intersection of layers 2 and 4. Below the bitmap representation in (B) are the delta values. For example, the overlap of the stitch is denoted as $\Delta x1 = |x2-x1|$. Its dimensions range from 1 to 290nm.

2.1.4 Exact patterns versus topological patterns for layout analysis and testing

To further clarify why both exact and topological patterns are important, the two types of patterns are compared in terms of analysis and testing applications. See Fig. 6.

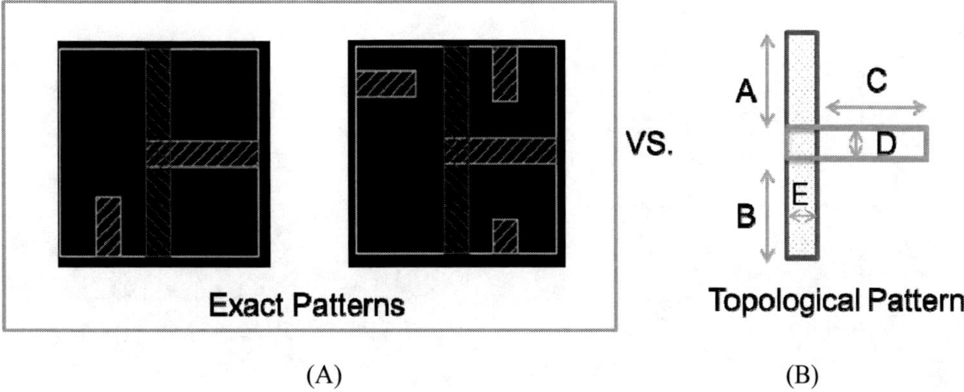

(A) (B)

Fig. 6: Exact patterns and topological patterns provide complementary analysis and testing.

Fig. 6 shows a cross-shaped stitch represented as an exact pattern versus a topological pattern. As shown in (A), exact pattern classification enables a simple and efficient way to extract layout clips across various contexts, preserving their exact dimensions. These patterns are suitable for context-aware testing, which is important for lithography and OPC. The reason is that lithography's optical radius of influence goes beyond that of the neighboring features.

On the other hand, topological patterns shown in Fig. 6(B) are suitable for dimensional analysis as well as parameterized testing. Topological pattern classification extracts similar root patterns. When classifying topologies from layout designs, the varying context and different neighboring features are not preserved. These patterns allow layout-induced manufacturing effects to be isolated to individual layout dimensions. They are used for automated layout fixing because individual edges of specific patterns could be easily defined and shifted to change dimensions.

2.2 Automated topological pattern-based layout optimizations

The stitching optimizations described in this work leverage a pre-existing Pattern Optimization (POP) flow for automated layout changes [9] [10]. The topological pattern-based automated layout fixing flow is called DFM-POP. See Fig. 7.

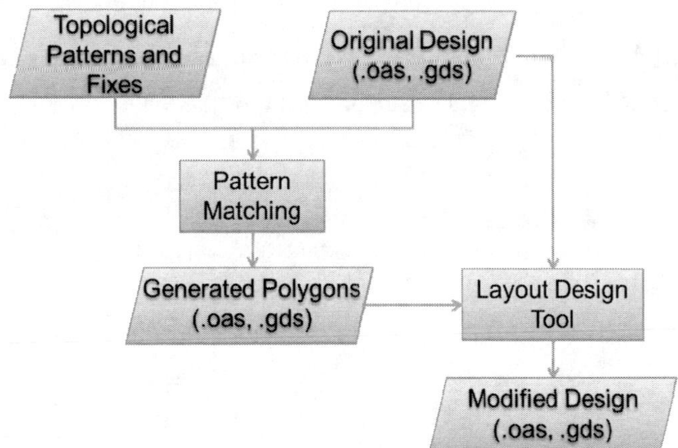

Fig. 7: The DFM-Pattern Optimization (POP) flow

The layout fixing flow shown in Fig. 7 uses a pattern matching engine to scan the original design for topological patterns. If a match is found, the pattern matching engine applies the fix by generating polygons on a temporary layer.

An external layout tool is then used to manipulate the original design with the generated polygons to create a modified design.

An example of how pre-characterized topological patterns are used for automated layout-fixing is shown in Fig. 8. Since topological patterns are defined based on scan lines that align to the edges of a pattern, these edges can be automatically extracted and associated with the generated shapes.

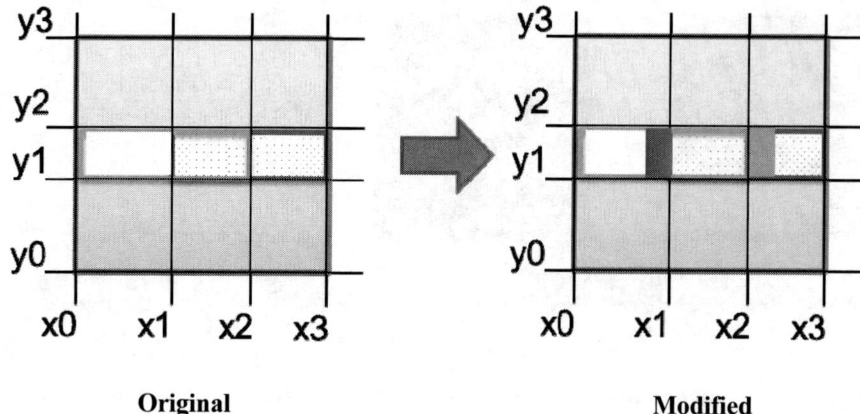

Original **Modified**

Fig. 8: The DFM-POP flow increases the overlay area by manipulating the original layout with generated shapes. These shapes are associated with the relevant edges in the original design. In the modified layout, two generated shapes (shown by the solid colors) are appended to the left and right edges of the stitch via using an external layout design tool.

Fig. 8 shows how DFM-POP increases the area of a stitch. The scan line definitions allow the ease of automated edge extraction and association with generated polygons. The dimensions of the generated polygons are specified based on the shifting of the scan lines using simple mathematical operations. For example, to shift the left edge of the stitch 5nm to the left, the pre-characterized fix is defined as $x1-5$, in which $x1$ is the scan line for the left edge of the stitch. The DFM-POP flow provides flexibility for fixing because an external layout design tool can use Booleans operations to post-process the original layout using the generated shapes.

3. AUTOMATED PATTERN-BASED LAYOUT CHARACTERIZATIONS AND OPTIMIZATIONS FOR STITCHES

Automated pattern-based stitching optimization involves characterizing and analyzing the layout patterns, building a topological pattern library, determining the associated fixes, and generating an optimized layout.

3.1 Characterizing stitches with DRCs, exact pattern classification, and topological pattern classification

The characterization flow described in Section 2.1 is demonstrated for the M1 layer stitches across four 20nm designs. Results are summarized in Table 2.

Design (M1 Layer Only)	Total counts	No. of Classes (Exact)	No. of Classes (Topological)
A	67 mil	2,890	85
B	2 mil	77	27
C	5k	28	7
D	8k	31	4

Table 2: The stitch usages for the M1 layer across four 20nm designs are summarized. Counts of unique patterns are compared with that of unique topologies for each design.

Table 2 shows that exact classification reduces the number of stitches, and topological pattern classification further reduces the number of exact patterns. For example, in Design A, design rules flag 67 million stitches. Exact pattern classification reduces the millions of stitches to 2,890 patterns. Topological pattern classification further reduces the thousands of patterns to 85 topologies.

In order to build a library of topological patterns and corresponding fixes, the top-occurring instances identified by each classification methodologies are examined. See Fig. 9.

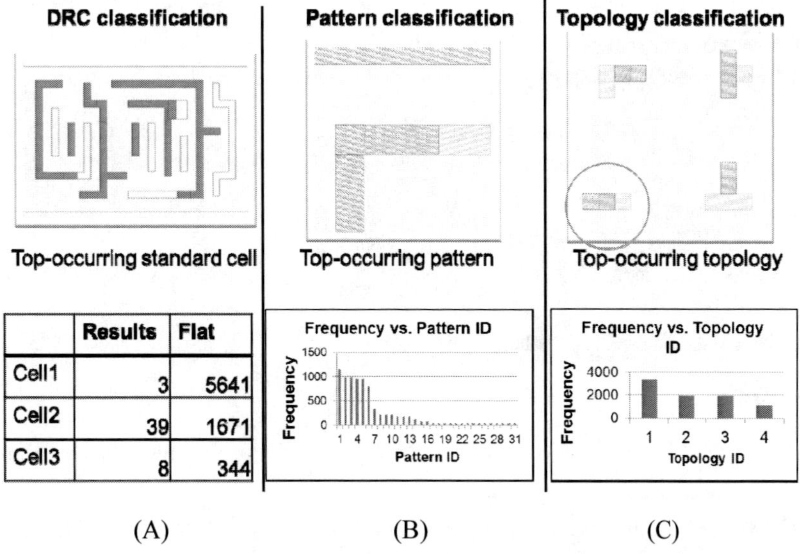

	Results	Flat
Cell1	3	5641
Cell2	39	1671
Cell3	8	344

Fig. 9: The top-occurring instances from each classification methodology are compared and contrasted. Rule-based classification, or DRCs, allows the study of the top-occurring standard cells that contained stitches. Exact pattern classification enables the study of the top-occurring patterns. Topological pattern classification identifies the top-occurring topologies.

Fig. 9 shows that each classification methodology provides complementary insights. In (A), rule-based classification identifies and ranks the standard cells that contribute the most number of stitches to the overall design. For example, Cell 1 has 3 stitches and contributes 5,641 stitches to the overall design. In (B), pattern classification enables the examining of the top-occurring patterns that are not obvious by studying the standard cells. In addition, the distribution of pattern usage in the overall design can be plotted and analyzed. For example, the top-occurring pattern shown in (B) with the Pattern ID 1 appears approximately 1,200 times in the overall design. In (C), topological pattern classification offers yet another perspective. For example, the top-occurring topology circled in (C) with the Topology ID 1 appears approximately 3,000 times in the overall design. The three top-occurring classified topologies are used for automated pattern-based auto fixing.

3.2 Using DFM-POP to improve layout regularity and robustness to mask-to-mask misalignment

The DFM-POP flow described in Section 2.2 is used to decrease the variability in stitch area. To do so, a layout design with 67 million M1 stitches is classified to 85 topologies. The three top-occurring topologies are studied, and each topology is coded with a pre-characterized fix. See Fig. 10.

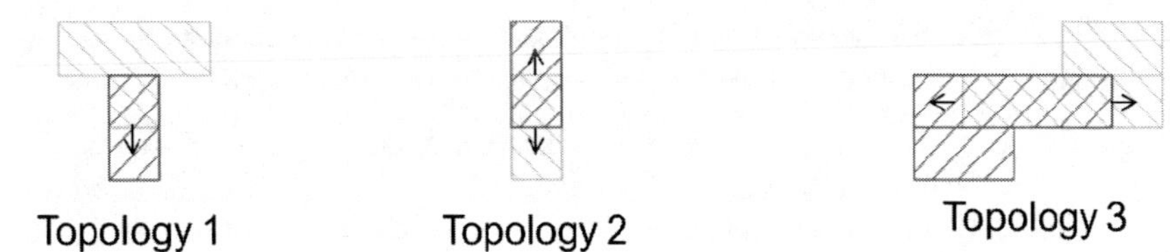

Fig. 10: The three top-occurring topologies used to improve layout regularity are shown. Each topology has a different pre-characterized fix as denoted by the arrows.

The three topologies and their pre-characterized fixes are shown in Fig. 10. Each topology employs a different style of fixing as denoted by the corresponding arrows. For example, the fix for Topology 1 is done by shifting the lower edge of the stitch. This way, no new topologies are introduced. For Topology 2 and 3, the fixes shift both edges of the stitch. These fixes reduce the number of unique stitch areas by regularizing the stitch area to a pre-determined dimension of 0.0028 um^2. First, the fixing flow detects the stitch areas that are smaller than 0.0028 um^2. Then, the flow adjusts the stitch overlay lengths so that the area meets the requirement. These fixes are opportunistic because any added shapes that generated new DRCs are dropped from the generated layout.

Table 3 shows the results of the applied fixes.

	Original	Post-Fixing
Mean (um^2)	0.0025	0.0029
Std_norm/ Mean (%)	3	0.7
Count	67 million	67 million

(A)

Topology	# of Fixes
1	750k
2	5.2 million
3	2800
Total	6 million

(B)

Table 3: Original and post-fixing results are shown in (A), and the number of fixes applied by each topology is summarized in (B).

Post-fixing analysis shows that the layout manufacturability improves in two ways. First, as shown in Table 3 (A), the mean area shifts from 0.0025 to 0.0029 um^2. This change shows that the layout manufacturability has improved because larger stitches are easier to manufacture. Secondly, the ratio of standard normal to mean of the stitch area is 3% in the original design and decreases to 0.7% post-fixing. This decrease indicates that the spread has decreased by approximately 4x. The manufacturability improves because a tighter spread indicates better layout regularity. Table 3(B) quantifies how many fixes are applied by each topology. Of the 67 million stitches, 6 million stitches are modified. Topology 2 applies the highest number of fixes, even though it is the second most frequently used topology in the original design.

The DFM-POP flow can also be used to improve the designs' robustness to mask-to-mask misalignment at the stitch. See Fig. 11.

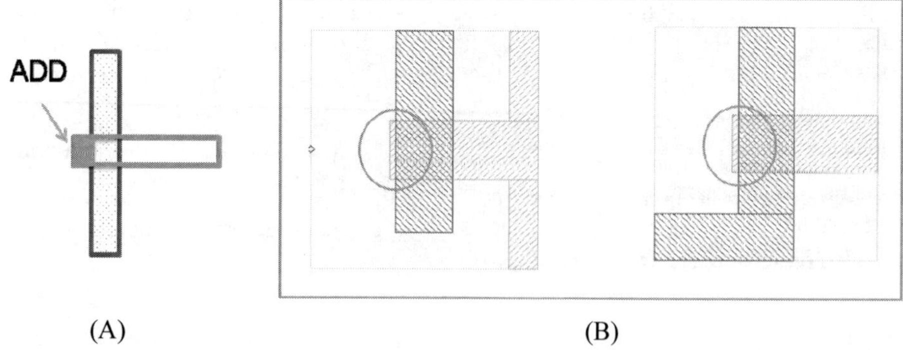

(A) (B)

Fig. 11: The DFM-POP flow is used to improve the stitch enclosure for a cross-shaped stitch topology. A topological representation of the fix is shown in (A), and the post-fixing layout clips are shown in (B).

A cross-shaped topological pattern is coded to add an additional 5nm enclosure at the stitch. To do so, DFM-POP adds a polygon at the associated edge of the topological pattern. See Fig. 11(A). By opportunistically adding a 5nm enclosure at the stitch, the robustness to mask-to-mask misalignment in the horizontal direction is improved. Post-fixing results are shown in Fig.11 (B). These layout clips show that the enclosures for the cross-shaped stitch topology is systematically improved (indicated by the circles).

3.3 Applying topological pattern-based fixes to complex patterns

Topological patterns can also be used to do more intrusive fixing for complex patterns. Furthermore, results can be confirmed with lithography simulations. See Fig. 12.

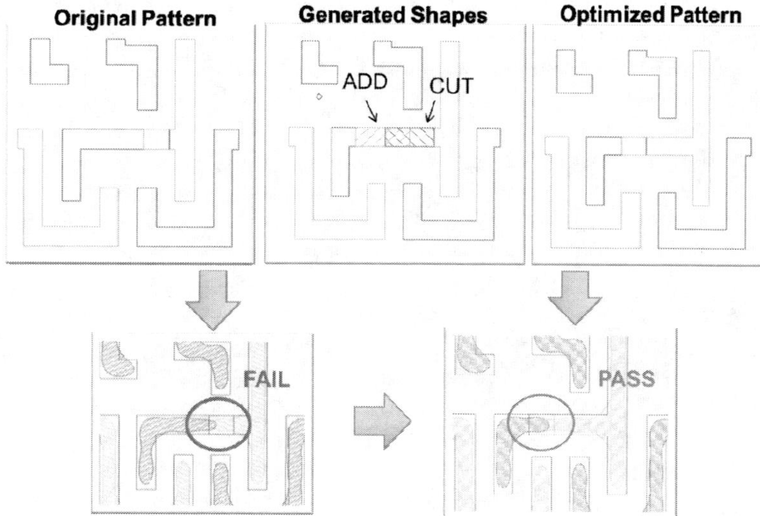

Fig. 12: The DFM-POP flow is used to shift stitches to improve lithography contours. The fix mitigates the lithography pinch in the original pattern.

The fixes applied by DFM-POP are validated using lithography simulations. In Fig. 12, the original pattern and its corresponding lithography simulations are shown. There is not enough overlap at the stitch, which causes a lithography pinch and an optical rule check failure. To address the issue, the DFM-POP flow generates two polygons, an add shape and a cut shape. These two shapes are then used by the external layout design tool to manipulate the original layout and generate the optimized pattern. This pattern shifts the original stitch location, and the corresponding lithography simulations show that the contours improved at the stitch. The original pinch is mitigated, as indicated by the passing of the optical rule check.

4. CONCLUSIONS

Stitching is a multi-patterning design technique that resolves conflicts with minimal area penalties. Current design methodologies do not guide where and how to stitch, which lead to a large variety of stitches being used in production designs. Pattern-based classification provides a complementary method for characterizing stitches, in addition to traditional rule-based approaches. An automated topological pattern-based fixing flow is applied to improve stitches. When optimizing based on stitching area, changing three topologies reduces the layout variability by approximately 4x.

REFERENCES

[1] Wong, A. K., "Resolution Enhancement Techniques in Optical Lithography," SPIE Publications (2001).

[2] Dusa, M., Finders, J., and Hsu, S., "Double patterning lithography: the bridge between low K1 ArF and EUV," Microlithography World (2008).

[3] Pan, D. Z., Yang, J.-S., Yuan, K., Cho, M., and Ban, Y., "Layout optimizations for double patterning lithography," IEEE 8th International Conference on ASIC (2009).

[4] Bencher, C., Dai, H., and Chen, Y., "Gridded design rule scaling: taking the CPU towards the 16nm node," Proc. SPIE 7274, 72740G-1-72740G-10 (2009).

[5] Rubinstein, J. and Neureuther, A. R., "Through-focus pattern matching applied to double patterning," Proc. SPIE 7274, 72741A-1- 72741A-11 (2009).

[6] Xie, P., Zou, Y., Capodieci, L., Dai, V., Muddu, S., and Wang, L.T.-N., "Methods for decomposing circuit design layouts and for fabricating semiconductor devices using decomposed patterns," USPTO, Patent 8,555,215 (2013).

[7] Dai, V., Yang, J., Rodriguez, N., and Capodieci, L., "DRC Plus: augmenting standard DRC with pattern matching on 2D geometries, "Proc. SPIE 6521, 65210A-1-65210A-12(2007).
[8] Dai, V., Lai, Y.-C., Gennari, F., Teoh, E., and Capodieci, L., "Systematic physical verification with topological patterns," Proc. SPIE 9053, 905304-1-905304-10 (2014).
[9] Wang, L.T.-N., Dai, V., and Capodieci, L., "Automated design layout pattern correction based on context-aware patterns," USPTO, Patent 8,924,896 (2014).
[10] Wang, L.T.-N., Dai, V., and Capodieci, L., "Methods for pattern matching in a double patterning technology-compliant physical design flow," USPTO, Patent 8,418,105 (2013).

Fast Detection of Manufacturing Systematic Design Pattern Failures Causing Device Yield Loss

Jean-Christophe LE DENMAT[1], Nelly FELDMAN[1], Olivia RIEWER[1],
Emek YESILADA[1], Michel VALLET[1], Christophe SUZOR[2], Salvatore TALLUTO[2]
(1) STMicroelectronics, 850 rue Jean Monnet, 38926, Crolles Cedex, France
(2) Synopsys, 700 E. Middlefield Rd, 94043 Mountain View, USA

ABSTRACT

Starting from the 45nm technology node, systematic defectivity has a significant impact on device yield loss with each new technology node. The effort required to achieve patterning maturity with zero yield detractor is also significantly increasing with technology nodes. Within the manufacturing environment, new in-line wafer inspection methods have been developed to identify device systematic defects, including the process window qualification (PWQ) methodology used to characterize process robustness. Although patterning is characterized with PWQ methodology, some questions remain: How can we demonstrate that the measured process window is large enough to avoid design-based defects which will impact the device yield? Can we monitor the systematic yield loss on nominal wafers? From device test engineering point of view, systematic yield detractors are expected to be identified by Automated Test Pattern Generator (ATPG) test results diagnostics performed after electrical wafer sort (EWS). Test diagnostics can identify failed nets or cells causing systematic yield loss [1],[2]. Convergence from device failed nets and cells to failed manufacturing design pattern are usually based on assumptions that should be confirmed by an electrical failure analysis (EFA). However, many EFA investigations are required before the design pattern failures are found, and thus design pattern failure identification was costly in time and resources. With this situation, an opportunity to share knowledge exists between device test engineering and manufacturing environments to help with device yield improvement.

This paper presents a new yield diagnostics flow dedicated to correlation of critical design patterns detected within manufacturing environment, with the observed device yield loss. The results obtained with this new flow on a 28nm technology device are described, with the defects of interest and the device yield impact for each design pattern. The EFA done to validate the design pattern to yield correlation are also presented, including physical cross sections. Finally, the application of this new flow for systematic design pattern yield monitoring, compared to classic in-line wafer inspection methods, is discussed.

Keywords: device yield, optical lithography, critical design pattern, wafer inspection, PWQ, systematic defectivity, electrical scan test, ATPG, test diagnosis, EWS, EFA.

Design-Process-Technology Co-optimization for Manufacturability IX, edited by John L. Sturtevant, Luigi Capodieci,
Proc. of SPIE Vol. 9427, 942705 · © 2015 SPIE · CCC code: 0277-786X/15/$18 · doi: 10.1117/12.2085307

1. Introduction

Starting from the 45nm technology node, systematic defectivity has a significant impact on device yield loss with each new technology node. In manufacturing environment, the effort required to achieve patterning maturity with zero yield detractor is also significantly increasing with technology nodes. Considering systematic defectivity, the objective for technology development is to drive process to maturity with no systematic defect causing device yield loss. One question remains: can we demonstrate that there is no residual systematic defect when the technology is considered as mature? From a device test engineering point of view, systematic yield detractors are expected to be identified by diagnostics of Automated Test Pattern Generator (ATPG) test performed after electrical wafer sort (EWS). Test diagnostics can identify failed nets or cells causing systematic yield loss [1],[2]. For a mature technology, typical device test diagnostics provides failure Pareto with no single net or cell as clear outlier from the general population. As a consequence, there is no possibility to select the major yield detractors for further analysis for process or product yield improvement. If a specific design pattern with a high failure rate is the root cause of failures on different nets, the time required to detect the defect with classic failure analysis methods is long and painstaking work because it must be repeated until the systematic defect is identified repeatedly. In this situation, there is no proof that the defect found is a major yield detractor.

The key idea of the proposed flow is to change the electrical failure Pareto provided by device yield analysis from net or cell based to a Pareto based on design patterns. With this new representation, the user can quickly select from the whole electrical failures population the critical design patterns which cause device yield loss. Once the critical design patterns are confirmed, this approach also allows to estimate the yield loss from the critical design patterns.

From the electrical failure analysis (EFA) point of view, convergence from device failed nets and cells to failed design patterns are usually based on assumptions that should be confirmed. However, many EFA investigations are required before the design pattern failures are found, and thus design pattern failure identification was costly in time and resources.

With the new electrical failure Pareto representation based on design patterns, users can provide to the EFA lab the reported critical design pattern to be checked, instead of the whole net or cell to analyze, to drastically reduce time to results for EFA operators.

From a product design point of view, unlike memory designs, systematic defects are difficult to detect within logic design, and the proposed methodology is developed to help systematic defect detection caused by critical design patterns in those logic areas.

2. Design pattern – Product Yield Correlation Flow description

The new flow aims to correlate critical design patterns for manufacturing to the observed device yield loss. After EWS test, the user is able to run ATPG diagnostics which results in an electrical failure Pareto with failures identified by net or cell. For devices in production, the yield analysis is performed on several production wafers in order to separate the systematic failures from global failures including random defectivity.

The new flow changes the electrical failure Pareto nets or cells to a new electrical failure Pareto based on design patterns. The additional information needed is a manufacturing design pattern library describing all known potential critical design patterns related to manufacturing process. This new Pareto is expected to highlight design pattern commonalities hidden in the classic net or cell Pareto results. In other words, several different failed nets or cells can be grouped together with a common design pattern, and identified as an outlier from the rest of the population. This allows the user to focus on the nets or cells with specific design pattern for electrical failure analysis. This representation provides visibility on which design patterns are involved in yield loss, and to estimate the yield loss impact of each design pattern.

2.1 Manufacturing design pattern library

The manufacturing design pattern library contains a list of all known design patterns which embeds a criticality toward a manufacturing process. Within the manufacturing environment, design patterning team composed by lithographers, etch engineers, metrology team and RET team are focused on improvement of device yield by decreasing design pattern criticalities.

Among several methods within manufacturing environment, in-line wafer inspection methods have been developed to identify device systematic defects. A popular method consists in characterizing design patterning robustness through dose and focus two important lithography process parameters to find out mask process window. This method well-known as process window qualification (PWQ) methodology is used intensively by many manufacturers today to screen out systematic defects from random defects on a wafer. This PWQ methodology is particularly efficient in giving list of design patterns limiting lithography dose and focus process window for a given design layer. This first source of critical design patterns was used to feed manufacturing design pattern library (Fig.1)

Design patterning is also characterized by OPC simulation which is used to simulate all new masks before mask fabrication in maskshop. Indeed, within Mask Data Preparation flow, one of main steps is mask sign-off with full chip mask data OPC simulation before mask fabrication in maskshop. This full chip mask data OPC simulation provides complete visibility on critical design patterns embedded on a device mask set (Fig.1).

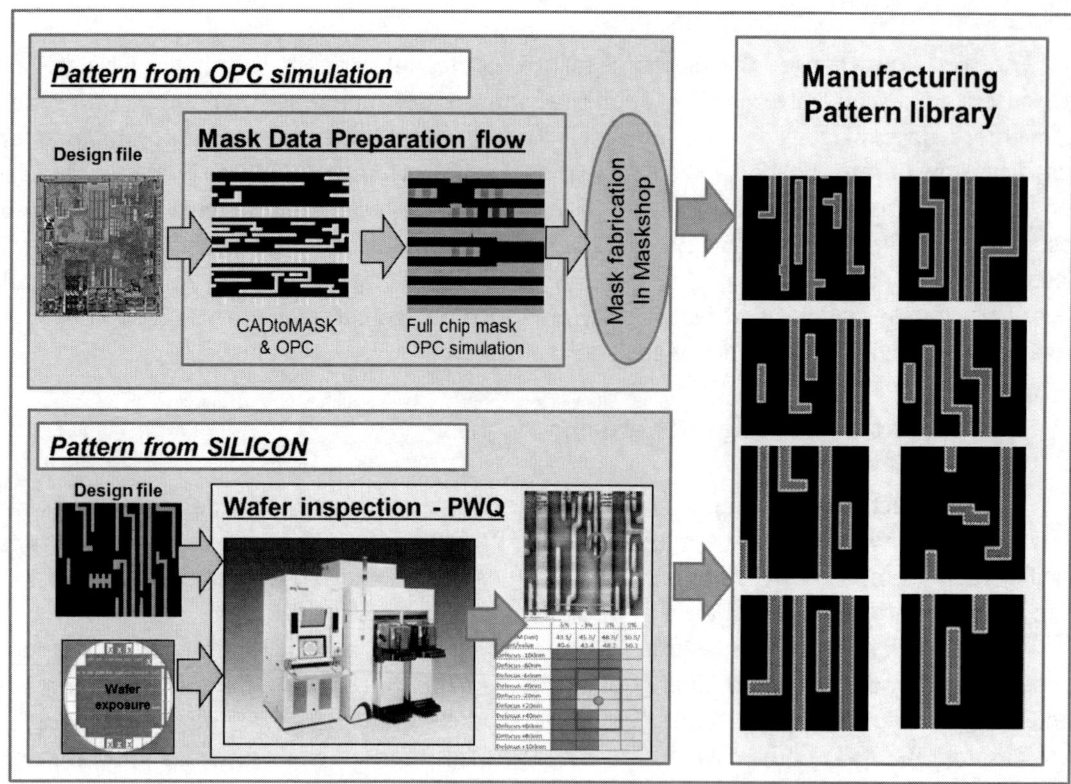

Fig.1 - Manufacturing design pattern library from Full chip mask data simulation & Systematic defect detection on Silicon.

2.2 Yield diagnosis flow with Manufacturing Design Pattern library

Manufacturing design pattern correlation to product yield is based on volume diagnostics results of ATPG failures detected during EWS testing, using statistical analysis to identify systematic failures related to critical design patterns. The volume diagnostics flow is setup with the logical netlist, the physical design, the ATPG test patterns, and the tester failure logs. The flow is performed in two steps: a) Diagnostics of the test failures to identify failure candidates by using logical netlist and physical design and fault simulation, and b) statistical analysis of the failure candidates correlated to the critical design patterns observed throughout the design. The flow is automated as shown below (Fig.2)

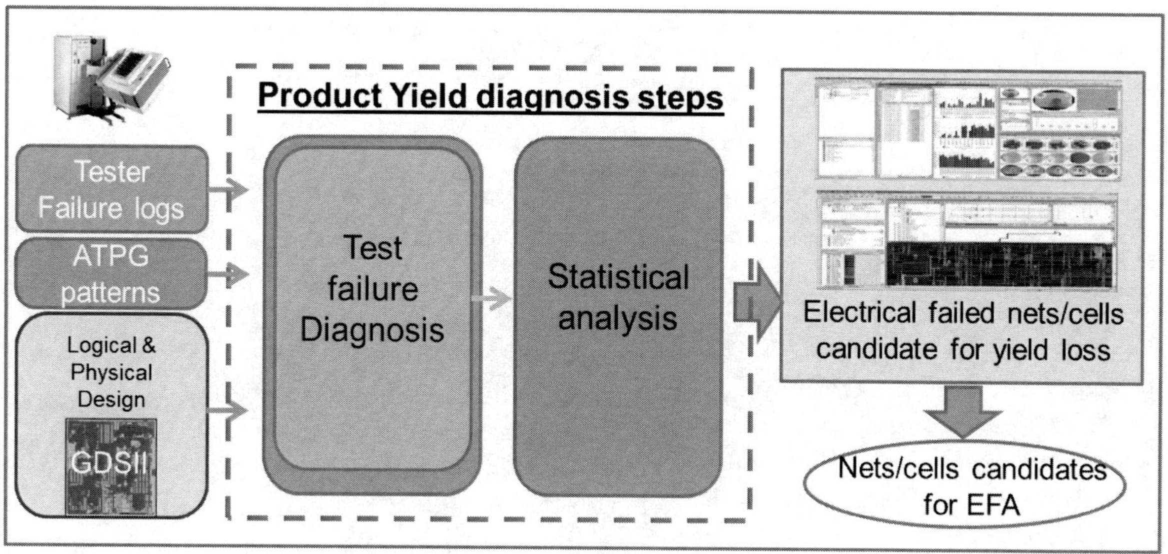

Fig.2 – Classic EWS volume diagnosis and statistical analysis.

To perform design pattern correlation to product yield, the first step is to run the design pattern matching engine on the device for this analysis. When all design patterns are identified on device under analysis, the correlation to the candidates from the diagnostics results is performed. This operation gives the critical design patterns correlated to measurable yield loss. With additional filtering, the user can determine the wafer signature associated to critical design patterns, and the failure statistics per design pattern, including wafer or lot trends associated to manufacturing equipment (Fig.3).

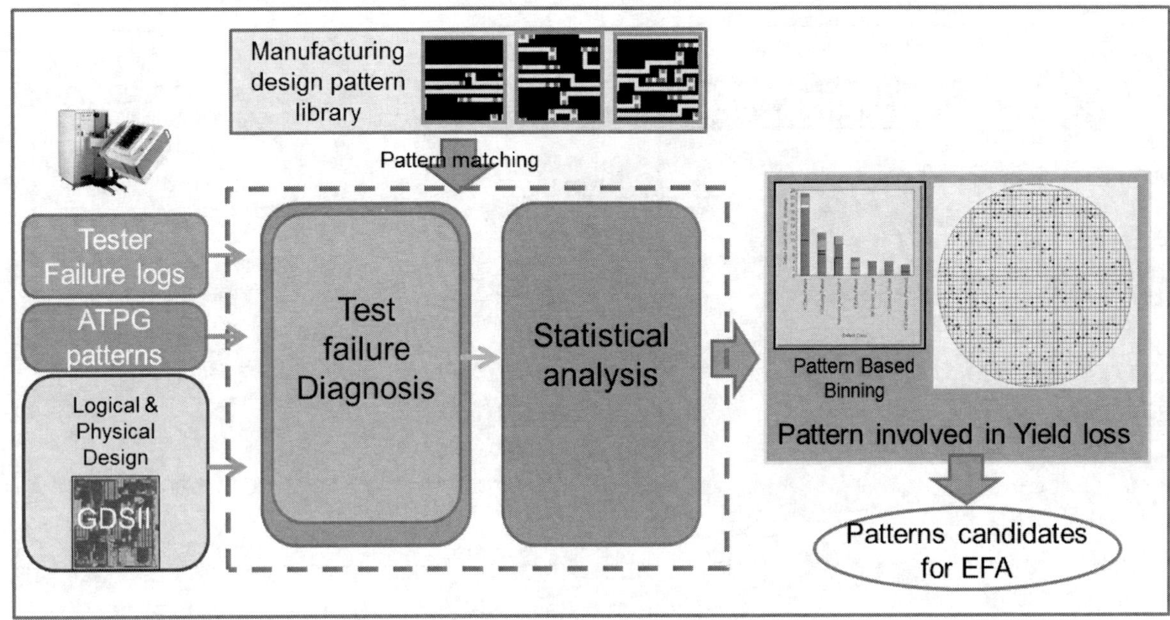

Fig.3 –Enhanced EWS volume diagnosis and statistical analysis with a manufacturing design pattern library.

This new flow is aimed to reduce time to result for EFA analysis, but also to provide better visibility to engineer concerning yield loss amount associated to a given pattern and found best solution to fix the problem like mask order, process change or even DFM library update. These results thus become an important tool for identifying manufacturing or design problems, and thus to improve the device yield.

3- Manufacturing Design pattern library - device Yield correlation results

3.1 Device, wafers and manufacturing design pattern library

The manufacturing design pattern correlation to product yield flow was exercised on a 28FDSOI device. 9 wafers from 3 lots were selected after EWS test, with their yield aligned with the current yield trend in order to avoid wafers with process excursion. The EWS tests used included ATPG tests covering 98.5% of product stuck-at fault library. Yield analysis was done with Yield Explorer, which is part of the Design-centric Yield management platform provided by Synopsys.

The manufacturing design patterns library was setup with back-end metal layers hotspots provided by OPC simulation, and with patterns found during design based wafer inspections.

OPC Hotspots were extracted from Metal line OPC simulation of Mask Metal line 2 (M2) to Mask Metal line 6 (M6) with the following defect detectors: Metal line open (*metal line - neck*), metal line short (*metal line - bridge*), *metal line - via coverage*, *metal line - via distance,* and *metal line min area*.

Design patterns from wafer inspection were provided on Metal line 2 layer after hard mask etch. The KLA-Tencor 2915 Broad Band Plasma wafer optical inspection tool coupled with the eDR-7000 e-beam review tool were used to perform this operation. A design-based approach for defect filtering was applied to select systematic defects amongst the global defectivity. Then we applied a pattern matching engine on device metal lines to detect all pattern occurrences for the design. For example, 11 defects of interest found by wafer inspection were pattern matched on device metal line 2. This resulted in the identification of 421 pattern instances, referred as *Inspected defect/Wafer inspection* in the table below (Tab.1) (for M2 only, since other layers were not analyzed).

Defect type	patt source	M2	M3	M4	M5	M6
Metal line - Neck	OPC simulation	395	1093	919	291	135
Metal line - Bridge	OPC simulation	998	2118	2072	1252	687
Metal line - Via Coverage	OPC simulation	997	11020	6901	4685	2979
Metal line - Via Distance	OPC simulation	490	904	876	286	530
Metal line MinArea	OPC simulation	99	38	8	2	0
Inspected defect	Wafer Inspection	421	0	0	0	0
TOTAL		**3400**	**15173**	**10776**	**6516**	**4331**

Tab.1 – Manufacturing design pattern library. Defect number per layer and per defect type

3.2 Design Patterns candidates for product yield loss

Once the manufacturing design pattern library was setup, and the EWS ATPG diagnostics analysis was done on 9 wafers, we could perform a spatial correlation between the design patterns (from OPC simulation and from wafer inspection) and the electrical failing nets and instances from the diagnostics results. The result is the identification of all the diagnostics candidates which are physical matched to critical design patterns, and thus potentially explaining the observed yield loss. This can be shown as a pareto chart by layer and defect Type (Fig.4).

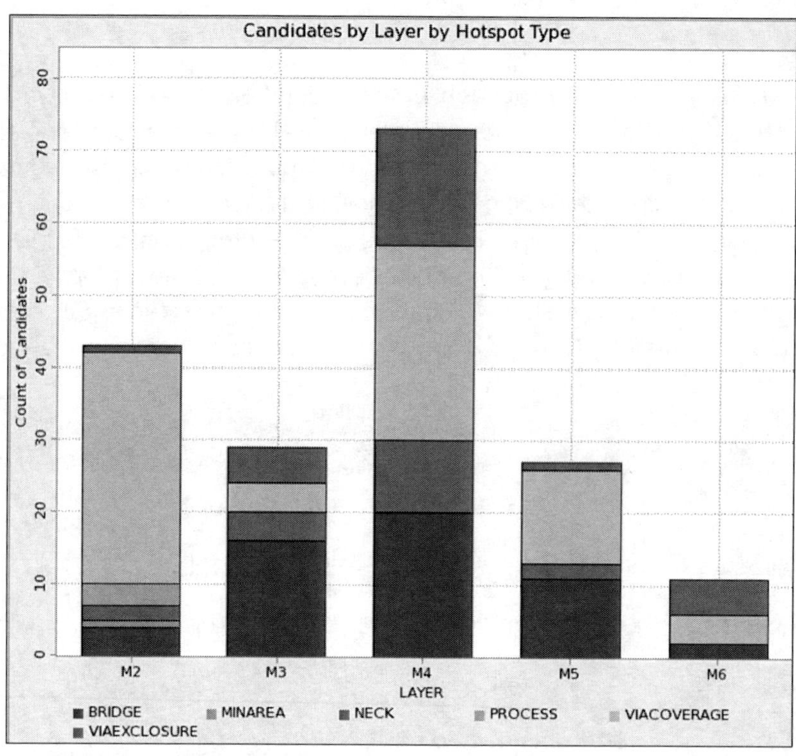

Fig.4 – Design pattern candidates for product yield loss

3.3 Result for Design patterns from OPC simulation

For design patterns provided by OPC simulation, a correlation to failing nets does not necessarily confirm that it is the root cause of the electrical net failure. Further investigation is needed to confirm the link between the design pattern and the electrical failure. For that purpose, an EFA campaign was done on several design patterns hitting failed nets. An EFA campaign on Metal line – bridge design patterns, Metal min area design patterns or Metal line-Via distance design patterns could not highlight any issue on cross sections and thus could not establish any link between the electrical failure and the manufacturing design pattern criticity. However, an EFA campaign on Metal line neck design patterns and Metal line - via coverage design patterns could identify a real defect on cross section and thus prove that the electrical failure was related to the manufacturing design pattern weakness.

For Metal line - Via coverage, 5 failure candidates with correlated design patterns were selected for EFA, and the root cause for all 5 failures were related to the design pattern failure on the net. After this achievement, a defect filtering could easily lead to wafer signature information with a clear center to edge wafer distribution. Also wafer-to-wafer and lot-to-lot correlation distribution could be extracted from Yield Explorer. The yield loss could be estimated at 0,9% for Metal line - via coverage design patterns (Fig.5)

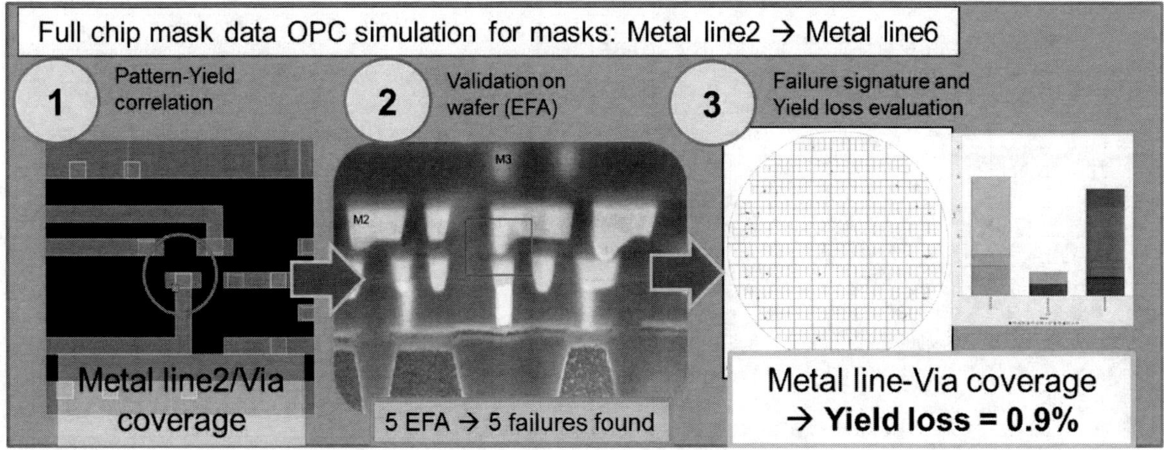

Fig.5 – Metal line - Via coverage correlation to product yield results

Wafer signature related to Metal line - via coverage could be correlated to Yield loss given by EWS on selected wafers. This achievement is particularly interesting as this type of defect is difficult to detect with classical in-line wafer inspection methods.

3.4 Result for Design patterns from Silicon

For results provided by wafer inspection method, the design pattern failure is already found on wafer. If a design pattern belonging to wafer inspection category in manufacturing design pattern library is correlated to electrical failed net, the probability to have a link between this design pattern and the electrical failure is high. As a consequence, there is no need to follow investigation by EFA analysis.

For Ushape defects in manufacturing design pattern library, some of them could be correlated to electrical failed nets. We could then extract from Yield Explorer the wafer signature information (no particular wafer signature in this case) and wafer-to-wafer and lot-to-lot defectivity distribution. Also yield loss could be estimated at 0,4% in this case (Fig.6).

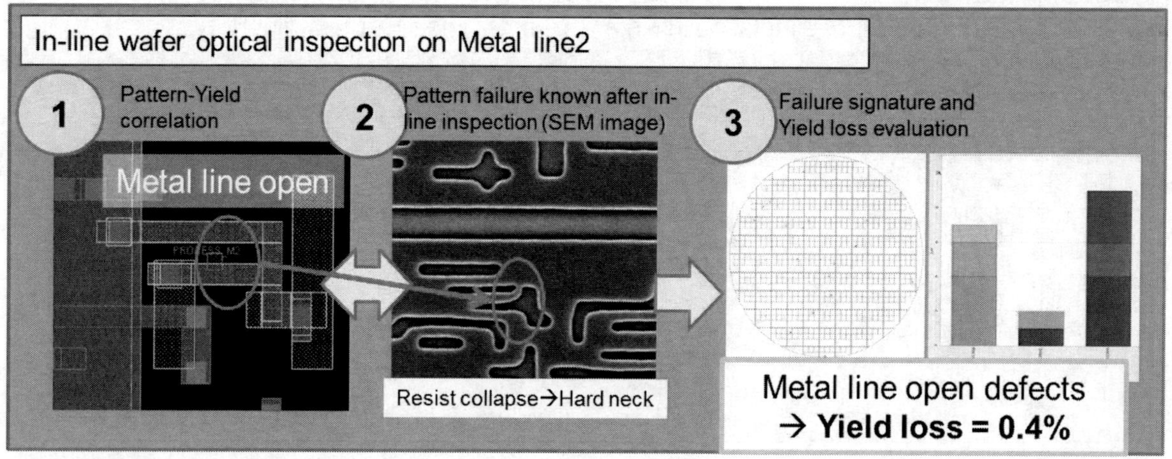

Fig.6 – Design pattern from Silicon (Inspected defect) correlation to product yield result

4. Discussion

4.1 Correlation method perimeter

It is important to note that this methodology has today some limitations. Product yield diagnosis is based among other data on product netlist to describe logical design. Design logic area is described by a Cell_ID for front-end design physical layers and by a net_ID for all design back-end layers (usually starting at metal2 design physical layer). Additional limitation is observable on memory blocks which IP is designed on whole front-end and large number of back-end layers. If an engineer wants to develop the proposed method also on memory design blocks, they need to provide electrical net description intra memory cell.

Another limitation is related to OPC modeling: if OPC simulation is not well calibrated to the process, it will result in wrong simulation of real defect of interest. Then the resulting manufacturing design pattern library will not embed the defects of interest coming from mask OPC simulation.

5. Conclusion

A new correlation methodology based on manufacturing design pattern knowledge and product yield was detailed in this paper. A manufacturing design pattern library is built on both full chip mask data OPC simulation and systematic defectivity found on silicon by in-line wafer inspection tool. This manufacturing design pattern library is then correlated to product yield loss through volume diagnostics and statistical analysis performed with Yield Explorer (Synopsys). The exercise done on a 28FDSOI product has shown very promising results, where defects such as "Metal line - Via coverage" and "Metal line open" were demonstrated with wafer signature and lot distribution information. An intensive EFA campaign confirmed the root cause for each defect type. The yield loss was estimated for each defect type: 0.9% yield loss for "Metal line – Via coverage", and 0.4% yield loss for "Metal line open".

This cross-domain analysis methodology aims to decrease the time to results for product debug, by focusing EFA analysis on manufacturing design patterns rather then failed nets. It also increases the visibility on product design pattern failures for the manufacturing process engineering teams. The methodology completes the manufacturing design pattern datamining efforts to speed-up the manufacturing process learning curve, and increase product yields.

References:

[1] Florent Garait, Stéphane Lecomte, Christophe Suzor, Aymen Lamine, Roberto Gonella, "Identification of systematic yield issues using volume diagnostics", SNUG June 2010

[2] Nelly Feldman, Vincent Robert, Christophe Suzor, Salvatore Talluto "Physical Data Loading to Improve Diagnosis Accuracy", SNUG June 2014

Topology and context-based pattern extraction using line-segment Voronoi diagram

Sandeep Kumar Dey[a], Panagiotis Cheilaris[a], Nathalie Casati[b], Maria Gabrani[b], Evanthia Papadopoulou[a]

[a]Università della Svizzera italiana (USI), Faculty of Informatics, Lugano, Switzerland
[b]IBM-Research Zurich, Säumerstrasse 4, 8803, Rüschlikon, Switzerland

Abstract

Abstract. Early identification of problematic patterns in real designs is of great value as the lithographic simulation tools face significant timing challenges. To reduce the processing time such a tool selects only a fraction of possible patterns,which have a probable area of failure, with the risk of missing some problematic patterns. In this paper, we introduce a fast method to automatically extract patterns based on their structure and context, using the Voronoi diagram of VLSI design shapes. We first identify possible problematic locations, represented as gauge centers, and then use the derived locations to extract windows and problematic patterns from the design layout. The problematic locations are prioritized by the shape and proximity information of design polygons. We performed experiments for pattern selection in a portion of a 22nm random logic design layout. The design layout had 38584 design polygons (consisting of 199946 line-segments) on layer Mx, and 7079 markers generated by an *Optical Rule Checker* (ORC) tool. We verified our approach by comparing the coverage of our extracted patterns to the ORC generated markers.

Keywords: Voronoi diagram, line-segment, photolithography, pattern selection, ORC, OPC.

1. Introduction

With the increase in miniaturization of current VLSI patterns, there is a significant rise in printability problems of such patterns, during the photolithography process. The analysis of patterns to find faults or error-prone locations, is of prime importance to the manufacturing process. There are mainly two kinds of faults that can occur during printing: a *pinch* and a *bridge*. A pinch corresponds to an open fault and it occurs due to incomplete printing of a shape or due to discontinuity in the printing of a shape. A bridge corresponds to a short fault and occurs when two printed shapes are touching each other.

The analysis of a complete layout for finding faults or error-prone locations is difficult and very time-consuming. The printability of a layout is related to the patterns it contains, thus, pattern selection should be done in a way that selected patterns suffice to asses the quality of the entire layout. In other words, it is important to identify patterns, which are prone to faults, known as *patterns of interest* (POI). A relatively small number of POIs that sufficiently cover the whole layout allow for a faster, yet effective layout analysis. Obtaining an optimal set of POIs is a big challenge. The success of printing a POI is verified by taking several measurements, such as critical distance, on potentially critical areas of *scanning electron microscope* (SEM) images of the printed pattern. The location of measurement is very important for the proper evaluation of a POI.

In each pattern, the measurement location is called a *gauge*.[1] The gauge is typically represented by a line segment in the VLSI pattern, around which a critical distance is measured. Gauge

Design-Process-Technology Co-optimization for Manufacturability IX, edited by John L. Sturtevant, Luigi Capodieci, Proc. of SPIE Vol. 9427, 942706 · © 2015 SPIE · CCC code: 0277-786X/15/$18 · doi: 10.1117/12.2085547

locations must be meaningful, that is, the critical distance measurement around the gauge location should be the correct measurement for the pattern. Current gauge suggestion techniques are rule-based or they are done manually by VLSI designers. The suggested gauges very often miss the location of the critical distance or the location of faults. The actual location of faults within a POI is known as a *hotspot*. Gauges are indicators that help to categorize a pattern as POI and to locate hotspots within the pattern. Good gauge suggestions improve the evaluation of a pattern, and contribute to the goal of achieving an optimal set of POIs.

In the literature, many variants of hotspot identification methods have been suggested, like, machine learning algorithms,[2,3] image recognition techniques,[1,4,5] a design based approach,[6] pattern matching techniques,[7,8] and topology oriented techniques.[9,10] For machine learning techniques, the learning time is high and there is a need for already available hotspots to be used as training sets. Image recognition techniques need to go over the entire layout for hotspot detection, which is very time consuming. Pattern matching techniques work on a predefined set of hotspot patterns, and thus, there is a limitation of detecting unseen hotspots. The design based approach has also been observed to be very time consuming as it needs to analyze the entire design. In general, the random nature of layout patterns is difficult to predict in all these methods. Topology oriented pattern extraction techniques[9] can be useful to handle random patterns, however, the existing technique[9] does not claim to be general enough to detect all the hotspots in a layout. In this paper, we introduce a new topology oriented technique for pattern extraction, based on the geometric information of layout shapes.

The set of POI and the predicted hotspots are of prime importance to address printability problems. After obtaining such a set, designers modify the mask designs with the goal of improving the level of yield. For verification, VLSI designers have developed several models based on POI, including the *model-based optical proximity correction* (MB-OPC).[1,5,11] Hotspots give feedback to the OPC process for better yield. An optimized MB-OPC demands an optimal set of problematic patterns, but identifying such a set in a time efficient manner is very hard. Currently, reliable OPC models are mostly based on image parameters of the test patterns,[1,5,11] however, techniques involving image parameters are computationally expensive. Another problem is the automatic identification of problematic patterns. In many cases, the expertise of lithographers and design engineers provides the problematic patterns by manual inspection of the layout. Some new automatic approaches[12,13] use a combination of parameters. These techniques sample the full spectrum of patterns, and thus, tend to be computationally expensive.

To provide feedback to OPC models, VLSI designers have developed tools to identify optical rule violation in the simulated lithographic patterns. An Optical rule checker (ORC)[14] is a program that encodes and verifies the rules for ideal simulated lithographic patterns. Given the lithographic processing conditions, an ORC run generates markers on the violation of an ORC rule. Therefore, the problematic patterns in the layout are around the ORC markers. These sets of patterns can be feedback to OPC for better yield in manufacturing.

In this paper we provide a fast automatic approach to derive a near optimal set of problematic patterns for a layout. These sets of patterns can also potentially be used for calibration and verification of MB-OPC. We first find gauge locations, using the line-segment Voronoi diagram of layout shapes, and give priority to the gauges depending upon the shape and proximity information of the design polygons. The gauge locations are then used to extract windows from the design layout.

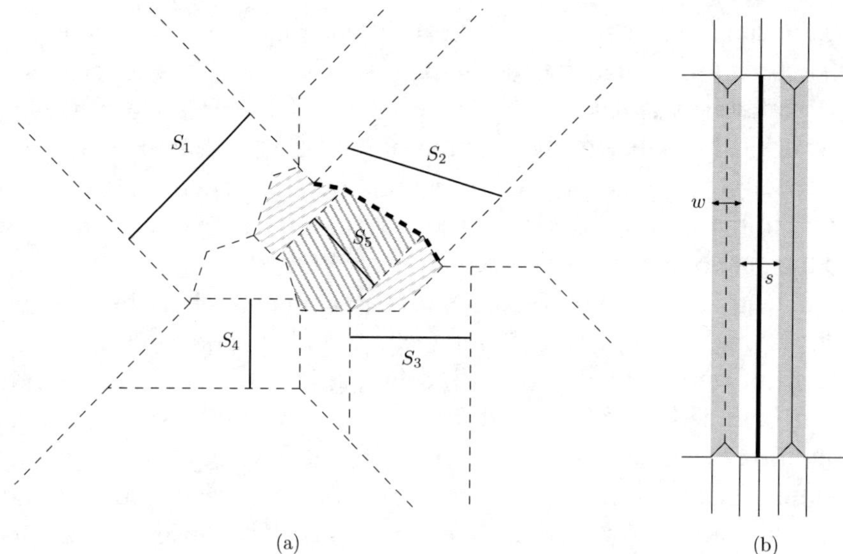

Fig 1 (a) Segment Voronoi diagram under L_∞ metric, with five distinct sites S_1, S_2, S_3, S_4, S_5, for interiors of segments and their endpoints. The lightly shaded portion is the Voronoi region of the endpoints of the segment S_5, dark shaded portion is the Voronoi region of the interior of the segment S_5. The thick black dashed line is the Voronoi edge separating the faces defined by segments S_5 and S_2, (b) Voronoi diagram (in black) of two design polygons (in grey), w is the width parameter and is encoded by a Voronoi edge shown in black dashed line, and s is the space parameter encoded by a Voronoi edge shown in thick black line.

We extract one window per gauge location. The windows contain patterns which are potentially problematic. Finally, we verify the usefulness of the extracted windows by comparing the coverage of the problematic patterns with respect to the ORC generated markers. We observe that the set of patterns extracted by our tool covers all the ORC generated markers for the given layout.

The rest of this work is organized as follows. In Section 2, we introduce the line-segment Voronoi diagram in the L_∞ metric as a tool for VLSI pattern analysis. We describe the gauges and their scoring method in Section 3. We describe our method to detect potentially critical locations in the VLSI design layout using the segment Voronoi diagram and then describe our pattern selection procedure in Section 4. We discuss experimental results on the given design layout in Section 5.

2. The L_∞ line-segment Voronoi diagram

Let S be a set of simple geometric objects, called sites, such as points, line-segments, or simple polygons. The Voronoi diagram[15,16] of S is a subdivision of the plane into regions such that the region of a site $s \in S$ is the locus of points closer to s than to any other site in S. The *distance* of a site s from a point q in the plane is $d(s,q) = \min_{p \in s} d(p,q)$, where the interpoint distance $d(p,q)$ can be the Euclidean (L_2) distance, the Manhattan (L_1, L_∞) distance, or any other metric. In this paper, we use the L_∞ metric (known also as the maximum norm), where the distance between two points p and q is given by $d(p,q) = d_\infty(p,q) = \max(|p_x - q_x|, |p_y - q_y|)$.

The Voronoi diagram of S is a planar graph of linear complexity, linear on the total complexity of the input sites. Each *face* corresponds to the Voronoi region of a site $s \in S$: $reg(s) = \{q \in \mathbb{R}^2 \mid d(s,q) < d(s',q), \forall s' \in S \setminus \{s\}$. Each region contains its defining site. Figure 1(a) illustrates the Voronoi diagram of a set of line-segments $\{S_1, S_2, S_3, S_4, S_5\}$. The shaded region is the Voronoi region of line-segment S_5 and it contains S_5. The boundary between two neighboring faces is an *edge* of the diagram (see e.g., the thick black dashed line in Figure 1(a)). A Voronoi edge is the locus of points equidistant from two defining sites and borders their corresponding neighboring faces. Edges meet at *vertices* of the diagram. At least three Voronoi edges meet at a Voronoi vertex. The Voronoi diagram encodes proximity information of the input sites. It can be constructed in $O(n \log n)$ time,[15,16] where n is the total complexity of the input sites. For more information on Voronoi diagrams see e.g., the books of Aurenhammer et al.[16] or Okabe et al.[17] For the Voronoi diagram of segments and polygons in the L_∞ metric, see.[18,19]

A VLSI layout is composed of design polygons which are predominantly rectilinear. There are two important parameters, *width* and *space*, that describe patterns in a VLSI layout. The width parameter is defined as the distance between two parallel edges of a design polygon. A Voronoi edge internal to a design polygon encodes this width parameter (see e.g. the thin dashed line in Figure 1(b)). The space parameter is defined as the distance of separation of two design polygons. A Voronoi edge induced by edges of two different design polygons encodes the space parameter (see e.g. the thick black line in Figure 1(b)).

For point sites, the Voronoi diagram is available through MATLAB[20] and for line segments in the Euclidean plane it is available through the CGAL library.[21,22] CGAL is an open source C++ library that provides easy access to efficient and reliable geometric algorithms. In the L_∞ metric, the segment Voronoi diagram has various desirable properties; for example, Voronoi edges are straight line segments and Voronoi vertices are on rational coordinates, unlike the corresponding diagram in the Euclidean metric. In the VLSI environment, where shapes can be assumed rectilinear, Voronoi edges are also rectilinear or 45-degree and Voronoi vertices appear on a grid only slightly finer than the coordinates of input shapes. Thus, the $L\infty$ version of the diagram is very well suited for modeling proximity on a VLSI layer. See e.g., the *critical area* extraction problem.[18,19,23,24] We have developed the L_∞ segment Voronoi diagram in the CGAL environment;[10,25] the code is currently under review for inclusion in the library.

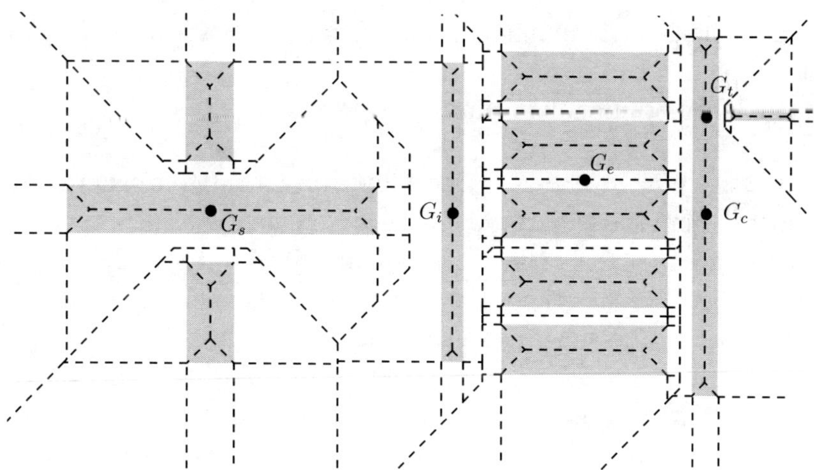

Fig 2 Different gauge suggestions.

3. Gauge suggestion using the segment Voronoi diagram

We use the L_∞ segment Voronoi diagram to suggest good gauge locations, based on proximity information among the shapes of a pattern, such as the space between shapes and the extent of interaction between neighboring shapes. We suggest five types of gauges, *internal, external, sandwich, comb* and *T*, as illustrated in Figure 2. Figure 3 illustrates different gauges in a portion of a design layout. Following is the description of the five gauge types.

1. *Internal gauge* (inside a shape), G_i: This gauge lies on the center of a Voronoi edge in the interior of the polygonal shape of minimum width in the pattern (see G_i in Figure 2). The position of G_i is a probable location for a pinch, when printing the pattern.

2. *External gauge* (between different neighboring shapes), G_e: This gauge lies on the center of the Voronoi edge between the two shapes that are closest in the pattern (see G_e in Figure 2). The position of G_e is a probable location for the formation of a bridge between the two corresponding shapes, when printing the pattern.

3. *Sandwich gauge,,* G_s: This gauge lies on the center of the Voronoi edge inside a polygonal shape P_1 that is "sandwiched" between two other shapes P_2 and P_3 for which the distance between P_2 and P_3 is the minimum in the pattern (see G_s in Figure 2). There is a probability of a pinch happening at P_1 around G_s because of the influence of P_2 and P_3.

4. *Comb gauge*, G_c: This gauge lies on the center of the Voronoi edge inside a long polygonal shape, which is the base of the comb and it has close to it and on one side of it a number of polygonal shapes which are the teeth of the comb. We report the gauge for the configuration where the base of the comb shape is closer to the teeth in the pattern (see G_c in Figure 2). The position of G_c is dangerous for a pinch, when printing the pattern.

5. *T gauge*, G_t: A comb gauge at a minimum must contain one tooth and a base. We call such a minimal comb gauge a *T gauge* (see G_t in Figure 2). The position of G_t is a probable location for a pinch, when printing the pattern.

We have validated the usefulness of these gauges by experiments on patterns of small size presented in a recent paper.[10] For each pattern we first suggested gauges. Then, for each suggested gauge we measured the distance in pixels in the corresponding SEM image and took the minimum. In most of the patterns we improved the critical distance measurement and in some cases we were able to locate hotspots missed by the suggested gauges of conventional methods. The ability to generate gauges may find additional uses, in particular, in the model generation of MB-OPC.

We introduce a scoring method for gauges, which is used for prioritizing the gauges to determine problematic patterns efficiently. The score associated with a gauge determines its affinity towards the printing problem. It indicates the potential of failure around the location of the gauge, when printing the related pattern.

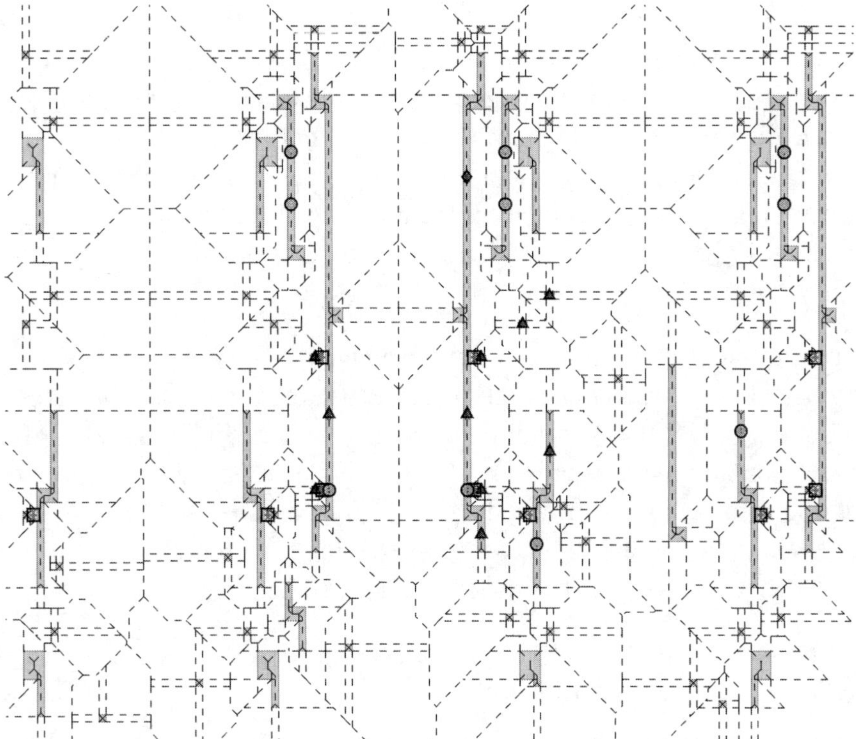

Fig 3 Showing different gauge locations in a portion of design layout: shapes in grey are design polygons, Voronoi diagram of design polygons is shown by dashed lines.

3.1 Scoring method for gauges

For each type of gauge we defined scores as follows:

- Score of an internal gauge: Let P_i be a shape in the design layout. The score of an internal gauge is the minimum width value in P_i (see w in Figure 4 (a)). In case there are shapes with the same width, we break ties using the *extension* parameter. The extension parameter is the length of the Voronoi edge associated with an internal gauge. The extension parameter does not change the value of the score; it is just used to change the order of the gauge in the priority list in case of ties. The gauge associated with the longer shape that is having a greater value of associated extension parameter gets more priority and will be ranked higher in the ordered list of internal gauges. Lower width implies a thinner and a greater extension implies a longer shape; a thin long shape has more probability to give rise to a pinch.

- Score of an external gauge: Let P_a and P_b be two neighboring shapes in the design layout. The score of an external gauge associated with P_a and P_b is the separating distance (s in the Figure 4 (b)) between P_a and P_b. This is encoded by the associated Voronoi edge between P_a and P_b. When there are gauges with the same score, we break ties using the extension parameter. The gauge whose associated Voronoi edge is longer gets higher priority, as the longer edge implies more interaction between the shapes, and thus, higher probability of a bridge.

- Score of a sandwich gauge: Let P_x, P_y, and P_z be three shapes in a design layout such that P_y is sandwiched between P_x and P_z. We define the score for the sandwich gauge equal to the $0.5\times\ d$ (see Figure 4 (c)), where d is the distance between the Voronoi edges E_{xy} and E_{yz} (see Figure 4 (c)). If there are (P_x, P_y, P_z) triples with the same score, we use the *overlap* parameter to break ties; a sandwich gauge with a greater overlap parameter gets higher priority. The overlap parameter is the measure of the length of the overlapping portion of Voronoi edges E_{xy}, E_y, and E_{yz} associated with the sandwich configuration (see Figure 4 (c)).
- Score of a comb gauge: For the score of comb gauges, we first define the distance d_{tb} between the tooth (P_t) and the base (P_b) as the distance between the Voronoi edges E_{tb} and E_b (as shown in Figure 4 (d)). The Voronoi edge E_{tb} is associated with an edge of the base and an edge of the tooth, and the Voronoi edge E_b is an edge associated with the base of the comb. The score of a comb gauge is then defined as $0.75 \times d_{tb}$. In case, we have comb gauges with same score, we break ties by the measure of overlap between E_{tb} and E_b. We give priority to the comb gauge where tooth has more overlap with the base.
- Score of a T gauge: For T gauges the score is defined as $0.80 \times d_{tb}$. The ties for T gauges are broken in the same way as for comb gauges.

The lower the score of a gauge, the higher is the probability of getting a problematic pattern around that gauge. When gauges of the same type get the same score, we break ties by the extension and overlap parameters described above. A higher value of these parameters gives a higher priority to the gauges. If two different types of gauges have the same score, we break ties using the *context* parameter, which reflects complexity of the pattern associated with the gauge. We have simply considered a hard coded order of context parameter, $C_s > C_c > C_t > C_e > C_i$, where

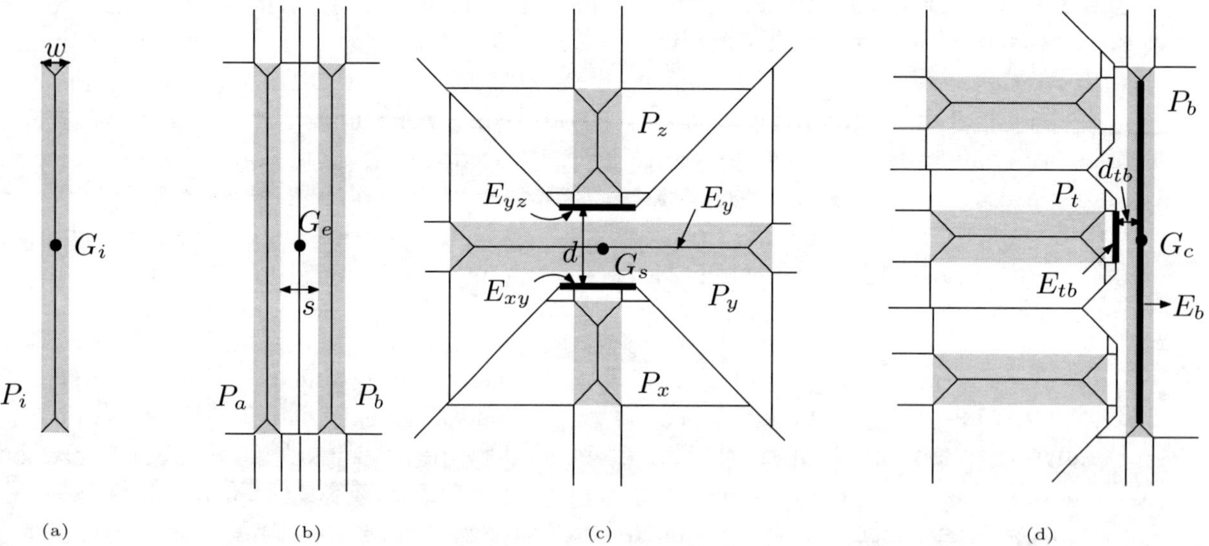

Fig 4 An example showing different gauges with their scoring formula: (a) internal gauge with score = w, (b) external gauge with score = s, (c) sandwich gauge with score = $0.5 \times d$, and (d) comb gauge with score = $0.75 \times d_{tb}$, the score of a T gauge is computed similarly with score = $0.80 \times d_{tb}$.

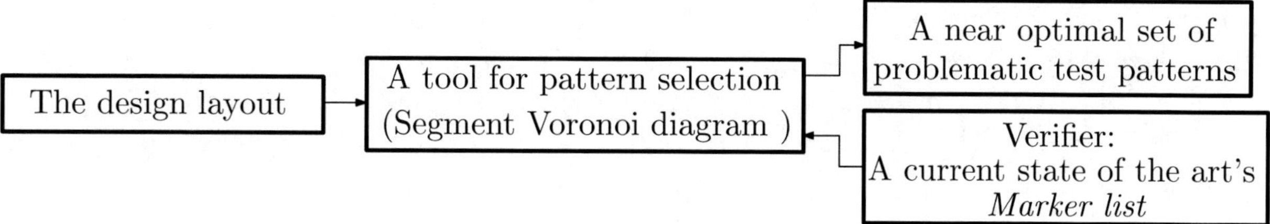

Fig 5 Block diagram of flow for the pattern selection using Segment Voronoi diagram.

C_i, C_e, C_s, C_c, and C_t is the context parameter of the internal, external, sandwich, comb and T gauges respectively. For example, if a sandwich gauge G_s and an internal gauge G_i have the same score, then G_s has higher priority, because $C_s > C_i$. The scoring method along with the parameters, provide us a priority-based ordered list of gauges. Figure 4 illustrates each type of gauge with its scoring formula.

4. Pattern selection based on the scoring method

We use the gauge locations as derived in Section 3 to extract windows and patterns from the design layout. We feed our pattern selection tool with two inputs (see Figure 5): (1) A design layout, (2) A set of *markers* for the design layout. A marker is a region in the design that indicates a problematic area that has high probability of faults. It is generally given in the form of rectangles. For a given design layout, we first obtain the ordered list of gauge locations according to their priority, and we traverse it to extract patterns, one pattern per gauge. Following are the steps of our Voronoi based pattern selection tool:

1. Compute all possible gauge locations in the given layout. Sort the gauges according to the scoring function.
2. For each gauge, consider a window of $5 - 8$ pitches with the gauge location as the center of the window.
3. For each window, check if it covers any marker. Select a window only if it covers some marker that has not already been covered by a previous window; otherwise discard it. A marker is considered covered in three different ways (A,B,C) as defined below.

As we do not desire overlapping windows, we prune any windows that cover markers that have already been covered, and thus, we obtain a set of disjoint windows for a layout. In step 3 we consider a marker to be covered in three different ways: (A) when the geometric center of the marker is strictly inside the corresponding window of a gauge, in which case we say that the gauge covers the marker; (B) when the whole marker rectangle is completely inside the corresponding window of the gauge; and (C) when the marker rectangle overlaps with the gauge window. The results of our experiment for the three different methods (A), (B), and (C) of marker covering are shown in Section 5, Table 1, Table 2, and Table 3 respectively. Note that method (B) may miss many markers because a long or a wide marker may not completely fit inside the windows we considered.

To analyze the quality of our gauges we further investigated the number of windows required to cover all the markers. We tried to put gauge locations at close proximity into one window, by using the following heuristic. The heuristic takes a window size and goes over the layout to cover the markers, while it discards unnecessary windows. The input to our heuristic algorithm is a set of points P, derived by the set of marker centers or the set of gauge locations, and the output is a set of windows, which covers the input set of points. The description of the heuristic algorithm follows:

1. Find the bottommost point in P. Let this point be (b_x, b_y). Construct a window W with $(b_x - 1, b_y - 1)$ as its bottom left corner; remove from P all points that lie inside W.
2. If $P \neq \emptyset$, increment the window counter by 1 and repeat step 1; otherwise report the window counter as the number of windows required to cover all points in the given set.
3. Output the windows generated by step 1.

To further reduce the number of patterns, we apply this heuristic to the obtained gauge locations. The next section provides experimental results.

5. Experimental results

We have done experiments for pattern selection on portion of a 22nm random logic design layout, provided with a state of the art markers for the corresponding layout. The 22nm layout had 38584 design polygons, and 7079 markers. The experiments are executed on a MacBook Pro 2.2 GHz Intel i7 with 4 GB RAM.

5.1 Marker coverage

We check the quality of the gauges by counting the number of gauges needed to cover all the markers. We considered four window sizes (in *pitches* $\times$ *pitches*), 5×5, 6×6, 7×7, and 8×8. We observed that the smaller windows were not able to cover all the markers. As we increased the window size the covering of markers increased. Our results are summarized in Tables 1, 2, and 3, and graphs shown in Figures 6, 7, and 8.

The notation in Tables 1, 2, and 3 are as follows: W_s = window size = *pitches* $\times$ *pitches*, M_c = Number of markers covered, G_u = Number of gauges used, r_u = Normalized range of scores of useful gauges (normalized gauge score = $\frac{g_s}{H_s}$, where g_s is the score of a gauge, and H_s is the highest recorded score among all considered gauges), G_f = Number of gauges failed to detect any marker, r_f = Normalized range of scores of failed gauges, p_d = Probability of detection of markers by the provided gauge set = $\frac{M_c}{7079} \times 100$, p_m = Probability that markers will be missed by the given gauge set will be $1 - p_d$, u_g = percentage of useful gauges those detect at least one marker (this basically evaluates the scoring function) = $\frac{G_u}{G_u+G_f} \times 100$, and T = Time taken to run the experiment on a MacBook Pro 2.2 GHz Intel i7 with 4 GB RAM.

In Table 1, we show results of our experiment, implementing the method (A) of marker covering. Recall that for method (A), a marker is considered covered, if the geometrical center of the marker is within some window. The gauge utilization, that is, the percentage of gauges that successfully covered some marker, increases with the increase in the window size. The larger window has greater chance to cover more marker centers. Windows in our experiments are chosen with priority, based on the priority of the gauges, which in turn depends on shapes and the proximity

Table 1 Covering of marker centers by windows generated by gauge set

W_s (*pitches* × *pitches*)	M_c	G_u	r_u	G_f	r_f	p_d	u_g	T
5×5	6968	5708	[0.05 - 1.0]	1638	[0.05 - 0.1]	98.43%	77.70 %	18.208 sec
6×6	7070	5306	[0.05 - 0.55]	675	[0.05 - 0.1]	99.87 %	88.71%	17.666 sec
7×7	7079	5029	[0.05 - 0.2]	383	[0.05 - 0.1]	100%	92.92%	16.768 sec
8×8	7079	4767	[0.05 - 0.16]	354	[0.05 - 0.1]	100%	93.08%	16.056 sec

Table 2 Covering of marker rectangles by windows generated by gauge set

W_s (*pitches* × *pitches*)	M_c	G_u	r_u	G_f	r_f	p_d	u_g	T
5×5	6101	5202	[0.05 - 0.6]	2278	[0.05 - 0.1]	86.18%	69.54%	68.63sec
6×6	6711	5263	[0.05 - 0.55]	691	[0.05 - 0.1]	94.8%	88.39 %	61.93sec
7×7	6908	5126	[0.05 - 0.4]	334	[0.05 - 0.1]	97.58%	93.88%	59.21sec
8×8	7014	4899	[0.05 - 0.2]	370	[0.05 - 0.1]	99.08%	92.97%	57.12sec

information of the shapes in the design. The probability of marker coverage increases with the increase in the window size. We observe 100% marker coverage for window sizes 7×7 and 8×8. The run time of the experiment decreases with the increase in the window size. For all the different window sizes the run time is within 20 seconds.

Table 2 shows the result for method (B) of marker covering, in which a marker is considered covered, if the marker area is completely within some window. The marker coverage increases with the increase in window size. The gauge utilization first increases with the increase in window size (from 5×5 to 7×7), and then decreases as we further increase the window size (from 7×7 to 8×8). This is because a bigger window has potential to cover more markers but we need to check with a larger number of windows, as there are many windows which do not completely contain any marker. In this case also we observe that the probability of marker coverage increases with the increase in window size. We were not able to cover 100% of markers, mainly due to the fact that some markers were very long or wide and were not fitting in within any acceptable window size. The best case was 99.08% for the window size of 8×8. For all the different window sizes the run time is within 70 seconds.

Table 3 shows the result for method (C) of marker covering, in which a marker is considered covered, if the marker area is overlapping within some window. The probability of marker coverage increases with the increase in window size. We observe 100% marker coverage for the window size of 7×7 and 8×8. For all the different window sizes the run time is within 4 minutes. The time taken in this case is more compared to the other two cases as the predicate to determine overlap between window and design shapes takes more time that the predicate that determines if a point is inside a window or a rectangle is completely inside a window.

We observe that the window size of 7×7 and 8×8 gives fairly acceptable results in terms of marker coverage in all cases. All types of gauges have the most critical gauge score of highest priority (0.05). A clear observation from the range of scores for G_u is that, with the increase of

Table 3 Covering (overlapping) of marker rectangles by windows generated by gauge set

W_s (*pitches* × *pitches*)	M_c	G_u	r_u	G_f	r_f	p_d	u_g	T
5×5	7011	5607	[0.05 - 0.6]	1627	[0.05 - 0.1]	99.03%	77.50%	3m 57s
6×6	7073	5180	[0.05 - 0.4]	341	[0.05 - 0.1]	99.91%	93.82%	3m 44s
7×7	7079	4910	[0.05 - 0.2]	319	[0.05 - 0.1]	100%	93.89%	2m 25s
8×8	7079	4686	[0.05 - 0.16]	339	[0.05 - 0.1]	100%	93.25%	1m 28s

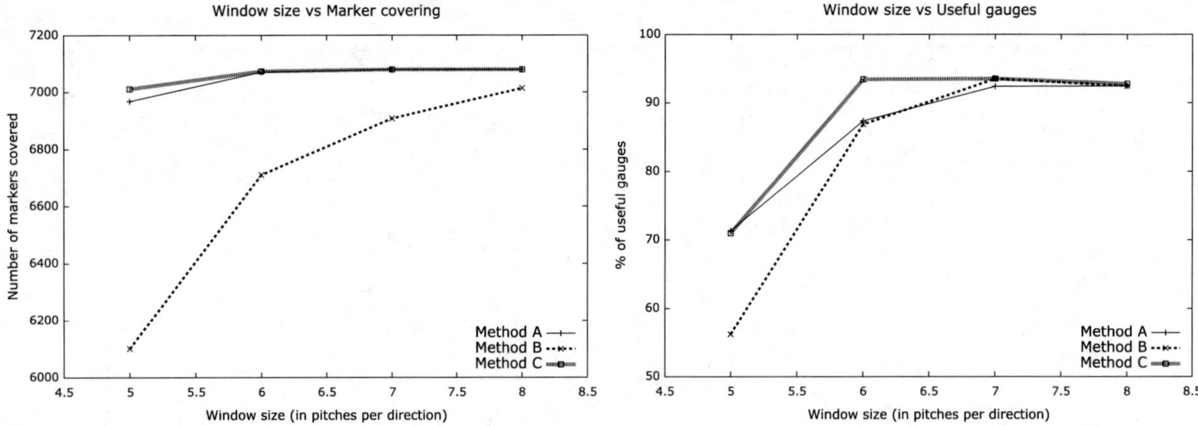

Fig 6 Variation of markers covered w.r.t the window size. **Fig 7** Variation of useful gauges w.r.t the window size.

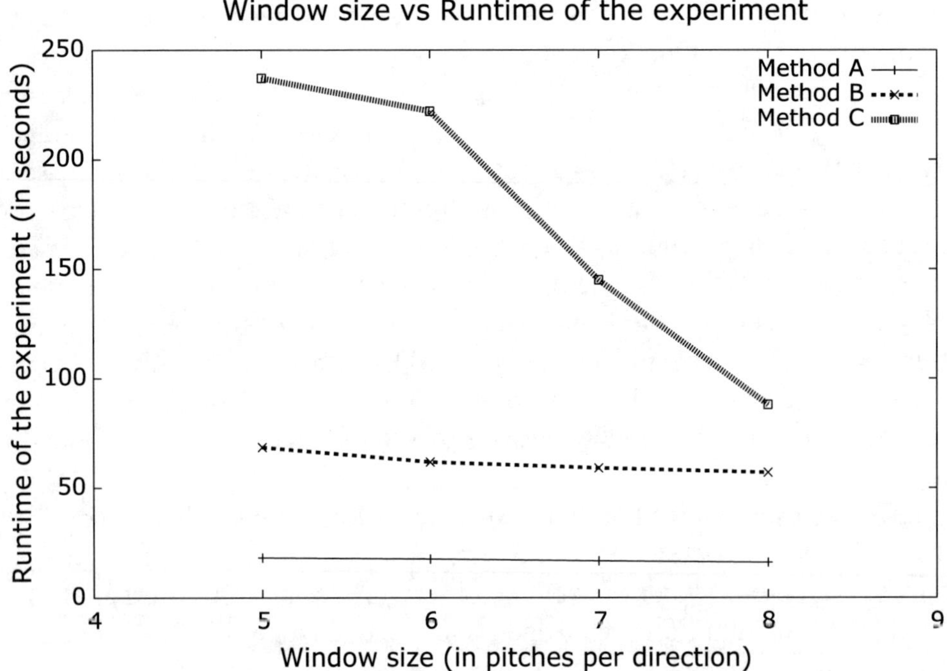

Fig 8 Variation of runtimes of the experiments w.r.t the window size.

window size, there is a decrease in the requirement of low priority gauges. In all cases the range of score for G_f is [0.05 - 0.1], which suggests that the gauge (which belongs to G_f) locations are critical and there is a possibility of finding a problematic pattern around these gauges, although in this specific layout they did not cover any ORC marker. The range of scores for G_f in all cases indicates that the scoring method captures the pattern criticality and is also able to capture more patterns that may be needed to minimize the probability of missing problematic patterns.

The variation of marker coverage with window size for all cases are shown in Figure 6. The variation of useful gauges with window size for all cases are shown in Figure 7. We compare the runtime of the experiments based on three different ways of marker covering in Figure 8.

Table 4 Gauge distribution for marker covering: A, B, C are gauge distribution for three different ways of covering markers that is, marker center inside the window, markers are completely inside the window respectively, and markers are overlapping with windows. Notations in the table: r_i, r_e, r_s, and r_t are normalized range of scores of internal, external, sandwich and T gauges respectively. N_i, N_e, N_s, and N_t are number of internal, external, sandwich and T gauges respectively.

	W_s	N_i	r_i	N_e	r_e	N_s	r_s	N_t	r_t	Total
A	5×5	30	[0.2 - 0.4]	38	[0.05 - 1.0]	5388	[0.075 - 0.3]	252	[0.16 - 0.32]	5708
	6×6	6	[0.2 - 0.4]	13	[0.05 - 0.55]	5240	[0.075 - 0.1]	47	[0.16 - 0.16]	5306
	7×7	1	[0.2 - 0.2]	7	[0.05 - 0.1]	5005	[0.075 - 0.1]	16	[0.16 - 0.16]	5029
	8×8	0	[0 - 0]	6	[0.05 - 0.05]	4753	[0.075 - 0.1]	8	[0.16 - 0.16]	4767
B	5×5	57	[0.2 - 0.4]	56	[0.05 - 0.6]	4686	[0.075 - 0.3]	403	[0.16 - 0.16]	5202
	6×6	12	[0.2 - 0.4]	18	[0.05 - 0.55]	5141	[0.075 - 0.2]	92	[0.16 - 0.16]	5263
	7×7	4	[0.2 - 0.4]	9	[0.05 - 0.4]	5078	[0.075 - 0.2]	35	[0.16 - 0.16]	5126
	8×8	1	[0.2 - 0.2]	7	[0.05 - 0.1]	4880	[0.075 - 0.1]	11	[0.16 - 0.16]	4899
C	5×5	15	[0.2 - 0.4]	29	[0.05 - 0.6]	5424	[0.075 - 0.2]	139	[0.16 - 0.32]	5607
	6×6	4	[0.2 - 0.4]	11	[0.05 - 0.4]	5135	[0.075 - 0.1]	30	[0.16 - 0.16]	5180
	7×7	1	[0.2 - 0.2]	7	[0.05 - 0.1]	4888	[0.075 - 0.1]	14	[0.16 - 0.16]	4910
	8×8	0	[0 - 0]	6	[0.05 - 0.05]	4673	[0.075 - 0.1]	7	[0.16 - 0.16]	4686

Table 5 Distribution of gauges do not covering any markers.

	W_s	N_i	r_i	N_e	r_e	N_s	r_s	N_t	r_t	Total
A	5×5	0	[0 - 0]	3	[0.05 - 0.05]	1635	[0.1 - 0.1]	0	[0 - 0]	1638
	6×6	0	[0 - 0]	3	[0.05 - 0.05]	672	[0.1 - 0.1]	0	[0 - 0]	675
	7×7	0	[0 - 0]	3	[0.05 - 0.05]	380	[0.1 - 0.1]	0	[0 - 0]	383
	8×8	0	[0 - 0]	3	[0.05 - 0.05]	351	[0.1 - 0.1]	0	[0 - 0]	354
B	5×5	0	[0 - 0]	3	[0.05 - 0.05]	2275	[0.1 - 0.1]	0	[0 - 0]	2278
	6×6	0	[0 - 0]	3	[0.05 - 0.05]	688	[0.1 - 0.1]	0	[0 - 0]	691
	7×7	0	[0 - 0]	3	[0.05 - 0.05]	331	[0.1 - 0.1]	0	[0 - 0]	334
	8×8	0	[0 - 0]	3	[0.05 - 0.05]	367	[0.1 - 0.1]	0	[0 - 0]	370
C	5×5	0	[0 - 0]	3	[0.05 - 0.05]	1624	[0.1 - 0.1]	0	[0 - 0]	1627
	6×6	0	[0 - 0]	3	[0.05 - 0.05]	338	[0.1 - 0.1]	0	[0 - 0]	341
	7×7	0	[0 - 0]	3	[0.05 - 0.05]	316	[0.1 - 0.1]	0	[0 - 0]	319
	8×8	0	[0 - 0]	3	[0.05 - 0.05]	336	[0.1 - 0.1]	0	[0 - 0]	339

5.2 *Gauge distribution*

We performed experiments to find out the distribution of gauges covering the markers. Results are reported in Tables 4 and 5. In Table 4, we give detailed results on the marker coverage by the different types of gaugesWe observe that all types of gauges are useful, as there are at least some gauges of each type covering some markers. For this particular design layout, the sandwich gauge turned out to cover most of the markers. In Table 5, we give the distribution of gauges which fail to cover any marker. We observe that the failed gauges are of the extra or the sandwich type. The internal and T type of gauges have no failure.

We have not included information on comb type of gauges in the tables. This is because comb gauges were not used to cover any marker and no comb gauge failed to cover a marker. The range of normalized gauge score for comb gauge is [0.15 - 0.15]. The score range of comb gauges suggests that they are useful. They did not appear in the distribution of useful gauges because they cover markers, which have already been covered by gauges of higher priority (normalized score ≤ 0.15).

Table 6 Covering of gauge locations and marker centers by our heuristic algorithm. Notations in the table: W_b = Number of windows before using heuristic, W_a = Number of windows obtained after using heuristic, W_m = Number of windows required by heuristic to cover marker centers, R_w = Reduction in windows from W_b to W_a.

W_s (pitches × pitches)	W_b	W_a	R_w	W_m
5 × 5	5709	4102	28.14%	4442
6 × 6	5306	3555	33.00%	4146
7 × 7	5029	3221	35.95%	3934
8 × 8	4767	3013	36.79%	3783

5.3 *Reduction of patterns by a simple heuristic*

We tried to obtain a lower bound on the number of windows required to cover all the marker centers, using the simple heuristic algorithm of Section 4. For this we took a list of geometrical centers of all the markers as an input to the heuristic. It would be desirable to reach the bound mentioned in the W_m column of Table 6 with our gauge locations.

We tried to further reduce the number of patterns using the simple heuristic algorithm of Section 4. We took the set of gauge locations corresponding to the set of windows obtained in our experiment using method (A), as an input to the heuristic algorithm. We observe a reduction in the number of patterns by around 30%. We have not evaluated the usefulness of reduced set of gauges. We report the results in Table 6.

6. Conclusions

We discussed a new method to select a set of problematic patterns, which is based on topological information extracted from the Voronoi diagram of layout shapes. Our method is fast and gives an automatic way to discover the potentially problematic areas when printing VLSI layout patterns. We verified our windows by covering ORC-generated markers. Using our pattern selection tool, we covered 7079 ORC markers for a design layout of 38584 design polygons using nearly 5000 extracted patterns. The patterns extracted covered all the ORC markers. Applying a simple heuristic algorithm we further reduced the number of patterns by approximately 30%. The variety of gauges has a potential to extract topology and context based interesting features of patterns. This lays a foundation for future work to obtain a set of patterns for calibration and verification of MB-OPC.

References

1 G. Viehoever, B. Ward, and H. J. Stock, "Pattern selection in high-dimensional parameter spaces," in *Proc. SPIE*, **8326**, 832618 (2012).

2 D. Ding, J. A. Torres, and D. Z. Pan, "High performance lithography hotspot detection with successively refined pattern identifications and machine learning," *IEEE Transactions on Computer-Aided Design of Integrated Circuits and Systems* **30**(11), 1621–1634 (2011).

3 J. R. Gao, B. Yu, and D. Z. Pan, "Accurate lithography hotspot detection based on pca-svm classifier with hierarchical data clustering," in *Proc. SPIE*, **9053**, 90530E–90530E–10 (2014).

4 H. Nosato, H. Sakanashi, E. Takahashi, M. Murakawa, T. Matsunawa, S. Maeda, S. Tanaka, and S. Mimotogi, "Hotspot prevention and detection method using an image-recognition technique based on higher-order local autocorrelation," *Journal of Micro/Nanolithography, MEMS, and MOEMS* **13**(1), 011007 (2014).

5 A. Abdo and R. Viswanathan, "The feasibility of using image parameters for test pattern selection during OPC model calibration," in *Proc. SPIE*, **7640**, 76401E (2010).

6 G. Yoo, J. Kim, T. Lee, A. Jung, H. Yang, D. Yim, S. Park, K. Maruyama, M. Yamamoto, A. Vikram, and S. Park, "OPC verification and hotspot management for yield enhancement through layout analysis," in *Proc. SPIE*, **7971**, 79710H–79710H–11 (2011).

7 J. Y. Wuu, F. G. Pikus, A. Torres, and M. Marek Sadowska, "Rapid layout pattern classification," in *Proceedings of the 16th Asia and South Pacific Design Automation Conference, ASPDAC '11*, 781–786, IEEE Press, (Piscataway, NJ, USA) (2011).

8 M. C. Simmons, J. H. Kang, Y. Kim, J. I. Park, S. W. Paek, and K. S. Kim, "A state-of-the-art hotspot recognition system for full chip verification with lithographic simulation," in *Proc. SPIE*, **7974**, 79740M–79740M–9 (2011).

9 S. Shim, W. Chung, and Y. Shin, "Synthesis of lithography test patterns through topology-oriented pattern extraction and classification," in *Proc. SPIE, Design-Process-Technology Co-optimization for Manufacturability VIII*, **9053**, 905305–905305–10 (2014).

10 P. Cheilaris, S. K. Dey, M. Gabrani, and E. Papadopoulou, "Implementing the L_∞ segment Voronoi diagram in CGAL and applying in VLSI pattern analysis," in *Proc. 4th International Congress on Mathematical software (ICMS'14), Lecture Notes in Computer Science* **8592**, 198–205, Springer (2014).

11 Y. Sun, Y. M. Foong, Y. Wang, J. Cheng, D. Zhang, S. Gao, N. Chen, B. I. Choi, A. J. Bruguier, M. Feng, J. Qiu, S. Hunsche, L. Liu, and W. Shao, "Optimizing OPC data sampling based on orthogonal vector space," in *Proc. SPIE*, **7973**, 79732K (2011).

12 N. Casati, M. Gabrani, R. Viswanathan, Z. Bayraktar, O. Jaiswal, D. L. DeMaris, A. Y. Abdo, J. Oberschmidt, and R. A. Krause, "Automated sample plan selection for OPC modeling," in *Optical Microlithography XXVII, Proceedings of SPIE*, **9052** (2014).

13 R. Viswanathan, O. Jaiswal, M. Gabrani, N. Casati, A. Y. Abdo, J. Oberschmidt, and J. Watts, "Experiments using automated sample plan selection for OPC modeling," in *Optical Microlithography XXVIII, Proceedings of SPIE*, **9426** (2015).

14 M. Mukherjee, Z. Baum, J. Nickel, and T. G. Dunham, "Optical rule checking for proximity-corrected mask shapes," in *Proc. SPIE, Optical Microlithography XVI*, **5040**, 420–430 (2003).

15 F. Aurenhammer, "Voronoi diagrams - a survey of a fundamental geometric data structure," *ACM Computing Surveys (CSUR)* **23**(3), 345–405 (1991).

16 F. Aurenhammer, R. Klein, and D. T. Lee, *Voronoi Diagrams and Delaunay Triangulations*, World Scientific Publishing Company, Singapore (2013).

17 A. Okabe, B. Boots, K. Sugihara, and S. Chiu, *Spatial Tessellations: Concepts and Applications of Voronoi Diagrams*, Wiley Series in Probability and Statistics., second ed. (2000).

18 E. Papadopoulou and D. T. Lee, "Critical area computation via Voronoi diagrams," *IEEE Transactions on Computer-Aided Design of Integrated Circuits and Systems* **18**(4), 463–474 (1999).

19 E. Papadopoulou and D. T. Lee, "The L_∞ Voronoi diagram of segments and VLSI applications," *International Journal of Computational Geometry and Application* **11**(5), 502–528 (2001).

20 MATLAB, *version 8.4 (R2014b)*, The MathWorks Inc., Natick, Massachusetts (2014).

21 "CGAL, Computational Geometry Algorithms Library." http://www.cgal.org.

22 M. Karavelas, "2D segment Delaunay graphs." http://doc.cgal.org/latest/Segment_Delaunay_graph_2/index.html.

23 "Voronoi CAA: Voronoi Critical Area Analysis." IBM CAD Tool, Department of Electronic Design Automation, IBM Microelectronics Division, Burlington, VT. Initial patents: US6178539, US6317859.

24 E. Papadopoulou, "Net-aware critical area extraction for opens in VLSI circuits via higher-order Voronoi diagrams," *IEEE Transactions on Computer-Aided Design of Integrated Circuits and Systems* **30**(5), 704–716 (2011).

25 P. Cheilaris, S. K. Dey, and E. Papadopoulou, "L_∞ Segment Delaunay graphs." http://compgeom.inf.usi.ch/cgal_doc/Segment_Delaunay_graph_Linf_2/. Code available upon request.

A Systematic Framework for Evaluating Standard Cell Middle-Of-Line (MOL) Robustness for Multiple Patterning

Xiaoqing Xu[a], Brian Cline[b], Greg Yeric[b], Bei Yu[a] and David Z. Pan[a]

[a]ECE Dept. University of Texas at Austin, Austin, TX 78712 USA
[b]ARM Inc., Austin, TX 78735 USA
Email: {xiaoqingxu, bei, dpan}@cerc.utexas.edu; {brian.cline, greg.yeric}@arm.com

ABSTRACT

Multiple patterning (triple and quadruple patterning) is being considered for use on the Middle-Of-Line (MOL) layers at the 10*nm* technology node and beyond.[1] For robust standard cell design, designers need to improve the inter-cell compatibility for all combinations of cells and cell placements. Multiple patterning colorability checks break the locality of traditional rule checking and N-wise checks are strongly needed to verify the multiple patterning colorability for layout interaction across cell boundaries. In this work, a systematic framework is proposed to evaluate the library-level robustness over multiple patterning from two perpectives, including illegal cell combinations and full chip interactions. With efficient N-wise checks, the vertical and horizontal boundary checks are explored to predict illegal cell combinations. For full chip interactions, random benchmarks are generated by cell shifting and tested to evaluate the placement-level efforts needed to reduce the quadruple patterning to triple patterning for the MOL layer.

Keywords: Multiple Patterning, Middle-Of-Line, Cell Compatibility Check

1. INTRODUCTION

Multiple patterning (MP) has been widely adopted in industry for technology scaling due to the resolution limits of the 193*nm* lithography tools.[1–4] To enable MP lithography, restrictive design rules or constraints have been introduced into the back-end design flow. To deal with these complex constraints, a wide range of research efforts have been focusing on standard cell (SC) design,[2,5] placement,[2,6,7] routing,[8–11] layout migration and decomposition.[12–16] From the library design perspective, it is important to provide quick feedback on the cell layout robustness over MP constraints so that SC designers can improve the cell layout design accordingly. However, there has been few works related to efficient SC library evaluation over MP design constraints. For SC library design, inter-cell compatibility is essential for achieving a robust library that can be used in any design implementation, no matter what kind of placement is implemented for that design. Historically, inter-cell placement compatibility could be guaranteed by checking all 2-cell combinations, because the lithographic interactions were small in comparison to the cell sizes and limited to single mask layers.

With the industry extending 193*nm* lithography via MP, the lithographic interactions now include complex layout design constraints that extend beyond 2-cell interactions. MP coloring and decomposition - especially on middle of line (MOL) layers like the Contact-to-Active (CA) layer[17] that routinely introduces interactions across cell boundaries - breaks the locality of traditional rule checking and makes historical pair-wise cell compatibility checking obsolete. Fig. 1 illustrates the cell interactions introduced by MP. Fig. 1(a) shows a 3-cell interaction that requires triple patterning (TP) to decompose the pattern at the boundary while Fig. 1(b) illustrates a 3-cell interaction example that requires quadruple patterning (QP). Specifically, for the dimensions in the representative 10*nm*/15*nm* technologies used in this work, TP or even QP is needed on the CA layer, depending on the SC boundary conditions and the strength of color-aware placement tools. In cutting-edge technology nodes like 10*nm* and beyond, N-wise checks are now strongly-needed to verify the MP colorability for cell interactions across boundaries, as demonstrated in Fig. 1. This means that what once was a tractable problem in pair-wise cell checking now becomes an exponential problem in the exhaustive N-wise checking. Fortunately, the extremely regular layout structure of layers that interact at the cell boundary leads to large amounts of redundancy among

Design-Process-Technology Co-optimization for Manufacturability IX, edited by John L. Sturtevant, Luigi Capodieci, Proc. of SPIE Vol. 9427, 942707 · © 2015 SPIE · CCC code: 0277-786X/15/$18 · doi: 10.1117/12.2085918

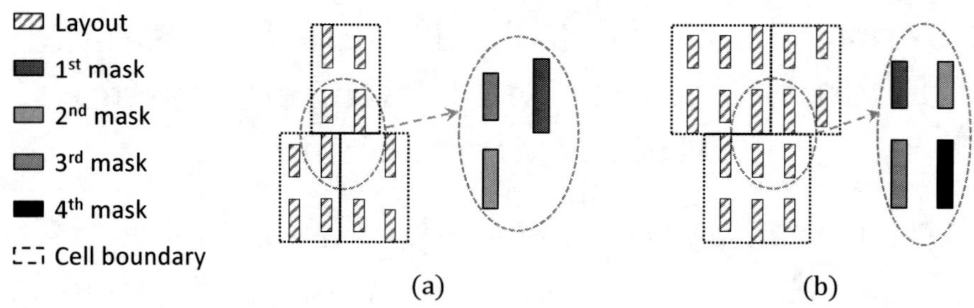

Figure 1. 3-cell interactions, (a) triple patterning, (b) quadruple patterning.

exhaustive inter-cell compatibility checks. For a typical SC library, it becomes worthwhile to explore and remove the redundancy among these inter-cell compatibility checks to make N-wise checks more tractable.

For the 10*nm* technology, the MOL layers, including CA and CB, are essential for the intra-cell routing.[18, 19] This work mainly focuses on the library robustness evaluation on the CA layer due to its complex MP design constraints. Specifically, for the representative 10*m*/15*nm* technologies used in this study, the CA layer has small dimensions/pitches that requires TP even QP. Compared to other MP layers, such as Metal-1, stitching is not allowed for the CA layer (which is enforced by a technology's Design Rule Checks - DRC's). In addition, the CA layer routinely introduces interactions across rows under MP constraints, which makes N-wise checks particularly important in the library evaluation stage. However, our evaluation framework can be easily adapted to other MP layers, such as Metal-1.

For library robustness evaluation, the inter-cell compatibility checks should report all the illegal cell combinations to the designers. Then, designers can improve the library cell layout and reduce the number of illegal cell combinations as much as possible. For the representative 10*nm*/15*nm* libraries we used, the inter-cell compatibility checks on the CA layer reveal that the TP DRC violations originate from the local coloring conflicts. Therefore, we further evaluate the robustness of the library over TP from the perspective of full chip interactions. In this work, the concepts of "cell sequence" and "multi-row cell compatibility checking" are proposed to enable efficient N-wise inter-cell compatibility checks. We manage the exponential growth of the N-wise checks by exploiting the pattern regularity of the CA layer and remove redundancy among various cell sequences. Vertical and horizontal boundary checks are explored in order to predict illegal cell combinations. Then, multiple rows of independent cell sequences are randomly combined to evaluate the full chip interactions on the CA layer over MP. Our main contributions are summarized as follows:

- A systematic framework is proposed to evaluate the library-level MOL robustness for MP.
- To enable efficient N-wise checks, the cell sequence size reduction problem is proposed and solved efficiently with a graph-based approach and divide-and-conquer technique.
- The illegal cell combinations are predicted with efficient horizontal and vertical boundary checks for a given SC library.
- For specific SC libraries, the full chip interactions are explored to evaluate the placement-level efforts in reducing the QP to TP for the MOL layer.

The rest of this paper is organized as follows: Section 2 introduces the related background information, basic definitions and overall framework. Section 3 presents the formulation and solution of the cell sequence size reduction problem. Section 4 introduces the multi-row cell compatibility checking to enable efficient horizontal and vertical boundary checks with the elementary checker. Section 5 demonstrates the effectiveness of our framework with the results on two sets of benchmarks, including ARM 10*nm* representative library and modified Open NanGate 15*nm* library.[20] Several key observations are discussed in detail.

2. PRELIMINARIES AND OVERALL FLOW

2.1 Middle-Of-Line and Row Structure

In advanced technology nodes, the complex lithography and process constraints prevent simply using Metal-1 layer for intra-cell routing. Hence, the MOL layers have been adopted for intra-cell signal-wiring in the 20nm technology node and beyond.[18,19] As shown in Fig. 2(a), typical MOL layers include CA and CB.[17] In particular, there may be cross-coupled connections, i.e. special constructs, in the CA layer depending on the technological choices. In the placement row structure shown in Fig. 2(b), SC's are aligned horizontally and share the same height. The mirrored structure of the power and ground rails goes from the left to right of the design. For practical placement, there might be white spaces among neighboring cells on the same row. However, in the library design or evaluation stage, the exact amount of white spaces between any cell pair is unknown without placement information. A typical way to enable robust SC design is to estimate the white spaces conservatively. As illustrated in Fig. 2(c), the white spaces among cells on a single row are removed and cells are abutted horizontally. Thus, we have the following definition.

Definition 1 (Cell Sequence). *A set of cells abutting each other in the horizontal direction without overlapping or white space is called a cell sequence. If the sequence consists of n SC's, it is defined as an n-cell sequence.*

An example of the 4-cell sequence is shown in Fig. 2(c).

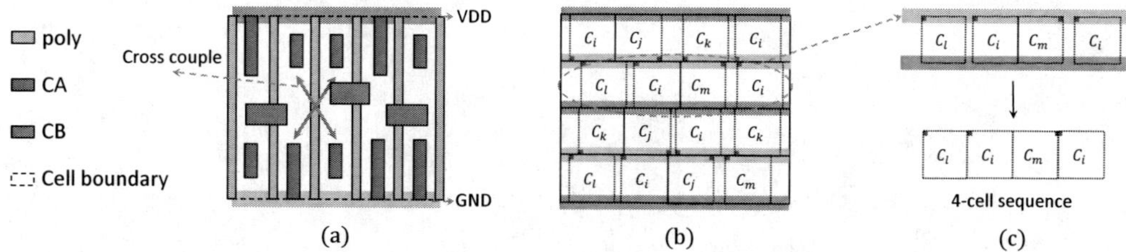

Figure 2. Standard cell and row structure, (a) MOL structures, (b) multiple rows, (c) one single row and 4-cell sequence.

2.2 Conflict Graph

To determine whether the target layout is free of MP DRC violations, i.e. k-patterning friendly, the first step is to built the conflict graph.[13] Fig. 3(a) and (b) shows a typical layout in the CA layer and the corresponding conflict graph, respectively. Then, the result of the k-patterning DRC is determined by the k-colorability of the conflict graph. For example, the target layout passes the quadruple ($k = 4$) patterning coloring check in Fig. 3(b) and the mask assignment is demonstrated in Fig. 3(c). The k-patterning design rule check is similar to the problem of layout decomposition. The MP layout decomposition problem is well-studied for double patterning,[13,21–23] triple patterning[14,24–28] and quadruple patterning.[16] Due to the timing criticality of the MOL layers for the SC design, we assume no stitches are allowed for the feasible coloring solution. In addition, MP DRC aims at deciding the existence of some coloring solution instead of finding an optimal one in layout decomposition. Then, an elementary DRC checker is presented to determine the k-colorability of the conflict graph for the CA layer.

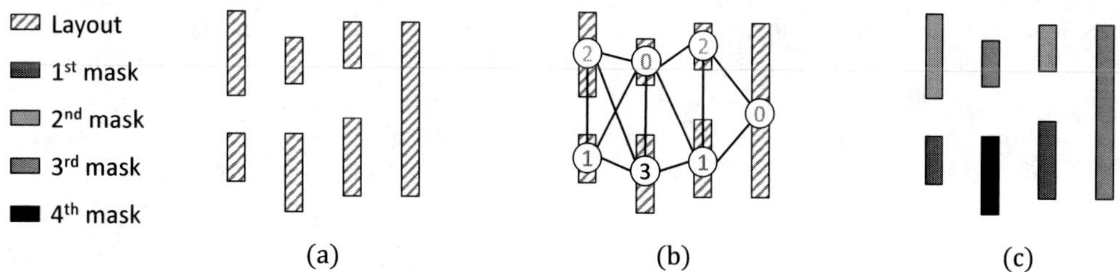

Figure 3. Layout graph for MP layout decomposition, (a) target layout, (b) graph and coloring, (c) mask assignment.

2.3 Overall Flow

The overall flow for our framework in demonstrated in Fig. 4. At the placement level, the TP or QP DRC involves 3-cell interactions as shown in Fig. 1. Thus, we need to enable N-wise checks across multiple rows to predict illegal cell combinations. Given a SC library, the framework starts with the cell sequence construction, which produces a set of independent k-cell sequences. Then, the two-row cells are exhaustively built from the independent cell sequences to compute the illegal cell combinations. For representative libraries we used, the cell compatibility checks reveal that the violations of TP DRCs come from the local conflicts on the CA layer. Random benchmarks are constructed to explore full chip interactions. This further produces the cell shift histogram to evaluate the placement-level efforts needed to reduce QP to TP. Reducing from QP to TP is potentially attractive because it may allow us to save one mask.

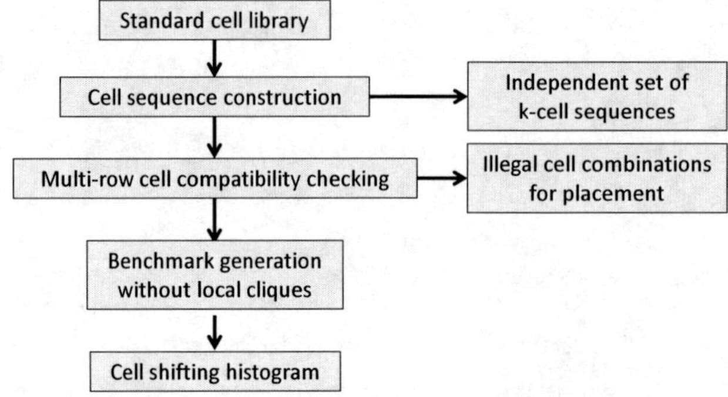

Figure 4. The overall flow for our framework.

3. CELL SEQUENCE CONSTRUCTION

In this section, we first define the cell sequence redundancy given a pair of cell sequences. Then, the redundancy graph and divide-and-conquer technique are presented to remove the redundancy and compute the independent set of cell sequences. In the end, we demonstrate the efficient cell sequence construction algorithm.

3.1 Cell Sequence Size Reduction

For the k-patterning DRC, the pass or violation is decided by the k-colorability of the conflict graph constructed from the layout.[13, 14, 16] For the k-colorability of a graph, we have the following theorem.

Theorem 1. *If a graph is k-colorable, a subset of the graph is also k-colorable. The subset of a graph is achieved by removing some nodes or edges of the given graph.*

Proof. A graph is k-colorable means a feasible coloring solution exists for the graph. Pick one feasible solution and assign color to each node. Then, we have legal color assigned to each node of the subset of graph, which means the subset of the graph is also k-colorable. □

With the regular layout structure on the CA layer, the conflict graph is determined by a set of rectangles and relative positions among them. Fig. 5(a) illustrates the layout of two 1-cell sequences , namely C_i and C_j, on the CA layer. As shown in Fig. 5(b), the layout of C_j is a **superset** of the layout of C_i on both the left and right boundary of C_j. Specifically, the left boundary layout of C_j exactly matches the layout of C_i. Moreover, the line ends of the right boundary layout of C_j are extended vertically compared to the layout of C_i, which may induce extra conflict edges during conflict graph construction. This means that the conflict graph for C_i is the subset of the conflict graph for C_j on both the left and right boundary. According to Theorem 1, the k-colorability of C_j guarantees that of C_i. Then, the k-patterning check for C_i will be redundant to that for C_j. As mentioned earlier, the MP DRC involves N-wise checks, which brings the necessity of k-cell sequence construction. Then, the redundancy of C_i to C_j is further demonstrated in Fig. 5(c). When we build the 2-cell sequences introducing another cell C_k, the layout of C_kC_j is a superset of the layout of C_kC_i and a similar relationship exists between C_jC_k and C_iC_k. If C_j is included in the independent set of cell sequences, C_i should be excluded from that independent set for both MP DRC and cell sequence construction, since including C_i would be redundant.

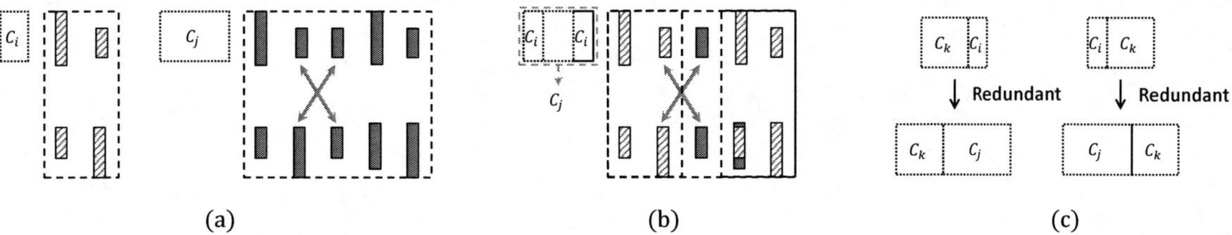

Figure 5. Sequence cell redundancy, (a) cell C_i and C_j, (b) C_i redundant to C_j, (c) C_kC_i redundant to C_kC_j, C_iC_k redundant to C_jC_k.

Redundancy Graph The ultimate goal for the cell sequence size reduction is to reduce the number of N-wise checks for MP. Given a set of cell sequences, we introduce the redundancy graph for removing redundancy and compute an independent set of cell sequences. The redundancy graph is a directed graph with each node representing one cell sequence. As shown in Fig. 6, there are n nodes and node C_i represents the cell sequence C_i. Now, we may use C_i to represent the cell sequence and the corresponding graph node interchangeably. In addition, an edge from node C_i to node C_j denotes that cell sequence C_i is redundant to cell sequence C_j. For example, in Fig. 6, an edge from C_3 to C_6 means cell sequence C_3 is redundant to C_6. In the end, with the redundancy graph constructed from the input set of cell sequences, the independent set consists of those nodes without successors in the graph. For instance, the dashed nodes $(C_2, C_5, C_6, C_{n-3}, C_{n-1})$ in Fig. 6 belong to the independent cell sequence set.

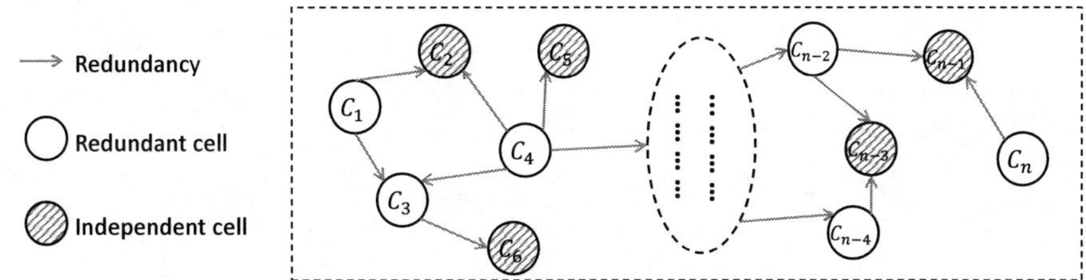

Figure 6. Cell sequence redundancy graph.

Given n cell sequences, the time complexity to build a redundancy graph is $O(n^2)$ because we need to determine the redundancy between each pair of cells. For practical implementations, the input size can be as large as $n \geq 10^6$, which would incur large runtime. Thus, a divide and conquer technique is deployed to optimize the set within an affordable runtime. Algorithm 1 illustrates the details on the recursive divide and conquer technique with the redundancy graph construction. Lines 5-10 explain the redundancy graph construction and independent set computation if the input size is within the graph node size bound (n_0). Otherwise, the input cell sequence set is segmented into various subsets and recursively computed. It can be observed that the inter-subset redundancy among cell sequences on different subsets are ignored in Algorithm 1. To account for that, Algorithm 2 demonstrates the iterative approach to implement the cell sequence size reduction. For each iteration from line 4 to line 9, the cell sequence set is optimized using the divide and conquer technique and the set is randomly shuffled for the next iteration. The exit condition is that no redundant cell is found or the iteration count exceeds the pre-set bound (m_0). A more intuitive interpretation of Algorithm 1 and Algorithm 2 is illustrated in Fig. 7. The input set is segmented into several subsets in Fig. 7(a) and each subset is optimized based on redundancy graph in Fig. 7(b) and the results are shown in Fig. 7(c). To prepare for the next iteration, the optimized set is randomly shuffled in Fig. 7(d).

3.2 Cell Sequence Construction

Given a standard cell library consisting of m cells, i.e. m 1-cell sequences, the exhaustive construction scheme gives the number of n-cell sequences as $(2 * m)^n$ (the factor of 2 accounts for horizontal cell flipping). A simple calculation with $m = 100$ yields that the number of 4-cell sequences will be 1.6 billion. To manage the size to an acceptable level, cell sequences are built in a bottom-up manner as shown in Algorithm 3. We iterate through

Algorithm 1 Divide and Conquer

```
Require: A set of cell sequences (CS), graph node size bound (n0);
Define ICS as the independent cell sequence set;
Define RG(CS) as the redundancy graph for set CS;
function DIVIDECONQUER(CS)
    Compute size = size of CS;
    if size ≤ n0 then
        for each node v_i in RG(CS) do
            if v_i has no successors then
                add v_i to ICS;
            end if
        end for
    else
        Divide CS into √size subsets;
        for each subset for CS do
            DIVIDECONQUER(subset);
        end for
    end if
end function
```

Algorithm 2 Cell Sequence Size Reduction (CSSR)

```
Require: A set of cell sequences (CS), iteration bound (m0);
Define iteration count m = 0;
Define ICS as the independent cell sequence set;
while m ≤ m0 do
    ICS = DIVIDECONQUER(CS);
    if size of ICS < size of CS then
        CS = random-shuffle(ICS);
    else
        break;
    end if
    m = m + 1;
end while
```

the number of cells, denoted as n, in one sequence in line 4. The independent n-cell sequences are computed in line 5. Then, in lines 6-9, the $(n+1)$-cell sequences are constructed from the independent sets, including the 1-cell sequence set to the n-cell sequence set. The iteration ends with the pre-set maximum number of cells (n_0) in one sequence. With Algorithm 3, the n-cell sequences are constructed from the independent cell sequences in a bottom up manner. For practical implementations, this method helps to achieve the complete set of 3-cell sequences in reasonable amount of runtime despite of the exponential growth of the cell sequences.

Algorithm 3 Cell Sequence Construction

```
Require: The set of 1-cell sequences (CS_1) in the library and the maximum number of cells in one sequence (n0);
 1: Define CS_n as the set of n-cell sequences;
 2: Define ICS_n as the set of independent n-cell sequences;
 3: Define initial number of cells in one sequence as n = 1;
 4: while n <= n0 do
 5:     ICS_n = CSSR(CS_n);
 6:     for m = 1; m ≤ ⌊(n + 1)/2⌋; m = m + 1 do
 7:         Construct the (n + 1)-cell sequence set (set_{n+1}) from ICS_m and ICS_{n+1−m};
 8:         CS_{n+1} = CS_{n+1} ∪ set_{n+1};
 9:     end for
10:     n = n + 1;
11: end while
```

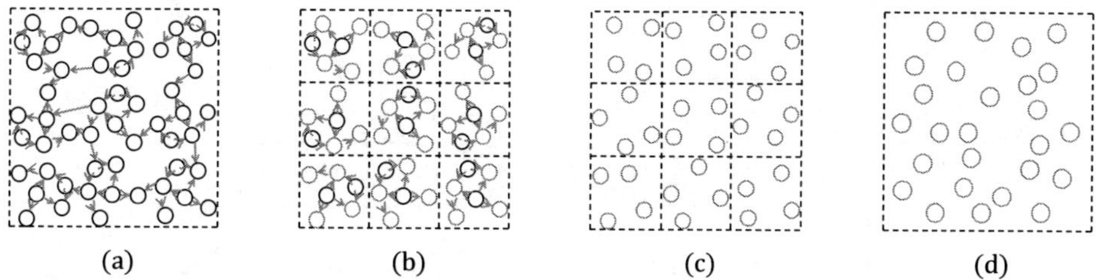

Figure 7. Divide and conquer, (a) divide the cell sequences into several subsets, (b) remove redundancy in each subset, (c) random shuffle, (d) for next iteration.

4. MULTI-ROW CELL COMPATIBILITY CHECKING

In this section, we build two-row cells and multi-row cells, to predict illegal cell combinations and explore full chip interactions, respectively. An elementary checker is presented predict the local conflicts and non-local conflicts for TP.

4.1 Two-Row Cells

Ideally, we need to exhaustively build cell combinations across as many rows as possible to explore long-range interactions for the MP DRC. However, exhaustive two-row cell checks are sufficient for the local conflict detections for QP and TP. In addition, the exponential growth of the cell combinations beyond 2 rows makes it unaffordable in practice. Hence, we only exhaustively build two-row cells from the independent cell sequences. In addition, we also enable the cell sequence shifting to account for the relative position changes of two cell sequences on neighboring rows. As shown in Fig. 8(a), one type of two-row cell consists of an independent 1-cell sequence and an independent 3-cell sequence on top of it, which is denoted as ics_1 on ics_3. For practical placement, the 1-cell sequence in Fig. 8(a) can be anywhere on top of the 3-cell sequence cell. Therefore, we allow for the cell sequence shifting from left to right in the unit of placement pitch and the k-pattering colorability is checked for each placement pitch. Moreover, Fig. 8(b) shows another type of two-row cell denoted as ics_2 on ics_2 and the cell shifting is also supported for the DRC. For practical implementations, we build three types of two-row cells, including ics_1 on ics_2, ics_1 on ics_3 and ics_2 on ics_2.

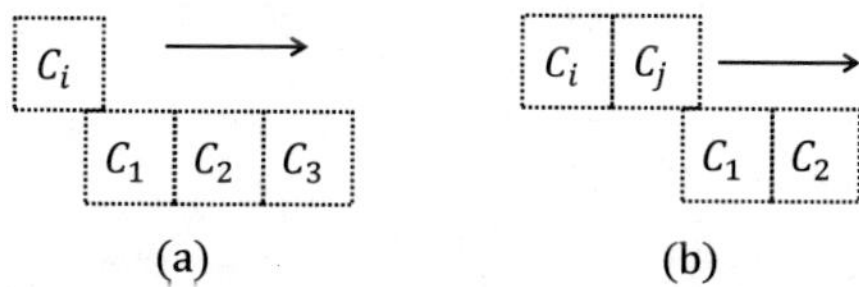

Figure 8. Two-row cells, (a) independent 1-cell sequence on independent 3-cell sequence (ics_1 on ics_3), (b) independent 2-cell sequence on independent 2-cell sequence (ics_2 on ics_2).

4.2 Multi-Row Cells

Next, we randomly build multi-row cells to explore the full chip interactions. Specifically, benchmarks with multiple rows are randomly generated from the independent 1-cell sequences as illustrated in Fig. 9. We pre-set a bound for the design width and height for placement. Independent 1-cell sequences are randomly chosen and abutted horizontally and vertically to fill in the placement region. In Fig. 9(a), 1-cell sequences are abutted horizontally without gaps among their placement and routing boundaries. However, for Fig. 9(b), the cell shifting is enabled to greedily remove the local conflicts for MP, which creates white spaces among cells in the same row. As discussed in Section 5, these types of multi-row cells help to evaluate the placement-level efforts required to reduce QP to TP. In particular, there exists white spaces at the end of each placement row when the random cell can not be inserted due to the fixed design width.

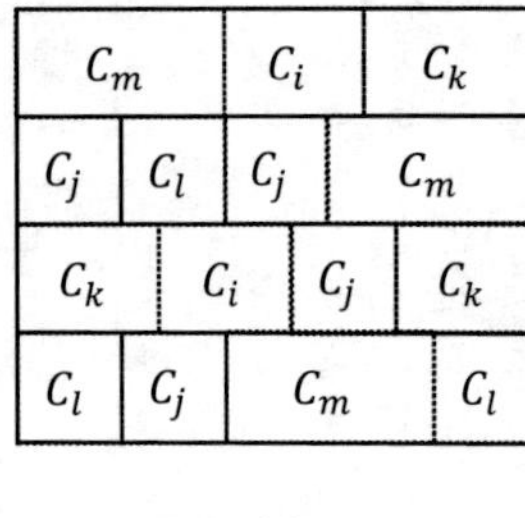

(a)

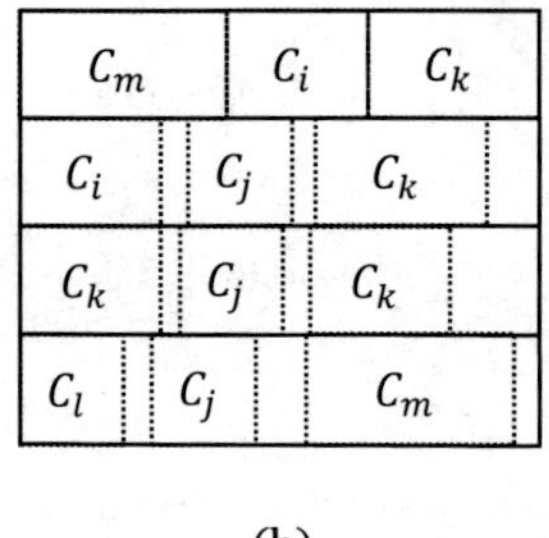

(b)

Figure 9. Multi-row cells, (a) Multiple rows of 1-cell sequences, (b) Multiple rows of 1-cell sequences with local conflict removed.

4.3 Multiple Patterning Compatibility Check

Given a cell combination, our elementary k-patterning checker starts with the conflict graph construction. With the conflict graph, the k-patterning DRC consists of three stages, including iterative vertex removal, clique detection and backtrack coloring. The iterative vertex removal means that any conflict graph node with degree $< k$ can be removed without impacting the colorability of the graph.[16,24] The clique detection searches for a $(k+1)$-clique, which is not k-colorable. For the complex conflict graph without $(k+1)$-cliques, the backtrack coloring is deployed to determine the k-colorability of the conflict graph in a brute-force manner. Our elementary checker works for small scale conflict graph within reasonable runtime.

Local conflict and non-local conflict Next, we discuss why QP, instead of TP, is needed for the CA layer in the $10nm$ representative technology. With the cell combinations constructed and our elementary checker, we can enable the efficient 3-patterning checks. In general, we find that there are two types of conflicts, including local conflicts and non-local conflicts, which lead to the TP DRC violations. As show in Fig. 10(a) and (b), the local conflict originates from the local 4-clique involving neighboring rows. The non-local conflict is due to the coloring failure of the layout across three rows or beyond. An example is illustrated in Fig. 10(a), (c) and (d). For the 3-coloring of the graph, if we have adjacent 3-circles sharing one edge, the two nodes without edge connection should share the same color. As shown in the dashed rectangle in Fig. 10(d), the two nodes without edge connection share the same color, denoted as "1". The same color constraint propagates across the entire graph and introduces a coloring conflict in the dash ellipse in Fig. 10(d).

Furthermore, the graph nodes due to the special constructs, as shown in Fig. 10(c), play a critical role in the non-local conflicts. If we forbid or remove the special construct in the library design, non-local conflicts never happen in our 3-patterning checks. This means the TP DRC violations are simply due to the local conflicts, i.e. local 4-cliques for TP. This also leads to the possibility of removing the local 4-cliques in the placement level and reducing the QP to TP for the CA layer. A local 4-clique consists of 4 rectangles, which at most involves 4 standard cells neighboring each other, and greedy local 4-clique removal for multi-row cells can be achieved by shifting cells and adding white spaces, as shown in Fig. 9(b).

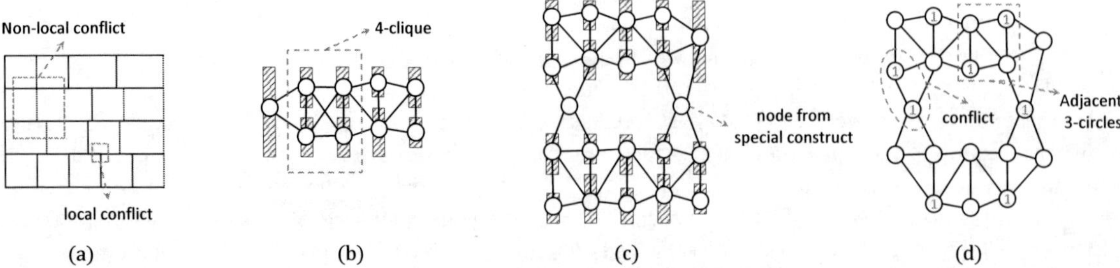

Figure 10. Violations of TP DRCs, (a) the local and non-local conflicts (b) the local conflict, (c) the non-local conflict, (d) the coloring failure for the non-local conflict.

5. EXPERIMENTAL RESULTS

We have implemented our algorithms in C++ and tested them with two sets of benchmarks. One is an ARM representative 10nm standard cell library consisting of 116 logic cells. The other is the NanGate 15nm open cell library[20] consisting of 76 logic cells. All experiments on ARM representative 10nm library are performed on a Linux machine with 3.33GHz Intel(R) Xeon(R) CPU X5680. For experiments on the modified NanGate 15nm library, they are performed on a Linux machine with 3.4GHz Intel(R) Core and 32GB memory. For the NanGate 15nm open cell library, there exits three MOL layers for intra-cell routing. The CA layer is pre-designed as being explicitly decomposed into two masks on separate layers. In order to test our methodology, we recombine these two MOL layers into one single layer, and then run our algorithms on the modified layout using the combined MOL layer. In Algorithm 1, the graph size bound (n_0) is set as 100 and in Algorithm 2, the iteration bound (m_0) is set as 10. We first show the benefit of our cell sequence size reduction. After identifying independent cell sequences, the size reduction of two-row cells are demonstrated. Then, we discuss a set of observations from the multi-row cell compatibility checking. In particular, the cell shifting histogram is introduced to demonstrate the placement-level efforts required to reduce QP to TP for the CA layer with the representative libraries we used.

5.1 Cell Sequence Size Reduction

The cell sequence size reduction helps remove the redundancy among cell sequences and enables efficient cell sequence construction as discussed in Section 3. However, from the theoretical perspective, the number of cell sequences still grows exponentially with the length of the sequence. As shown in Fig. 11, the maximum number of cells in one sequence we can achieve is only 3. However, the 3-cell sequences are sufficient to explore the two-row cell interactions for TP and QP. Meanwhile, we see orders of magnitude reduction on the number of cell sequences from the brute-force construction for both the ARM and NanGate representative libraries. In Fig. 11, "Initial" denotes the brute-force construction while the "CSSR" denotes the cell sequence size reduction. Meanwhile, for each 3-cell sequence built from the library, it is either included in the independent set or redundant to some sequence within the independent set. Moreover, we discuss the necessity of the divide and conquer approximation and speedup. As mentioned earlier, the redundancy graph construction has the time complexity as $O(n^2)$, where n denotes the graph node size. Here, we give an example on the runtime for practical implementations. For ARM representative library, the number of 3-cell sequences before redundancy removal is $n = 120960$ and the average amount of time to compute the redundancy between a pair of 3-cell sequences is $t_0 = 0.36s$. Then, we can estimate the amount of runtime to compute the redundancy graph without the speedup will be $t = 0.5 * n^2 * t_0$, which is approximately 83 years. With the divide and conquer speedup, we achieve the results in Fig. 11 within several hours.

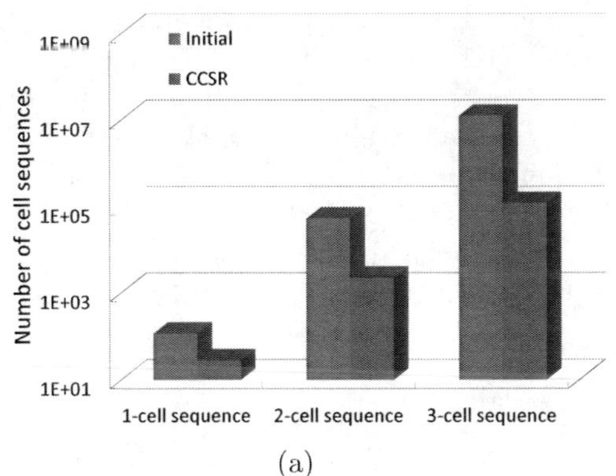

(a)

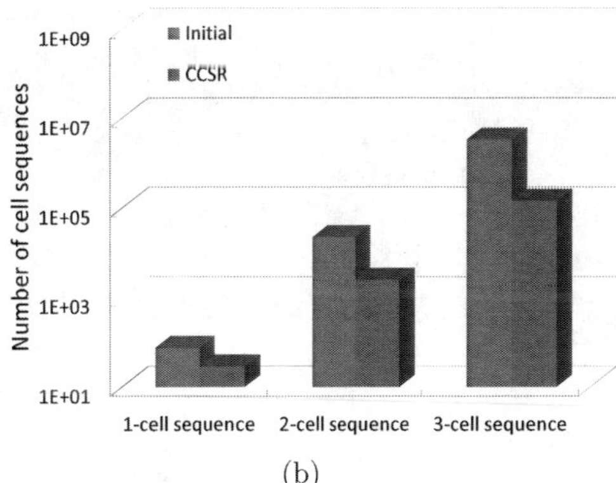

(b)

Figure 11. Cell sequence size reduction, (a) ARM 10nm representative library, (b) Modified NanGate 15nm library.

5.2 Size Reduction of Two-Row Cells

Fig. 12 demonstrates the size reduction of two-row cells. For the horizontal axis, the "ics_i on ics_j" denotes one type of two-row cells, which has an i-cell sequence on top of a j-cell sequence. As discussed in Section 4, the two-row cells are only exhaustively built from the independent cell sequences based on the cell sequence size reduction. Then, the size of "ics_i *on* ics_j" $=$ $2 *$ the size of i-cell sequences $*$ the size of j-cell sequences, where the factor of 2 accounts for the vertical flipping of cell sequences. Actually, the orders of magnitude reductions for two-row cells are due to the cell sequence size reduction in Fig. 11. For example, in Fig. 12(a), we observe that billions of "ics_1 *on* ics_3" are reduced to millions of them. In addition, any two-row cell constructed from the cell library is either included in the reduced sets or redundant to some two-row cell in the reduced sets.

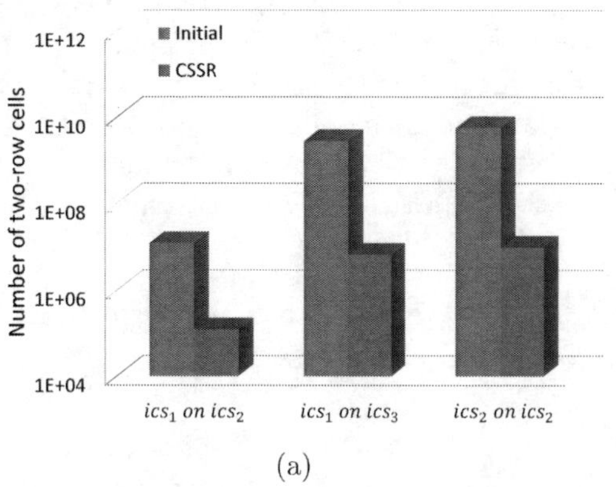

(a)

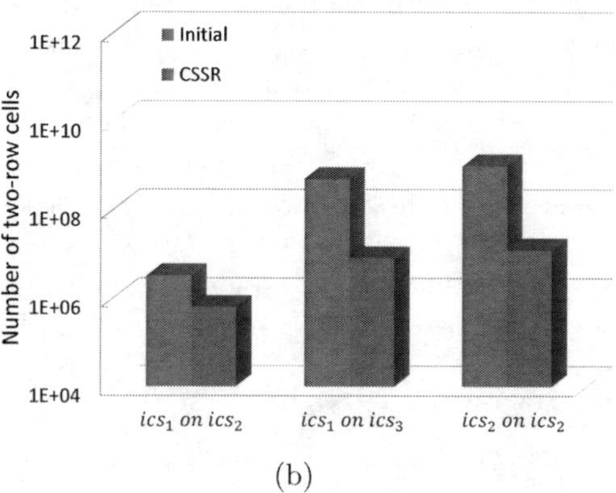

(b)

Figure 12. Size reduction of two-row cells from the cell sequence size reduction, (a) ARM $10nm$ representative library, (b) Modified NanGate $15nm$ library.

5.3 Full Chip Interactions

Table 1 and Table 2 demonstrate the multi-row cells randomly generated from the set of independent 1-cell sequences for the NanGate $15nm$ library and the ARM $10nm$ representative library, respectively. Within the tables, the "Bench" denotes the benchmark name, the "size" is the area for placement, the "Util(%)" is the utilization rate representing the percentage of cell area over total area of the design, the "Cell#" is the number of cells, the "Rect#" is the number of rectangles, "Pass" denotes the DRC result and "CPU" denotes the amount of runtime. We have the utilization rate (Util(%)) less than 100% because white space is left at the end of a row when there is not enough to insert another random 1-cell sequence. As shown in both tables, all the multi-row cells pass the QP DRCs. In contrast, TP DRCs either lead to violations or can not be finished within an acceptable amount of runtime. As mentioned earlier, the violations of TP DRCs originate from two types of conflicts, including local conflicts and non-local conflicts. The non-local conflicts are due to the propagation of same color constraints across three rows or beyond as shown in Fig. 10(d). The conflict graph nodes from the special constructs prevents simplifying the graph into small components. Then, we may encounter large graph node sizes for the multi-row cells, which makes the 3-coloring problem intractable. For instance, the TP DRCs for the multi-row cells from ARM_4 to ARM_6 in Table 2 can not be finished within a reasonable amount of runtime due to the large graph sizes. Actually, the use of special constructs varies from technology to technology and the NanGate $15nm$ library does not have this cross-coupled structure. Thus, the 3-patterning checks for all the multi-row cells from the NanGate library can be achieved in Table 1 (without the special construct that merges the graph across a row, the graph naturally partitions itself by row, which leads to much more manageable graph sizes and runtimes).

Table 1. NanGate $15nm$ multi-row cells with local 4-cliques

					TP Checks		QP checks	
Bench	Size(um^2)	Util(%)	Cell#	Rect#	Pass	CPU(s)	Pass	CPU(s)
NanGate_1	0.8 x 0.8	91.0	77	1870	No	0.07	Yes	0.04
NanGate_2	1.5 x 1.5	96.5	288	7875	No	0.35	Yes	0.29
NanGate_3	31 x 31	98.0	1257	31783	No	3.37	Yes	2.83
NanGate_4	61 x 61	99.0	4948	12864	No	52.7	Yes	39.6
NanGate_5	115 x 115	99.5	17382	454671	No	584.9	Yes	449.1
NanGate_6	154 x 154	99.6	31057	809015	No	1823.9	Yes	1403.4

Table 2. ARM 10nm representative multi-row cells with local 4-cliques

					TP Checks		QP checks	
Bench	Size(um^2)	Util(%)	Cell#	Rect#	Pass	CPU(s)	Pass	CPU(s)
ARM_1	17.2 x 17.2	73.6	29	937	No	0.04	Yes	0.02
ARM_2	34.6 x 34.6	86.3	124	4411	No	24.4	Yes	0.15
ARM_3	69.1 x 69.1	93.4	444	19381	No	915.6	Yes	1.36
ARM_4	138 x 138	95.4	1831	79287	No	308605.0	Yes	16.8
ARM_5	259 x 259	97.5	6362	284933	n/a	n/a	Yes	200.0
ARM_6	346 x 346	98.3	11261	511888	n/a	n/a	Yes	638.4

For the NanGate library without special constructs, the TP DRC violations of multi-row cells are simply due to the local 4-cliques. As shown in Fig. 13(a), the conflict graph on the CA layer are split into independent components in a row by row manner. The local 4-cliques come from the cell interactions between neighboring rows. The multi-row cells without local 4-cliques are still built horizontally and vertically from the random 1-cell sequences. When local 4-cliques are detected, the current cell being inserted is shifted from left to right by one placement pitch until all local 4-cliques are eliminated.

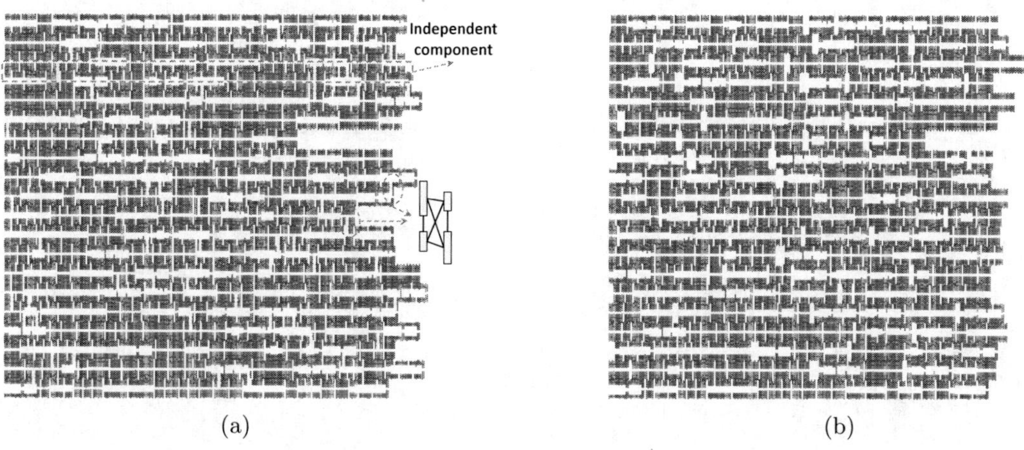

Figure 13. Benchmarks with multiple rows, (a) with local 4-cliques, (b) without local 4-cliques.

The non-local conflicts for TP are inevitable if special constructs are included in the SC design as discussed in Section 4.3. If the special constructs are excluded, it is possible to resolve the local conflicts in the placement level and reduce QP to TP. Therefore, we build two sets of multi-row cells, one is from the NanGate library in Table 3 and the other is obtained by modifying the ARM library and removing the special constructs within the standard cells in Table 4. We first discuss the **cell shifting histogram** in Fig. 14. In the cell shifting histogram, the horizontal axis is the number of placement pitches each inserted cell is shifted before achieving a placement site that is free of local 4-cliques and the vertical axis is the number of cells. In general, the cell shifting histogram is an initial attempt to capture the effort that might be needed by a placement algorithm in order to achieve a legal placement given the same scenario of cells. For most benchmarks, the number of cells is a non-increasing function of the number pitches shifted. To the first order, we prefer that cells are shifted less than 3 placement pitches, as shown in the shadowed region marked, "Preferred region", in Fig. 14. In the representative libraries we used, a minimum-sized inverter occupied 3 placement pitches (in width), so we chose to use the same width as the upper bound of our "Preferred Region". For each benchmark, we further calculate the percentage of cells in the preferred region and the utilization rate, which are two metrics to evaluate the placement-level efforts to reduce QP to TP. If TP on the CA layer for practical designs is desired, we expect to require that a high percentage of cells would be in the preferred region while the utilization rate of the placement is also high. A high percentage of cells in the preferred region means that minimum placement-level efforts will be needed to remove the local 4-cliques and achieve TP friendly design. The high utilization of each benchmark means that the cell shifting in the placement stage will introduce less area penalty. To improve the robustness at the standard cell library level, SC designers need to increase the utilization rate and the percentage of cells in the preferred region for the multi-row cells.

In the end, the multi-row cell compatibility checking yields a unique set of observations for the representative 10nm technology. (1) The 4-clique exists due to the row to row interactions, which means QP needs to be deployed to enable implicit coloring for standard cells; (2) There is no local five-clique for the CA layer in the representative technology used; (3) For all the two-row cells and multi-row cells generated, the colorability for QP can be easily proved with our elementary checker.

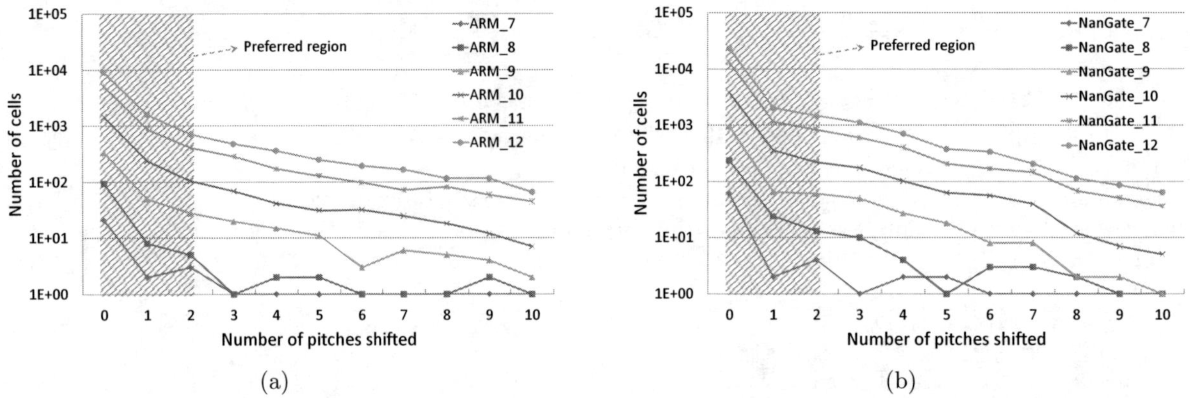

Figure 14. The cell shifting histogram denoting the number of pitches shifted to remove local 4-cliques, (a) ARM 10nm representative library (special constructs removed), (b) Modified NanGate 15nm library.

Table 3. NanGate 15nm multi-row cells without local 4-cliques

						TP Checks	
Bench	Size(um^2)	Util(%)	Cell#	Rect#	C.i.p(%)	Pass	CPU(s)
NanGate_7	0.8 x 0.8	91.3	72	1865	91.7	Yes	0.02
NanGate_8	1.5 x 1.5	92.8	295	7499	92.5	Yes	0.19
NanGate_9	31 x 31	94.6	1191	30561	90.3	Yes	2.53
NanGate_10	61 x 61	95.6	4800	123621	90.4	Yes	37.9
NanGate_11	115 x 115	95.5	16757	433639	90.0	Yes	461.8
NanGate_12	154 x 154	95.7	29870	771964	89.9	Yes	1475.8

Table 4. ARM 10nm representative multi-row cells (special constructs removed) without local 4-cliques

						TP Checks	
Bench	Size(um^2)	Util(%)	Cell#	Rect#	C.i.p(%)	Pass	CPU(s)
ARM_7	17.2 x 17.2	89.7	29	913	89.7	Yes	0.09
ARM_8	34.6 x 34.6	86.9	117	4443	91.5	Yes	1.00
ARM_9	69.1 x 69.1	84.7	480	17300	86.3	Yes	2.23
ARM_10	138 x 138	92.3	2028	75405	88.4	Yes	1237.6
ARM_11	259 x 259	93.2	7338	267291	87.1	Yes	4904.2
ARM_12	346 x 346	93.5	13162	476000	86.8	Yes	1502.1

6. CONCLUSION

This paper proposes a comprehensive framework to evaluate the MOL robustness for a standard cell library over multiple patterning. To reduce the number of N-wise checks, the issues of cell sequence construction and size reduction are proposed and solved efficiently. The multi-row cell compatibility checking and an elementary checker are presented to enable efficient N-wise checks. For representative 10nm/15nm libraries we used, the cell shifting histogram is proposed to quantify the placement-level efforts required to reduce QP to TP for the CA layer.

REFERENCES

[1] R. Merritt, "Intel sees quad-patterned path to 10-nm chips." http://www.eetimes.com/document.asp?doc_id=1262512, September 2012.

[2] L. Liebmann, D. Pietromonaco, and M. Graf, "Decomposition-aware standard cell design flows to enable double-patterning technology," in *Proc. of SPIE*, **7974**, 2011.

[3] D. Z. Pan, B. Yu, and J.-R. Gao, "Design for manufacturing with emerging nanolithography," *IEEE Transactions on Computer-Aided Design of Integrated Circuits and Systems (TCAD)* **32**(10), pp. 1453–1472, 2013.

[4] K. Lucas, C. Cork, B. Yu, G. Luk-Pat, B. Painter, and D. Z. Pan, "Implications of triple patterning for 14 nm node design and patterning," in *Proc. of SPIE*, **8327**, 2012.

[5] X. Xu, B. Cline, G. Yeric, B. Yu, and D. Z. Pan, "Self-aligned double patterning aware pin access and standard cell layout co-optimization," in *ACM International Symposium on Physical Design (ISPD)*, pp. 101–108, 2014.

[6] K. B. Agarwal, C. J. Alpert, Z. Li, G.-J. Nam, and N. Viswanathan, "Multi-patterning lithography aware cell placement in integrated circuit design," July 23 2013. US Patent 8,495,548.

[7] B. Yu, X. Xu, J.-R. Gao, and D. Z. Pan, "Methodology for standard cell compliance and detailed placement for triple patterning lithography," in *IEEE/ACM International Conference on Computer-Aided Design (ICCAD)*, pp. 349–356, 2013.

[8] M. Cho, Y. Ban, and D. Z. Pan, "Double patterning technology friendly detailed routing," in *IEEE/ACM International Conference on Computer-Aided Design (ICCAD)*, pp. 506–511, 2008.

[9] M. Mirsaeedi, J. A. Torres, and M. Anis, "Self-aligned double-patterning (SADP) friendly detailed routing," in *Proc. of SPIE*, **7974**, 2011.

[10] Q. Ma, H. Zhang, and M. D. F. Wong, "Triple patterning aware routing and its comparison with double patterning aware routing in 14nm technology," in *ACM/IEEE Design Automation Conference (DAC)*, pp. 591–596, 2012.

[11] Y.-H. Lin, B. Yu, D. Z. Pan, and Y.-L. Li, "TRIAD: A triple patterning lithography aware detailed router," in *IEEE/ACM International Conference on Computer-Aided Design (ICCAD)*, pp. 123–129, 2012.

[12] C.-H. Hsu, Y.-W. Chang, and S. R. Nassif, "Simultaneous layout migration and decomposition for double patterning technology," in *IEEE/ACM International Conference on Computer-Aided Design (ICCAD)*, pp. 595–600, 2009.

[13] A. B. Kahng, C.-H. Park, X. Xu, and H. Yao, "Layout decomposition for double patterning lithography," in *IEEE/ACM International Conference on Computer-Aided Design (ICCAD)*, pp. 465–472, 2008.

[14] B. Yu, K. Yuan, B. Zhang, D. Ding, and D. Z. Pan, "Layout decomposition for triple patterning lithography," in *IEEE/ACM International Conference on Computer-Aided Design (ICCAD)*, pp. 1–8, 2011.

[15] B. Yu, J.-R. Gao, and D. Z. Pan, "Triple patterning lithography (TPL) layout decomposition using end-cutting," in *Proc. of SPIE*, **8684**, 2013.

[16] B. Yu and D. Z. Pan, "Layout decomposition for quadruple patterning lithography and beyond," in *ACM/IEEE Design Automation Conference (DAC)*, pp. 1–6, 2014.

[17] M. Rashed, N. Jain, J. Kim, M. Tarabbia, I. Rahim, S. Ahmed, J. Kim, I. Lin, S. Chan, H. Yoshida, *et al.*, "Innovations in special constructs for standard cell libraries in sub 28nm technologies," in *IEEE International Electron Devices Meeting (IEDM)*, pp. 248–251, 2013.

[18] G. Northrop, "Design technology co-optimization in technology definition for 22nm and beyond," in *IEEE Symposium on VLSI Technology (VLSIT)*, pp. 112–113, 2011.

[19] M. Rashed, S. Soss, J. Kye, I. Lin, J. Gullette, C. Nguyen, J. Kim, M. Tarabbia, Y. Ma, Y. Deng, *et al.*, "Semiconductor device with transistor local interconnects," Nov. 12 2013. US Patent 8,581,348.

[20] "NanGate FreePDK15 Open Cell Library." http://www.nangate.com/?page_id=2328.

[21] K. Yuan, J.-S. Yang, and D. Z. Pan, "Double patterning layout decomposition for simultaneous conflict and stitch minimization," in *ACM International Symposium on Physical Design (ISPD)*, pp. 185–196, 2009.

[22] Y. Xu and C. Chu, "A matching based decomposer for double patterning lithography," in *ACM International Symposium on Physical Design (ISPD)*, pp. 121–126, 2010.

[23] X. Tang and M. Cho, "Optimal layout decomposition for double patterning technology," in *IEEE/ACM International Conference on Computer-Aided Design (ICCAD)*, pp. 9–13, 2011.

[24] S.-Y. Fang, W.-Y. Chen, and Y.-W. Chang, "A novel layout decomposition algorithm for triple patterning lithography," in *ACM/IEEE Design Automation Conference (DAC)*, pp. 1185–1190, 2012.

[25] H. Tian, H. Zhang, Q. Ma, Z. Xiao, and M. Wong, "A polynomial time triple patterning algorithm for cell based row-structure layout," in *IEEE/ACM International Conference on Computer-Aided Design (ICCAD)*, pp. 57–64, 2012.

[26] J. Kuang and E. F. Young, "An efficient layout decomposition approach for triple patterning lithography," in *ACM/IEEE Design Automation Conference (DAC)*, pp. 69:1–19:6, 2013.

[27] B. Yu, Y.-H. Lin, G. Luk-Pat, D. Ding, K. Lucas, and D. Z. Pan, "A high-performance triple patterning layout decomposer with balanced density," in *IEEE/ACM International Conference on Computer-Aided Design (ICCAD)*, pp. 163–169, 2013.

[28] Y. Zhang, W.-S. Luk, H. Zhou, C. Yan, and X. Zeng, "Layout decomposition with pairwise coloring for multiple patterning lithography," in *IEEE/ACM International Conference on Computer-Aided Design (ICCAD)*, pp. 170–177, 2013.

Self-Aligned Quadruple Patterning-Compliant Placement

Fumiharu Nakajima[1], Chikaaki Kodama[2], Koichi Nakayama,
Shigeki Nojima and Toshiya Kotani
Toshiba Corp. Semiconductor & Storage Products Company, Sakae-ku, Yokohama 247-8585, Japan

ABSTRACT

Self-Aligned Quadruple Patterning (SAQP) will be one of the leading candidates for sub-14 nm node and beyond. However, compared with triple patterning, making a feasible standard cell placement has following problems. (1) When coloring conflicts occur between two adjoining cells, they may not be solved easily since SAQP layout has stronger coloring constraints. (2) SAQP layout cannot use stitch to solve coloring conflict. In this paper, we present a framework of SAQP-aware standard cell placement considering the above problems. When standard cell is placed, the proposed method tries to solve coloring conflicts between two cells by exchanging two of three colors. If some conflicts remain between adjoining cells, dummy space will be inserted to keep coloring constraints of SAQP. We show some examples to confirm effectiveness of the proposed framework. To our best knowledge, this is the first framework of SAQP-aware standard cell placement.

Keywords: Multiple patterning, Self-Aligned Quadruple Patterning, SAQP, Optical lithography, Placement, DFM

1. INTRODUCTION

Due to shrinking of semiconductor technology node, Self-Aligned Quadruple Patterning (SAQP) will be one of the leading candidates for sub-14 nm node and beyond. Despite its robustness against overlay, SAQP is a challenging process since predicting wafer image instantly is almost impossible and we must follow stricter constraint than in triple patterning process in order to design feasible layout.

A feasible layout drawing for SAQP process has been studied so far. Nakayama et al. analyzed SAQP construction process and revealed that patterns could be distinguished into three kinds[1]. Kodama et al. proposed the first SAQP layout method using pre-colored grid[2] which can obtain resultant layout painted into three colors, but strict constraints are imposed to draw a feasible layout. Nakajima et al. relaxed a part of the routing constraints and developed a standard cell layout method without using pre-coloring grid[3]. Therefore, SAQP-aware standard cell placement is natural issue for further SAQP application.

Triple patterning-aware standard cell placement was studied by Yu et al.[4] They presented the first systematic study for the triple patterning-aware ordered single row placement. It can solve cell placement and color assignment simultaneously. However, we cannot easily apply their method to SAQP process since there are following problems for standard cell placement. (1) When coloring conflicts occur between two adjoining cells, they may not be solved easily since SAQP layout has stronger coloring constraints. In triple patterning layout, color of patterns can be exchanged each other flexibly under no coloring conflicts, but in SAQP, only two of three colors are permitted to exchange each other. (2) SAQP layout cannot use stitch to solve coloring conflict like triple patterning.

In this paper, we present a framework of SAQP-aware standard cell placement considering the above problems. When standard cell is placed, the proposed method tries to solve coloring conflicts between two cells as many as possible by exchanging two of three colors. If some conflicts remain between adjoining cells, dummy space will be inserted and they are connected keeping coloring constraints (Fig. 1). Trim mask is derived according to feasible cell placement. Some examples will be shown to confirm effectiveness of the proposed framework. To our best knowledge, this is the first framework of SAQP-aware standard cell placement.

---

[1] Fumiharu Nakajima: fumiharu.nakajima@toshiba.co.jp
[2] Chikaaki Kodama: chikaaki1.kodama@toshiba.co.jp

Design-Process-Technology Co-optimization for Manufacturability IX, edited by John L. Sturtevant, Luigi Capodieci,
Proc. of SPIE Vol. 9427, 942708 · © 2015 SPIE · CCC code: 0277-786X/15/$18 · doi: 10.1117/12.2085705

2. SAQP PROCESS AND LAYOUT PRINCIPLES

2.1 SAQP process overview

In SAQP process, sidewall spacer is formed twice and the final pattern pitch becomes a quarter of a mandrel pattern pitch. There are two types of SAQP processes, positive tone (also called Spacer-Is-Metal) and negative tone (also called Spacer-Is-Dielectric). The second sidewall spacer becomes the final pattern in positive tone process and becomes dielectric in negative tone process. In this paper, we utilize negative tone SAQP (nSAQP) for metal layer because design flexibility is higher and the number of required masks is fewer than positive tone process.

Figure 1 shows nSAQP process with two masks, mandrel and trim masks. "F" represents half pitch of the final pattern. (a) Mandrel pattern of 8F pitch is exposed by a mask in the first lithography process. (b) The first masking material is deposited to form the first sidewall spacer. (c) The mandrel pattern is selectively removed. The second sidewall spacer process from (d) to (e) is the same as from (b) to (c). (f) Trim pattern is covered by the second lithography. (g) The substrate is etched using the remaining second sidewall spacer and trim pattern as a hard mask. The spacer and trim pattern are removed. (f) The etched trench is filled with conductive material and the final pattern is formed.

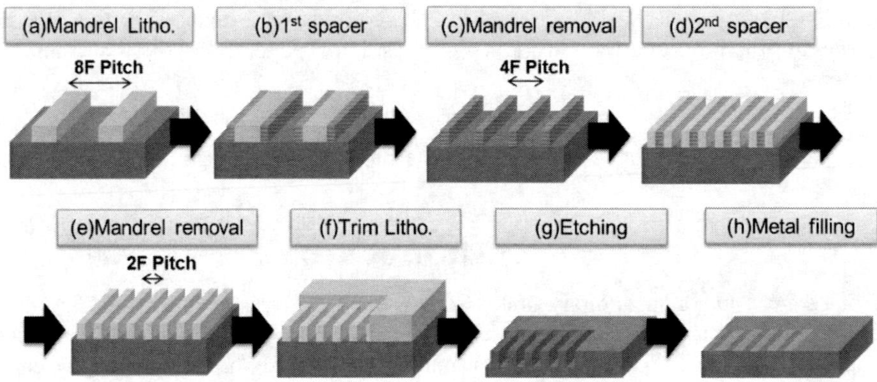

Figure 1: Negative tone SAQP process.

2.2 Feasible SAQP layout principle

In SAQP process, a feasible SAQP layout is necessarily decomposed into three patterns, **primary**, **secondary** and **tertiary** patterns[1]. As shown in Figure 2, primary patterns are formed on mandrel positions, and secondary patterns on gap between mandrels. Therefore the minimum pitch between primary patterns and that between secondary patterns becomes 8F which is the same as mandrel pattern pitch shown in Figure 1. The minimum pitch between primary and secondary patterns is 4F because it is half pitch of mandrel pattern. Tertiary pattern is formed on the position of first sidewall spacer and it is located between primary and secondary pattern, so the minimum pitch between tertiary patterns is 4F. The minimum pitch between tertiary and primary or tertiary and secondary patterns is 2F, the same as the final pattern pitch. We can introduce three-color mapping for SAQP layout design (Figure 2(d)).

Space width between each pattern is constant in the final pattern as is defined by the constant width of the second sidewall spacer. The width of the second sidewall spacer is determined based on the deposited quantity of material and it is impossible to partially control the quantity, so space width of the final wafer image becomes constant. Tertiary pattern is formed like a loop shape because its shape is the same as the first sidewall spacer, so T-shape is not allowed for tertiary pattern. Any stitch like LELELE is not allowed since the tertiary pattern takes a role of layout separator between primary and secondary patterns. Different type of patterns cannot "connect" each other. It means one color should be assigned to one polygon.

Accordingly, if a certain layout is given and it can be decomposed into three colors keeping the following conditions by some kinds of layout coloring technique, the layout is possible to produce by SAQP process. The following is a summary of the above mentioned conditions of feasible SAQP pattern.

- Patterns can be decomposed into three colors like Figure 2.
- Each of primary, secondary and tertiary patterns has specific minimum pitch (2F, 4F, or 8F).
  - primary-primary : 8F
  - secondary-secondary : 8F
  - primary-secondary : 4F
  - tertiary-tertiary : 4F
  - primary-tertiary : 2F
  - secondary-tertiary : 2F
- Space of the final pattern is constant.
- Different type of patterns cannot connect each other. Stich pattern is not allowed.
- Tertiary pattern forms a loop shape. T-shape is not allowed.

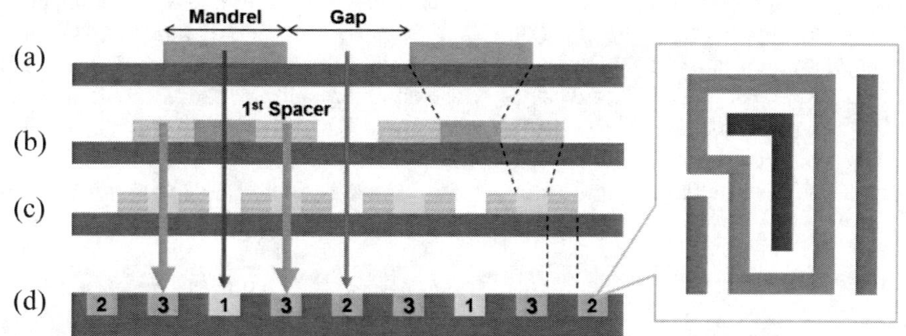

Figure 2: Relation of mandrel, sidewall spacer and the final pattern.
(a) Mandrel lithography. (b) First spacer process. (c) Second spacer process. (d) Final pattern.

3. COMPLIANT SAQP-AWARE PLACEMENT

In this paper, we propose a SAQP-compliant cell placement method. In order to obtain a feasible cell placement for SAQP, we focus on pre-coloring approach where coloring is performed in the cell level design and the coloring information is used at placement stage because of shorter design TAT compared with re-coloring approach after placement. The method comprises five steps as shown in Figure 3. In Step 1, we prepare feasible cell layout for SAQP by our SAQP-aware routing method[3]. In initial cell placement, each cell is placed without any extra space and is checked whether it is compliant for SAQP or not. If it is not, some solutions are performed to remove coloring conflict in cell boundary in Step 4. Mask patterns are extracted from the resultant layout. In this section, we explain each layout step in detail and show an example of SAQP process with obtained masks.

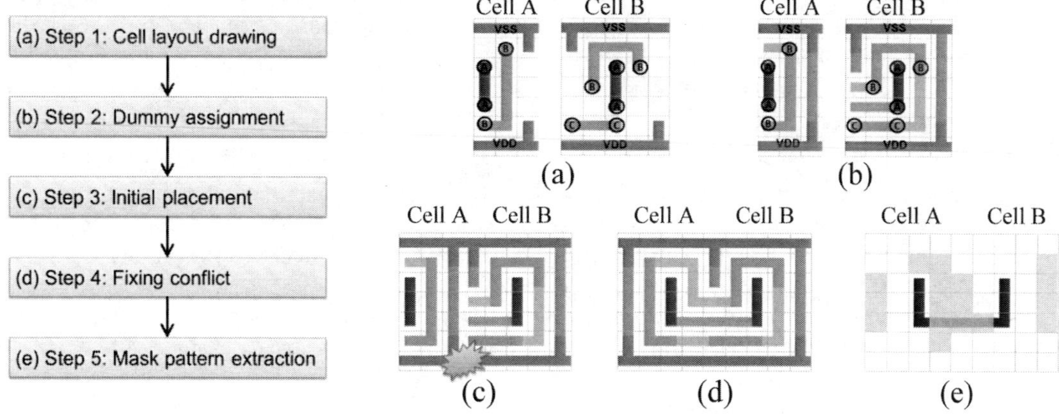

Figure 3: Flow of the proposed method.

3.1 Step 1: Cell layout drawing and Step 2: Dummy pattern assignment

In Step 1, a feasible cell layout for SAQP keeping three color constraints is prepared according to SAQP-aware routing method[3]. A grid structure called "base grid" is prepared for routing (Figure 4(a)). Each grid size is set as the target metal pitch. We call a grid used by routing "occupied" and a grid not used "vacant" in this paper. Each pin is connected by grid routing and each grid is allowed to be occupied by one net only. The width of path keeps the target half pitch multiplied by odd number in the final pattern. Power and ground rails are located on top and bottom of the each cell. Since it is difficult to solve the coloring conflicts of power and ground nets without stitch insertion on cell boundary, keeping their consistent coloring for all cells is preferable.

We show an example of drawing paths between pins "A1" to "A2," "B1" to "B2," and "C1" to "C2" in Figure 4(b) and (c). "VDD" and "VSS" each is also connected to corresponding power or ground net. Red color is assigned to both power and ground nets in this example. To make feasible SAQP layout, each path is drawn under the proposed SAQP-aware routing rule[3]. First we connect "VSS," "VDD," "A" and "B" by blue and red paths. For blue and red paths, different color path can be routed on every other or more than every other grid and same color path on every fourth or more than fourth grid. Therefore, we cannot connect "C1" and "C2" by blue or red path since they are placed on adjacent grid to blue or red path. So, they must be connected by green path. Different from blue and red paths drawing, a strict constraint called "green pattern routing rule" must be imposed on green path for SAQP manufacturability. Details of this rule were explained in our previous paper[3]. Due to this rule, green path must make a detour as shown in Figure 4(c). After path drawing, if vacant grids remain, dummy patterns must be assigned to all of them to make the layout feasible for SAQP process (Figure 4(d)). The drawing procedure of Step 1 and 2 can be carried out by designers or EDA tools.

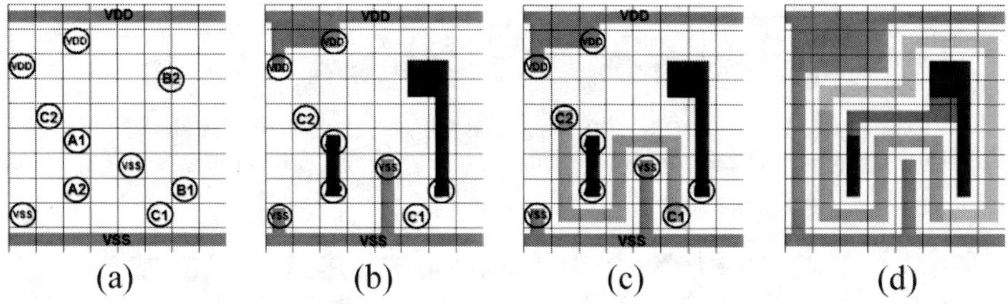

Figure 4: Drawing example of cell layout keeping three color constraints. (a) Base grid. (b) Blue and red pattern drawing. (c) Green pattern drawing. (d) Dummy pattern assignment.

3.2 Step 3: Initial placement and Step 4: fixing conflict

Our framework assumes a row-structure layout, where cells in each row are located at the same height. In Step 3, we obtain initial cell placement with pre-coloring cell layout prepared in Step 1 and 2. After each cell is placed side by side without any extra space, some conflicts will happen on the cell boundaries even though each cell is compliant for SAQP. If there are some conflicts, we can apply some solutions to fix the coloring conflicts in Step 4. Table 1 shows their comparisons between LELELE and SAQP at placement stage. In general, there are four major solutions in LELELE[4], but stitch insertion and flipping green pattern into other color are not allowed in SAQP due to conditions mentioned in Section 2.

We show an example of initial placement and fixing conflict. As shown in Figure 5(a), there are three cells A, B and C. Cell B is the same pattern of the example of Step 1 (Figure 4(d)). For easy understanding, all dummy patterns are hidden in Figure 5(b) and (c). When these cells are placed without any extra space, two coloring conflicts E1 and E2 happen on the cell boundaries (Figure 5(c)). For E1, there is one grid interval between two blue patterns on the boundary between cell B and C. Since the same blue patterns must be drawn on every fourth grid or more for a feasible SAQP layout, E1 is coloring conflict. For E2, blue pattern is adjacent to red pattern on the boundary between cell A and B. However, blue and red patterns must not be drawn adjacently for a feasible SAQP layout, so E2 is also coloring conflict.

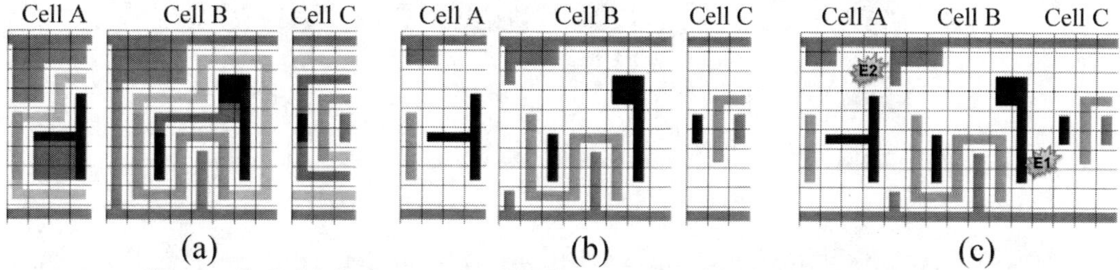

Figure 5: Initial placement. (a) Three cells with dummy patterns. (b) Three cells without dummy patterns. (c) Initial placement.

We can solve conflict E1 if cell C is flipped in left-right direction or blue and red patterns in cell C are flipped each other. We adopted flipping of blue and red patterns as shown in Figure 6(a) and E1 was solved. For conflict E2, if we flip blue and green patterns as shown in Figure 7(a), the conflict will disappear. However, T-shape pattern cannot become green (tertiary) pattern due to SAQP process. Green pattern is formed based on the shape of the first sidewall spacer but sidewall spacer process cannot form T-shape pattern. Therefore, this flipping operation is not acceptable. If we can insert a stitch as shown in Figure 7(b), the conflict will disappear but stitch insertion is not originally allowed in SAQP process. As a result, we have to insert extra space between two cells and SAQP-compliant cell placement (target design) is obtained (Figure 6(b)). Figure 8(a) shows target design with dummy pattern. As described in Section 2, dummy pattern must be assigned to all vacant grids to satisfy constant space width, so we assign dummy pattern to extra space region in the same way as Step 2 (Figure 8(b)).

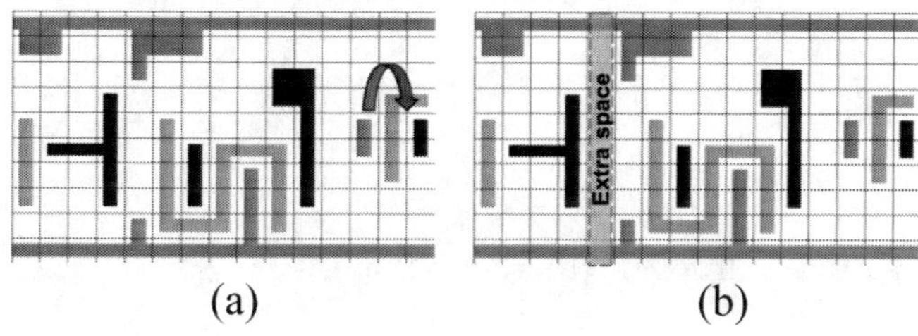

Figure 6: Example of fixing conflicts. (a) Blue and red pattern flipped. (b) Layout after extra space insertion between cell A and B (target design).

		LELELE	SAQP
Flip the color.	Blue	✓	✓
	Red	✓	✓
	Green	✓	–
Flip the one cell in right-left direction.		✓	✓
Insert stitch.		✓	–
Add a extra space between two cells.		✓	✓

✓: allowed

Table 1: Comparison of the fixing solutions.

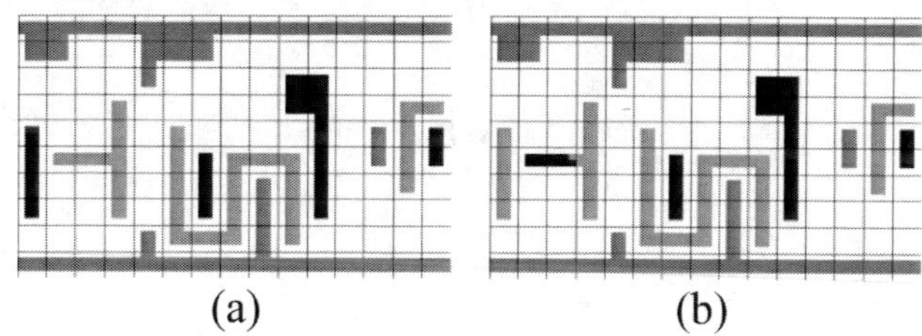

Figure 7: Examples of improper fix. (a) Blue and green pattern flipped. (b) Stitch insertion.

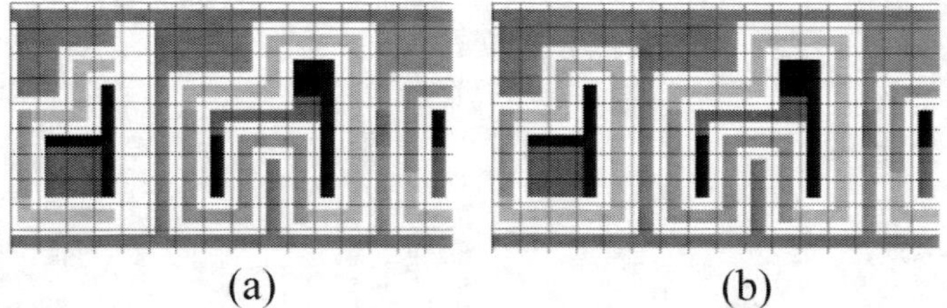

Figure 8: Example of SAQP-compliant cell placement after fixing coloring conflicts. (a) Layout fixed two conflicts. (b) Dummy pattern assignment to extra space region.

3.3 Step 5: Mask pattern extraction

After fixing conflicts in Step 4, mandrel and trim mask patterns are extracted from the resultant layout in Step 5. Mandrel mask pattern can be obtained from the layout by selecting either blue or red pattern but not green pattern. In Figure 9(a), blue pattern was extracted from Figure 8(b) and expanded to make the mandrel mask. All dummy patterns in Figure 8(b) were extracted as trim mask pattern shown in Figure 9(b). Figure 10(a) shows an example of SAQP process using mandrel and trim masks as shown in Figure 9. The final wafer image obtained by SAQP process (Figure 10(b)) is the same as the target design of Figure 6(b).

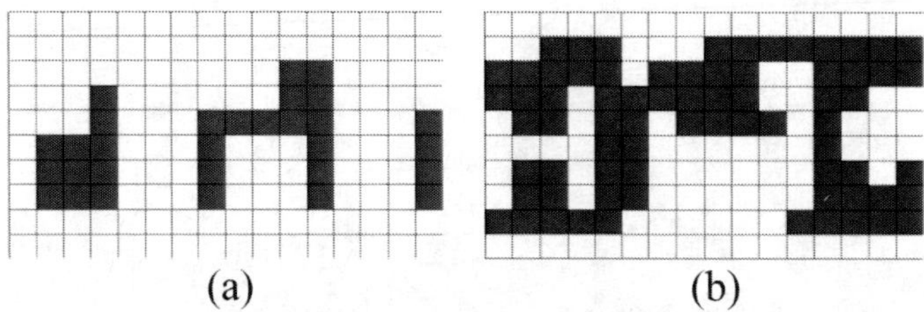

Figure 9: Mask pattern extraction. (a) Mandrel mask pattern. (b) Trim mask pattern.

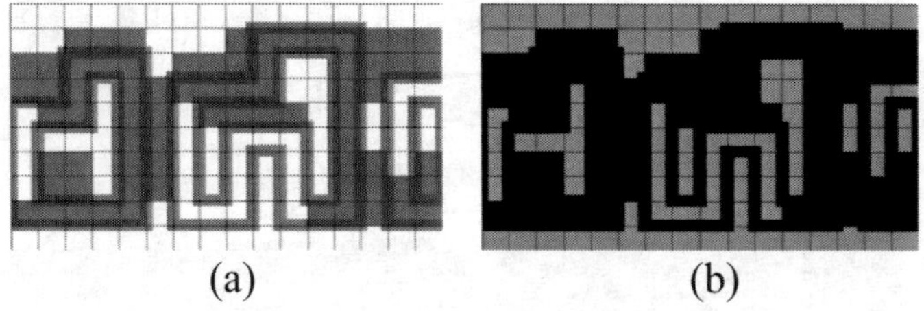

Figure 10: Example of SAQP process with mandrel and trim masks in Figure 9. (a) Spacer and trim patterns. (b) Final wafer image.

4. EXPERIMENTS AND RESULTS

In order to confirm effectiveness, we applied the proposed method to a metal layer of logic stand cell layout and carried out lithography and spacer sidewall process simulations. We utilized 45 nm node standard cell layout of NanGate open cell library (PDKv1_3_v2010_12)[9]. NanGate standard cell is not designed for SAQP and some contacts were initially placed on off-grid, so we slightly shifted them to on-grid. The minimum width and space of the standard cell were scaled down to 20 nm. Figure 11 shows one of the experimental results. Pins were connected manually according to the procedure mentioned in Section 3.1 (Figure 11(a)) and dummy patterns were also manually drawn at cell level layout stage (Figure 11(b)). When cells were placed without any extra space, some conflicts happened on the cell boundaries

(Figure 11(c)), so we tried to fix them by flipping and extra space insertion (Figure 11(d)). Mandrel and trim masks for SAQP process were extracted from the resultant layout after dummy pattern assignment (Figure 11(e)).

After mask pattern extraction, we carried out lithography and sidewall spacer process simulations. Since the minimum pitch of mandrel pattern is 160 nm, the simulation conditions, NA = 1.35 and a cross pole illumination, were used for lithography of mandrel. Trim pattern will be formed by EUV or double patterning technique because its minimum pitch becomes double pitch of the final pattern. In this experiment, we adopted double patterning process simulation to form trim pattern of 80 nm pitch. In the spacer process simulation, sidewall with 20 nm width was deposited twice. Figure 11(f) shows the result of mandrel lithography and Figure 11(g) shows trim pattern and the first and the second sidewall spacers. As shown in Figure 11(h), the final pattern was obtained after metal filling. The proposed method will be utilized for standard cell design effectively.

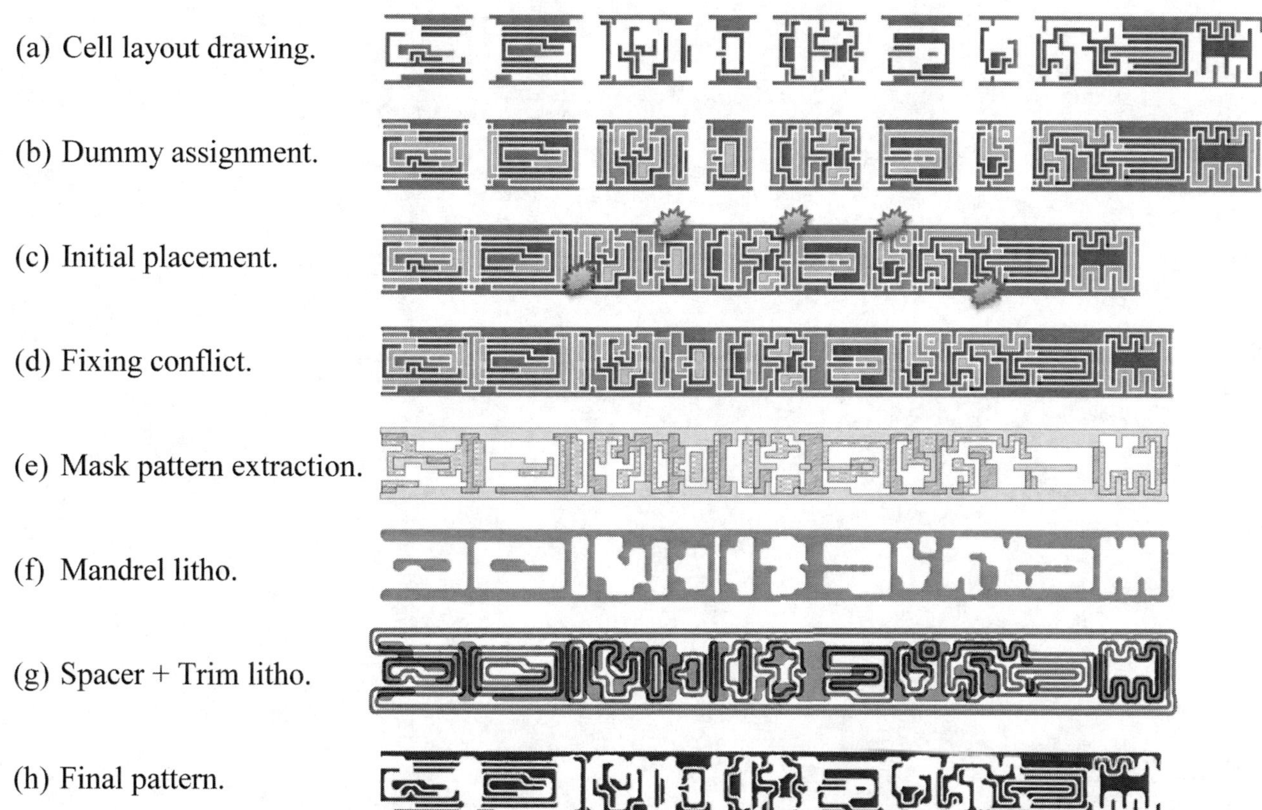

Figure 11: Layout obtained by the proposed method.

5. CONCLUSIONS

To our best knowledge, we proposed the first SAQP-compliant cell placement method. Before cell placement, we prepare SAQP-feasible cell layout by utilizing our SAQP-aware layout method[3]. After each cell is placed, some fixing solutions are performed to remove coloring conflicts on cell boundary. In order to confirm effectiveness, we applied the proposed method to logic standard cell layout. Our experimental results show the validations of the SAQP-compliant cell placement. As future work, relative research on placement-friendly standard cells will be carried out for smaller area penalty.

REFERENCES

[1] K. Nakayama, et al., "Self-aligned double and quadruple patterning layout principle," Proceedings of SPIE Vol. 8327, 83270V (2012)

[2] C. Kodama, et al., "Self-aligned double and quadruple patterning-aware grid routing with hotspots control," Proc. in ACM/IEEE ASP-DAC 2013, pp.267-272, (2013).

[3] F. Nakajima, et al., "Self-aligned quadruple patterning-friendly routing," Proc. SPIE 9053, 9053C-7 (2014)

[4] B. Yu, et al. "Methodology for standard cell compliance and detailed placement for triple patterning lithography," Proc. in ACM/IEEE ICCAD 2013, pp.349-356, 2013.

[5] NanGate Open Cell Library. http://www.si2.org/openeda.si2.org/projects/nangatelib

Impact of a SADP flow on the design and process for N10/N7 Metal layers

W. Gillijns, S. M. Y. Sherazi, D. Trivkovic, B. Chava, B. Vandewalle[§], V. Gerousis°, P. Raghavan, J. Ryckaert, K. Mercha, D. Verkest, G. McIntyre, K. Ronse
imec, Belgium; °Cadence, San Jose, USA; [§] Brion, USA.

Abstract

This work addresses the difficulties in creating a manufacturable M2 layer based on an SADP process for N10/N7 and proposes a couple of solutions. For the N10 design, we opted for a line staggering approach in which each line-end ends on a contact. We highlight the challenges to obtain a reasonable process window, both in simulation as on based on exposures on wafer. The main challenges come from a very complex keep mask, consisting of complicated 2D structures which are very challenging for 193i litho. Therefore, we propose a solution in which we perform a traditional LELE process on top of a mandrel layer. Towards N7 we show that a line staggering approach starts to break down and design needs to allow better process window for lithography by having metal lines ending in an aligned fashion. has many challenges and we propose to switch to a line cut approach. A more lithography friendly approach is needed for design where the lines end at aligned points so that the process window can be enhanced.

1. Introduction

Self-aligned double patterning (SADP) is an important patterning technique that is needed for allowing scaling of metal pitch down to the 42nm pitch. In order to achieve successful decomposition, an SADP-compliant design becomes a necessity. Compared to Litho-Etch-Litho-Etch (LELE) double patterning lithography (DPL), spacer defined patterning techniques have clearly advantages with respect to overlay tolerance and line-width roughness (LWR). However the key challenge of SADP or SAQP approaches is how the line ends. This pushes the problem to keep or block masks that are used to terminate lines. This paper evaluates for different metal pitches ranging from 48nm down to 32nm, different metal patterning techniques that can be used and which give the largest process window to be eventually usable in industry

The paper is organized as follows: the next section describes standard cell template that has been used for the rest of the paper. Section 3 describes the different place and route options that can be used for the tight metal layers. As pitches move to 48nm for N10 and 32-28nm for N7 are considered the most viable option for tight pitch Mx layers.

2. Standard cells

For evaluating M2 it is crucial that the right design context is considered and this needs to be done post place and route of a complete design. For this purpose we have designed standard cells for both N10 as well as for N7 technologies with similar templates. The ground rules used for these designs are summarized in the table below:

Layer	N10 Patterning	N10 Pitch	N7 Pattering	N7 Pitch
Vertical M1	SADP	64nm	SADP	42nm
Horizontal M2	SADP	48nm	SAQP	32nm
V1	LELE		LELELE(LE)	
Fin	SADP	36nm	SAQP	24nm

The design of standard cells template was made generic to both N10 and N7 technology nodes. As described in paper [x], we have explored different standard cell styles, and evaluated the pros and cons of each of them. Given SADP/SAQP assumptions on the metal layers, a strictly unidirectional routing has been used to all metal layers. For the place and route experiments we have pursued the style with M1 vertical to the gate at the gate pitch. This template also forms a gear ratio of ¾ between the fins and the metal allowing building 8 active fins in the 9 track cell. A standard cell library consisting of about 50 cells with 9 metal tracks has been developed. A few snapshots of these standard cells are illustrated in Figure 1. More information about these templates can be found in [1].

Design-Process-Technology Co-optimization for Manufacturability IX, edited by John L. Sturtevant, Luigi Capodieci,
Proc. of SPIE Vol. 9427, 942709 · © 2015 SPIE · CCC code: 0277-786X/15/$18 · doi: 10.1117/12.2085923

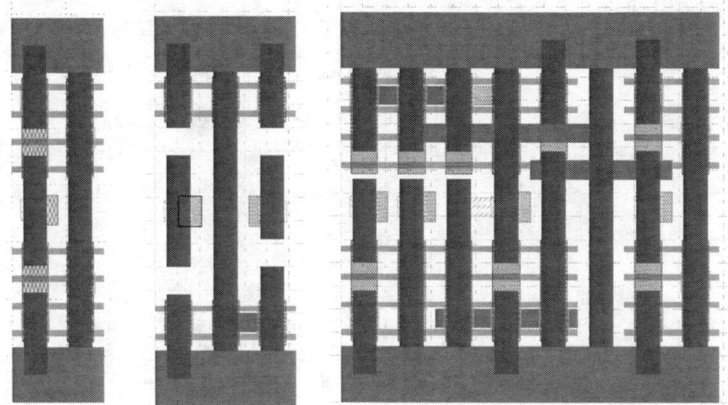

Figure 1: 1D Standard Cell with Vertical M1 (blue) and Horizontal M2 (red)

3. Place and route options for Metal

In order to take benefit of an SADP/SAQP flow, N10/N7 Mx layers need to be kept strictly 1D. For initial trials we performed place and route with 2D freedom to the router, while remaining SADP and keep mask compliant in terms of decomposition. With the 2D freedom for the router, there were very few places in the die where the router used the 2D freedom to land on vias. However given the limited use of the 2D constructs and no area overhead of restricting to 1D routing, we constrained the router to perform strictly 1D routing.

Within the constraint of 1D routing, place and route can be done with two styles: i) line and cut or ii) line staggering, both with strict 1D routing. This is illustrated in the figure below. The first style is called line staggering, this is the style where the line ends where the via makes the contact with the metal above or below (shown in the right side of Figure 2). The second scheme is known as the line and cut scheme. This is the scheme where the line is extended to the edge of so that cuts (or line ends) are aligned as much as possible. This is illustrated in the left side of the Figure 2 below.

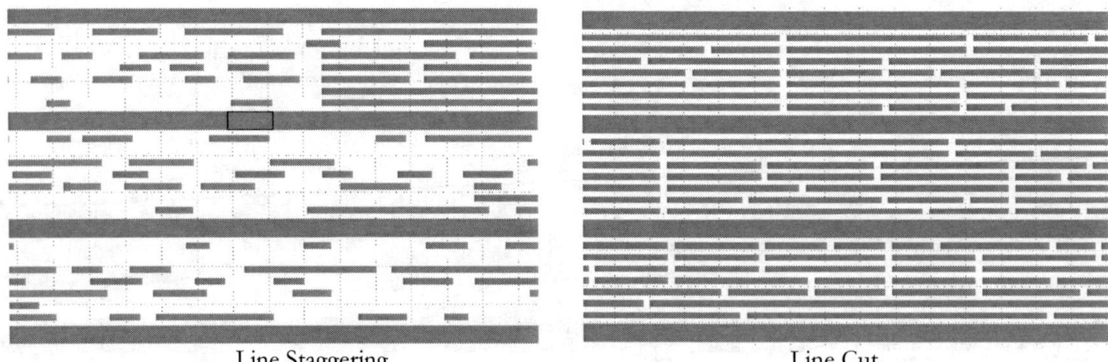

Figure 2: Place and route techniques

The line staggering approach can be decomposed with keep or block mask fashion. The number of keep-masks needed depend on the pitch and the process window. However given that the line cut approach has most of the die covered with metal, keep-mask based printing would not work and a block based approach makes more sense to follow. Figure 3 shows the decomposition of the intent metal to one keep mask or two keep masks. The coloring for the two keep mask is relatively simple where the color is alternated between odd and even lines. The multiple keep solution works provided there is no coloring conflict within a single row. At tight metal pitches (N7), as the tip to tip scales down this approach also break down and would require a more complex coloring on the keep mask based approach.

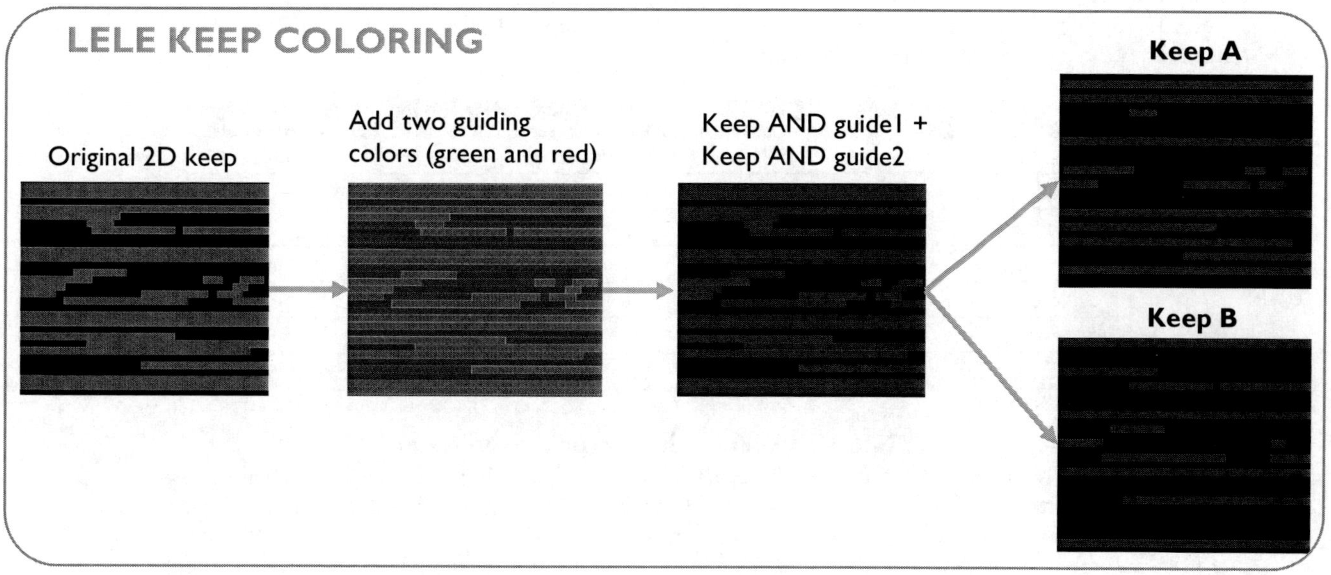

Figure 3: Keep mask coloring

The alternative method of coloring is by the block mask based approach. Figure 4 shows the decomposition options for block masks for both the design solutions (line cut and line staggering). It can be seen from the figure that the coloring for the line staggering scheme requires the blocks to be stitched at tight pitches. Also when the pitch for the metal is reduced further down to 28nm, the pitch of the block mask will also decrease, therefore this approach is more sensitive to metal pitch. However in case of the line cut, the coloring of the block is coupled to the metal pitch a bit more loosely. For metal pitch of 32nm, the line cut version requires 2 colors (with limited violations that can be fixed), whereas the line staggering requires 3 colors in the block. For a tighter pitch of 28nm, the line cut requires 3 colors whereas the line staggering will require 4 colors.

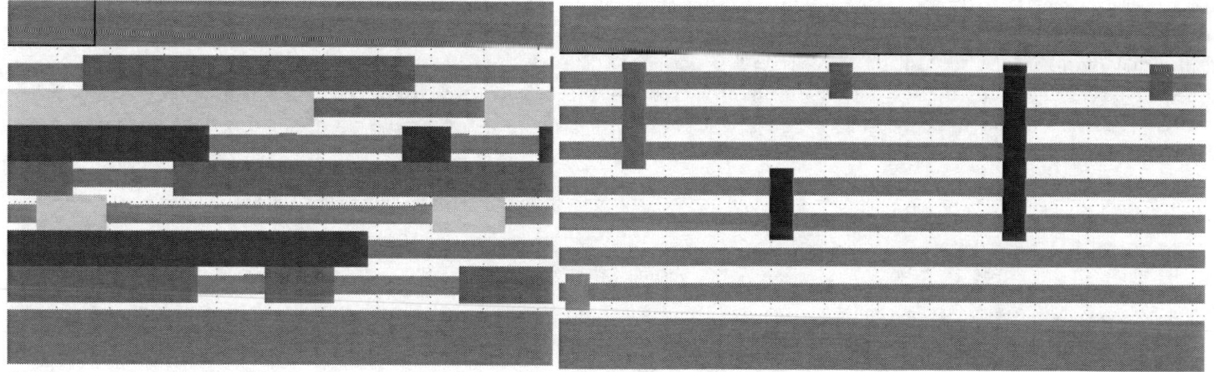

(a) Block mask decomposition for line staggering approach (b) Block mask decomposition for line cut approach

Figure 4: Block mask coloring

Using the line cut approach increases the average wire length of each net by about 30% on average. This increase translates to increased capacitance in the wires and therefore due to increased power also in the design. However critical paths of the design are not affected much and has no/limited impact in terms of timing of the design. However the line cut has the overwhelming advantage in terms of the block mask requirement. In a metal pitch of 32nm (N7), the line cut approach needs 2 block masks (with limited hotspots). However the line staggering needs 3 block masks at a metal pitch of 32nm. At N10 like dimensions (48nm pitch), the line staggering approach can possibly be used with one or two block or keep masks. The different options for this are further explored in the following sections. Furthermore the block masks need to be stitched in case of the line staggering scheme to ensure no metal is filled. The other key advantage of the line cut approach is the average metal density in the design is 50% (assuming the metal width is 50% of the pitch). However there is no difference in terms of area between the two options. Table below also summarizes the differences between the two techniques of routing for the block mask scheme.

Line Staggering	Line Cut
Equal Area	Equal Area
Reduced wire length	Increased wire length ~30%
Block mask stitching needed	Simple block mask scheme
Metal density design dependent	Near perfect uniform metal density
For pitch 28nm, 4 block masks needed	For pitch 28nm, 3 block masks needed
For pitch 32nm, 3 block masks needed	For pitch 32nm, 2 block masks needed
Lower capacitance in the BEOL	Increased capacitance (increased power)

4. N10 Metal 48nm pitch: exposure on wafer

In this section we will describe the patterning challenges and propose a solution for an N10 M2 layer. Since the line staggering approach has the advantage of being compatible with an LELE approach and has the added benefit of lower capacitance, we assumed line staggering as our front-up approach for our N10 design.

After place and route of the design we decomposed the final metal into a core and a keep layer. At this point we started our computational work to assess the printability of the design. For both the Core as the keep layer source-mask-optimization (SMO) was performed. Following this, we calibrated 2 OPC models and tuned the OPC recipes.

As could be expected the core layer did not show many problems, as it is mostly a 1D grating defining the metal CD. The OPC script was optimized with a tight spec on the CD, while the line ends were less strictly controlled. The reason is that the line ends of the final metal are define by the Keep layer.

Problems did start to appear in the keep layer. Some problems could easily be traced back to P&R settings. A typical example is the min tip-to-tip distance. As can be seen in Figure 5 (a) the keep cannot cut two metal lines below a certain value. In our case the min tip-to-tip distance needed to be ~80nm in order to resolve this with the keep layer. This rule has a strong impact on chip area and may need to be unacceptably high in order to have a stable process. A second type of defect we found are related to "staircase" configurations [see Figure 5 (b)]: in these positions the contours are all diagonal, which will lead to sharp corners on the final metal line ends. These can be quite severe, but we have initial data on an SADP+Block process that shows that those sharp line ends are not transferred during etch (see Figure 6). Apart from those, other defects (mainly bridging) were appearing due to some challenging 2D configurations [see Figure 5(c)]. Even by using very aggressive OPC, they were proving to be impossible to print with some process window. Even more problematic was the fact that it was not possible to relate those configurations back to the rules imposed to the P&R giving us very little control on the design side to try to improve the process window. In terms of places and route the restrictions needed for enabling these constructs (staircase, min tip-tip and extension from opposite sides) can be imposed, however these cause immediate non-acceptable increase in area of the chip and therefore were not considered.

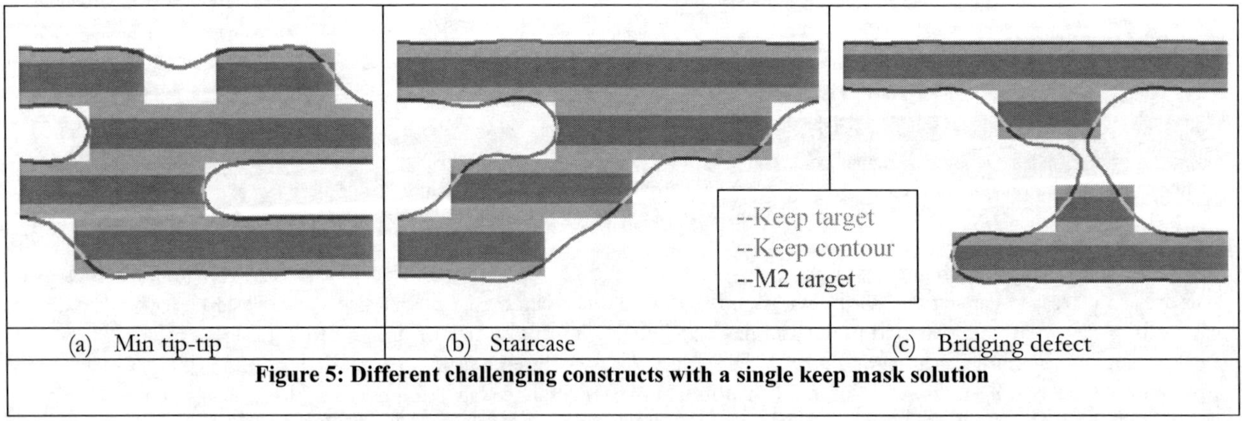

(a) Min tip-tip (b) Staircase (c) Bridging defect

Figure 5: Different challenging constructs with a single keep mask solution

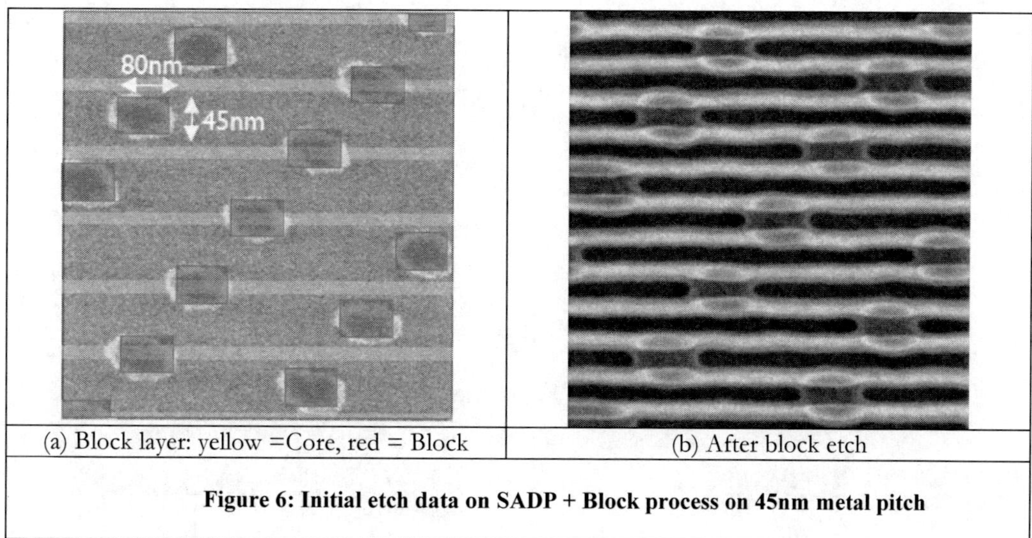

(a) Block layer: yellow =Core, red = Block	(b) After block etch

Figure 6: Initial etch data on SADP + Block process on 45nm metal pitch

To verify our simulations we exposed a focus-energy-matrix (FEM) wafer and compared the resist to the target design (see Figure 7) and the simulated contour (Figure 8)

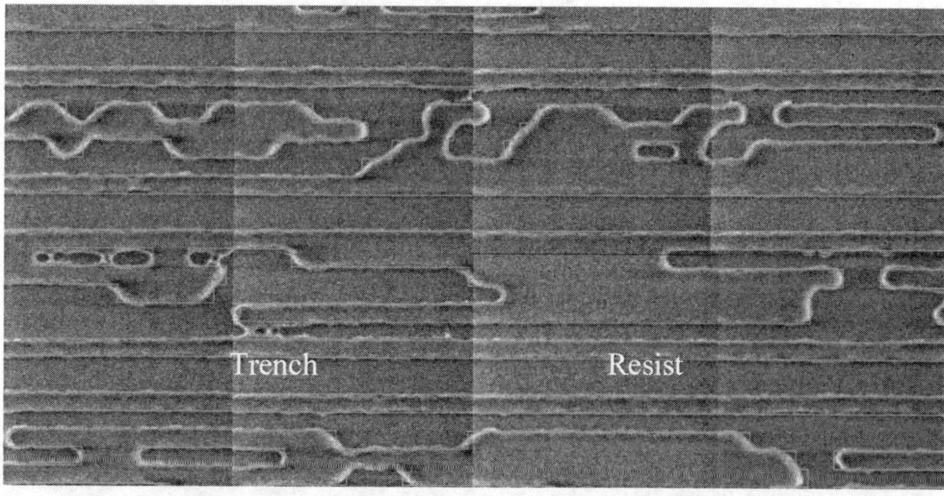

Figure 7: Target design of the keep layer overlapped with SEM images

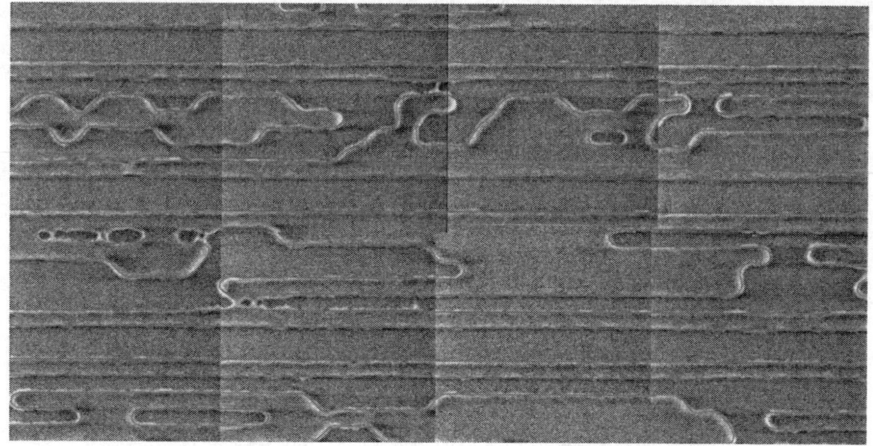

Figure 8: NC simulation contour overlapped with SEM images

It became clear that a good compromise between patternability and chip area was difficult to achieve. Therefore we proposed to split the keep layer and use a spacer assisted LELE approach for this layer. The decomposition itself is easy because all the M2 tracks are on the multiple of the 48nm pitch. Therefore it is sufficient to create a uniform grating of 2 color 96nm pitch lines shifted by ½ pitch and then applying a simple "AND" Boolean operation with original keep (Figure 3).

In Figure 10 we show the difference between the single keep and the LELE-keep approach. The main conclusion is that the final metal has higher pattern fidelity, line ends are better controlled and a larger process window is obtained. Importantly the min tip to tip can be reduced to ~50nm.

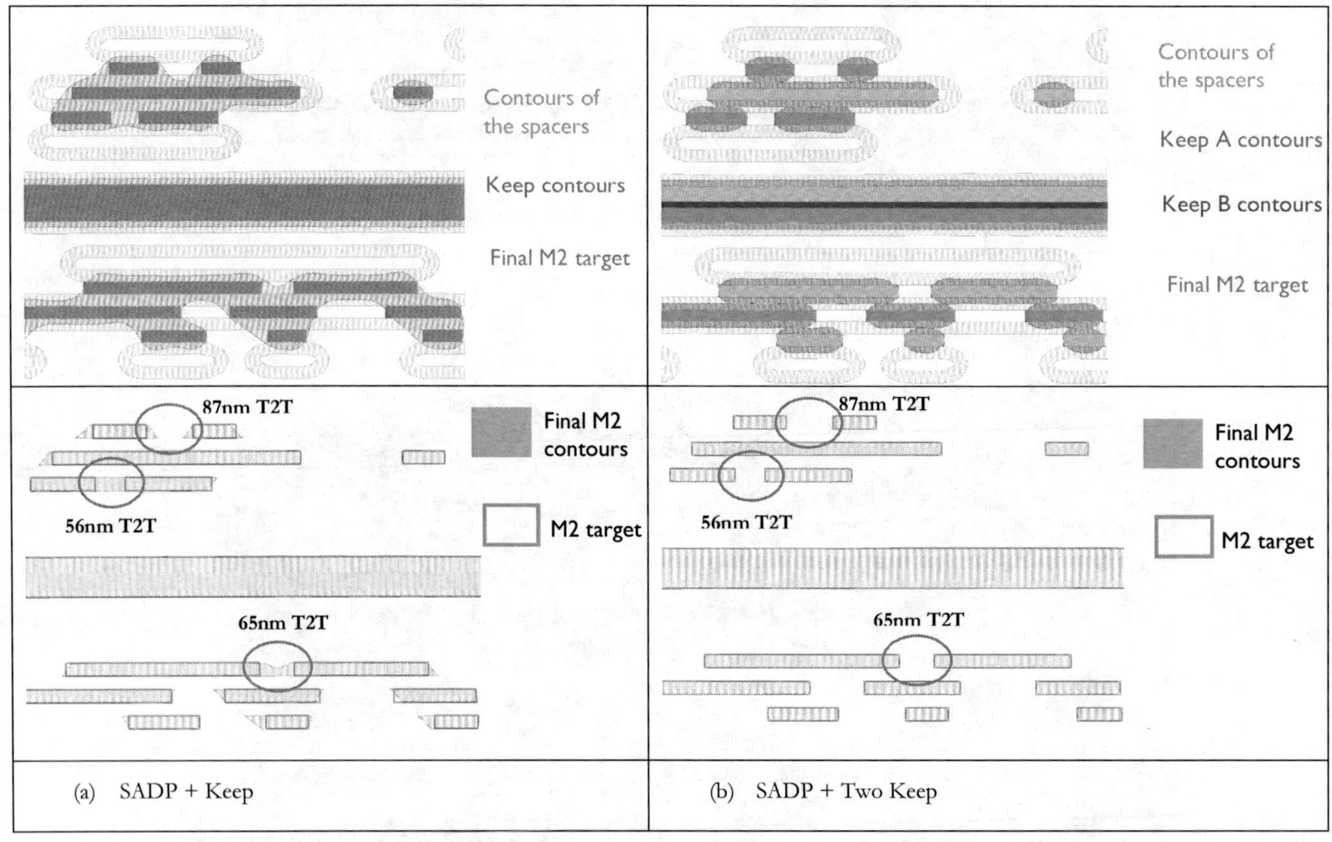

Figure 9: SADP + Keep vs. Two Keep simulation results

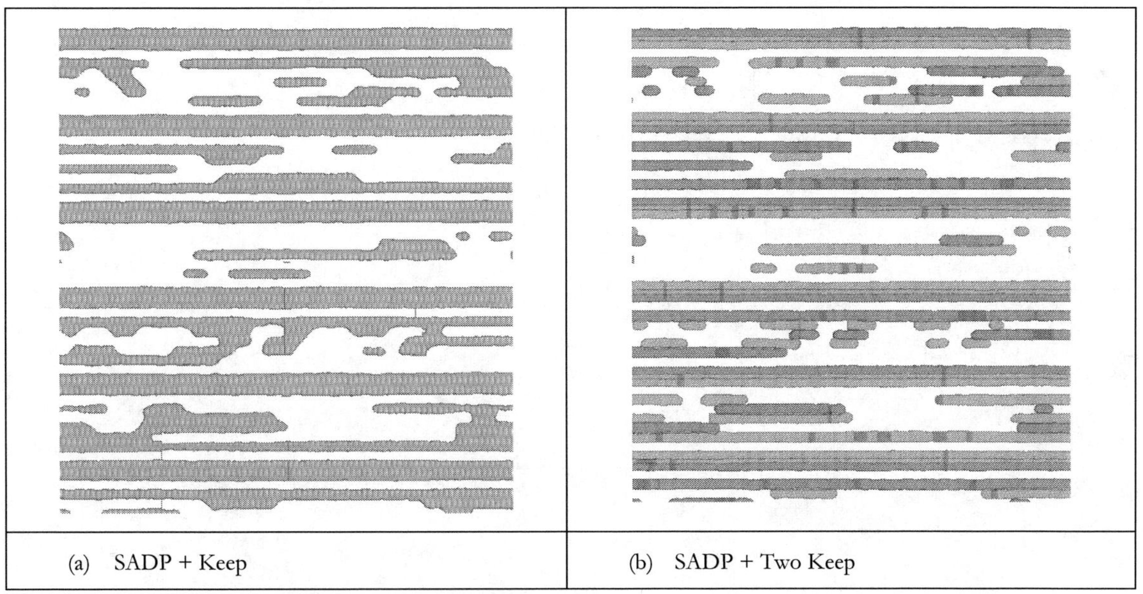

(a) SADP + Keep

(b) SADP + Two Keep

Figure 10: SADP+Keep vs. SADP+ Two Keep SEM extracted contours

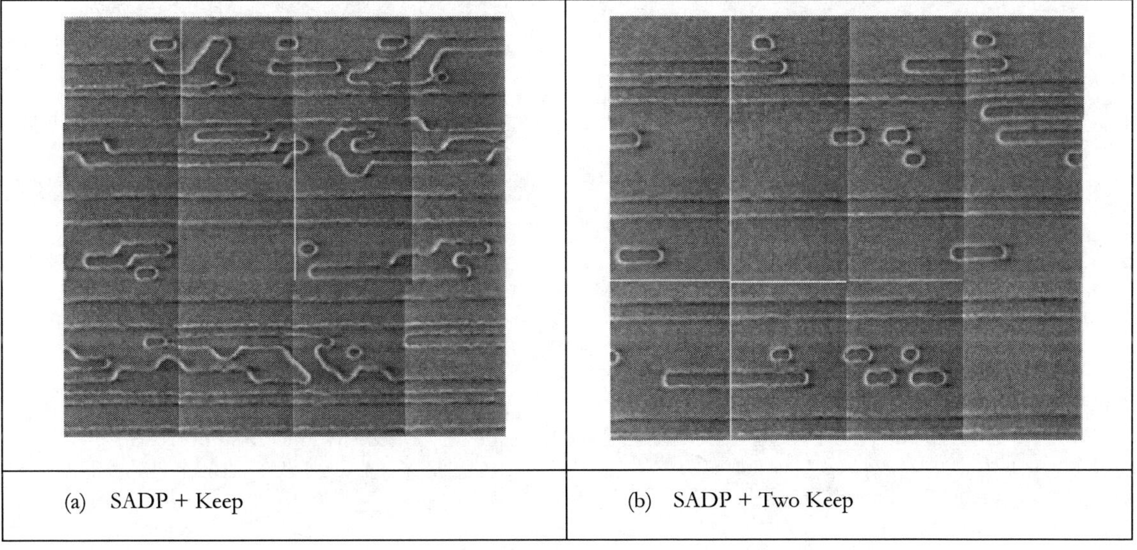

(a) SADP + Keep

(b) SADP + Two Keep

Figure 11: SADP+Keep vs. SADP+Two Keep SEM images

5. N7 Metal 32nm pitch: simulation results

A first consequence of the reduced metal pitch target of 28nm or 32nm for an N7 technology, there is a need to use an SAQP process instead of an SADP process. Assuming the mandrels are created well, our main problem is the block/keep layer once more. At these pitches a keep based strategy would not work. This would give raise to conflicts of coloring to allow good tip to tip printability and also odd cycles which may not allow 'colorability' in 3 colors or less for the keep masks.

Now going with the block strategy, we need to find the decomposition for the block mask. Figure 12 shows a typical decomposed block layer for the line staggering approach. Whether a triple block is feasible or not depends on the exact metal pitch that is targeted. We believe that this approach could work for a 32nm pitch, but when moving to 28nm pitch a fourth color would be needed for the line stagger approach.

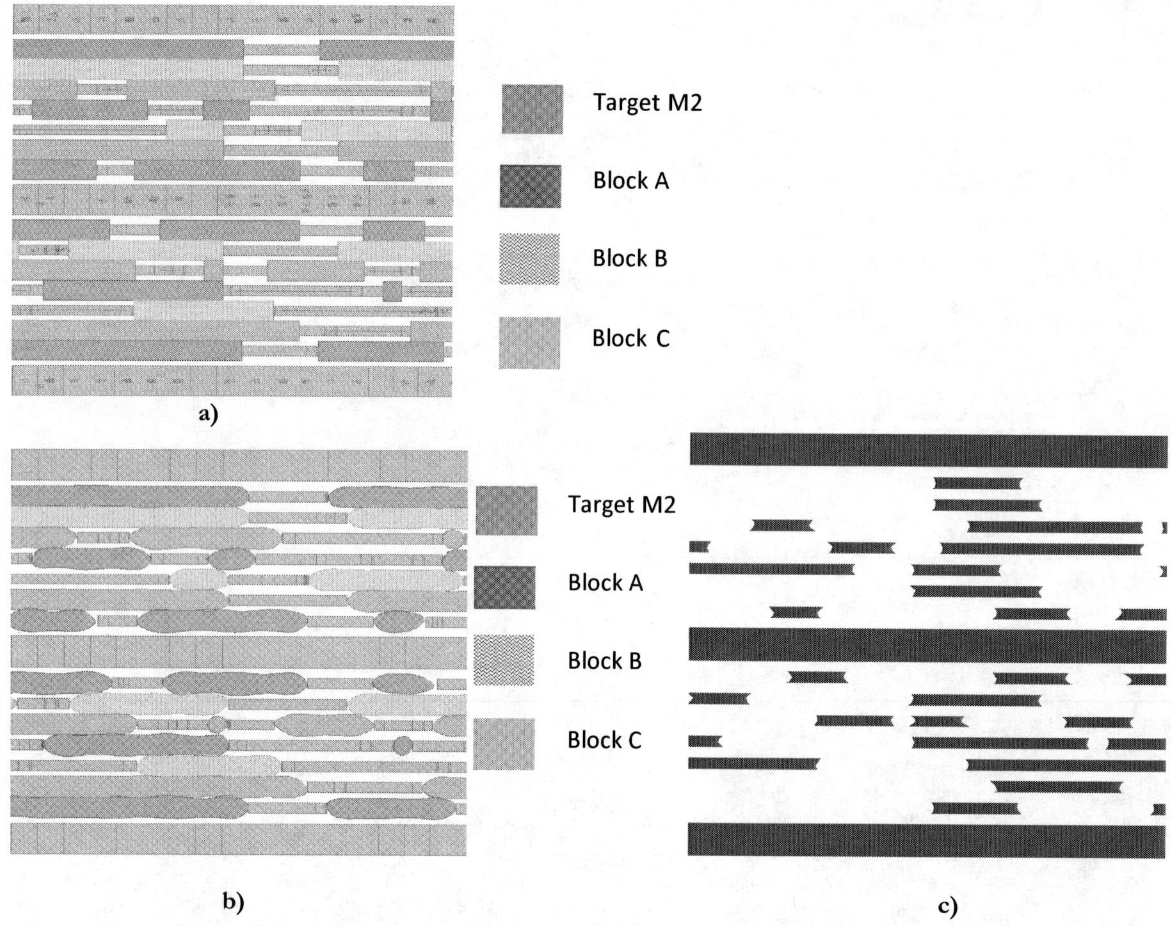

Figure 12: Staggered line approach a) Triple LE colored block layer, b) Simulated NC block, c) Simulated M2 nominal contours after full SADP + Block

Figure 13 shows the decomposition of the block masks and also computational litho simulation for the line cut approach of design where the line ends are aligned. For metal pitch 32nm, given that the line-cut approach can allow 2 colors for the block and is equal in process window with respect to line stagger, it is a superior approach even though it comes with the disadvantage of increased capacitance on the lines.

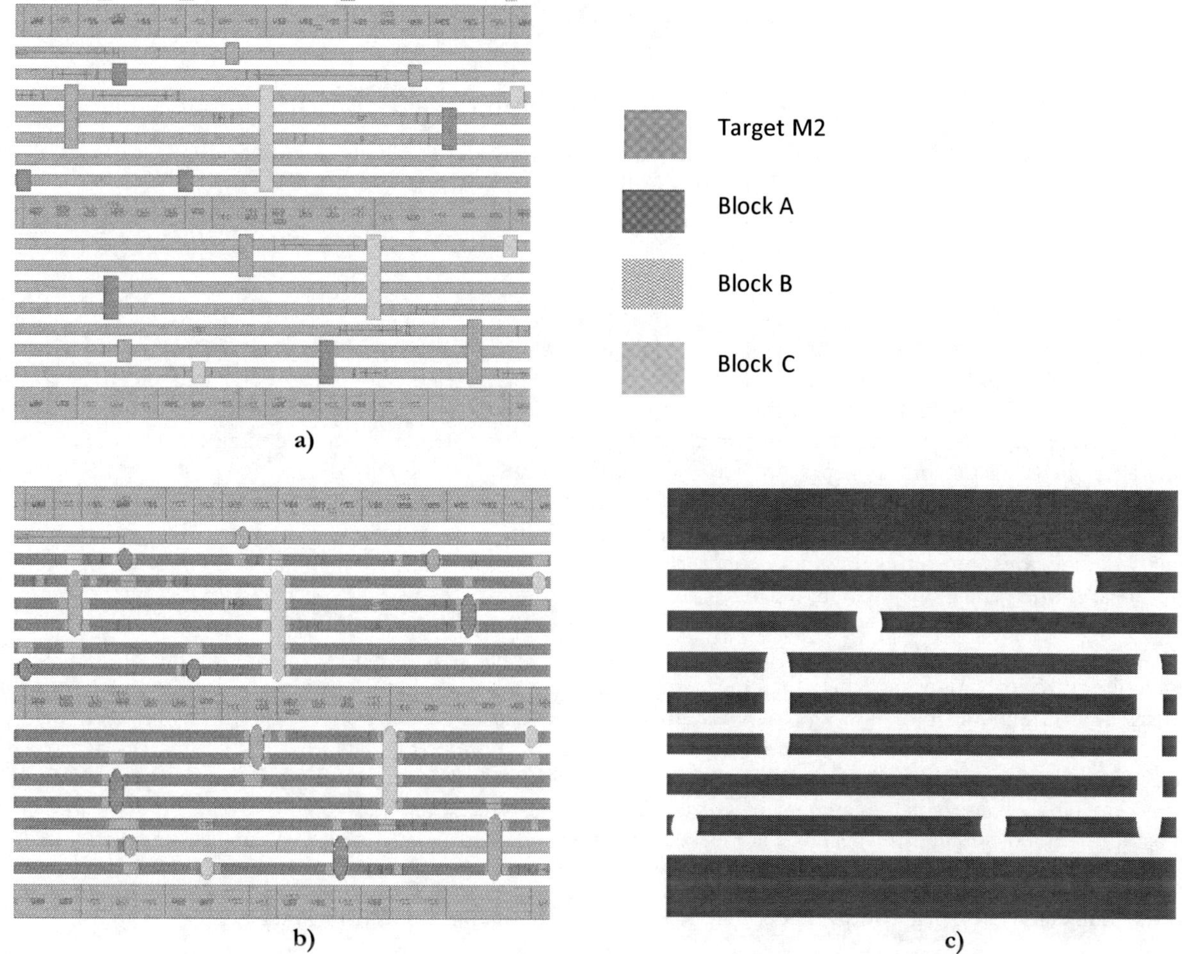

Figure 13: Line Cut approach a) LELELE colored block layer, b) Simulated NC Block, c) Simulated M2 nominal contours after full SADP + block

Conclusions

In this paper we discussed the different decomposition approaches and design approaches for tight pitch metal layers used for routing. From N10 metal pitches of 48nm down to N7 metal pitch down to 28nm we have done an analysis of the process window of different approaches. For N10, we showed on wafer that an SADP with two keep approach would allow to print in high fidelity giving the design freedom for all different constructs needed to reach high density logic place and route. We also showed the different decomposition approaches for tight pitch metals for an N7 technology. We showed that a block approach would work better than a keep approach. From design a line cut approach would be needed to allow lower number of colors needed for the block mask compared to the regular design style of ending metal where the via lands.

References

[1] **Standard cell design in 7nm node: EUV versus immersion**, Bharani Chava, Peter Debacker, Yasser Sherazi, Praveen Raghavan, Werner Gillijns, David Rio, Ahmad Elsaid, Mircea Dusa, Julien Ryckaert, SPIE 2015

[2] **DTCO at N7 and beyond: patterning and electrical compromises and opportunities** Diederik Verkest, Julien Ryckaert, Praveen Raghavan, Arindam Mallik, Sushil S. Sakhare, Bharani Chava, Yasser Sherazi, Philippe Leray, Abdelkarim Mercha, Jürgen Bömmels, Gregory R. McIntyre, Kurt G. Ronse, Aaron Thean, Zsolt Tőkei, An Steegen, SPIE 2015

An Efficient Auto TPT Stitch Guidance Generation for Optimized Standard Cell Design

Nagaraj Chary Samboju[1], Soo-Han Choi[2], Srini Arikati[2], Erdem Cilingir[2]

[1]Synopsys India Pvt. Ltd., Block B, Divya Sree Omega, Hyderabad, Telangana 500032, India; [2]Synopsys Inc., 700 E. Middlefield Rd, Mountain View, CA 94043, USA

ABSTRACT

As the technology continues to shrink below 14nm, triple patterning lithography (TPT) is a worthwhile lithography methodology for printing dense layers such as Metal1. However, this increases the complexity of standard cell design, as it is very difficult to develop a TPT compliant layout without compromising on the area. Hence, this emphasizes the importance to have an accurate stitch generation methodology to meet the standard cell area requirement as defined by the technology shrink factor. In this paper, we present an efficient auto TPT stitch guidance generation technique for optimized standard cell design. The basic idea here is to first identify the conflicting polygons based on the Fix Guidance [1] solution developed by Synopsys. Fix Guidance is a reduced sub-graph containing minimum set of edges along with the connecting polygons; by eliminating these edges in a design 3-color conflicts can be resolved. Once the conflicting polygons are identified using this method, they are categorized into four types [2] - (Type 1 to 4). The categorization is based on number of interactions a polygon has with the coloring links and the triangle loops of fix guidance. For each type a certain criteria for keep-out region is defined, based on which the final stitch guidance locations are generated. This technique provides various possible stitch locations to the user and helps the user to select the best stitch location considering both design flexibility (max. pin access/small area) and process-preferences. Based on this technique, a standard cell library for place and route (P&R) can be developed with colorless data and a stitch marker defined by designer using our proposed method. After P&R, the full chip (block) would contain the colorless data and standard cell stitch markers only. These stitch markers are considered as "must be stitch" candidates. Hence during full chip decomposition it is not required to generate and select the stitch markers again for the complete data; therefore, the proposed method reduces the decomposition time significantly.

Key Words: Standard Cell Design, Triple Patterning, Place & Route, Categorized Stitch marker

1. INTRODUCTION

The biggest challenge for semiconductor industry today is to manufacture chips at sub-10nm technology node with current limitations of ArF immersion lithography. To overcome this, several advanced lithography techniques such as extreme ultraviolet (EUV) lithography and electron beam (e-beam) lithography have been proposed. However, these advanced lithography techniques are not yet ready at production level. This means the chip manufacturers would still have to rely on current 193nm ArF immersion lithography with multi-patterning for the sub-10nm nodes and beyond. Immersion lithography with Double patterning (DPT) is very successful for manufacturing chips at 20nm and 14nm node. The same technique can be extended for the sub-10nm node for most of the layers except for the dense layers such as Metal1 (M1) and interconnects. These layers are denser and can have very complicated structure, which makes it impossible to split them into two masks without conflicts. This problem can be solved by splitting the layers into three masks instead of two, which is referred as Tipple patterning (TPT). Unlike DPT, TPT is more complicated as triple patterning decomposition is to solve a graph three-coloring problem – a known NP-complete problem. In addition TPT color conflict cannot be represented in a single loop, so reporting these errors in a way that user can effectively fix them is challenging. This issue has been studied and addressed in [1]; the solution provides a minimal fix guidance to report color conflict errors. Along with this solution, it is also important to have a good stitch generation algorithm. A stitch generation algorithm should be able to generate the stitches automatically or report all possible stitch locations to the user, so that the user has flexibility to select a stich candidate based on design methodology. In this paper, we present an efficient auto TPT stitch guidance generation technique to help user identify all possible stitch candidates and thereby aid them to optimize their standard cell design. . The techniques described in this paper are protected under pending patents [1][2]. The rest of the paper is organized as follows. Section 2 discusses previous studies and associated problem. The

Design-Process-Technology Co-optimization for Manufacturability IX, edited by John L. Sturtevant, Luigi Capodieci, Proc. of SPIE Vol. 9427, 94270A · © 2015 SPIE · CCC code: 0277-786X/15/$18 · doi: 10.1117/12.2085834 proposed stitch generation algorithm is explained in Section 3. Section 4 provides a standard cell design methodology based on the proposed technique. The experimental results are provided in Section 5 and Section 6 will summarize our work.

2. PRELIMINARIES

2.1 Problem formulation

The major challenge while considering TPT decomposition for sub-10nm is increase in complexity of layout design, particularly of Standard Cells. As the technology shrinks, the standard cell area has to be scaled down appropriately. However, it is very difficult to develop a TPT complaint layout without compromising on area. As the M1 layer is extensively used in both directions in the standard cells, developing a conflict free layout is less probable unless the user has flexibility to increase the area by spacing the M1farther. The other solution to resolve a TPT conflict is to split a pattern into two touching parts, which is called a stitch. Considering the complexity of TPT decomposition and error reporting, it is essential to have an accurate TPT stitch generation methodology to meet the standard cell requirements.

2.2 Previous study

Some of the previous studies [3][4][5][6][7] focused on the stitch generation for TPT decomposition. However, none of these techniques could propose a method to identify and report all the possible stitch locations. The previous algorithms could identify either only one or very few stitch locations out of all the possible locations in a layout. By using these methods the user will have limited choices to fix a layout. Instead, if all the possible stitch locations are identified, then the user has more flexibility in choosing the right stitch candidate based on the design methodology and process preference.

The Figure 1 explains a case study proposed in [3] & [4]. This method could identify only two stitch candidates for this layout as shown in Figure 1.d. However, this layout configuration can have four possible stitch candidates shown in Figure 1.e. This methodology is missing two legal stitch candidates, and hence the user does not have much flexibility to fix this layout.

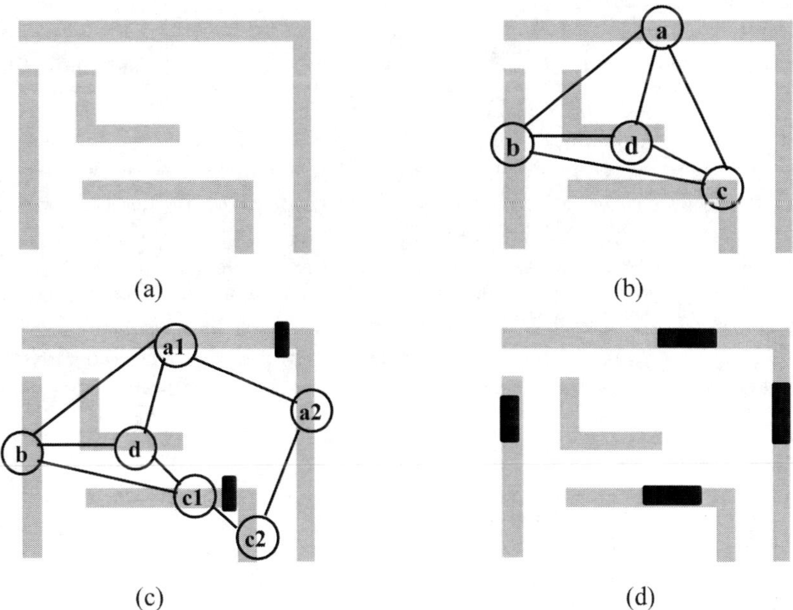

Figure 1: Previous case study 1. (a) Original colorless Layout (b) Original layout represented as graph; (c) Generated stitch location as per previous study, only 2 stich locations identified; (d) All possible stitch candidates for this layout.

A similar analysis is shown in Figure 2 for the case study proposed in [6]. With this technique only one stitch location can be identified, shown in Figure 2.c.Whereas this layout configuration can have 3 legal stitch candidates as shown in Figure 2.d

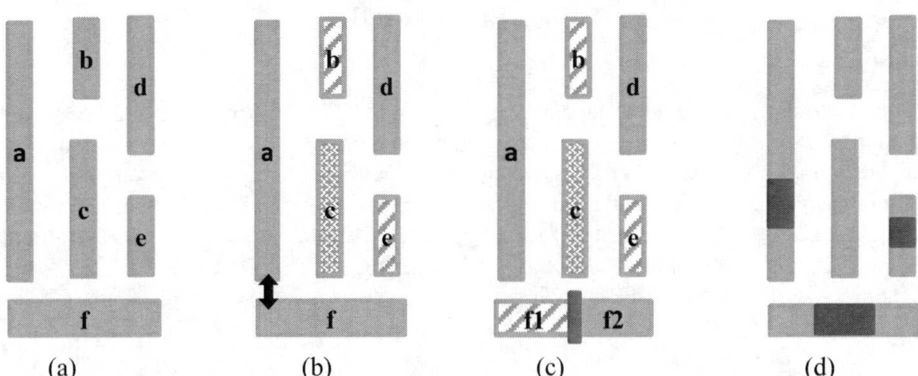

Figure 2: Previous case study 2. (a) Original colorless Layout (b) Possible color conflict while decomposition; (c) Generated stitch location as per previous study, only 1 stich location identified; (d) All possible stitch candidates for this layout.

As explained above, the existing solutions are not accurate enough to identify all the legal stitch candidates for a design. This problem has been addressed in our proposed stitch guidance generation technique explained in Section 3. The proposed technique can identify and report all the stitch candidates shown in figure 3.d & figure 4.d.

3. STITCH GUIDANCE GENERATION

In this section we explain our proposed stitch guidance generation technique which is based on the fix guidance solution [1]. The figure 3 shows the complete flow chart.

In step1 a colorless design is decomposed using TPT solver. The TPT solver colors the design into three colors and reports any color conflicts as Fix guidance. In step 2 the conflicting polygons identified by fix guidance are then categorized in to four different types. The categorization is based on the number of interactions a conflict polygon has with the coloring links and the triangle loops of the fix-guidance. In step 3 the keep-out region for each conflicting polygon is derived using spacing constraint. The criterion for generating keep-out region is different for each type of polygon categorized in step2. Using these keep-out regions the stitch markers are generated in step 4. In step5 the best possible stitch solution is automatically selected based on least number of stitches for each conflict type. Instead of auto stitching, all the possible stitching solution for each conflict type can be reported such that user can select optimum stitch candidate based on design preference and flexibility.

The following sub-sections explain each step in detail.

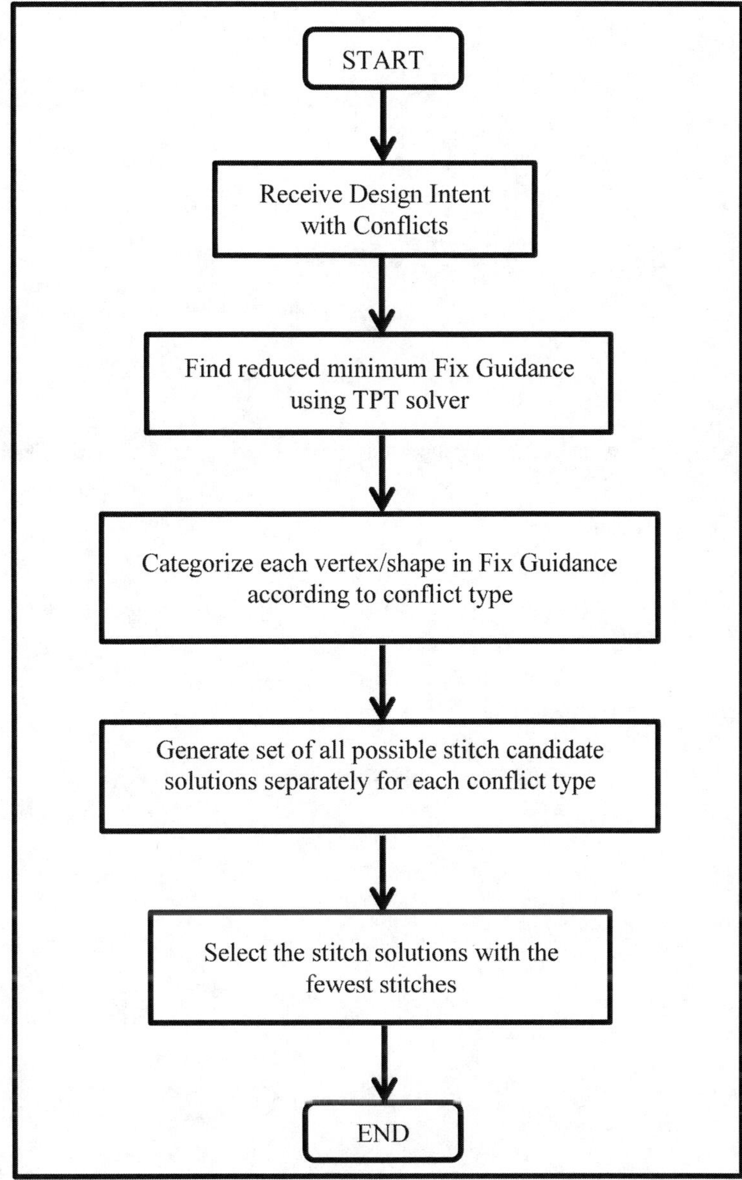

Figure 3: Flow chart of proposed stitch guidance generation algorithm.

3.1 Fix Guidance

To identify the color conflicting polygons, the original colorless data is three colored using TPT solver. The input to TPT solver is coloring links and the original data. Coloring links are defined as the distance between two patterns less than the minimum space. The TPT solver colors the input data and reports any color conflicts as a Fix Guidance. Unlike DPT, in TPT a color conflict cannot be reported as a single loop and it is difficult to report conflicts in way such that the user can understand to fix the conflicts. To resolve this issue, Synopsys has developed an algorithm to detect and report the multi-patterning conflicts as a smaller graph called Fix Guidance [1]. The Fix Guidance is a reduced sub-graph containing minimum set of edges along with the connecting polygons; by eliminating these edges in a design, 3-color conflicts can be resolved. Figure 4 shows an example for the Fix Guidance output. By eliminating the edges reported in Figure 4.b, the layout can be three colored. The edges can be eliminated either by increasing the space between polygons or by splitting the polygon into 2 masks by introducing stitches.

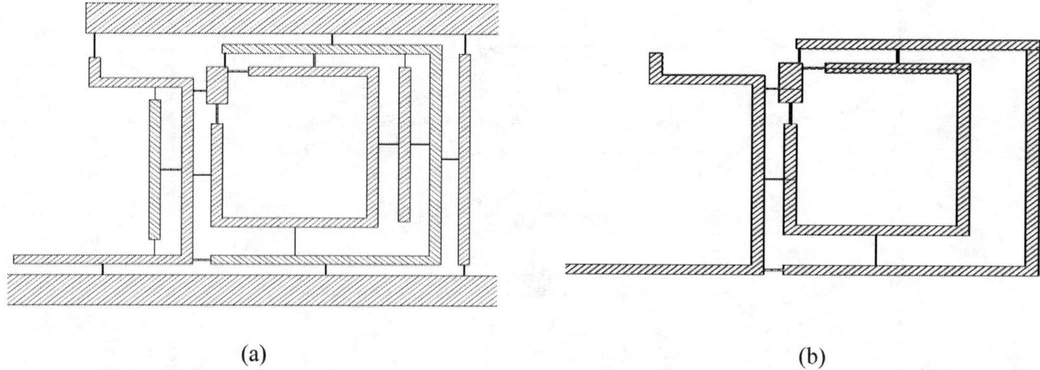

Figure 4: Example for Fix Guidance (a) Original conflict layout with coloring links; (b) Fix Guidance output from TPT solver.

3.2 Categorization of conflict polygon

Once the conflicting polygons are identified, the main idea is to categorize them into four different types. The categorization is based on the conflict polygon interaction with the coloring links and triangle loops of Fix Guidance as shown in Figure 5. Each type is defined as follows

Type 1: Conflict polygons interacting with less than or equal to two coloring links
Type 2: Conflict polygons interacting with one triangle-loop and an additional coloring link
Type 3: Conflict polygons interacting with two triangle-loops
Type 4: Conflict polygons interacting with more than two triangle-loops

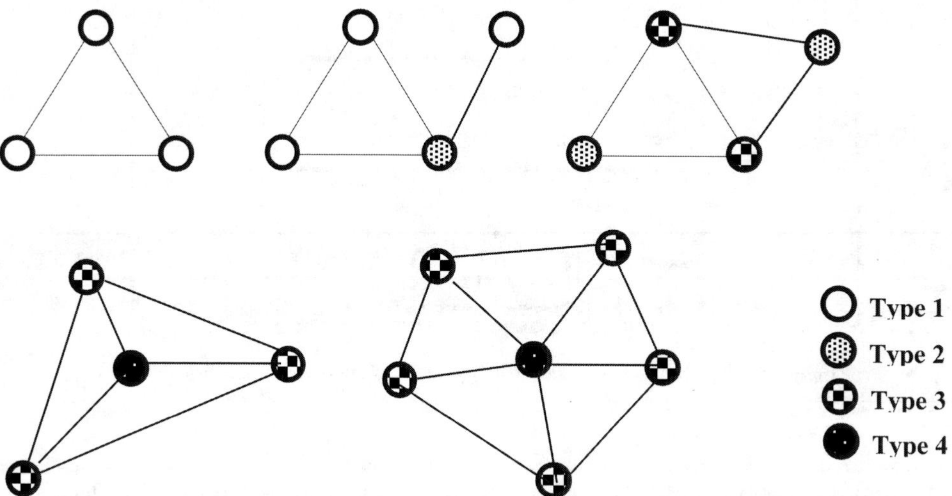

Figure 5: Categorization of conflicting polygons.

The categorization for the example provided in Figure 4 is shown in Figure 6. Polygons B, C & D represented as graph in Figure 6.b are categorized as Type 3, as they interact with exactly two triangle loops. Whereas polygon A interacts with three triangle loops and hence it is categorized as Type 4. Figure 6.c shows the categorized polygons in layout format.

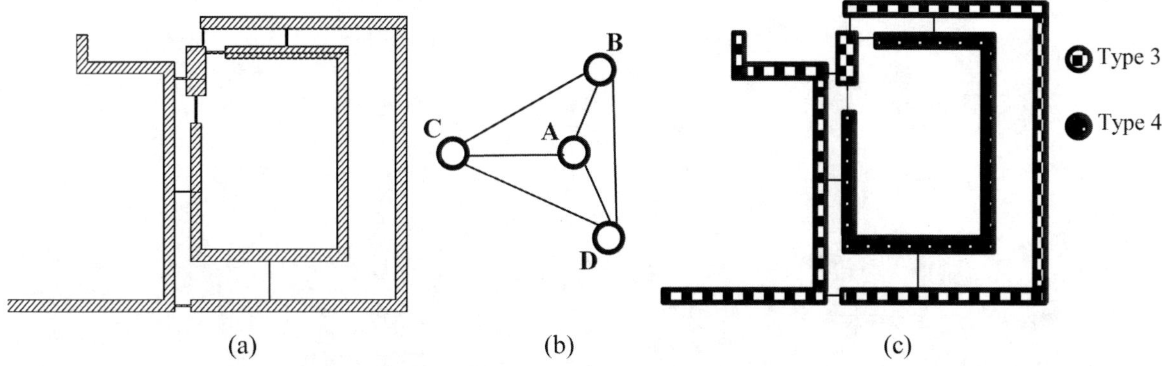

Figure 6: Categorization of polygons for example shown in Figure 4. (a) Fix guidance output from TPT solver; (b) Fix Guidance represented as graph; (c) Categorized conflicting polygons.

3.3 Keep-out region & Stitch generation

After the conflicting polygons are categorized, the next step is to identify the possible stitch locations for every conflict polygon. Stitch locations can be identified by first deriving the keep-out regions and then excluding those regions. Keep-out regions are defined using the minimum design rule violation area. In our algorithm, for each conflict type, we define different criteria for keep-out region generation. By excluding these keep-out regions, we can generate a set of all possible stitch candidate solutions for each categorized conflict type. The proposed stitch guidance generation criterion for each type is mentioned below.

Type 1: Stitching guidance is defined by keep-out region defined with respect to another Type 1 polygon
Type 2: Stitching guidance is defined by keep-out of region defined with respect to another Type 1 and Type 3
Type 3: Stitching guidance is defined by keep-out of region defined with respect to Type 2 and another Type 3
Type 4: The whole polygon face with tip-to-side link will be the stitching candidate.

Using our proposed method, the creation of the stitch locations for the conflict layout provided in Figure 4.b is explained in Figure 7. In this figure, the stitch candidates are derived for type 3 conflict polygons. Generation of keep-out region using design rule violation is explained in Figure 7.b & c. The final possible stitch locations for type 3 conflict are shown in Figure 7.d with black marker. One of these three locations can be an optimal stitch candidate and the user can select any one based on design flexibility. Similarly, for the same layout the stitch location for Type 4 conflicts is shown in Figure 7.e with black marker. Both of these stitch candidates form optimal solution.

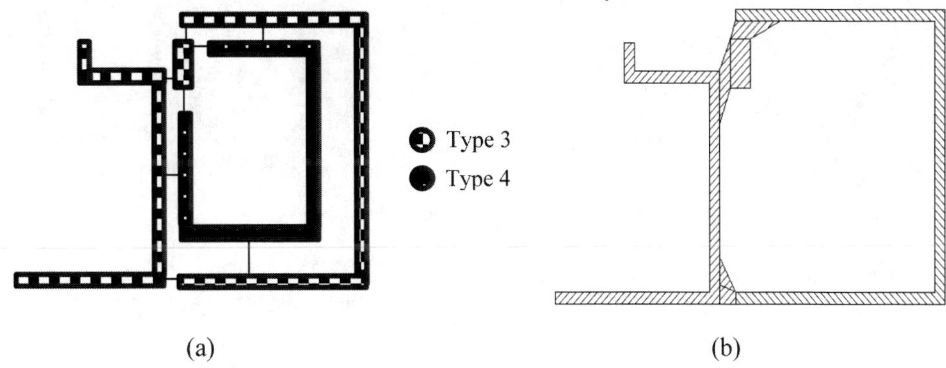

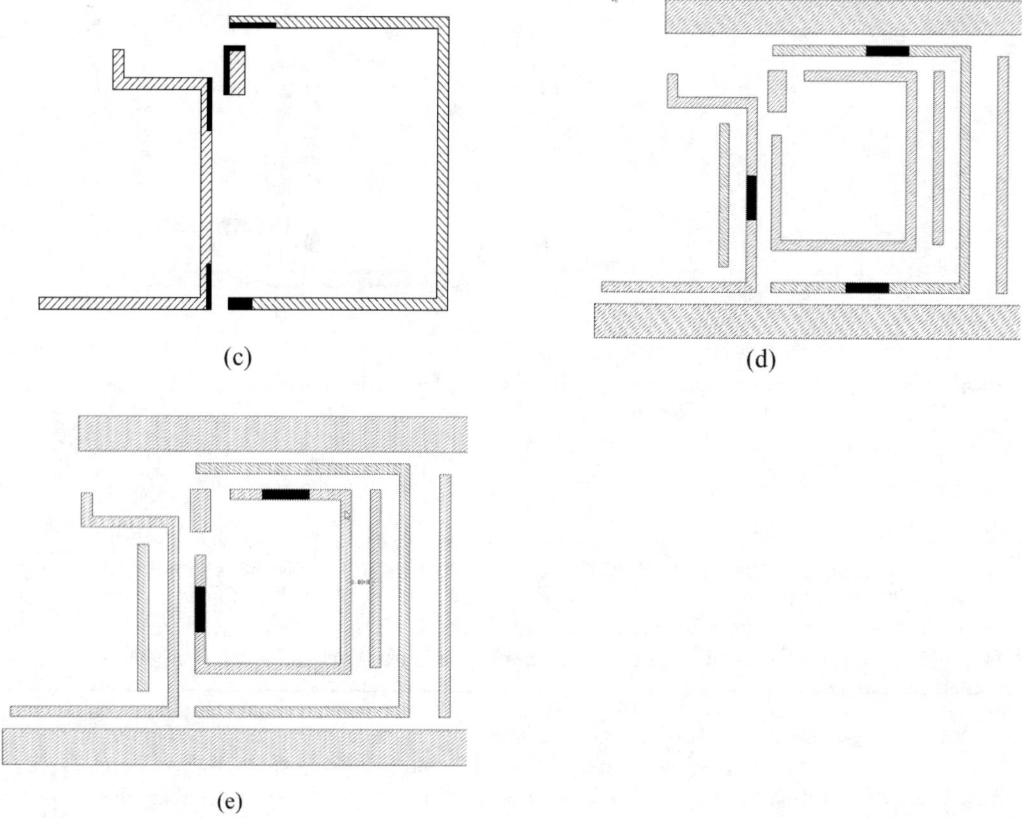

Figure 7: Stitch guidance generation for example shown in Figure 6. (a) Categorized conflicting polygons; (b) Design rule violation for Type 3 conflict; (c) Keep-out region defined for Type 3 conflict; (d) All possible stitch candidates for the type 3 conflicts; (e) All possible stitch candidates for Type 4 conflicts.

Another example is shown in Figure 8 for categorized polygons of Type 1 & 2.

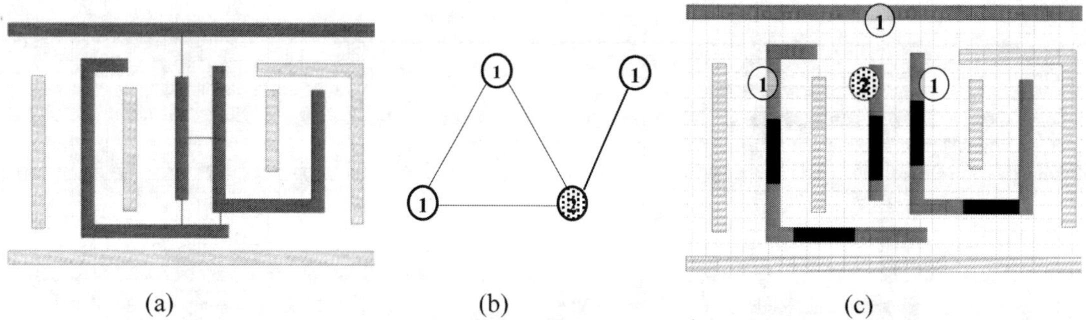

Figure 8: Example to explain categorization of Type 1 & Type 2 Polygons. (a) Original Decomposed Layout with Fix Guidance (in blue); (b) Conflicting polygons of fix Guidance represented as graph; (c) Categorized polygons with all possible stitch locations for Type 1 & Type 2.

3.4 Automatic stitch selection

Once all the legal stitch candidates are identified using the above algorithm, the final stitch candidates can be automatically selected by using following steps.

Step1: Check the number of stitches used to solve color conflicts for each categorized conflict type.

Step2: Among the stitching solution of each categorized conflict type, select the stitch solutions of one conflict type with the fewest stitches.

Instead of automatic stitching, user can also select the stitch candidates based on their preference.

Figure 9 shows the possible decomposed solution with stitches using our proposed method. Figure 9.a uses stitch candidate of type 3 conflicts and Figure 9.b uses stitch candidates of type 4 conflicts.

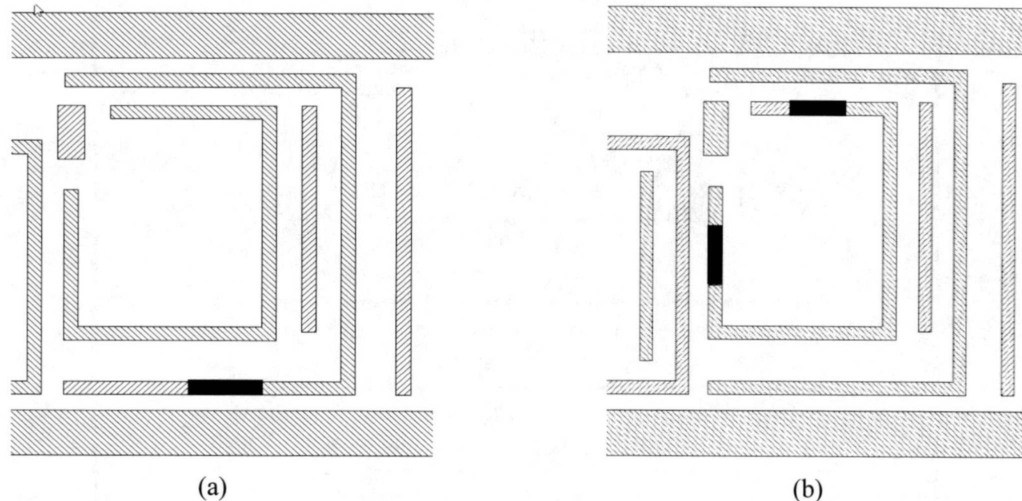

Figure 9: TPT compliant decomposed layout with stitches generated using our proposed method for example shown in Figure 4. (a) Decomposed result using type 3 stitch; (b) Decomposed result using type 4 stitch.

4. PROPOSED DESIGN METHODOLOGY

Based on the stitch guidance technique discussed in section 3, we propose a new optimized standard cell design methodology. This methodology can help users design a TPT complaint layout without compromising on area, also this can help reduce decomposition runtime at chip level thereby improving performance. Figure 10 shows the flow chart of the proposed design methodology.

A colorless Standard Cell design is decomposed using Synopsys TPT compliance checker. The output design contains three color data along with coloring conflicts reported as Fix Guidance. The proposed categorized stitch guidance generation algorithm is applied to identify all the possible legal stitch candidates. The optimal stitch candidate are selected and added to the original colorless design. The stitch candidate can be either selected manually or using automatic stitch selection method. The final standard cell libraries can be released with colorless data and the stitch marker. During place and route these standard cell libraries are used for full chip (block) design. At this stage, after the completion of full chip P&R, the design will contain only colorless data and the standard cell stitch markers. The full chip data can be decomposed using Synopsys TPT compliance checker. During decomposition of full chip data, the standard cell stitch markers are considered as "Must be Stitch" candidates. Hence it is not required to generate and select the stitch markers again for the complete data; therefore it reduces the decomposition time significantly.

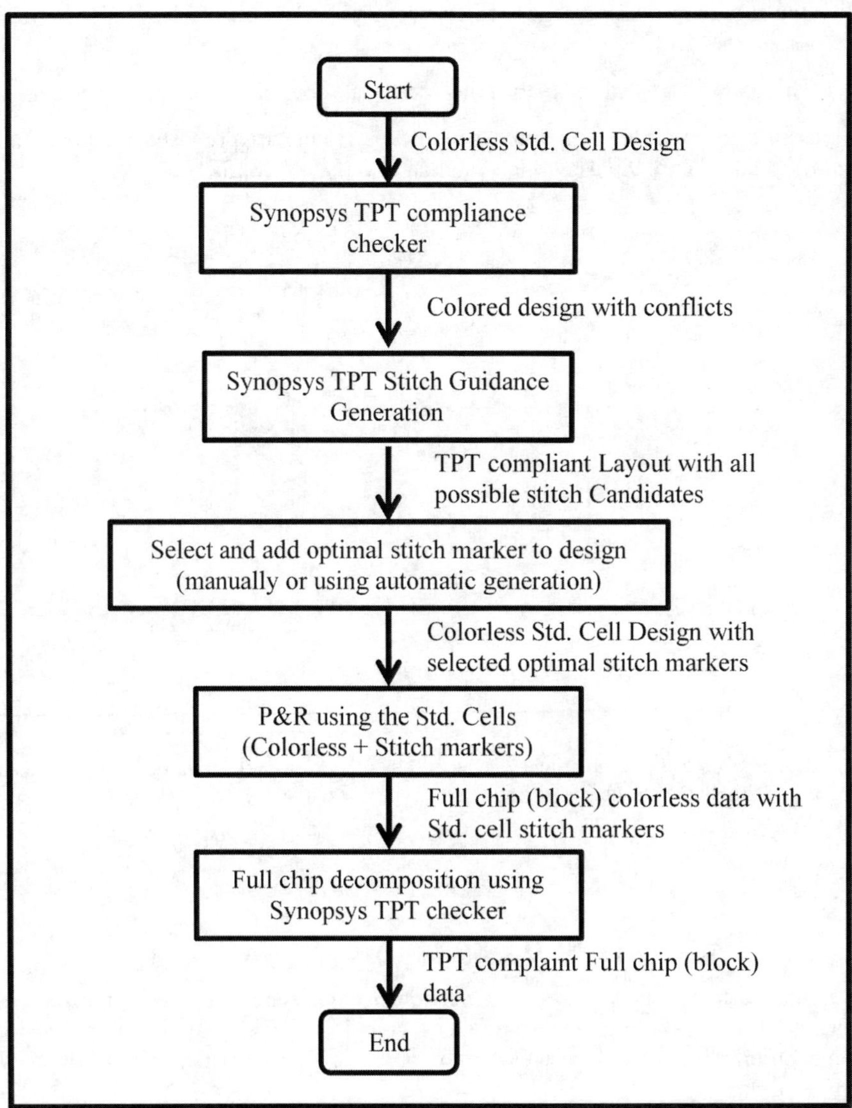

Figure 10: Proposed design flow chart for efficient and optimized TPT decomposition at full chip.

5. EXPERIMENTAL RESULTS

We implemented the proposed TPT stitching algorithm on the layout configurations provided in previous study and compared the results. The Figure 11 shows the results for the first case study. The polygons ***a, b & c*** are categorized as Type 3 whereas polygon ***d*** is categorized as Type 4. There are four stitch location (shown in Figure 11.d) identified by the proposed method, compare to the previous study [3][4] which identifies only two stitch location shown in Figure 11.c. The stitch location missed by previous study is highlighted in Figure 11.d.

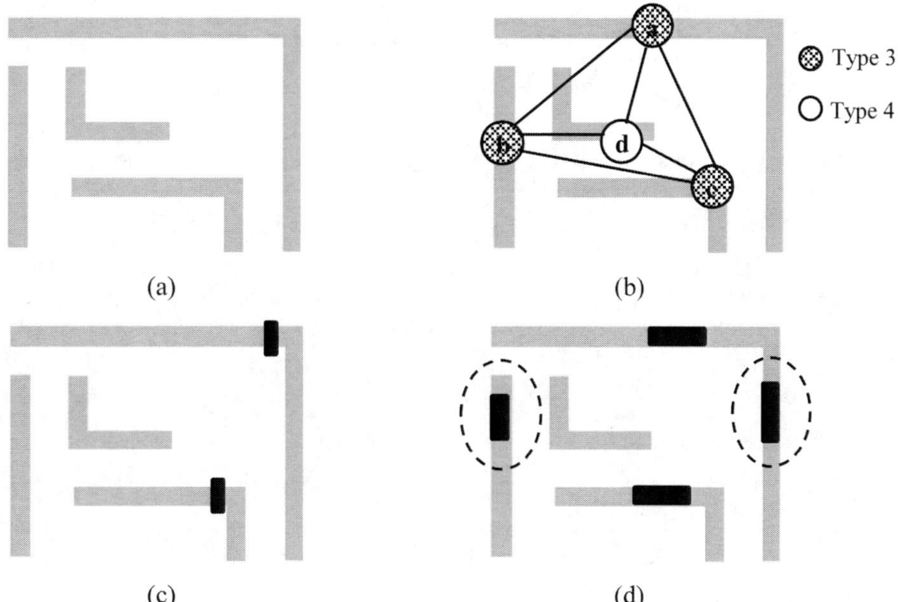

Figure 11: Case study 1. (a) Original colorless Layout (b) Original layout with categorization of polygons represented as graph (a, b & c – Type 3, d – Type 4); (c) Generated stitch location as per previous study[3][4], only 2 stich locations identified; (d) All four stitch locations identified with the proposed method, the encircled stitches are missed by previous study.

Result for another case study is shown in Figure 12. In this example, the polygons ***a, b, d, e & f*** are categorized as Type 3 and polygon ***c*** is categorized as Type 4. Based on this totally there are three stitch candidates identified (shown in Figure 12.d) by proposed algorithm when compared to only one solution provided in previous study [6] (Figure 12.c).

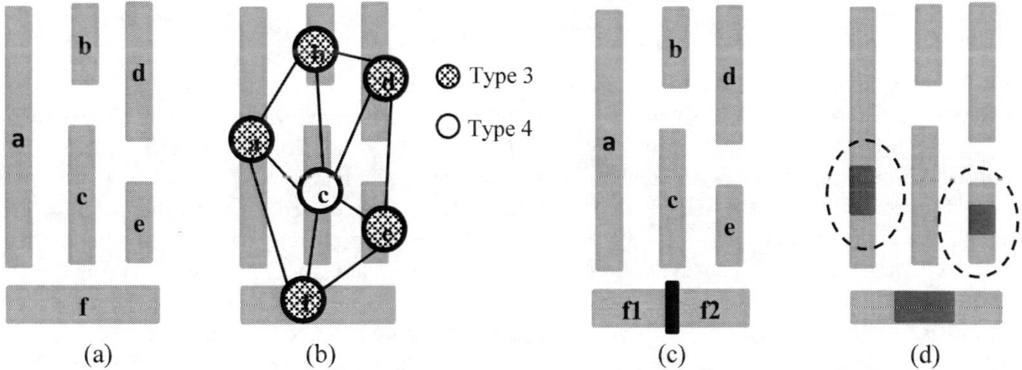

Figure 12: Case study 2. (a) Original colorless Layout (b) Categorized polygons represented as graph (a, b, d, e & f – Type 3, c – Type 4); (c) Generated stitch location as per previous study [6], only 1 stich location identified; (d) All three stitch locations identified with the proposed method, the encircled stitches are missed by previous study.

From case studies 1 & 2 it is evident that the proposed stitch guidance generation algorithm is efficient in identifying and reporting all the possible stitch candidates. This gives the user more flexibility to fix a TPT conflict by selecting the best suitable stitch candidate based on their design & process preference. At the same time using the proposed stitch solution with the recommended design methodology, the full chip decomposition time can be improved significantly. The Table 1 shows the runtime comparison for full chip decomposition using the proposed technique versus conventional technique. In the proposed technique, the stitch markers are already introduced at Standard cell level and these stitch markers are considered as "must be stitch candidates". Hence, we can skip Step 1 to Step 3 for full chip decomposition from the below table, thereby reducing the final decomposition time and improving the runtime performance from 4.7x to 1x approximately.

Table 1: Approximate runtime comparison between conventional and proposed stitch generation technique & design methodology.

Step	**Operation**	**Conventional Method (Colorless Design + Full-Chip automatic stitching)**	**Proposed Method (Colorless Design + Standard cell stitching markers)**
1	Color conflict check @ full chip	Required (**3.0x**)	Not Required (**0x**)
2	Stitching candidate generation @ full chip	Required (**0.2x**)	Not Required (**0x**)
3	Stitching optimization @ full chip	Required (**0.5x**)	Not Required (**0x**)
4	Decomposition @ full chip	Required (**1.0x**)	Required (**1.0x**)
x	**Estimated total runtime @ full chip**	**4.7x**	**1.0x**

6. CONCLUSION

To adopt Triple patterning for sub-10nm node, it is important to have a good TPT stitch generation technique. In this paper we presented an efficient auto TPT stitch generation technique. The technique is based on categorization of conflicting polygons reported as Fix guidance by a TPT solver. For each categorized type the stitch locations are identified by defining certain keep-out region criteria. Using an automated flow, the best possible stitch location can be selected and reported. In addition, all the possible stitch locations can also be reported such that the user has flexibility to select the best stitching candidate depending on their design and process preference. Based on this technique we also presented an optimized standard cell design methodology. With this, the standard cell libraries can be developed and released with colorless data and a stitch marker. After place & route, the full chip (block) would contain the colorless data and the standard cell stitch markers only. These stitch markers are considered as "must be stitch" candidates for final decomposition. Hence during full chip decomposition it is not required to generate and select the stitch markers again for the complete data; therefore, it reduces the decomposition time significantly.

REFERENCES

[1] Weiping Fang; Srini Arikati; Erdem Cilingir; Marco A. Hug; Peter De Bisschop; Julien Mailfert; Kevin Lucas; Weimin Gao, "A fast triple patterning solution with fix guidance", Proc. SPIE 9053, Design-Process-Technology Co-optimization for Manufacturability VIII, 90530A (March 28, 2014); doi:10.1117/12.2046487

[2] Soo Han Choi; Srini Arikati; Erdem Cilingir, "Categorized Stitching Guidance for Triple-Patterning Technology", United States Patent, provisional application no. 61/975,701

[3] B. Yu, "Layout decomposition for triple patterning lithography", Proc. ICCAD, p. 1-8, 2011. doi:10.1109/ICCAD.2011.6105297

[4] Tian, H., Zhang, H., Ma, Q., Xiao, Z., Wong, M.D.F.: "A polynomial time triple patterning algorithm for cell based row-structure layout"; In ICCAD(2012)57-64

[5] Kuang, J., Young, E.F.Y.: "An efficient layout decomposition approach for triple patterning lithography." ;In DAC(2013)69-69

[6] S. Fang, Y. Chang, and W. Chen, "A Novel Layout Decomposition Algorithm for Triple Patterning Lithography"; presented at IEEE Trans. on CAD of Integrated Circuits and Systems, 2014, pp.397-408.

[7] B. Yu, Y. Lin, G. Luk-Pat, D. Ding, K. Lucas, and D.Z. Pan, A High-Performance Triple Patterning Layout Decomposer with Balanced Density; In Proceedings of CoRR. 2014.

Yield-aware Mask Assignment using Positive Semidefinite Relaxation in LELECUT Triple Patterning

Yukihide Kohira[a], Chikaaki Kodama[b], Tomomi Matsui[c], Atsushi Takahashi[c], Shigeki Nojima[b] and Satoshi Tanaka[b]

[a]The University of Aizu, Aizu-Wakamatsu, Japan
[b]Toshiba Corporation, Yokohama, Japan
[c]Tokyo Institute of Technology, Tokyo, Japan

ABSTRACT

LELECUT type triple patterning lithography is one of the most promising techniques in the next generation lithography. To prevent yield loss caused by overlay error, LELECUT mask assignment which is tolerant to overlay error is desired. In this paper, we propose a method that obtains an LELECUT assignment which is tolerant to overlay error. The proposed method uses positive semidefinite relaxation and randomized rounding technique. In our method, the cost function that takes the length of boundary of features determined by the cut mask into account is introduced.

Keywords: Triple Patterning, LELECUT, Design for Manufacturability, Positive Semidefinite Relaxation

1. INTRODUCTION

Multiple patterning technique enables us to fabricate small features without using advanced technologies such as extreme ultra violet (EUV) lithography. Triple patterning lithography (TPL) is one of the most promising techniques in 14 nm logic node and beyond. In order to realize a target pattern, various types of techniques including design for manufacturability such as LELE type double patterning lithograph,[1–7] LELELE type TPL,[8–18] LELECUT type TPL,[19,20] and side wall process,[21] are used in addition to a basic litho-etch process with optimized mask. These techniques are summarized in.[22,23]

Sidewall process[21] forms a wall feature with unique width so that it surrounds the prefabricated polygon. The sidewall process which is used in self-aligned double patterning enables us to fabricate finer pattern pitch by combining a slimming process, but the variety of target patterns that can be fabricated is limited.

Two types of TPL technologies are often discussed in literature. In LELELE, litho-etch process is repeated three times. However, it is difficult to achieve high yield due to native conflict and overlay problems. In LELECUT, the third mask called cut (or trim) process removes a part of a fabricated pattern. It is used to improve the quality of fabricated patterns as well as to enhance the flexibility of layout. However, it has overlay problems and lithographical limitations. In order to prevent yield loss caused by overlay error as much as possible, LELECUT mask assignment which is tolerant to overlay error is desired.

To our best knowledge, two LELECUT mask assignment methods have been proposed. In,[10] LELECUT mask assignment problem is formulated as an integer linear programming problem. Although it minimizes the weighted summation of the number of conflicts and stitches, the effect of cuts on layout quality is not taken into account. In,[20] LELECUT mask assignment problem is solved by positive semidefinite relaxation. Although it minimizes the weighted summation of the number of conflicts, stitches, and polygons in the cut mask, the yield of obtained layout is also not discussed. Fig. 1 shows mask assignments in LELECUT. A target pattern is shown in Fig. 1 (a). The layouts obtained by two LELECUT mask assignments which are represented by blue, magenta, and cut masks without overlay error are shown in Fig. 1 (b) and Fig. 1 (d). These mask assignments have no

Further author information: (Send correspondence to Yukihide Kohira, Chikaaki Kodama and Atsushi Takahashi)
Yukihide Kohira: E-mail:kohira@u-aizu.ac.jp, Telephone: +81-242-37-2536
Chikaaki Kodama: E-mail:chikaaki1.kodama@toshiba.co.jp, Telephone: +81-45-890-2818
Atsushi Takahashi: E-mail:atsushi@eda.ce.titech.ac.jp, Telephone: +81-3-5734-2665

Design-Process-Technology Co-optimization for Manufacturability IX, edited by John L. Sturtevant, Luigi Capodieci, Proc. of SPIE Vol. 9427, 94270B · © 2015 SPIE · CCC code: 0277-786X/15/$18 · doi: 10.1117/12.2085285

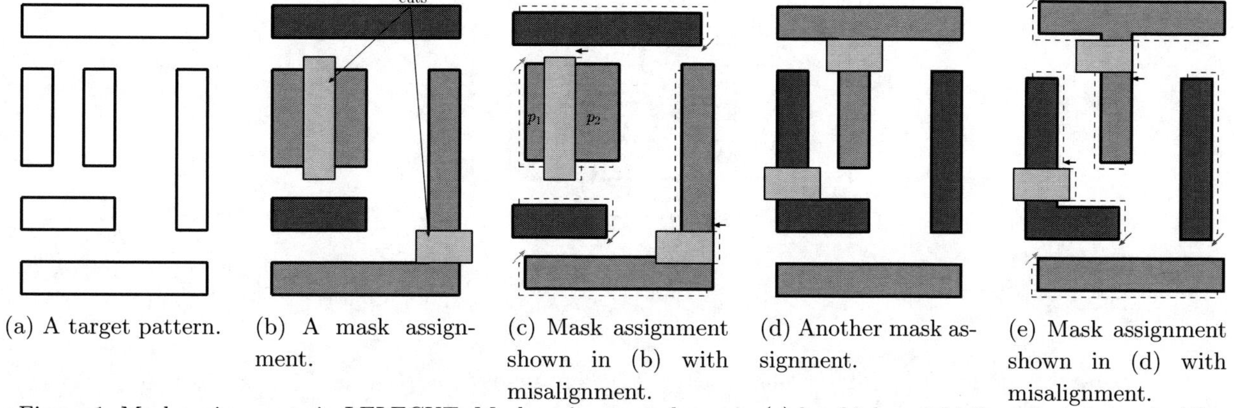

(a) A target pattern. (b) A mask assignment. (c) Mask assignment shown in (b) with misalignment. (d) Another mask assignment. (e) Mask assignment shown in (d) with misalignment.

Figure 1. Mask assignments in LELECUT. Mask assignment shown in (c) has higher yield than that shown in (d).

conflicts, no stitches, and the number of polygons in the cut mask is same. The layouts of them with overlay error in which blue, magenta, and cut masks move to the lower left, the upper right, and the left, respectively, are shown in Fig. 1 (c) and Fig. 1 (e), respectively. The former is expected to have lower yield than the latter since a longer dimension of features such as p_1 and p_2 is determined by the cut mask and is affected directly by the overlay error. The length of a boundary of a feature that is determined by the cut mask should be small enough to prevent the yield loss caused by overlay error.

In this paper, we propose a method that obtains an LELECUT assignment which is tolerant to overlay error. The proposed method is an enhancement of the method proposed in[20] and uses positive semidefinite relaxation and randomized rounding technique. In our method, the cost function that takes the length of boundary of features determined by the cut mask into account is introduced to obtain an overlay tolerant LELECUT assignment.

2. PRELIMINARIES

2.1 Problem Definition

Let $P = \{p_1, p_2, \ldots, p_n\}$ be the set of polygons in a target pattern. A polygon may represent a polygon decomposed by given stitch candidates. A stitch edge is defined between two polygons if and only if two polygons are decomposed by a stitch candidate. A polygon conflict edge is defined between two polygons if and only if two polygons are too close to assign the same mask. A cut candidate is defined between two polygons connected by a polygon conflict edge if and only if they can be cut by the cut mask when they are assigned to the same mask. A cut candidate c has a cost $l(c)$ which is defined by the length of boundary between polygons in the target pattern and the cut candidate. A cut candidate may not be independent of other cut candidates. A cut conflict edge is defined between two cut candidates if and only if they cannot be used simultaneously. Note that even if the distance between two cut candidates is not long enough, they can be used simultaneously if they can be merged into one. A cut conflict edge is not defined between two cut candidates if the distance between them is long enough or if they can be merged into one without affecting the critical dimension of pattern. Let S, C_P, T, and C_T be the set of stitch edges, the set of polygon conflict edges, the set of cut candidates, and the set of cut conflict edges, respectively. Note that both the set of stitch edges S and the set of polygon conflict edges C_P are families of unordered pairs of polygons in P, the set of cut candidates T is a sub-set of the set of polygon conflict edges C_P, and the set of cut conflict edges C_T is a family of unordered pairs of cut candidates in T.

Fig. 2 shows an example of problem. In this example, $P = \{p_1, p_2, p_3, p_4\}$, $S = \{\{p_3, p_4\}\}$, $C_P = T = \{c_1, c_2, c_3, c_4, c_5\}$, where $c_1 = \{p_1, p_3\}$, $c_2 = \{p_1, p_4\}$, $c_3 = \{p_1, p_2\}$, $c_4 = \{p_2, p_3\}$, and $c_5 = \{p_2, p_4\}$ with costs $l(c_1) = 2$, $l(c_2) = 2$, $l(c_3) = 1$, $l(c_4) = 2$, and $l(c_5) = 2$. The set of cut conflict edges is given by $C_T = \{\{c_1, c_2\}, \{c_1, c_3\}, \{c_2, c_3\}, \{c_3, c_4\}, \{c_3, c_5\}, \{c_4, c_5\}\}$.

In this paper, a polygon is assigned to one of two masks except the cut mask. The problem of finding an assignment of polygons, and/or two-coloring, is essentially equivalent to a maximum cut problem. We employ a

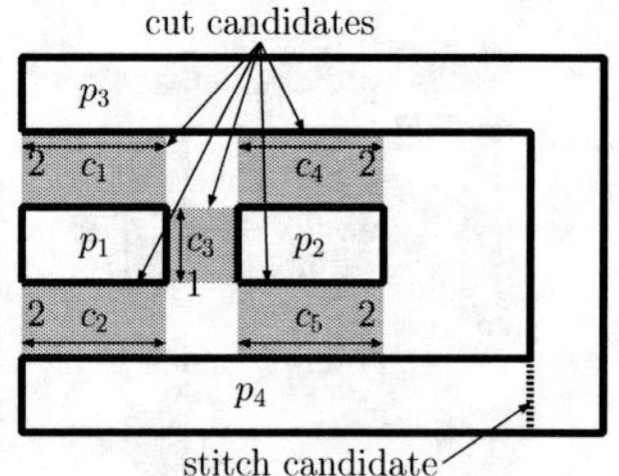

Figure 2. Polygons with stitch and cut candidates.

$\{-1,1\}$ formulation represented by function $x : P \to \{-1,1\}$, which is introduced by Goemans and Williamson in[24] for a max cut problem. This formulation naturally yields a positive semidefinite relaxation appearing in a later section. For two-coloring x, the set of polygon conflict edges that connect the same color polygons is represented by

$$C_P(x) \stackrel{\text{def.}}{=} \{\{p,q\} \in C_P \mid x(p) = x(q)\}.$$

Similarly, the set of stitch candidates that connect the different color polygons is represented by

$$S(x) \stackrel{\text{def.}}{=} \{\{p,q\} \in S \mid x(p) \neq x(q)\}.$$

The set of feasible cuts for two-coloring x is a subset of $C_P(x) \cap T$ and an independent set of cut graph (T, C_T). The problem we consider is given as follows:

P1: minimize

$$\begin{aligned} & \alpha_1 |C_P(x) \setminus T'| + \alpha_2 |S(x)| + \alpha_3 \sum_{c \in T'} l(c) \\ = \; & \alpha_1 |C_P(x)| + \alpha_2 |S(x)| - \alpha_1 \sum_{c \in T'} 1 + \alpha_3 \sum_{c \in T'} l(c) \\ = \; & \alpha_1 |C_P(x)| + \alpha_2 |S(x)| + \sum_{c \in T'} (\alpha_3 l(c) - \alpha_1) \end{aligned}$$

subject to

- $x(p) \in \{-1,1\} \quad (\forall p \in P)$,
- $T' \subseteq C_P(x) \cap T$,
- T' is an independent set of cut graph (T, C_T).

In this formulation, the weighted sum of the number of unresolved conflict edges, the number of caused stitches, and the total cost of used cut candidates is minimized. In the following, we assume that $\alpha_1 \geq \alpha_2 \geq 0$ and $\alpha_1 \geq \alpha_3 l(c) \geq 0$. According to this assumption, the total cost of used cut candidates is minimized under the condition that the number of conflict edges that connect the same polygons is minimized and the number of used cut candidates is maximized.

Fig. 3 shows examples of mask assignments. The mask assignment shown in Fig. 3 (a) has two cuts with total cost 4 and one stitch. On the other hand, that shown in Fig. 3 (b) has one cut with total cost 1. Obviously, the mask assignment shown in Fig. 3 (b) is better than that shown in Fig. 3 (a).

2.2 Maximum Independent Set with Minimum Total Cost Problem

For a given two-coloring $x : P \to \{-1,1\}$, the problem P1 is equivalent to a maximum independent set with minimum total cost problem MISMTCP1 since we assume $\alpha_1 \geq \alpha_3 l(c)$.

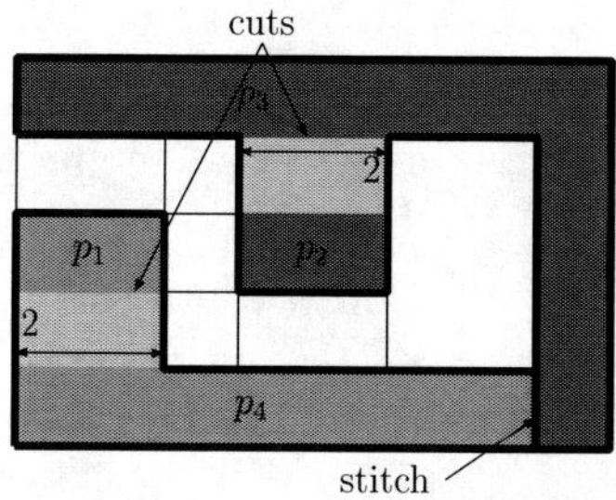

(a) Two cuts with total cost 4 and one stitch.

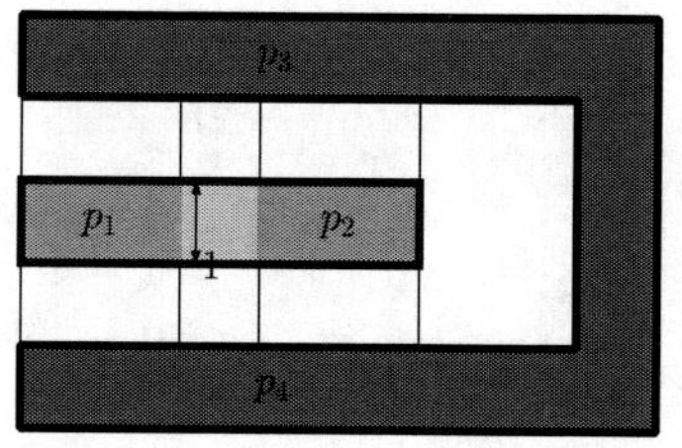

(b) One cut with total cost 1.

Figure 3. Mask assignments for layout in Fig. 2.

Step 1: Formulate a positive semidefinite relaxation SDP-L.

Step 2: Solve SDP-L by SDP solver.

Step 3: Obtain a mask assignment by randomized rounding algorithm with iterative improvement.

Figure 4. Outline of the proposed method.

MISMTCP1: maximize

$$\sum_{c \in T'} (1 - \alpha l(c))$$

subject to

- $\alpha l(c) < 1$,
- $T' \subseteq C_P(x) \cap T$,
- T' is an independent set of cut graph (T, C_T).

A maximum independent set with minimum total cost problem is known to be NP-hard. In this paper, the following 0-1 integer linear programming MISMTCP2 is formulated by introducing 0-1 variable $y(c)$.

MISMTCP2: maximize

$$\sum_{c \in T} y(c) \cdot (1 - \alpha l(c))$$

subject to

$$y(c) \in \{0, 1\} \qquad (\forall c \in T), \tag{1}$$

$$y(c) + y(c') \leq 1 \qquad (\forall \{c, c'\} \in C_T), \tag{2}$$

$$y(c) = 0 \qquad (\forall c \in T \setminus C_P(x)). \tag{3}$$

3. POSITIVE SEMIDEFINITE RELAXATION

3.1 Outline of Proposed Method

The outline of the proposed method is shown in Fig. 4. The proposed method is based on the method proposed in.[20] In the proposed method, a positive semidefinite relaxation of P1 called SDP-L is formulated and a mask assignment is obtained from an optimum solution of the relaxation by randomized rounding technique with iterative improvement. It is well-known that a positive semidefinite programming problem can be solved by interior point methods in polynomial time.

3.2 Our Semidefinite Relaxation

In this subsection, we introduce a positive semidefinite relaxation of P1. Our relaxation is an enhancement of the formulation proposed by Goemans and Williamson[24] for a max cut problem to handle a maximum independent set with minimum total cost.

First, we represent the objective function of P1 as a linear function. An arbitrary two-coloring $x : P \rightarrow \{-1, 1\}$, satisfies: $x(p) = x(q) \leftrightarrow x(p)x(q) = 1$ and $x(p) \neq x(q) \leftrightarrow x(p)x(q) = -1$. By using these properties, $|C_P(x)|$ and $|S(x)|$ are represented in terms of x as follows:

$$\begin{aligned} |C_P(x)| &= \sum_{\{p,q\} \in C_P} \left(\frac{x(p)x(q)}{2} + \frac{1}{2} \right) \\ |S(x)| &= \sum_{\{p,q\} \in S} \left(-\frac{x(p)x(q)}{2} + \frac{1}{2} \right). \end{aligned}$$

Let X be the $n \times n$ matrix whose (p,q)-th element is $X_{pq} = x(p)x(q)$ $(\forall p, \forall q \in P)$. Let C and S be matrixes that represent polygon conflicts and stitch candidates, respectively. That is, C and S are the $n \times n$ symmetric matrixes, and

$$C_{pq} = \begin{cases} \frac{1}{4} & (\{p,q\} \in C_P), \\ 0 & (\{p,q\} \notin C_P), \end{cases} \qquad S_{pq} = \begin{cases} \frac{1}{4} & (\{u,v\} \in S), \\ 0 & (\{u,v\} \notin S), \end{cases}$$

respectively. Then, we have

$$|C_P(x)| = \sum_{\{p,q\} \in C_P} \left(\frac{x(p)x(q)}{2} + \frac{1}{2} \right) = C \bullet X + \text{const},$$

$$|S(x)| = \sum_{\{p,q\} \in S} \left(-\frac{x(p)x(q)}{2} + \frac{1}{2} \right) = -S \bullet X + \text{const},$$

where $M \bullet M'$ is defined as $\sum_i \sum_j M_{ij} M'_{ij}$ for square matrixes M and M' of the same size. Let $y(c)$ be a 0-1 variable for a cut candidate c. c is assigned to the cut mask, if and only if $y(c) = 1$. Then, we have

$$\sum_{c \in T'} (\alpha_3 l(c) - \alpha_1) = \sum_{c \in T} y(c) \cdot (\alpha_3 l(c) - \alpha_1).$$

Therefore, the objective function of P1 is represented as

$$\alpha_1 (C \bullet X) - \alpha_2 (S \bullet X) + \sum_{c \in T} y(c) \cdot (\alpha_3 l(c) - \alpha_1) + \text{const}. \tag{4}$$

Note that C and S are constant matrixes.

Next, constraints of P1 are represented as linear functions in terms of X and y. Constraint (3) in MISMTCP2 is represented as

$$0 \leq y(c) \leq \begin{cases} 1 & (\text{if } c \in T \cap C_P(x)), \\ 0 & (\text{if } c \in T \setminus C_P(x)). \end{cases}$$

Then, constraint (3) is represented by using two-coloring x as follows:

$$0 \leq y(c) \leq \frac{x(p)x(q)}{2} + \frac{1}{2} \qquad (\forall c = \{p,q\} \in T).$$

The representations of other constraints are straightforward, and P1 is represented as follows:

P2: minimize

$$\alpha_1(C \bullet X) - \alpha_2(S \bullet X) + \sum_{c \in T} y(c) \cdot (\alpha_3 l(c) - \alpha_1)$$

subject to

$$
\begin{aligned}
X_{pq} &= x(p)x(q) && (\forall (p,q) \in P^2), && (5)\\
x(p) &\in \{-1,1\} && (\forall p \in P), && (6)\\
y(c) &\in \{0,1\} && (\forall c \in T), && (7)\\
y(c) + y(c') &\leq 1 && (\forall \{c,c'\} \in C_T), &&\\
0 \leq y(c) &\leq \frac{x(p)x(q)}{2} + \frac{1}{2} && (\forall c = \{p,q\} \in T). &&
\end{aligned}
$$

The objective function of P2 is obtained from Eq. (4) by removing constant term. Note that $X_{pp} = 1$ for all $p \in P$.

Since X in P2 is a positive semidefinite symmetric matrix, the problem P2 has a positive semidefinite programming relaxation as follows. Let $\mathcal{S}^n_+$ be the set of $n \times n$ positive semidefinite symmetric matrix. A positive semidefinite relaxation problem SDP-L is obtained from P2 by restricting X within $\mathcal{S}^n_+$ and ignoring 0-1 constraints for y, instead of constraints (5), (6), and (7).

SDP-L: minimize

$$\alpha_1(C \bullet X) - \alpha_2(S \bullet X) + \sum_{c \in T} y(c) \cdot (\alpha_3 l(c) - \alpha_1)$$

subject to

$$
\begin{aligned}
X_{pp} &= 1 && (\forall p \in P),\\
y(c) + y(c') &\leq 1 && (\forall \{c,c'\} \in C_T),\\
0 \leq y(c) &\leq \frac{1}{2}X_{pq} + \frac{1}{2} && (\forall c = \{p,q\} \in T),\\
X &\in \mathcal{S}^n_+.
\end{aligned}
$$

SDP-L is a positive semidefinite programming problem and can be solved by interior point methods in polynomial time.

3.3 Randomized Rounding for LELECUT

In this subsection, we propose a randomized rounding technique based on the *hyper-plane separation technique* proposed by Goemans and Wiiliamson,[24] which gives 0.878 approximation algorithm for a max cut problem. The randomized rounding technique is based on the method proposed in,[20] and the iterative improvement is applied as post-processing.

For any positive semidefinite symmetric matrix $X \in \mathcal{S}^n_+$, there exists a matrix Z satisfying $X = Z^\top Z$. This decomposition is called Cholesky decomposition.

We solve problem SDP-L by a SDP solver and obtain an optimal solution $(\widetilde{X}, \widetilde{y})$ for SDP-L. Let $\widetilde{Z}^\top \widetilde{Z}$ be the Cholesky decomposition of $\widetilde{X}$ and d be the number of rows of $\widetilde{Z}$. Here we note that columns of $\widetilde{Z}$ are indexed by polygons in P and the length of every column vector is equal to 1. For each polygon $p \in P$, vector $\widetilde{z}(p) \in \mathbb{R}^d$ denotes the corresponding column vector of $\widetilde{Z}$. Algorithm RR shown in Fig. 5 outputs a two coloring $\widetilde{x} : P \to \{-1,1\}$ and the set $\widetilde{T}$ of cuts. The mask assignment is modified by the greedy iterative improvement so that better solutions are obtained. The quality of the obtained mask assignment depends on the generated random unit vector and the runtime of Algorithm RR is very small. Therefore, Algorithm RR is repeated appropriate times and the best mask assignment is output. During the repetition, we might choose a comfortable mask assignment as well.

Algorithm RR

Step 1: Generate a random unit vector $\boldsymbol{u} \in \mathbb{R}^d$ (satisfying $||\boldsymbol{u}|| = 1$).

Step 2: For each polygon $p \in P$, set $\widetilde{x}(p) = \begin{cases} 1 & (\text{if } \boldsymbol{u}^\top \widetilde{\boldsymbol{z}}(p) > 0), \\ -1 & (\text{otherwise}). \end{cases}$

Step 3: Construct a subgraph $\widetilde{G}$ of cut graph (T, C_T) induced by vertex subset $C_P(\widetilde{x}) \cap T$. Find a maximal independent set $\widetilde{S}$ of $\widetilde{G}$ by employing a heuristic algorithm for maximum independent set with minimum total cost problem.

Step 4: Apply the greedy iterative improvement in which mask assignment of a polygon is changed and a heuristic algorithm for maximum independent set with minimum total cost problem is applied until the solution is not improved.

Figure 5. Randomized Rounding Algorithm.

Table 1. Experimental results. The obtained mask assignments have no polygon conflicts and cut conflicts in ILP-#, ILP-L, SDP-#, and SDP-L Imp. The units of cost and time are [nm] and [s], respectively.

circuit	$\|P\|$	# Seg	$\|C_P\|$	# comp total	# comp target	$\|T\|$	ILP-# $\|T'\|$	ILP-# cost	ILP-# time	ILP-L $\|T'\|$	ILP-L cost	ILP-L time	SDP-# $\|T'\|$	SDP-# cost	SDP-# time	SDP-L $\|T'\|$	SDP-L $\|C_P(x)\|$	SDP-L cost	SDP-L time	SDP-L Imp $\|T'\|$	SDP-L Imp cost	SDP-L Imp time
c432	850	4918	540	414	1	12	1	40	0.01	1	40	0.01	1	40	0	1	0	40	0	1	40	0
c499	1491	9518	1489	502	50	278	58	2720	0.42	58	2720	0.45	58	6030	0.12	62	0	2940	0.12	58	2720	0.15
c880	1872	10666	1422	717	168	934	172	11720	2.46	198	10390	6.25	172	20590	0.44	202	2	11105	0.35	198	10390	0.51
c1355	2656	15246	1514	1328	76	514	90	9760	1.10	122	8320	6.10	90	13230	0.27	130	2	9635	0.17	122	8320	0.37
c1908	4191	24370	3141	1733	182	1462	211	37700	5.77	373	30410	24.34	211	41140	0.73	411	13	36600	0.43	373	30410	1.37
c2670	6371	37564	5802	2056	585	4298	686	76950	10.52	958	64710	38.98	686	101975	2.32	1069	66	84540	1.39	952	65775	5.26
c3540	8188	47244	6897	2896	775	4794	821	73560	11.06	1044	63445	29.61	821	111330	2.24	1090	11	72630	1.47	1044	63445	3.02
c5315	11498	68476	10097	3926	1193	7552	1259	108290	17.52	1572	93965	53.23	1259	162510	3.59	1647	13	106280	2.33	1572	93965	4.11
c6288	11605	64762	5602	6259	256	1282	256	10320	2.39	256	10240	2.33	256	29840	0.66	256	0	10240	0.45	256	10240	0.66
c7552	17167	99526	14027	6258	1448	9325	1587	162980	20.02	2122	138905	89.62	1587	221025	4.47	2243	23	159505	3.01	2122	138905	6.12
ave.							0.82	1.12	0.50	(1)	(1)	(1)	0.82	1.77	0.09	1.05		1.12	0.07	1.00	1.00	0.12

$\|P\|$	the number of polygons
# Seg	the number of line segments
$\|C_P\|$	the number of polygon conflict edges
total	the number of components in the conflict graph (P, C_P)
target	the number of components in the conflict graph (P, C_P) in which cuts must be inserted
$\|T\|$	the number of cut candidates in the target components
$\|T'\|$	the number of inserted cuts
$\|C_P(x)\|$	the number of conflicts in the obtained mask assignment
cost	the total cost of cuts in the obtained mask assignment
time	computational time
ave.	the average normalized by ILP-L

4. EXPERIMENTS

Our proposed mask assignment method is implemented by using a SDP solver and C++ language. We compare the following five methods. ILP-# and ILP-L are ILP formulations based on[19] which minimizes the number of cuts and minimizes the total cost of cuts, respectively. SDP-# is the positive semidefinite relaxation proposed in[20] which minimizes the number of cuts. SDP-L Imp and SDP-L are the proposed methods. SDP-L Imp applies the greedy iterative improvement as post-processing and SDP-L does not. The methods are executed on a Linux machine with 12 GB memory by using Intel core i7-3770 of 3.40 GHz. In our implementation, SDP problems are solved by SDPA 7.3.8[25] which is a free tool. On the other hand, ILP problems are solved by CPLEX 12.6.1[26] which is one of the most famous commercial ILP solvers. In these methods, a speedup technique in which the conflict graph $(P, S \cap C_P)$ is decomposed into connected components is adopted. This speedup technique is discussed in many previous studies.[1–3,5,6,19,20] In this experiment, we do not prepare stitch candidates to focus on observing the total cost, the number of cuts, and the number of conflicts. The parameters in objective functions are set to $\alpha_1 = 10^6$ and $\alpha_3 = 1$. $\alpha_1 = 10^6$ is much larger than cut costs. Algorithm RR is applied 100 times in SDP-#, SDP-L, and SDP-L Imp.

ISCAS benchmarks which were used in[3,5] are used. The benchmarks are reproduced from the information given by authors of[3,5] and from figures in,[3] though we could not obtain the same data. We followed the parameters as in,[3,5] where the minimum polygon space in a mask is 54 nm. If stitches are allowed to be inserted, the mask assignment without cuts is obtained. Therefore, we also do not insert stitches in this experiment. Table 1 shows the results. Note that the mask assignments obtained by all methods except SDP-L have no polygon conflicts and cut conflicts. Since the minimization of the total cost is added into the objective function of ILP-L, the total cost obtained by ILP-L is optimum. Similarly, since the minimization of cuts is added into the objective function of ILP-#, the number of cuts obtained by ILP-# is optimum. Although ILP-# and SDP-# obtain mask assignments with the minimum number of cuts, the total cost of the mask assignment obtained by them is larger than that by ILP-L since the minimization of cuts does not corresponds to that of the total cost. Although SDP-L is fast, it obtains mask assignments with conflicts. The total cost of the mask assignment obtained by SDP-L Imp is the same as that by ILP-L in nine circuits of ten circuits. Moreover, SDP-L Imp is much faster than ILP-L. Consequently, SDP-L Imp obtains optimum solutions in the shortest computational time in almost all circuits.

5. CONCLUSIONS

In this paper, we propose a fast LELECUT mask assignment method to be tolerant to overlay error. The proposed method applies a positive semidefinite relaxation. The experimental results show the efficiency and the validity of the proposed method. We will take the mask density balance, stitch direction, and etc. into account to improve the quality of the mask assignment in our future works.

ACKNOWLEDGMENTS

This work was supported by JSPS KAKENHI Grant-in-Aid for Scientific Research (B) 25280013.

REFERENCES

[1] Kahng, A., Park, C.-H., Xu, X., and Yao, H., "Layout decomposition for double patterning lithography," in [*Proc. ICCAD*], 465–472 (2008).

[2] Yuan, K., Yang, J.-S., and Pan, D., "Double patterning layout decomposition for simultaneous conflict and stitch minimization," in [*Proc. ISPD*], 107–114 (2009).

[3] Yang, J.-S., Lu, K., Cho, M., Yuan, K., and Pan, D. Z., "A new graph-theoretic, multi-objective layout decomposition framework for double patterning lithography," in [*Proc. ASP-DAC*], 637–644 (2010).

[4] Chen, S.-Y. and Chang, Y.-W., "Native-conflict-aware wire perturbation for double patterning technology," in [*Proc. ICCAD*], 556–561 (2010).

[5] Tang, X. and Cho, M., "Optimal layout decomposition for double patterning technology," in [*Proc. ICCAD*], 9–13 (2011).

[6] Kohira, Y., Yokoyama, Y., Kodama, C., Takahashi, A., Nojima, S., and Tanaka, S., "Yield-aware decomposition for LELE double patterning," in [*Proc. SPIE*], **9053, 90530T** (2014).

[7] Yokoyama, Y., Sakanushi, K., Kohira, Y., Takahashi, A., Kodama, C., Tanaka, S., and Nojima, S., "Localization concept of re-decomposition area to fix hotspots for LELE process," in [*Proc. SPIE*], **9053, 90530V** (2014).

[8] Yu, B., Yuan, K., Zhang, B., Ding, D., and Pan, D. Z., "Layout decomposition for triple patterning lithography," in [*Proc. ICCAD*], 1–8 (2011).

[9] Fang, S.-Y., Chang, Y.-W., and Chen, W.-Y., "A novel layout decomposition algorithm for triple patterning lithography," in [*Proc. DAC*], 1181–1186 (2012).

[10] Ma, Q., Zhang, H., and Wong, M., "Triple patterning aware routing and its comparison with double patterning aware routing in 14 nm technology," in [*Proc. DAC*], 591–596 (2012).

[11] Tian, H., Zhang, H., Ma, Q., Xiao, Z., and Wong, M., "A polynomial time triple patterning algorithm for cell based row-structure layout," in [*Proc. ICCAD*], 57–64 (2012).

[12] Lin, Y.-H., Yu, B., Pan, D. Z., and Li, Y.-L., "TRIAD: a triple patterning lithography aware detailed router," in [*Proc. ICCAD*], 123–129 (2012).

[13] Kuang, J. and Young, E., "An efficient layout decomposition approach for triple patterning lithography," in [*Proc. DAC*], 69:1–69:6 (2013).

[14] Zhang, Y., Luk, W.-S., Zhou, H., Yan, C., and Zeng, X., "Layout decomposition with pairwise coloring for multiple patterning lithography," in [*Proc. ICCAD*], 170–177 (2013).

[15] Yu, B., Lin, Y.-H., Luk-Pat, G., Ding, D., Lucas, K., and Pan, D. Z., "A high-performance triple patterning layout decomposer with balanced density," in [*Proc. ICCAD*], 163–169 (2013).

[16] Tian, H., Du, Y., Zhang, H., Xiao, Z., and Wong, M. D., "Constrained pattern assignment for standard cell based triple patterning lithography," in [*Proc. ICCAD*], 178–185 (2013).

[17] Yu, B., Xu, X., Gao, J.-R., and Pan, D. Z., "Methodology for standard cell compliance and detailed placement for triple patterning lithography," in [*Proc. ICCAD*], 349–356 (2013).
[18] Matsui, T., Kohira, Y., Kodama, C., and Takahashi, A., "Positive semidefinite relaxation and approximation algorithm for triple patterning lithography," in [*Proc. ISAAC, LNCS 8889*], 365–375 (2014).
[19] Yu, B., Gao, J.-R., and Pan, D. Z., "Triple patterning lithography (TPL) layout decomposition using end-cutting," in [*Proc. SPIE*], **8684, 86840G** (2013).
[20] Kohira, Y., Matsui, T., Yokoyama, Y., Kodama, C., Takahashi, A., and Shigeki Nojima, S. T., "Fast mask assignment using positive semidefinite relaxation in lelecut triple patterning," in [*Proc. ASPDAC*], 665–670 (2015).
[21] Kodama, C., Ichikawa, H., Nakayama, K., Kotani, T., Nojima, S., Mimotogi, S., Miyamoto, S., and Takahashi, A., "Self-aligned double and quadruple patterning aware grid routing with hotspots control," in [*Proc. ASP-DAC*], 267–272 (2013).
[22] Yu, B., Gao, J.-R., Ding, D., Ban, Y., Yang, J.-S., Yuan, K., Cho, M., and Pan, D. Z., "Dealing with ic manufacturability in extreme scaling," in [*Proc. ICCAD*], 240–242 (2012).
[23] Takahashi, A., Awad, A., Kohira, Y., Matsui, T., Kodama, C., Nojima, S., and Tanaka, S., "Multi patterning techniques for manufacturability enhancement in optical lithography," in [*Proc. ICDV*], 117–122 (2014).
[24] Goemans, M. X. and Williamson, D. P., "Improved approximation algorithms for maximum cut and satisfiability problems using semidefinite programming," *Journal of the ACM* **42**, 1115–1145 (1995).
[25] "SDPA 7.3.8." `http://sdpa.sourceforge.net/`.
[26] "CPLEX 12.6.1." `http://www-01.ibm.com/software/commerce/optimization/cplex-optimizer/`.

DTCO at N7 and Beyond: Patterning and Electrical Compromises and Opportunities

Julien Ryckaert, Praveen Raghavan, Pieter Schuddinck, Huynh Bao Trong, Arindam Mallik, Sushil S. Sakhare, Bharani Chava, Yasser Sherazi, Philippe Leray, Abdelkarim Mercha, Jürgen Bömmels, Gregory R. McIntyre, Kurt G. Ronse, Aaron Thean, Zsolt Tőkei, An Steegen, Diederick Verkest
imec, Kapeldreef 75 3001 Leuven Belgium

ABSTRACT

At 7nm and beyond, designers need to support scaling by identifying the most optimal patterning schemes for their designs. Moreover, designers can actively help by exploring scaling options that do not necessarily require aggressive pitch scaling. In this paper we will illustrate how MOL scheme and patterning can be optimized to achieve a dense SRAM cell; how optimizing device performance can lead to smaller standard cells; how the metal interconnect stack needs to be adjusted for unidirectional metals and how a vertical transistor can shift design paradigms. This paper demonstrates that scaling has become a joint design-technology co-optimization effort between process technology and design specialists, that expands beyond just patterning enabled dimensional scaling.

Keywords: N7, FinFET, DTCO, 193i lithography, EUV lithography, SADP, SAQP

1. INTRODUCTION

From the 16nm node onwards, a Design-Technology Co-Optimization approach has become the most efficient scaling strategy [1-3]. Two nodes down the road with the arising of the 7nm node and looking forward to the nodes to come, it becomes evident that scaling based on pure lithography can no longer absorb the challenges of scaling. Designers must now contribute to a joint patterning exploration effort by identifying patterning processes that lead to the most optimal solution. Techniques such as fin depopulation, gridded designs, horizontal/vertical asymmetric pitch scaling, modified interconnect stack or alternative device architectures will all need to be used to achieve the 2x area scaling target without requiring design rules to scale by a factor 0.7. The pitch scaling target of the 7nm node can be projected by plotting the contacted poly pitch as function of the metal pitch keeping area constant for each technology node, as shown in Figure 1.

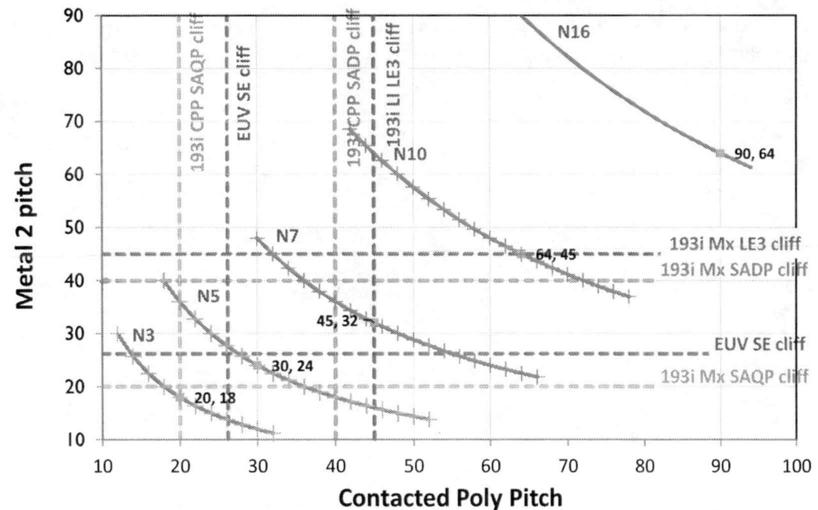

Figure 1: Iso-area scaling curves for 16nm technology to 3nm

Design-Process-Technology Co-optimization for Manufacturability IX, edited by John L. Sturtevant, Luigi Capodieci, Proc. of SPIE Vol. 9427, 94270C · © 2015 SPIE · CCC code: 0277-786X/15/$18 · doi: 10.1117/12.2178997

The iso-area curves for each node highlight the fact that metal pitch reduction can be traded for CPP depending on patterning or electrical considerations. For example, recognizing that device performance suffers from gate length and contact area scaling, one can adopt a scaling path where metal pitch is more aggressively scaled than the gate pitch as was done in the 14nm technology of [3]. Various patterning cliffs are also plotted on this graph to define critical cost and complexity boundaries for patterning decisions. For example, keeping the CPP above 40nm allows to keep the SADP assumption on gate. This directly translates into a metal pitch around 32nm pitch. From Figure 1, the 7nm node metal pitch lands in the range of 30-36nm with a contacted poly pitch around 40-45nm.

At these dimensions pitch quadrupling in 193i becomes a necessity and some features reach mean to sigma values rising up to few tens of percentages due to CD non-uniformity and overlay. Therefore patterning can no longer be done without compromising on layout templates leading to gridded and eventually unidirectional designs. Thereby, spacer-defined technology is best exploited and self-alignement becomes the only way to reduce sensitivities to the process variations. Furthermore, some design templates may enable patterning techniques that offer larger process windows. One then needs to consider different layout options in combination with different corresponding patterning techniques. The different options may lead to different electrical performances. Power performance area and cost trade-offs is then the right metric to define the process assumptions of a technology. Extreme-UV can obviously simplify patterning and alleviate some of the patterning complexity inherent to the usage of 193i lithography thereby allowing more design flexibility and relaxing some of the design rules. But the pitch targeted for 7nm get critically close to the limits of a single exposure in EUV. The 2D design flexibility in EUV at these pitches still remains to be demonstrated. Although this issue is being seriously addressed by tool and material suppliers, the urgency to lock down the process assumptions for 7nm may force industry to adopt a "safe" scaling path that rather relies on EUV insertion in a mix and match combination with 193i. This option offers cost benefits and better process windows but will require an all-193i solution to start from.

In this paper we will explore how design exploration co-optimized with patterning can lead to optimal solutions for the 7nm node. We will look at different cases starting with the SRAM in Section 2, followed by a fin depopulation study in the logic allowing standard cell scaling by increasing device drive capabilities in Section 3. Section 4 will explore 7nm standard cell architectures designed and optimized for unidirectional patterning. In Section 5 we will look ahead beyond 7nm to see how new device architectures can enable scaling. Section 6 concludes the paper.

2. MOL PATTERNING TO ENABLE 7nm SRAM

In terms of the FEOL and MOL, patterning is usually challenged by the SRAM rather than by the logic context. Indeed, in order to achieve a dense SRAM cell, devices need to be arranged in close proximity requiring dense and complex local interconnect patterns as shown in [4]. On the other hand, the metal layers are usually very regular in the SRAM hence the logic will challenge these layers more due to the multiple different interconnect requirements inherent to logic contexts. We will explore metal patterning in the next Sections.

The left of Figure 2 shows the MOL stack that was used in imec 10nm platform. The active contacts are partially self-aligned to the gate but still feature a non-self-aligned portion that will be subject to overlay errors

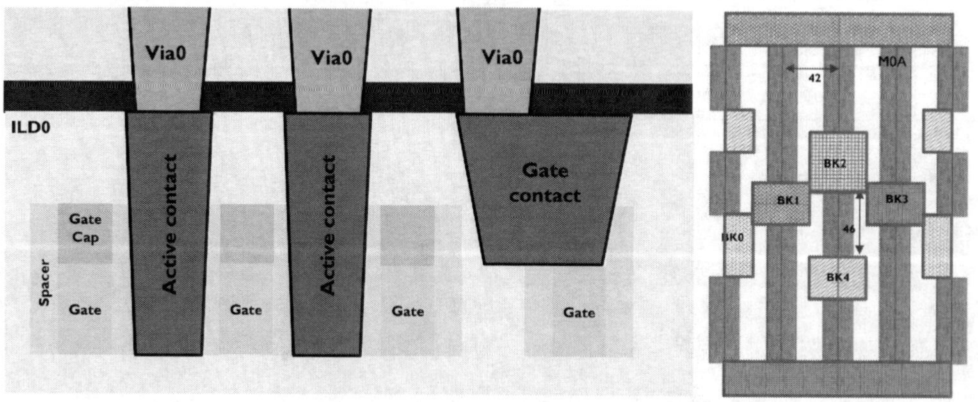

Figure 2: imec 10nm MOL stack and a corresponding SRAM layout

An active contact MOL pattern required to enable a dense 111 SRAM cell is shown on the right of Figure 2. Active contacts are aligned to the gate pitch which is around 42nm. Therefore, SADP combined with multiple blocks was assumed for this example. As can be seen the complex block arrangement in the 42nm CPP requires 5 different exposures to guarantee an optimal printing.

By adopting a fully self-aligned MOL scheme as depicted in the left of Figure 3, one can now adopt a different patterning scheme that features NTD Litho-Etch constructs continuous across the gates. This fully self-aligned scheme allows to merge multiple isolated active contacts at litho since they are self-aligned to the gate spacers. Using this approach in the 111 SRAM together with an optimization of the active contact layer [5], we obtain a 2 mask patterning solution instead of the 6 masks (1 mandrel + 5 blocks) in the previous solution.

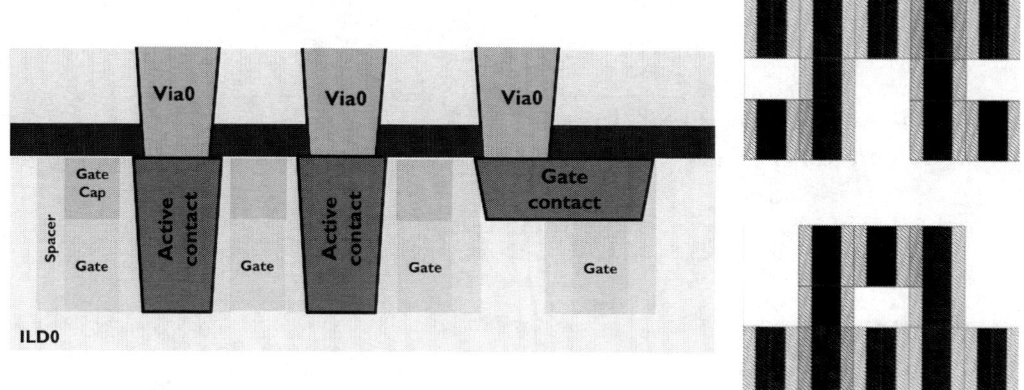

Figure 3: Suggested fully-self-aligned MOL scheme or 7nm and a corresponding SRAM layout

This experiment illustrates how by analyzing design requirements and optimizing the layout to the patterning as well as the integration scheme, one can optimize the technology to achieve the required scaling performance at acceptable cost.

3. FIN DEPOPULATION TO REDUCE LOGIC CELL AREA

In low-power SoC application where a majority of the standard cells do not require high current drive capabilities, standard cells are optimized for minimum area. This is usually done by building libraries with metal-2 track heights down to 7.5. In these type of cells the active area limits the minimum height a cell can achieve. In FinFET technology, since active area is quantized, one will seek an optimum between the metal-2 pitch, defining the cell height and the fin pitch [3], to achieve a target cell performance. A gear ratio of ¾ is a common practice in most applications leading to 4 fins per FET in a 9T and 3 fins per FET in a 7.5T library as shown in Figure 4. If we can increase the current drive capabilities of a fin, by for example increasing the fin height, the number of fins may be reduced and thereby the cell height. In this way area scaling can be achieved by increasing the electrical performance of the device without modifying the pitches of the technology. This approach is called fin depopulation.

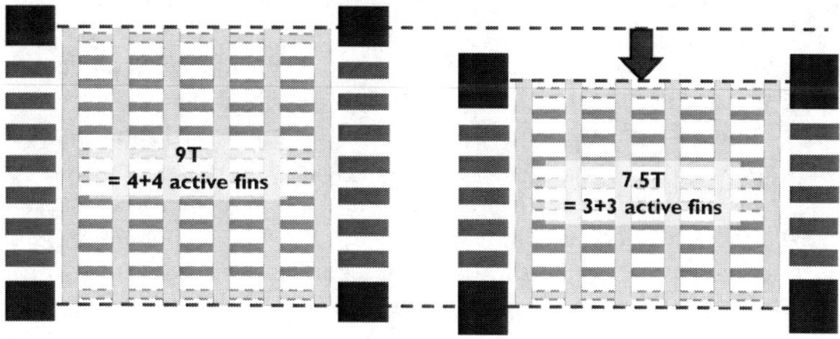

Figure 4: 9 track (left) standard cell featuring 4 fins per device and 7.5 track (right) with 3 fins per device

A possible fin depopulation strategy is described in Figure 5 for a fin depopulation from 4 to 3 fins per FET.

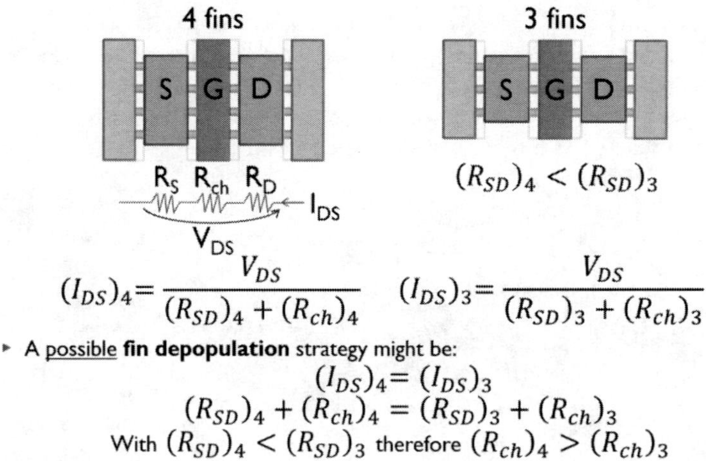

Figure 5: Fin depopulation strategy. Rch is the channel resistance and Rsd is the source-drain parasitic resistance.

Reducing the channel resistance (Rch) can be done by increasing fin height. The amount of fin height increase necessary to equalize the currents will depend on the evolution of the source-drain resistance. Figure 6 shows the Rsd increase extracted from parasitic simulations for 3 nodes (14nm, 10nm and 7nm). In 7nm, the limited contact area due to an aggressive CPP of 42nm increases the Rsd drastically at low number of fins. On the other hand, we could expect Rsd to reduce when fin height increases due to a larger fin cross-section but this reduction is rather limited as shown in the middle of Figure 6.

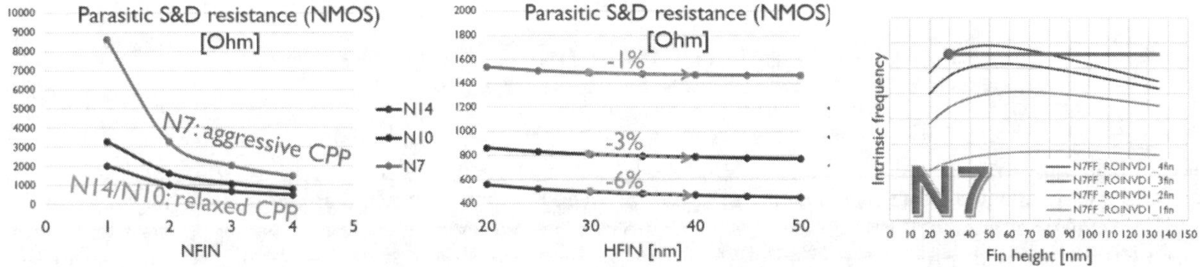

Figure 6: Parasitic Rsd as function of number of fins for 14nm, 10nm and 7nm technology (left). Parasitic resistance as function of the fin height (HFIN) for 14nm, 10nm and 7nm (middle). Ring oscillator intrinsic frequency as function of fin height for different number of fins (right).

The channel resistance reduction needs to balance the Rsd increase to enable fin depopulation. The right of Figure 6 shows the intrinsic frequency of a ring oscillator as function of fin height for different number of fins assuming a contact resistivity of 7e-9 Ohm.cm2. The fin height is 30nm in imec 10nm technology. From this graph, in order to apply fin depopulation starting from a 9 track/4 fin based library, we would need to obtain the same performance with a 3 fin device to enable a 7.5T library with equal performance. However, the high parasitic resistance in 7nm saturates the performance improvement at fin heights above 30nm jeopardizing any fin depopulation strategy. Lowering contact resistivity is the best way to reduce Rsd. Table 1 shows that the contact resistivity needs to be reduced down to 5e-9 Ohm.cm2 to reduce number of fins from 4 to 3 in 7nm technology.

Table 1: Fin depopulation strategy assuming different contact resistivities. 5e-9 enables 4 to 3 fin depopulation at 7nm.

Assumed contact resistivities		INV	NFIN	4	3	2	1
$\rho_{ct,total}$=12e-9 Ω.cm²	OK	N14	HFIN [nm]	(30)	45	80	X
$\rho_{ct,total}$=12e-9 Ω.cm²	OK	N10	HFIN [nm]	30 OK	(45)	X	X
$\rho_{ct,total}$= 7e-9 Ω.cm²	NOK	N7	HFIN [nm]	30 OK	(X) NOK	(X) OK	X
$\rho_{ct,total}$= 5e-9 Ω.cm²	OK	N7	HFIN [nm]	25	(35)	X	X
$\rho_{ct,total}$= 3e-9 Ω.cm²	OK	N7	HFIN [nm]	20	25	(50)	X

4. STANDARD CELL OPTIMIZATION FOR UNIDIRECTIONAL METALS

In order to pattern metals in 193i at the 7nm pitches, spacer-defined technology is required. This directly forces designers to adopt a unidirectional metal orientation. A key question is then the logic cell area increase that results from this constraint. To address this question we first need to define the metal stack that is essential to finish the internal connections of a standard cell. In most libraries, metal-1 allows to finish standard cells in 2D designs. A simple example is shown in Figure 7. Complex cells sometimes do require some metal-2 usage but usually do not congest the metal-2 layer, hence the latter can be used as 1st routing layer. In this scheme Metal 1 has a preferred vertical orientation and metal-2 a horizontal direction. This guarantees optimal port accessibility at metal-1 from the 1st routing layer at metal-2.

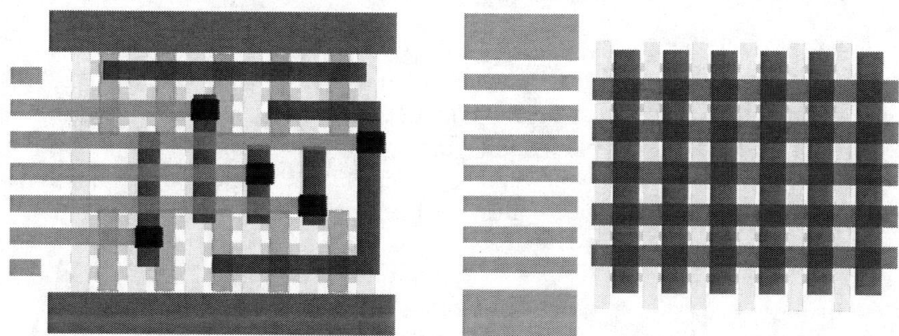

Figure 7: Example of a 10nm standard cell with 2D metal-1 and firs routing layer at metal-2 horizontal (left). Suggested scheme for a unidirectional metal-1 at 7nm introducing a horizontal layer under metal-1: Mint (right).

Although the vertical metal-1 orientation is necessary to guarantee port access, a horizontal metal-1 orientation is required to distribute power across rows of transistors in the horizontal direction, as well as to connect non-adjacent transistor nets. If metal-1 is defined to be in a single orientation, one of the above constraints will break. A straightforward way to solve this issue is to transfer all horizontal connections to metal-2. However, this creates high congestions in metal-2 that will therefore hamper any further routing. Recognizing that an extra layer must be used to compensate for the unidirectionality of metals, a more optimal solution is to add a horizontal layer under the vertical metal-1 as shown in Figure 7. This option helps in 2 ways. First it guarantees optimal port access at metal-1 with limited blockages at metal-2. Secondly since horizontal connections are mostly happening between transistor terminals, all horizontal connections in that additional layer would not necessarily need to transit through an metal-1 construct that usually creates heavy blockages hence simplifying the metal-1 pattern. In imec technology this additional layer is called the Mint layer.

A 7.5T standard cell template has been built optimizing the usage of the Mint layer and is shown in Figure 8. Metal-1 pitch is chosen to be aligned to the gate pitch to maintain the metal-1 grid after cell placement. The Mint pitch is chosen to be the same as the metal-2 pitch allowing 7 routing tracks on the Mint. It is essential for a Mint track to intercept the center of the cell where most gate connections are located. We see that each of the 7 track feature a preferred allocation. In this template 4 typical arrangements are shown to illustrate the track allocations. The arrangement A is used to shift the outbound metal-2 power rails to inbound power rails on the Mint. This track will then distribute locally the power to the transistors. The arrangement B is used to enable a gate connection at the border of the cells. This feature proves to be very useful since otherwise a single Mint track in the center would intercept all active gates creating high blockages for multiple consecutive gate connections. A typical usage of this construct is in pass-gate constructs. Note that the in-bound distribution of power on the Mint enables this extra gate connection allocation. Arrangement C is used for central gate connections and arangement D for pFET drain to nFET drain connections.

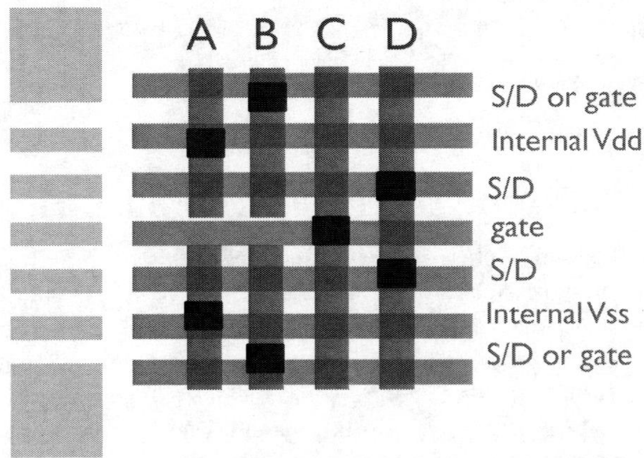

Figure 8: Four typical arrangements in imec 7nm technology standard cell library.

This template allows to build compact standard cells that are competitive in area with typical 2D libraries in previous technology nodes as discussed in [6]. Compared to a 9T library designed with a 2D metal-1 in 20nm, the amount of extra gate pitches necessary to finish the 7.5T cell is plotted in Figure 9 for a sample of 50 cells. As can be seen, a majority of the cell do not suffer from an increased width while designed with 7.5T in height. This example illustrates how design considerations allows choosing the right metal interconnect stack and associated patterning technique for the technology.

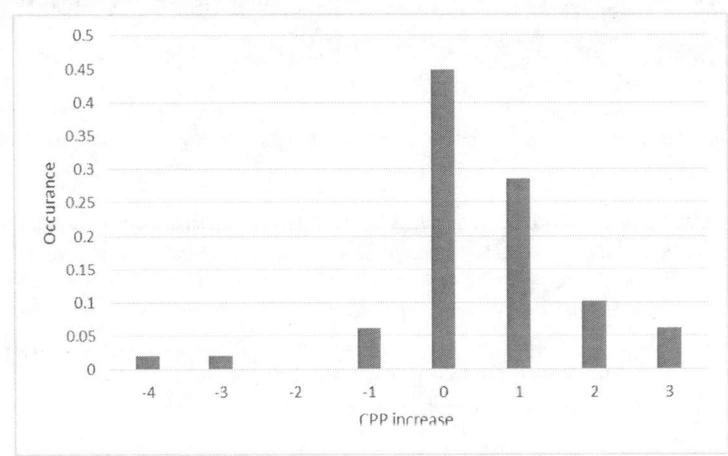

Figure 9: Amount of extra gate pitch consumed in a 1D 7.5T library compared to a 20nm 2D library for a library of 50 cells.

5. NEW DEVICE ARCHITECTURE TO ENABLE SCALING

As was discussed previously, contacted gate pitch (CGP) scaling reduces the source-drain contact area to the transistor largely affecting transistor performances. This issue is even amplified by a gate length that does not scale proportionally with pitch as shown in Figure 10.

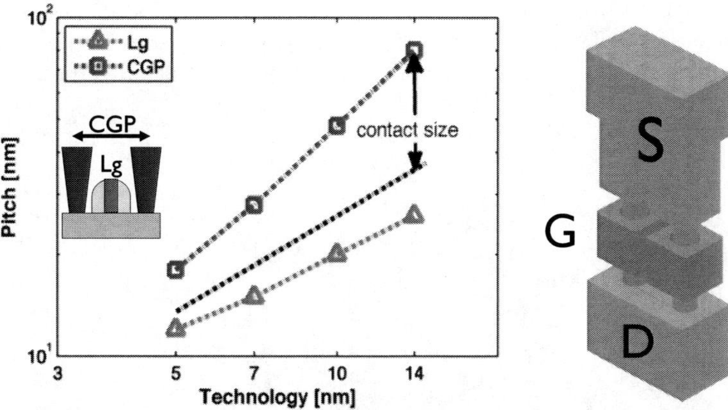

Figure 10: Contacted gate pitch and gate length scaling with recent technology nodes leaving limited space for device contacting (left). Vertically oriented devices (right) allows to break this constrain.

A way to circumvent this issue it by orienting the transistor in the vertical direction with source and drain on top of each other and gate wrapping around what would become a nanowire as shown in the right of Figure 10. Vertical transistors have unique properties. First gate length is now independent of lithography and can be scaled appropriately to achieve targeted transistor performances. Secondly, in compact 2D layouts, adjacent transistors merge their source/drain terminals with the adjacent devices. Isolation can only be made by introducing dummy gates in between. In vertical transistors, source and drain are on top and bottom of the gate hence transistor terminals do not need to be shared between adjacent devices and all transistors are now completely independent from each other. This allows very compact arrangements of transistors in the x-direction of the standard cell. Figure 11 shows two standard cells designed with VFET transistor architecture.

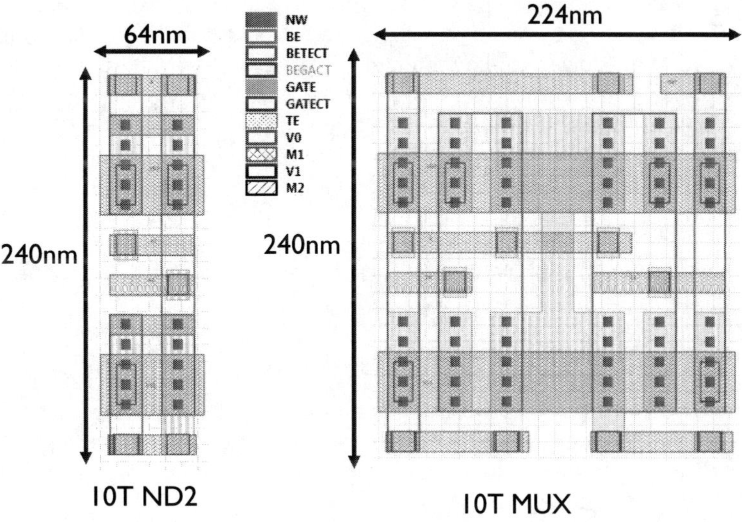

Figure 11: A 10T NAND2 and MUX designed with vertical transistors.

As can be seen, gates are placed at a very aggressive pitch of 32nm thanks to the absence of any source drain contact in between. However, source and drains terminals do need a path to escape to the interconnect layers. Therefore the upper side and the bottom side of the cell feature extra tracks that enable connections from the bottom terminal of the device to the upper layers. This means that inbound power rails are best for this architecture. It also implies that track heights will increase for this technology and scaling will happen because the width reduction of the cells thanks to the dense gate arrangements will compensate the track height increase of the cells.

The SRAM also benefits strongly from this natural isolation between devices. Indeed, as shown in Figure 12, The pull-up transistors in each part of the bit-cell can now be placed next to each other offering a straight 20% shrink advantage to the SRAM [7]. The layout of the bit-cell assumes a CGP of 32nm and a metal pitch of 24nm.

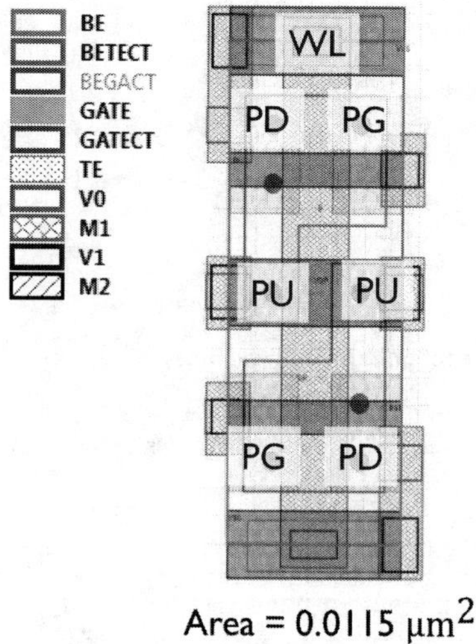

Figure 12: 112 SRAM cell designed with vertical transistors.

6. CONCLUSIONS

We have illustrated through various examples how a design perspective to scaling can support process technology experts in identifying the optimal process assumptions for advanced technology nodes such as 7nm and beyond. Revisiting the MOL stack and optimizing it to the requirements of the SRAM allowed a drastic reduction in masking layers. Improving device performance can enable logic cell area scaling without modifying printing dimensions. Optimizing the interconnect stack can enable unidirectional designs offering robust 193i scaling solutions competitive in area as their 2D counterparts. Finally changing the device architecture calls for a paradigm change in standard cell and SRAM design that has the potential to offer large area savings at constant patterning pitches.

REFERENCES

[1] J. Kye, et al., “Lithography and Design Interaction – new paradigm for the technology architecture development”, IEEE Proceedings of the Custom Integrated Circuits Conference (CICC), Sept. 2012

[2] Yeric, G. et al., “The past, present and future of design-technology co-optimization”, 2013 IEEE Proceedings of the Custom Integrated Circuits Conference pp 1-8, Sept. 2013.

[3] Ryckaert, J.et al., "Design technology co-optimization for N10," , 2014 IEEE Proceedings of the Custom Integrated Circuits Conference 1-8, Sept. 2014.

[4] Natarajan, S., et al. “A 14nm logic technology featuring 2nd-generation FinFET transistors, air-gapped interconnects, self-aligned double patterning and a 0.0588 μm2 SRAM cell size” IEEE International Electron Devices Meeting (IEDM), Dec. 2015

[5] Sakhare, S. et al., “Layout optimization and trade-off between 193i and EUV-based patterning for SRAM cells to improve performance and process variability at 7nm technology node”, Proc. SPIE 2015.

[6] Chava, B. et al., “Standard cell design in N7: EUV vs. Immersion”, Proc. SPIE 2015.

[7] Trong, H. B., “Circuit and process co-design with vertical gate-all-around nanowire FET technology to extend CMOS Scaling for 5nm and Beyond Technologies ” IEEE European Solid-State Design Research Conference, Sept 2014

Layout optimization with assist features placement by model based rule tables for 2x node random contact

Jinhyuck Jun [a], Minwoo Park [a], Chanha Park [a], Hyunjo Yang [a], Donggyu Yim [a], Munhoe Do [b], Dongchan Lee [b], Taehoon Kim [b], Junghoe Choi [b], Gerard Luk-Pat [b], Alex Miloslavsky [b]

[a] Memory Research & Development Division, SK hynix Semiconductor Inc., Korea
[b] Synopsys Inc.

ABSTRACT

As the industry pushes to ever more complex illumination schemes to increase resolution for next generation memory and logic circuits, sub-resolution assist feature (SRAF) placement requirements become increasingly severe. Therefore device manufacturers are evaluating improvements in SRAF placement algorithms which do not sacrifice main feature (MF) patterning capability. There are known-well several methods to generate SRAF such as Rule based Assist Features (RBAF), Model Based Assist Features (MBAF) and Hybrid Assisted Features combining features of the different algorithms using both RBAF and MBAF.

Rule Based Assist Features (RBAF) continue to be deployed, even with the availability of Model Based Assist Features (MBAF) and Inverse Lithography Technology (ILT). Certainly for the 3x nm node, and even at the 2x nm nodes and lower, RBAF is used because it demands less run time and provides better consistency.
Since RBAF is needed now and in the future, what is also needed is a faster method to create the AF rule tables. The current method typically involves making masks and printing wafers that contain several experiments, varying the main feature configurations, AF configurations, dose conditions, and defocus conditions – this is a time consuming and expensive process. In addition, as the technology node shrinks, wafer process changes and source shape redesigns occur more frequently, escalating the cost of rule table creation. Furthermore, as the demand on process margin escalates, there is a greater need for multiple rule tables: each tailored to a specific set of main-feature configurations.

Model Assisted Rule Tables(MART) creates a set of test patterns, and evaluates the simulated CD at nominal conditions, defocused conditions and off-dose conditions. It also uses lithographic simulation to evaluate the likelihood of AF printing. It then analyzes the simulation data to automatically create AF rule tables. It means that analysis results display the cost of different AF configurations as the space grows between a pair of main features. In summary, model based rule tables method is able to make it much easier to create rule tables, leading to faster rule-table creation and a lower barrier to the creation of more rule tables.

Keywords: Model Assisted Rule Tables(MART), RBAF(Rule Based Assist Feature), MBAF(Model Based Assist Feature), time consuming, process margin

Design-Process-Technology Co-optimization for Manufacturability IX, edited by John L. Sturtevant, Luigi Capodieci,
Proc. of SPIE Vol. 9427, 94270D · © 2015 SPIE · CCC code: 0277-786X/15/$18 · doi: 10.1117/12.2085460

1. INTRODUCTION

As the industry pushes to ever more complex illumination schemes to increase resolution for next generation memory and logic circuits, the process window for ArF lithography continues to decrease. An increasingly common problem is that we now have a set of features, which cannot be resolved by the imaging system through the entire process window (PW). The simplest solution to this problem is to completely avoid such features in the layout. Unfortunately, this requires restrictive design rules, and as RET becomes complex, it may not be possible to describe the restrictions in a set of layout design rules. The alternative is to apply sub-resolution assist features (SRAF) to non-resolving features to modify their optical behavior to enable the feature to resolve across the entire PW. SRAF also offer the advantage of contributing to dense mask environments that exhibit lower sensitivity to focus variations than isolated features.

There are several AF generation options such as Rule Based AF (RBAF) and Model Based AF (MBAF). In case of MBAF method, it delivers the ability to reliably place assist features for enhanced process window control across a wide variety of layout feature configurations. However, this MBAF approach has several demerits. OPC run time using MBAF is greatly increased compared to RBAF. This might be the problem sometimes, especially when mask revisions happen frequently due to design or litho process change, etc, OPC work including AF generation should be performed repeatedly. Thus, this issue caused by OPC run time with MBAF may impact on MTO (Mask Tape Out) schedule. Another possible issue is related to the consistency of MF (Main Feature) and AF (Assist Feature) location. MBAF placement technology has been used widely as resolution enhanced method and it almost reached maturity. However, the consistency issue of layout by MBAF has not been resolved completely yet as shown in Fig. 1 example.

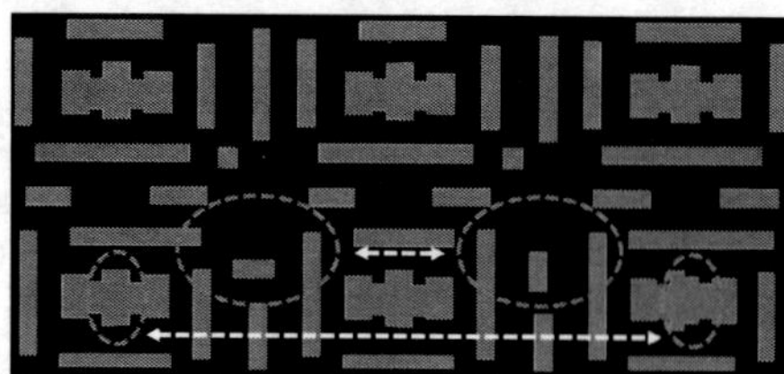

Figure 1. Broken consistency on array pattern out of the result of general MBAF method.

On the contrary, Rule Based Assist Feature (RBAF) provides more consistent AF treatment across repeated main-feature configurations. A particular RBAF advantage is fast run time for generating of AFs, which is with frequent MTOs. RBAF is the baseline AF placement method for many previous technology nodes. Although RBAF algorithm complexity limits its use with very extreme illumination, RBAF is still a powerful option in certain scenarios. However, RBAF method also has several demerits compared to MBAF. In case of RBAF method, rule table generation is time consuming and not flexible to changes in the process nodes and illumination conditions. Moreover, for advanced nodes, the time-consuming effort to extract a rule table is magnified by the increasing complexity of layouts. As the technology node shrinks, however, we have more frequent changes of the wafer process, and more re-designs of the illumination-source shape, escalating the cost of rule-table creation. Therefore, we need faster methods to create the AF rule tables. The traditional method typically involves making masks and printing wafers that contain a Design of Experiments (DOE), varying the main-feature configurations, AF configurations, dose conditions, and defocus conditions.

An alternative to the traditional method is to take advantage of lithography simulation to reduce the number of mask and wafer experiments needed to create AF rule tables. We call this alternative method, "MART," which stands for "Model Assisted Rule Tables." Not only does MART reduce the time and effort required to produce one AF rule table, it also makes it feasible to create multiple rule tables, tailored to specific layout patterns, which can improve the process-window (PW) performance. In case of MART approach, rule tables are created using lithographic models. They are automatically extracted from simulated lithography results. This MART concept allows user to save resources associated with mask making and inspection and to secure more optimized rule tables that are capable of covering the assorted pattern types. In effect, it combines the merits of RBAF and MART to improve on-wafer results. In this paper, we explore the manufacturing applications of MART approaches and compare them to the current Process Of Record (POR), which is RBAF with manual rule-tables generation, and no AFs.

2. MART FLOW

MART uses lithography simulation to reduce the number of mask and wafer experiments needed to produce AF-placement rule tables. An example of MART output is shown in Table 1. As the space, ‘s,’ between a pair of main features (MFs) increases, the total number of AFs between them, ‘Bin Type,’ increases. The parameters for some rows in this AF rule table are illustrated in Fig. 2.

Table 1. Example output from MART. As the space, ‘s,’ between a pair of main features increases, the total number of AFs between them, ‘Bin Type,’ increases. The primary AF is characterized by its width, ‘w1’, and position, ‘d1’. The secondary AF is characterized by its width, ‘w2’, and position, ‘d2’. The MF bias is given by ‘e’, which also denotes how much the AFs should be lengthened.

MART Summary Table (by space)

	Bin Type	AF Width w1	AF Pos. d1	AF Ext. e	AF2 Width w2	AF2 Pos. d2
0<= s <160	0	0	0	0	0	0
160<= s <220	1	23	0	4	0	0
220<= s <260	2	24	55	6	0	0
260<= s <270	2	24	55	5	0	0
270<= s <310	2	30	55	5	0	0
310<= s	3	27	54	4	33	0

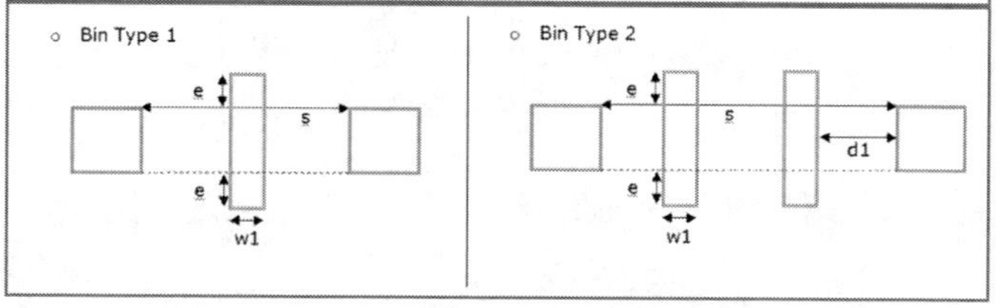

Figure 2: MART parameters for contact-hole MFs. The MFs are in green and the AFs are in blue. The MFs are separated by an edge-to-edge distance of ‘s’. (a) “Bin Type 1” means 1 AF centered between a pair of contacts; it has a width, ‘w1’. This AF’s length is extended by ‘e’ on each end to match the MF OPC. (b) “Bin Type 2” means 2 AFs between a pair of contacts. The distance between AF edge and MF edge is given by ‘d1’.

The data flow for MART is shown in Fig. 3. In Step 1, MF test patterns are generated. That is, for a given, MF CD, the space, ‘s’ between a pair of MFs is systematically varied. In Step 2, AF test patterns are generated. For a given MF pattern, several AF patterns are generated, varying the number of AFs, their widths, and their positions. In Step 3, OPC is performed. This can be as simple as systematically varying the biasing of the MFs, which adds test patterns. Or a conventional OPC tool can be used. In Step 4, the lithography performance of these test patterns is simulated under PW variation, measuring MF CDs and any AF printing. In Step 5, the AF-placement rules are extracted. For each MF configuration, we find the AF configurations that best print the MF of interest. Here, “best” can employ a cost function that includes DOF, NILS, MEEF, and AF-printing margin. After the best AF configurations are found, we find the AF-insertion points. For example, starting with 0 AFs, we find the smallest space that should receive 1 AF, as depicted in Fig. 4. Having found the AF-insertion points, we apply fitting in the AF-parameter space to produce AF-placement rules. An example is shown in Fig. 5 for the AF width.

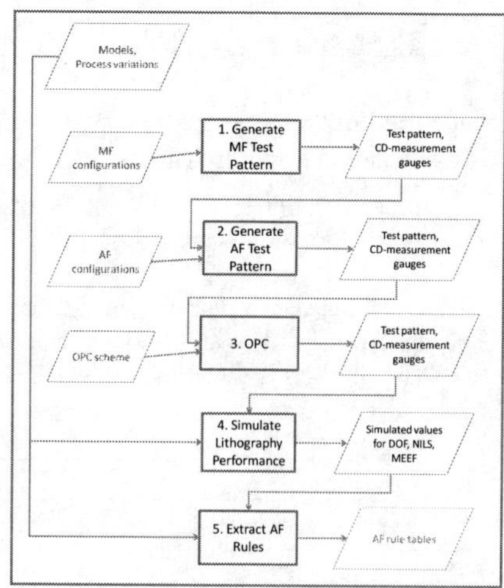

Figure 3: Data flow for MART. Inputs to the flow are shown in red, and outputs from the flow are shown in green. Here, "MF" denotes "Main Feature." MRC limits are an important part of the inputs, and are implicit in the "MF configurations" and "AF configurations."

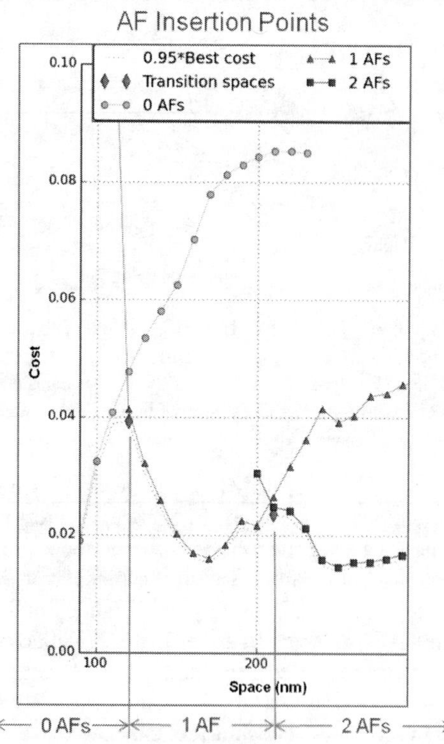

Figure 4: AF-insertion points found by MART are shown by a plot of cost (lower is better) vs. space (between a pair of MFs). The yellow plot is for 0 AFs, the green plot for 1 AF, and the blue plot for 2 AFs. The insertion point for 1 AF is at space 160, where the green plot first dips below the yellow plot. The insertion point for 2 AFs is at space 220, where the blue plot first dips below the green plot.

Choice of AF width
(for 2 AFs)

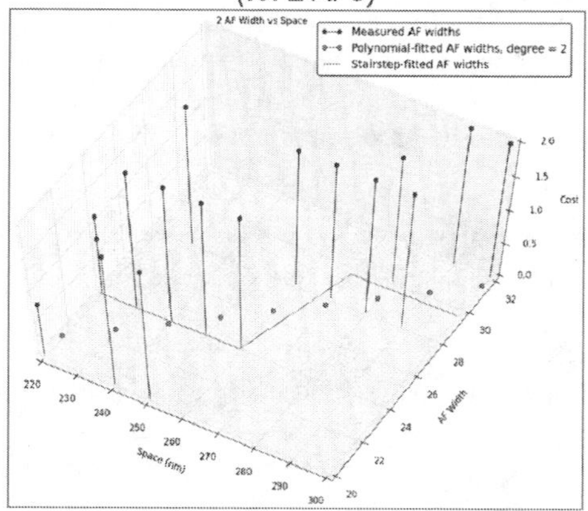

Figure 5: Fitting to find AF-placement rules. For a given space, the AF widths are plotted from the lowest-cost AF configurations. The pink dots show a quadratic fit, and the green lines show a stair-step fit. This stair-step fit prescribes an AF width of 24 for spaces between 220 and 260, and an AF width of about 30 for spaces between 280 and 300.

3. EXPERIMENT OF MART METHOD FOR RULE TABLES

We evaluated the performance of MART and POR on 2X node random contact hole. These flows are listed in Table 2 below. For the extraction of modeling data, we manufactured the test mask. This test mask consists of many kinds of pattern types to cover all random contact density and shape, characteristic, etc, for the experiment. After wafer evaluation with this mask, we generated the model for MART using wafer data from the test mask. Wafer data from the sample AF placed according to the size and distance to main features were also obtained to evaluate rule tables. The target specification for OPC results is no more than 10% CD variation from target, across a focus window of 100nm. Calibrated models for nominal focus, +60nm focus, and -40nm focus were used to verify the post-OPC results. OPC was co-optimized with AF placement to ensure optimal results. Assist features in both POR and MART were Manhattan-only (no 45-deg orientations), and limited to rectangular shapes (no merging).

Table 2. Rule Tables generation flows that were compared.

	FLOW	Method for AF Rules	# of Rule tables	Method for OPC
1	POR	Manual	Single	Auto OPC with AF
2	MART+RBAF	MART	Multiple	Auto OPC with AF

For MART, we created multiple AF-placement rule tables. Specifically, we created one rule table for each of these layout-pattern variations. Altogether, we used 10 rule tables as shown in Table 3.

Table 3. Rule Tables contents for MART.

Rule Table 1	hole type A	short side	.
Rule Table 2	hole type A	long side	.
Rule Table 3	hole type B	short side	.
Rule Table 4	hole type B	long side	.
Rule Table 5	hole type C (array)	short side	pitch 1
Rule Table 6	hole type C (array)	short side	pitch 2
Rule Table 7	hole type C (array)	short side	pitch 3
Rule Table 8	hole type C (array)	long side	pitch 1
Rule Table 9	hole type C (array)	long side	pitch 2
Rule Table 10	hole type C (array)	long side	pitch 3

3.1. MART TAT

For MART, the turn-around time (TAT) for creating the first rule table is typically 1 working day. While the actual computation time is about 1-2 hours, mostly for the lithography simulation, the TAT is dominated by the need to iterate between rule extraction and test-pattern creation. For example, if the final cost-summary plot (Figure 4 is an example) shows that we need more data to find the AF-insertion point for 1 AF. For subsequent rule tables, the TAT is more like ½ of a working day, since the flow is very similar to the first table's.

3.2. AF Printability on Simulation Results

We simulated the AF printability under these conditions, and obtained the results shown in Fig. 6. The mask was dark-field.

- Condition 1: (Nominal Model with 75% Threshold)
- Condition 2: (Nominal Model with 80% Threshold) + (Mask Bias of 1 nm)

While the POR flow shows AF printing in simulation, the MART flow does not show any AF printing. We note that wafer experiments show no AF printing for the POR flow, so our simulation results are conservative.

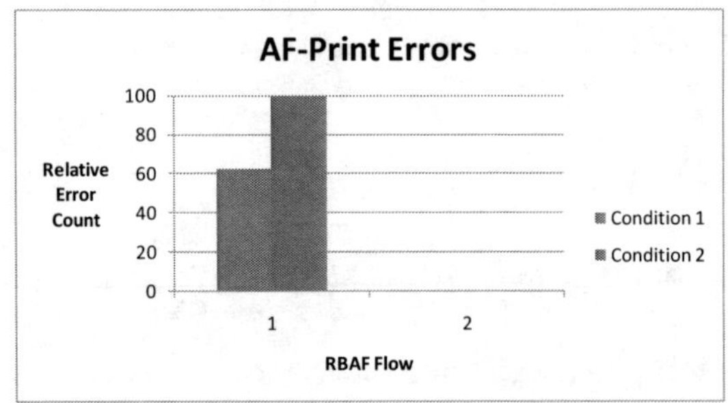

Figure 6: AF-print errors for the RBAF flows listed in Table 2.

3.3. CD Variation on Simulation Results

We simulated the CD variation, and obtained the results shown in Fig.7. A CD error is regarded as any location whose CD variation is greater than 10% from target. There no errors on the short edges of our rectangular contact. On the long edges, the error count decreases by about 10x from the POR flow to the MART flow.

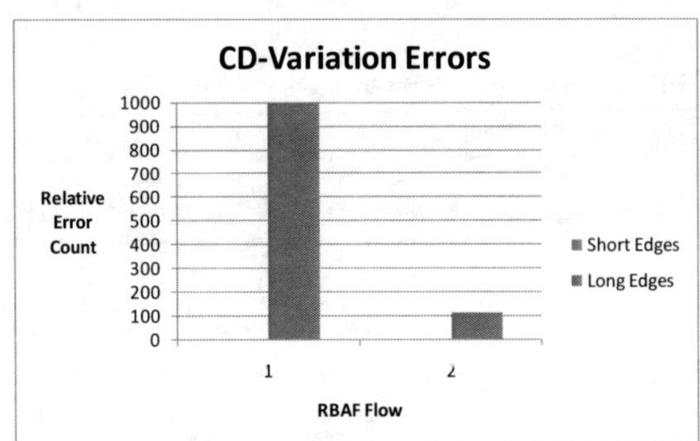

Figure 7: CD-variation errors for the RBAF flows listed in Table 2. A CD error is recorded at any location whose CD variation is greater than 10% from target.

3.4. NILS on Simulation Results

We simulated the NILS, and obtained the results shown in Fig. 8. A NILS error is regarded as any location whose NILS is less than '1.0'. For the short edges of our rectangular contacts, the NILS errors decrease by about 30% from the POR to the MART flow. For the long edges, the error-count change is more dramatic, falling about 8x from the POR to the MART flow.

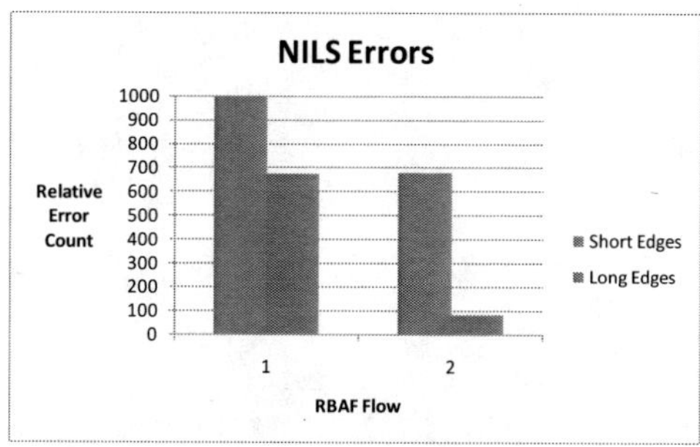

Figure 8: NILS errors for the RBAF flows listed in Table 2. A NILS error is recorded at any location whose NILS is less than '1.0'.

The error distributions for the long-edge errors are shown in Fig. 9, for the POR flow and the MART flow. The median of the distribution is shifted towards the higher NILS values for the MART flow. The low-NILS tail of the distribution is more populous (circled in red) for the POR flow.

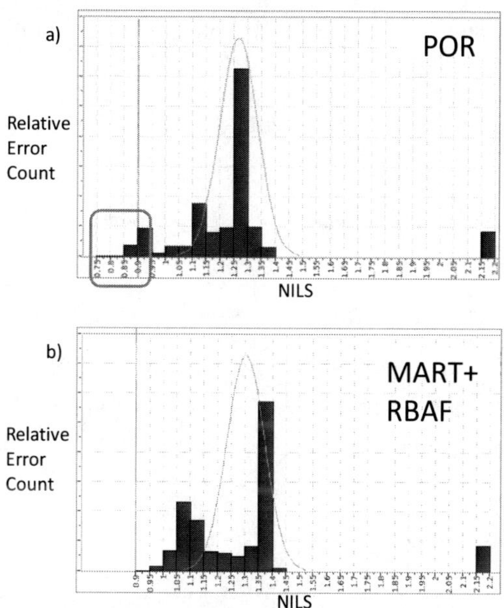

Figure 9: NILS error distributions for the RBAF flows listed in Table 2.

4. WAFER EVALUATION RESULT AND ANALYSIS

We manufactured a photo-mask to evaluate the performance of different rule tables generation methods in terms of process margin and CD-targeting, variation. Fig. 10 shows how this option mask is organized: it has 4 x 4 distributed chips as AF types.

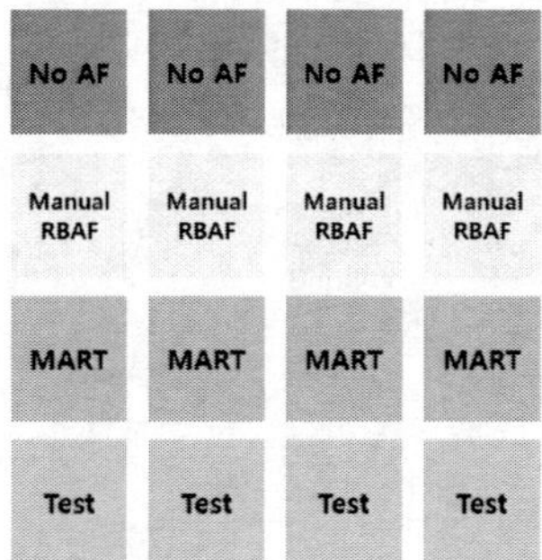

Figure 10. Option mask design for inspection of full chip level

4.1. Brief simulation test preceding the wafer evaluation.

We selected the most vulnerable pattern on process margin to be used for simulation. This was used to predict the process margin for the three types of chip before the fabricated wafer was available. The lithography process for this evaluation has a bright field 6% attenuated phase shifted mask and an optical system with 0.9x NA. The determined illumination condition by source optimizer is an annular source. Wavelength is 193nm, and the source is not polarized. Mask simulation was carried out using a single wave length exposure source. For simulation, a simple stack configuration was used with commercial-volume production ArF resist on top of BARC/Si substrate, with optimum thicknesses for multi-film reflectivity. Figure 11 shows the layout and simulation contours for the weakest pattern from a previous experiment. These contours are at nominal conditions after OPC, and compare AF generation by manual RBAF and by MART.

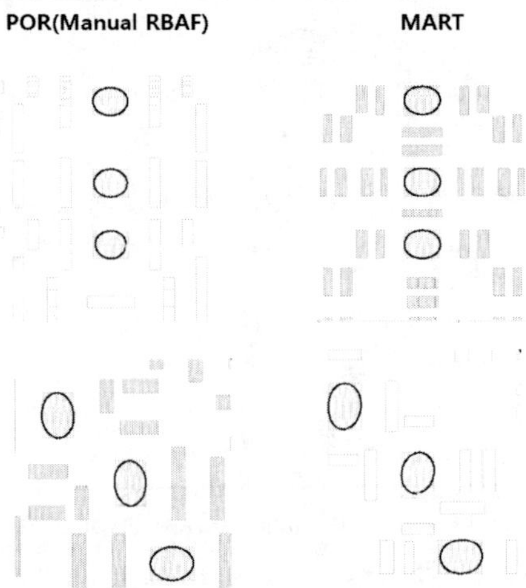

Figure 11. Layout and post-OPC simulation contours after AF generation by POR and MART

It can be seen from Fig. 11 that simulation CD and contours are not significantly different each other at nominal conditions. However, at defocus-offset and dose-offset conditions, there are differences in the number of defects, where a defect is any location falling outside the CD tolerance, as shown in Fig. 12. With MART, the number of defects decreased by more than 50% when compared to POR.

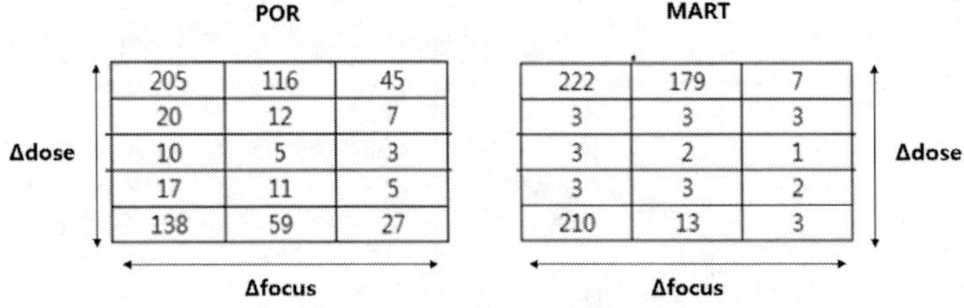

POR (rows: Δdose; columns: Δfocus)

205	116	45
20	12	7
10	5	3
17	11	5
138	59	27

MART (rows: Δdose; columns: Δfocus)

222	179	7
3	3	3
3	2	1
3	3	2
210	13	3

Figure 12. The number of defect (CD tolerance ±10%)

4.2. Wafer verification.

Focus Exposure Matrix (FEM) wafers were used to experimentally verify the process window and CD variation for no AFs, POR and MART. The following conditions were used for these experiments:

* Scanner : ArF /1400E @ASML
* Illumination setting
  - No AF : 0.9xNA, Conventional
  - POR & MART : 0.9xNA, Annular
* Device & Layer : 2X node, Dram memory random contact-hole layer
* MASK : Bright-field, 6% attenuated phase shifted

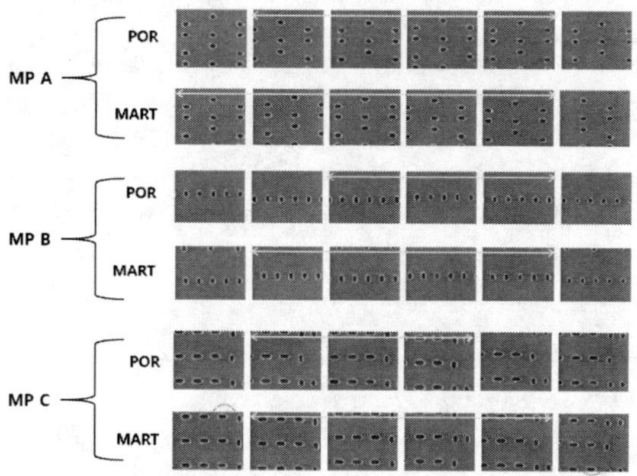

Figure 13. Wafer results for DoF margin on 3 weak points

The FEM was performed in dose steps (column steps) and in focus steps (row steps). Hitachi High-Technologies CD-SEM CG4000 was used for all the measurements. The target CD of the anchor contact hole was around 50nm after additional CD shrink process. The acceptable range of CDs for process-window analysis was within $\pm 10\%$ of target CD. Note that, mask-write variation across the mask does contribute to slight variations of the wafer CD. Although tuning can be done to further improve results, slight improvement of DoF by MART was demonstrated, as shown in Fig. 13. One of the common issues with AFs is the potential for printing sidelobes. The wafer was inspected for printing AFs on the most complex area with aggressive AF by MART, and none were found through-process as shown in Fig. 14 below.

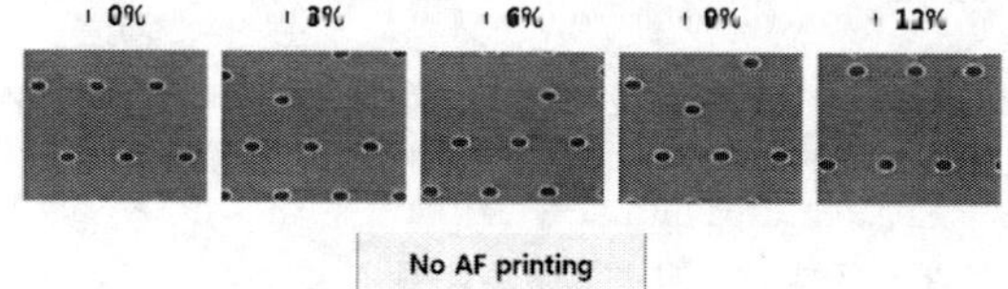

Figure 14. AF printing inspection result

We selected around 9 monitoring points that are the most critical patterns on the chip in terms of CD targeting, and evaluated CD variation according to defocus. Figure 16 indicates wafer-evaluation results for comparing rule-tables generation methodologies using CD process margin. For the POR (manual RBAF) case, although the common PW is not increased, individual process window improvement based on the CD tolerance criteria in defocus conditions can be observed compared to the No-AF case. Still comparing DoF to the No-AF case, the MART case has the most improvement while the POR case has moderate improvement. One of the monitoring points was analyzed with wafer images in detail to realize how the wafer profile changes with defocus conditions. As shown in Fig. 15, it confirmed that the small CD was found in defocus conditions for both the No AF and POR cases but not for the MART case. This demonstrates that AF generation by MART was more optimized than for other cases, and that the MART method for rule tables has good coverage for the various main feature types.

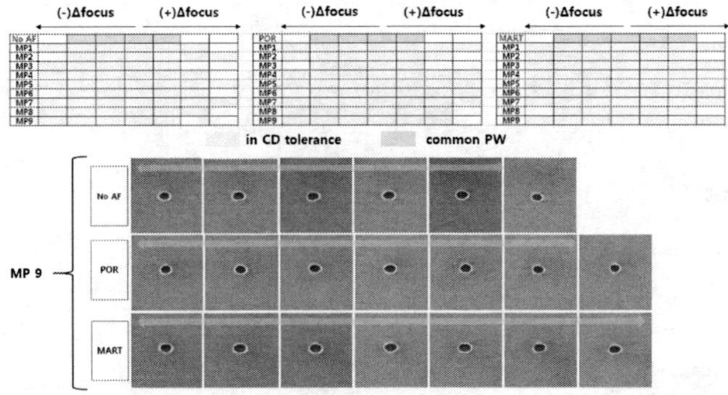

Figure 15. Wafer results for DoF margin on CD and profile.

Figure 16 shows distribution plots at nominal conditions for 53 monitoring points, and compares the POR and MART rule tables generation methods. As analyzed in Fig. 16, the overall CD error level between POR and MART does not show a significant difference. However, for some points, there is an improvement of CD error with MART. Some patterns, such as MP26 and MP30 have a large CD error on POR. This large CD error caused by insufficiently optimized AF locations and sizes. While there are still several patterns with CDs out of tolerance, these can be fine-tuned through additional OPC, AF, and process optimization.

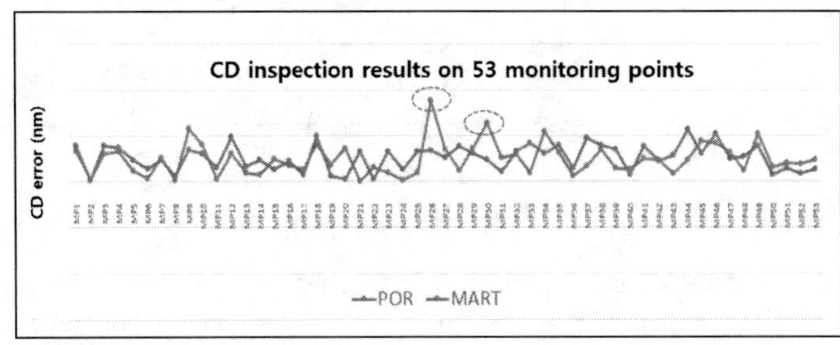

Figure 16. CD error of the random contacts comparison results on 53 monitoring points. MP=Measurement Points.

The results of the process window analysis on wafer conclude that rule tables generated by MART improves the process margin and CD targeting compared to the POR.

5. CONCLUSIONS

RBAF is still used today, even at advanced nodes, because it has fast runtimes, and offers more consistent AF treatment across repeated input-layout geometries. We introduced the MART rule-table generation method and demonstrated its benefits versus no assist features and versus manual rule-table generation. For RBAF with manual rule-table generation, much trial and error is needed to create the rule tables. As an alternative, MART uses lithography simulation to greatly reduce the required number of mask and wafer experiments for creating RBAF rule tables. It provides greater ease of use and a shorter time for creating rule tables. With MART reducing the cost of rule-tables creation, we created multiple rule tables to improve PW performance. We improved the CD variation and pattern profile in defocus conditions with extra rule tables aimed at commonly occurring input-layout patterns.

REFERENCES

[1] Levi D. Barnes, *"Model-based Placement and Optimization of Sub-resolution Assist Features"*, Proc. SPIE 6154, (2006)

[2] Ji Li, *"Process Window and Integration Results for Full-Chip Model-Based Assist-Feature Placement at the 32nm Node And Below"*, Proc. SPIE 7640, (2010).

[3] Jinhyuck Jeon, *"Process Window analysis of algorithmic AF placement options"*, Proc. SPIE 8684, (2013).

Standard cell design in N7: EUV vs. Immersion

Bharani Chava[a], David Rio[b], Yasser Sherazi[a], Darko Trivkovic[a], Werner Gillijns[a], Peter Debacker[a], Praveen Raghavan[a], Ahmad Elsaid[b], Mircea Dusa[b], Abdelkarim Mercha[a], Julien Ryckaert[a], Diederik Verkest[a]

(a). IMEC, Kapeldreef 75, Heverlee 3001, Belgium

(b). ASML US Inc, 4211 Burton Dr. Santa Clara, CA, USA

(c). ASML US Inc, 4211 Burton Dr. Santa Clara, CA, USA

ABSTRACT

While waiting for EUV lithography to become ready for adoption, we need to create designs compatible with both EUV single exposures as well as with 193i multiple splits strategy for technology nodes 7nm and below needed to keep the scaling trend intact. However, the standard approach of designing standard cells in two-dimensional directions is no more valid owing to insufficient resolution of 193-i scanner. Therefore, we propose a standard cell design methodology, which exploits purely one-dimensional interconnect.

Keywords: 193-i, DTCO, standard cell

1. INTRODUCTION

Multiple patterning based on 193-i sustained scaling of LOGIC and SRAM till date with a conventional 2D routing style in M1 and Mx layers. Migrating designs from N10 to N7 with next generation of FinFET technology proves challenging from physical design and process option path finding perspective. A 193-i based design with Litho-Etch-Litho-Etch-Litho-Etch (LELELE) Metal1 (M1) was proposed as a process option for N10 in [1]. The complexity of design rules in 2D design starts to put too many restrictions on LOGIC cells and triple patterning is not feasible for Mx pitches below 45nm due to the constraints on overlay and time-dependent dielectric breakdown (TDDB). For pitches below 45nm, the only patterning options feasible with DUV immersion lithography are self-aligned double patterning (SADP) and quadruple (SAQP) according to the scaling roadmap of IMEC as shown in Figure 1. We expect the SADP cliff at 42nm while SAQP can be scalable to 24nm. Both SADP and SAQP are scalable to these tight pitches given the constraint they force on design to be purely 1D as shown in [2]. Earlier work focused [1] & [2] on the architecture for ≥ 9Track standard cells. This work exclusively deals with complexities of looking at enabling 9 Track and 7.5 Track LOGIC libraries used for low power applications, which require reduced drive strength compared to 12 Track libraries.

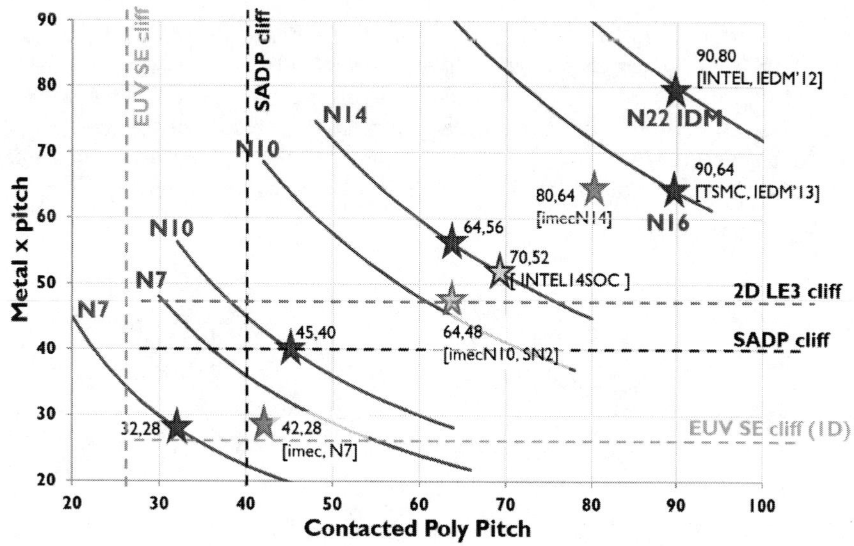

Figure 1.IMEC scaling roadmap

Design-Process-Technology Co-optimization for Manufacturability IX, edited by John L. Sturtevant, Luigi Capodieci, Proc. of SPIE Vol. 9427, 94270E · © 2015 SPIE · CCC code: 0277-786X/15/$18 · doi: 10.1117/12.2085739

The first solution in 1D design is to find an architecture with the same number of Interconnect layers as N10 which means that the N7 standard cells have to be completed in M1 - M2 with each of them being purely 1D. That solution means that M2 is more aggressively used in completing the connections inside the standard cells which creates congestion in place and route (PNR) level. An example of a 9T INVD4 is shown in Figure 2, where 3/7 M2 tracks are used to complete the routing within the cell. This solution could still be feasible for ≥ 9Track standard cell libraries with minor adjustments to Logic cells by addition of dummy gates inside the cells, which have congestion problems at PNR level. However, this assumption breaks as we look at denser standard cell libraries e.g.7.5Track.

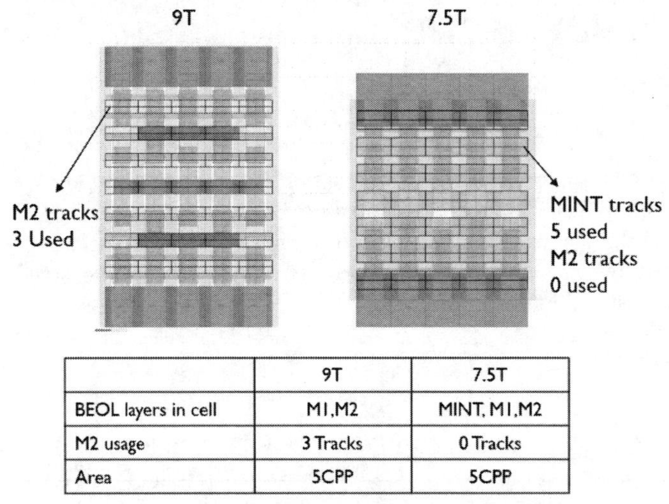

	9T	7.5T
BEOL layers in cell	M1,M2	MINT, M1,M2
M2 usage	3 Tracks	0 Tracks
Area	5CPP	5CPP

Figure 2. N7 INVD4 in 9T and 7.5T libraries

A method to enable 7.5 Track libraries is to introduce additional routing layer underneath M1 to alleviate the burden on M2 and PNR resources as can be seen from Figure 2. We will call this new layer as MINT, because we introduce this layer in-between the Local Interconnect and M1. In the 7.5T INVD4 under consideration, we did not utilize any M2 tracks, which prevent routing congestion. Finally, we try to migrate the fully immersion based architectures to EUV without the need for a complete redesign of the cells. The M1 layer is evaluated to be compatible with both EUV and 193-i process flows and we present some PV band simulations.

2. 1D DESIGN STYLE FOR N7

The first question to be answered in 1D design is the choice of M1 being Horizontal versus Vertical while in the traditional design M1 is purely 2D. Each choice has consequences on the cell design and PNR. The options of M1 horizontal and vertical are shown in Figure 3. For the discussion consider M2 is the level at which PNR is performed. The vertical connections have to be pitch matched to the gate to avoid vertical constructs out of grid after PNR is performed which is an advantage for M1 vertical option as it allows M2 to be at a much tighter pitch compared to M1 horizontal option. The advantage of having M2 at a tighter pitch is having additional routing resources at PNR level.

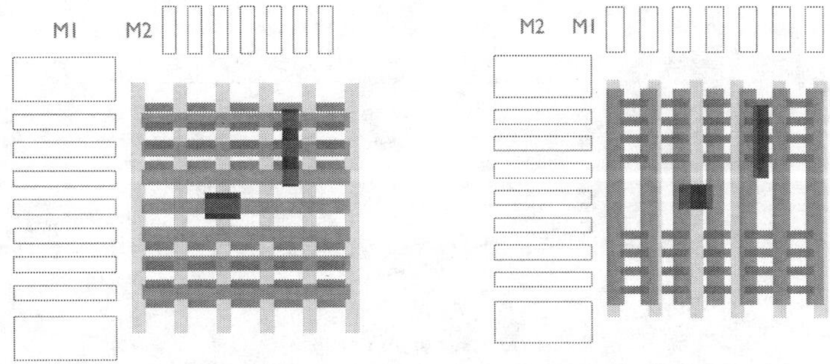

Figure 3. M1 Vertical vs. M1 Horizontal options

In Figure 4, the cross-sections of both options are shown. Although M1 Vertical option has an advantage of higher PNR resources, the M0 connecting gate to M1 becomes complicated and has to be offset from the gate grid to connect to M1. A qualitative comparison of both these process options is presented in Table 1. As one of the goals of standard cell design is to make the cells PNR friendly, our choice is to explore the design style where M1 is vertical and M2 is horizontal.

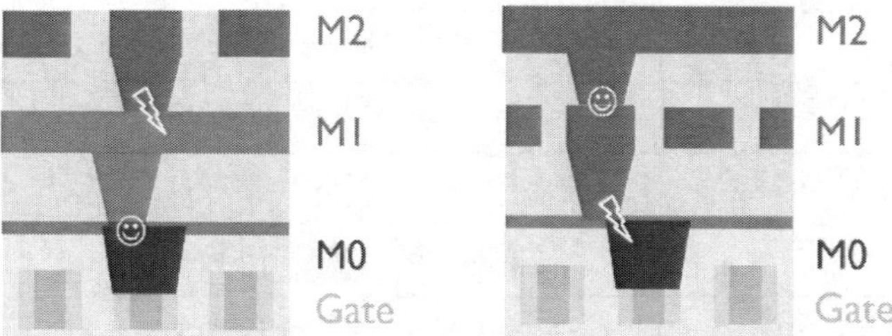

Figure 4. Cross-section of M1 Horizontal (left) vs. M1 Vertical (right)

	M1 Horizontal	M1 Vertical
PNR resources	Poor PNR resources compared to conventional design style	Similar PNR resources compared to conventional design style
M0 complexity	M0 aligned to gate grid to make connection to M1	M0 has to be offset from Gate grid to make connection to M1
M2 pitch	M2 pitch has to be equal to gate pitch to avoid pitch walking	M2 pitch can be chosen independent of gate pitch

Table 1 . Comparison between M1 Horizontal vs. M1 Vertical

2.1 Architectures for ≥9Track libraries

Given the regularity of 1D design, all the cells in a standard cell library can be built based on a combination of a set of 13 basic templates. Only a selection of eight basic templates is shown in Figure 1to explain the concept. The templates are based on M1 vertical option and minimum M1 area requirements for M1 islands. We isolate the M1 islands to enable multiple M1 nets in a single gate pitch. In a 9 Track library, this yielded a maximum of 3 M1 nets within a gate pitch.

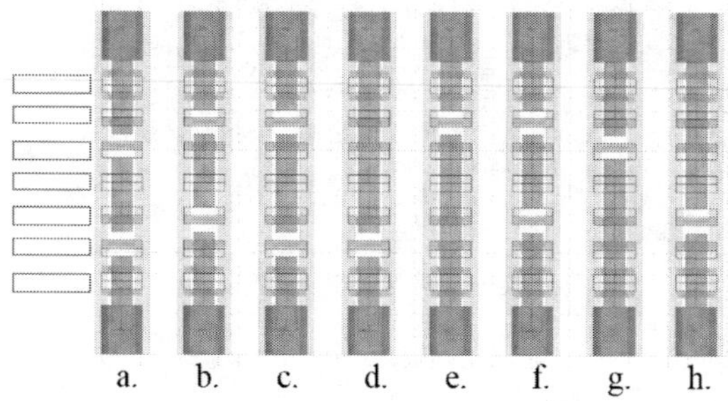

Figure 5. N7 9Track basic templates

The basic design rules, which we follow in this template, are the tip-tip (24nm) and minimum M1 area (1500nm^2). These basic rules are translated into the complete template. The description of the basic templates is given in Table 2.

Template	Definition	Usage	M1 port intersection
a.	P-G-VSS	Pfin, Gate and VSS connections	2,2,1
b.	VDD-G-N	VDD, Gate and Nfin connections	1,2,2
c.	VDD-G-VSS	VDD, Gate and VSS connections	1,3,1
d.	P-VSS	Pfin and VSS connections	5,1
e.	VDD-N	VDD and Nfin connections	1,5
f.	VDD-G-VSS-UP	Longer VSS connections	1,2,2
g.	P-N-UP	Longer Nfin connections	2,4
h.	P-N-DOWN	Longer Pfin connections	4,2

Table 2.

The advantage of having the templates d, e, g & h is to have better intersection of M1 and M2 with longer M1 islands. Basic standard cells can be built from these templates by choosing the appropriate template for every gate pitch. The vertical connections are completed in M1 and the horizontal routing inside the standard cells is performed with the help of M2 as shown in the example cell shown in. From Figure 1, our choice for M2 pitch is 32nm and M1 pitch is 42nm and aligned to the diffusion regions to make vertical connections as discussed in 2. The choice of M2 pitch of 32nm establishes the 9 Track standard cell height to be 288nm, VDD/VSS rail width to be 48nm and a maximum 4 Fins/FET. The gate connections are achieved by offsetting the gate Local interconnects to the adjacent M1 grid. This is highlighted in Figure 6. The VDD/VSS taps to the Fin are achieved by a combination of Local interconnect and M1. The VDD/VSS connections are performed with M1 and not the interconnect layers.

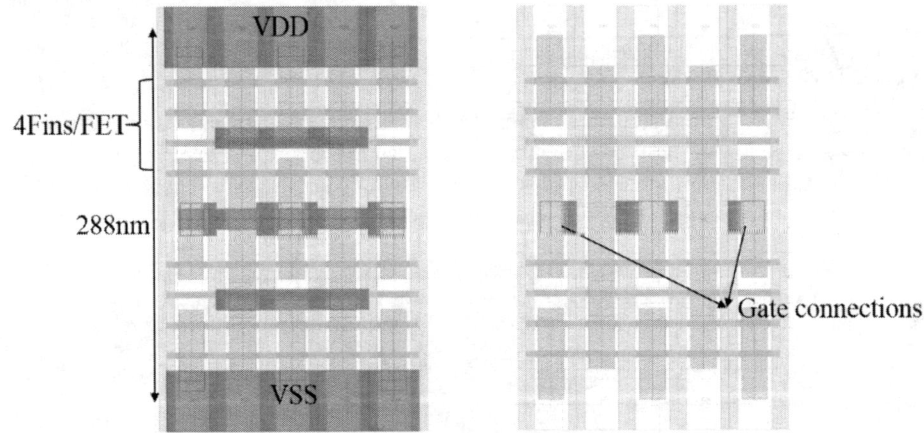

Figure 6. N7 9Track INVD4

2.2 Process options

The process options for N7 $\geq$ 9 Track are closely tied to the tight pitch requirement to sustain area scaling and the design space under consideration. We use the same number of interconnect as N10 for our 9Track libraries, which means that the libraries are completed in M2.

Layer	Pitch requirement	Process option
FIN	24nm	SAQP+LELE CUT
M1	42nm	SADP+LELELE CUT
M2	32nm	SAQP+LELELE CUT
GATE	42nm	SADP+CUT

Table 3. Process assumptions for N7 9Track libraries

3. MINT INTRODUCTION

The 9 Track library is feasible in the same number of interconnect layers as N10. We have developed ~50 cells in varying complexity to check the complexity on patterning and the congestion it imposes on M2. The more complex cells can be designed with the penalty on usage of more M2 tracks. However, to enable a 7.5 Track library, with the same number of interconnect layers is not possible due to the limited number of M2 tracks 5 vs. 7 in the case of 9 Track cells. Although, the simple cells can be designed with the same process options, it is not possible to design more complex cells in the 7.5Track architecture.

3.1 Architectures for $\leq$ 9 Track libraries (7.5 Track library)

As described in Section 2.1, the technique used in building the standard cells is employed again for 7.5 track cell library, where basic templates are designed first. Figure 7, shows the basic templates used in the design of the cells. These templates show the arrangement where in the number of Fins are 3, equivalent basic templates are used for the case of 2-fins. Basic standard cells can be built from these templates by choosing the appropriate template for every gate pitch. In the design of these cells the pitch of MINT is considered same as M2, which is 32nm. The tip-to-tip spacing between the two MINT islands is 18nm; the minimum area is considered 1452nm^2. Furthermore, the choice of M2/MINT pitch of 32nm establishes the 7.5 Track standard cell height to be 240nm, VDD/VSS rail width to be 4CD (64nm) and a maximum 3 Fins/FET. The vertical connections are completed in M1 and the horizontal routing inside the standard cells is performed with the help of MINT as shown in the example cell shown in Figure 8. This is equivalent INVD4 cell in 7.5 Track. Here the cell is free of M2 usage for signal routing within the cell, all the internal routing has been performed with MINT and M1. This allows the automated routers to route the cells easily using Metal layers from M2 and above. However, there are rare cases of very complex gates such as a Full Adder cells where, the designer is compelled to use M2 in internal routing of signal as shown in Figure 9. However, it is worth a mention that the usage of the M2 is minimal and only 2/5 M2 tracks are used to complete the cell while a 9 Track cell without the additional MINT uses 7/7 M2 tracks for internal cell routing. Although, all MINT tracks 5/5 are utilized in the complex cells in a 7.5 Track library, their usage does not have an impact on PNR resources and thus the additional layer is justified.

Layer	Pitch requirement	Process option
FIN	24nm	SAQP+LELE CUT
MINT	32nm	SAQP+LELELE CUT
M1	42nm	SADP+LELELE CUT
M2	32nm	SAQP+LELELE CUT
GATE	42nm	SADP+CUT

Table 4. Process options for N7 7.5 Track libraries

Template	Definition	Usage	M1 port intersection
a.	VDD-SIG-VSS	VDD, SIGNAL, and VSS connections	2,1,2
b.	VDD-SIG	VDD, and SIGNAL connections	2,4
c.	VDD-VSS	VDD, and VSS connections	3,3
d.	SIG-VSS	SIGNAL and VSS connections	4,2
e.	DUMMY	Free of M1	0

Table 5. N7 7.5T templates

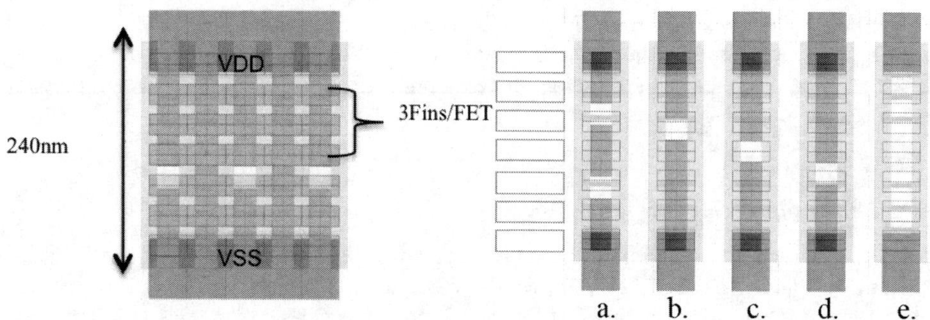

Figure 8. INVD4 N7 7.5 Track library **Figure 7. N7 7.5Track basic templates**

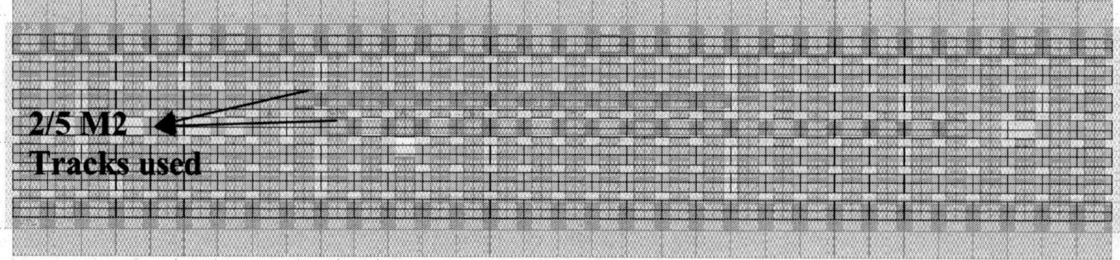

Figure 9. FA1D1 N7 7.5 Track cell. Only 2 M2 tracks are used for cell routing

3.2 Process options

The additional horizontal routing resources required define the process options for N7 $\leq$ 9 Track. The M2 and MINT are pitch matched in our exploration to fully exploit the benefits of a SAQP based design. The process options are shown in Table 4.

3.3 Litho simulations

Because we expect EUV to intercept within the N7 node for mass production, 7.5 tracks M1 architecture has been verified by imaging simulation to assess its feasibility for production in both DUV immersion and EUV. M1 target is presented in Figure 10. M1 is at a pitch of 42nm, 24nm CD target and 24nm Line-End-to-Line-End (LE-to-LE) gaps. We will focus our analysis on this critical gap in this study.

To print this M1 layer using a DUV immersion scanner, we investigated an SADP flow with an LELE CUT layer. CUT layer lithography printability is our concern. Spacer deposition and etch processes are out of the scope of this study. We consider ideal processes. Our simulations assume a LELE compatible wafer stack, a block Litho-etch bias of 10nm per edge, a bright field 6% attenuated phase shift mask and the use of NXT: 1950i immersion DUV scanner. The second process considered is an EUV single exposure with NXE: 3300B scanner. Realistic M1 wafer stack and mask stack were assumed. Lithography simulations are based on Tachyon SMO (Source Mask Optimization) tool. For both block layer and EUV single exposure test cases SMO tool optimized the mask as well as freeform illumination shapes, based on optical models aware of mask topography (M3D+ models). PV Bands were calculated to compare EUV and DUV immersion results. Default PV band settings for DUV in Tachyon SMO are delta focus ±40nm, delta dose ±3%, mask error ±0.5nm. For EUV it is delta focus ±40nm, delta dose ±5%, and mask error ±0.25nm, constant flare of 1.2%.

For DUV immersion test case, M1 cell was decomposed to comply with SADP CUT flow. CUT layer is shown in Figure 10. SMO tool was run on CUT layer, split A and split B. Then contours after etch were built thanks to Boolean operations using Tachyon LMC's Multi-Patterning Technology (MPT) capability, assuming ideal SADP and CUT layer trimming processes.

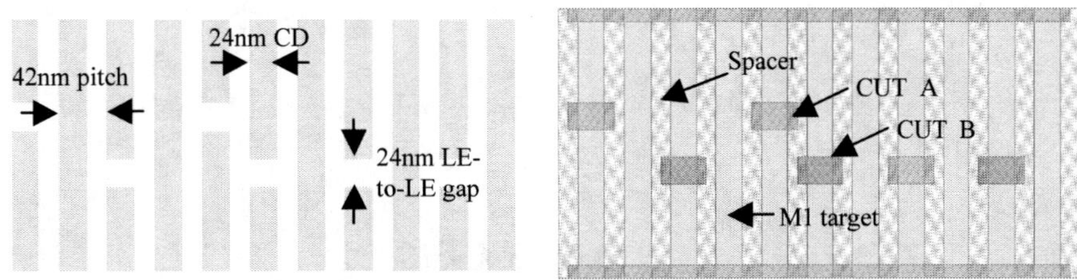

Figure 10. SADP LELE CUT decomposition

Simulation flow is simpler for EUV. M1 target is loaded in SMO tool. Lithography target matches etch target. No further work on lithography contour shaping is required. Figure 11 shows simulated PV bands overlapping with M1 target for both DUV (on the left) and EUV (on the right) test cases. DUV immersion simulation informs us on LE-to-LE gap printability. Variability of vertical edges will depend on SADP process. PV bands up to 4.8nm are found on LE-to-LE gaps. Intra-layer overlay between spacer, CUT A and CUT B is a concern. CUT-spacer overlay error in horizontal direction could create sever peaks in M1 layer; while good metal-contact overlay will depend more on vertical overlay error. EUV simulation computed the entire layer PV bands. Line ends (LE) have PV bands up to 4.6nm while vertical edges PV bands is 2.4nm.

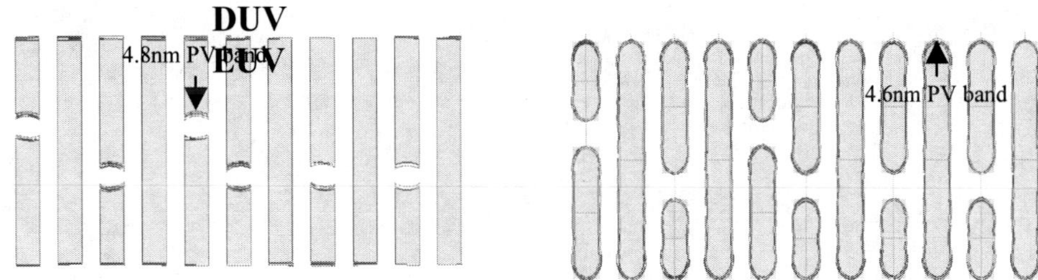

Figure 11. PV band simulations of M1

For this M1 test case, with our simulation settings M1 LE-to-LE gap lithography printability leads to comparable PV bands. But obviously EUV has clear advantages with respect to process complexity. It avoids SADP and LELE process steps and avoids intra-layer overlay.

For this M1 test case, with our simulation settings M1 LE-to-LE gap lithography printability leads to comparable PV bands. But obviously EUV has clear advantages with respect to process complexity. It avoids SADP and LELE process steps and has no intra-layer overlay. This also eases design scalability. EUV simulations show that M1 cell scalability down to 80% is possible, though process variations become significant (see Table 6). Design technology co-optimization will then be required to ensure a reasonable printability.

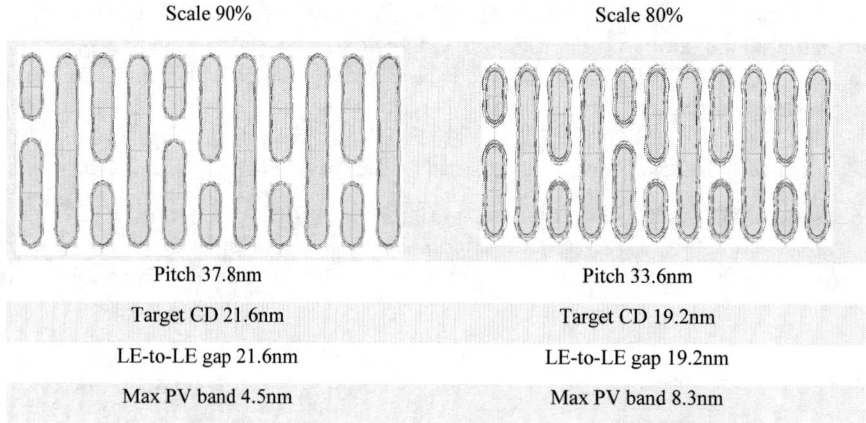

Scale 90%	Scale 80%
Pitch 37.8nm	Pitch 33.6nm
Target CD 21.6nm	Target CD 19.2nm
LE-to-LE gap 21.6nm	LE-to-LE gap 19.2nm
Max PV band 4.5nm	Max PV band 8.3nm

Table 6. EUV PV band simulations after scaling cells

4. CONCLUSIONS

In our study, we have designed ~50 cells in each of the 9Track and 7.5Track flavors. We refrain from making actual area comparison. However, we compare cell size based on number of CPP required to complete the cell. Our conclusions are that we would face penalty in cell area when we try to migrate design from 2D design in N20 to purely 1D design in N7. In our experiment, ~ 26 cells suffer from additional CPP caused by 1D design. The 7.5Track library also suffers from additional CPP in ~16 cells due to the limited routing resources inside the cell compared to a 9 Track library. Some example cells are shown in Table 7.

However, the total cell area will be much smaller due to smaller cell height. Our proposed cell architecture is structured in a fashion that it is firstly fully DUV compatible and can be readily migrated to EUV for critical layers when it becomes available for mass production.

Cells	No. Of Polys (CPP)			Cost of migration in CPP	
	N20	N7 9T	N7 7.5T	N7 9T -> N7 7.5T	N20 -> N7 7.5T
AOI21D1	7	9	10	1	3
AOI22D0	7	10	10	0	3
INVD0	3	3	3	0	0
INVD1	3	3	3	0	0
INVD2	4	4	4	0	0
INVD4	6	6	6	0	0
INVD8	10	10	10	0	0
ND2D0	4	4	4	0	0
ND2D1	4	4	5	1	1
ND3D0	5	6	5	-1	0
ND3XD0	5	6	5	-1	0
ND4D0	6	8	6	-2	0
NR2D0	4	4	4	0	0
NR2D1	4	4	5	1	1
NR3D0	5	6	5	-1	0
NR3D1	5	0	6	6	1
NR4D0	6	8	6	-2	0
XOR2D1	12	14	12	-2	0

Table 7. Table comparing cell areas

5. REFERENCES

[1] B. Vandewalle, B. Chava, S. Sakhare, J. Ryckaert, and M. Dusa, "Design technology co-optimization for a robust 10nm Metal1 solution for logic design and SRAM," *Proc. SPIE*, vol. 9053, p. 90530Q–90530Q–13, 2014.

[2] K. Vaidyanathan, R. Liu, L. Liebmann, K. Lai, A. Strojwas, and L. Pileggi, "Rethinking ASIC design with next generation lithography and process integration," *SPIE Adv. Lithogr.*, vol. 8684, p. 86840C–86840C–15, 2013.

Layout Dependent Effects Analysis on 28nm Process

Helen Li[1], Mealie Zhang[1], Waisum Wong[1], Huiyuan Song[2], Wei Xu[2], Philippe Hurat[2], Hua Ding[2], Yifan Zhang[2], Michel Cote[2], Jason Huang[2], Ya-ch Lai[2]

1 Semiconductor Manufacturing International Corporation, No. 18 Zhangjiang Road, Pudong New Area, Shanghai 201203, People's Republic of China
2 Cadence Design Systems, Inc., 2655 Seely Ave., San Jose, CA 95134, USA

ABSTRACT

Advanced process nodes introduce new variability effects due to increased density, new material, new device structures, and so forth. This creates more and stronger Layout Dependent effects (LDE), especially below 28nm. These effects such as WPE (Well Proximity Effect), PSE (Poly Spacing Effect) change the carrier mobility and threshold voltage and therefore make the device performances, such as Vth and Idsat, extremely layout dependent. In traditional flows, the impact of these changes can only be simulated after the block has been fully laid out, the design is LVS and DRC clean. It's too late in the design cycle and it increases the number of post-layout iteration.

We collaborated to develop a method on an advanced process to embed several LDE sources into a LDE kit. We integrated this LDE kit in custom analog design environment, for LDE analysis at early design stage. These features allow circuit and layout designers to detect the variations caused by LDE, and to fix the weak points caused by LDE. In this paper, we will present this method and how it accelerates design convergence of advanced node custom analog designs by detecting early-on LDE hotspots on partial or fully placed layout, reporting contribution of each LDE component to help identify the root cause of LDE variation, and even providing fixing guidelines on how to modify the layout and to reduce the LDE impact.

Keyword: Layout Dependent Effect, stress, Well Proximity Effect, Length OF Diffusion, Poly Spacing Effect

1. INTRODUCTION

With the development of advanced nodes, the manufacturing process becomes more complex. Layout Dependent Effects (LDE) significance is growing and that is a category of physical phenomena through which layout structure located near. At the 28nm process technology and below, the impact of LDE changes transistor characteristics typically by 20% and up to 80%, such as WPE (Well Proximity Effect), PSE (Poly Spacing Effect), LOD (Length OF Diffusion) and others [1]. These effects are great enough to degrade the design performance.

Design-Process-Technology Co-optimization for Manufacturability IX, edited by John L. Sturtevant, Luigi Capodieci, Proc. of SPIE Vol. 9427, 94270F · © 2015 SPIE · CCC code: 0277-786X/15/$18 · doi: 10.1117/12.2087443

As shown on Figure1 for WPE example, Vth increases if the distance to the well edge is too close due to dopant ions scattering off resist sidewall into active area during well implants [2]. So the delta value adjustment of Vth depends on the transistor channel distance to well mask edge, and it can be mitigated in layout by increasing well enclosure.

Because LDE impact device electrical performance, the effectiveness of doing ideal simulation without layout effects is problematic. We need a solution to consider LDE effects as early as possible during design stage.

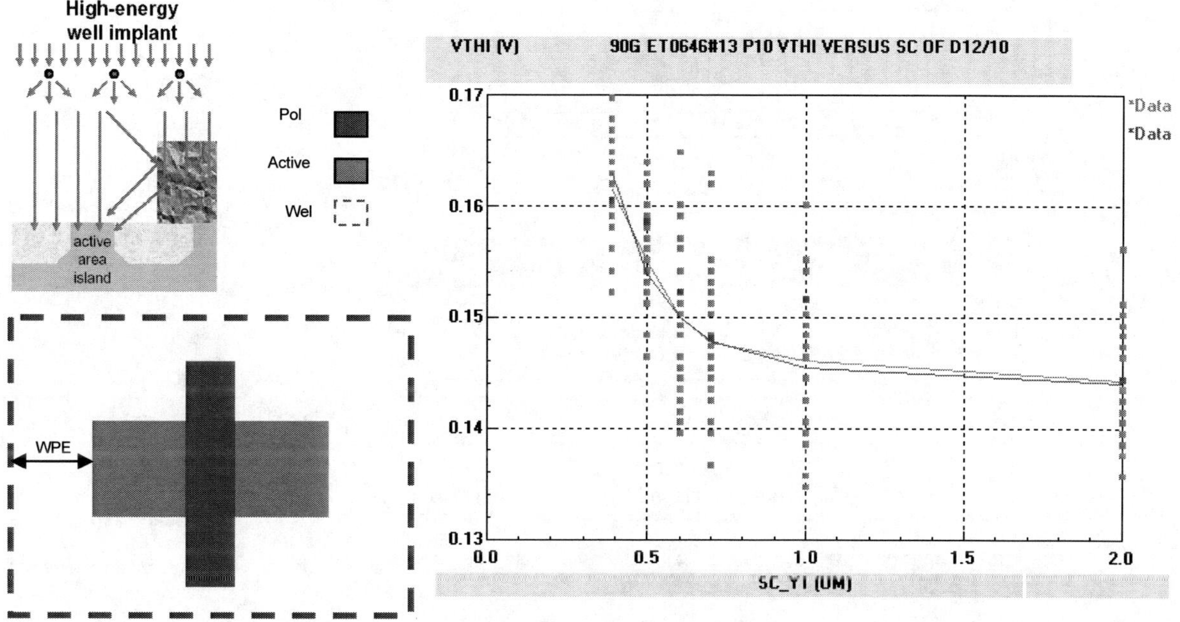

Figure 1: Well Proximity Effect (WPE)

2. CUSTOM VIRTUOSO LDE-KIT AND FLOW

In traditional flow as shown on Figure 2, issues detection is very late in the design stage due to the simulation just can be accessed after the design has been completely laid out and DRC is clean, LVS has been run and a post-layout netlist has been generated.

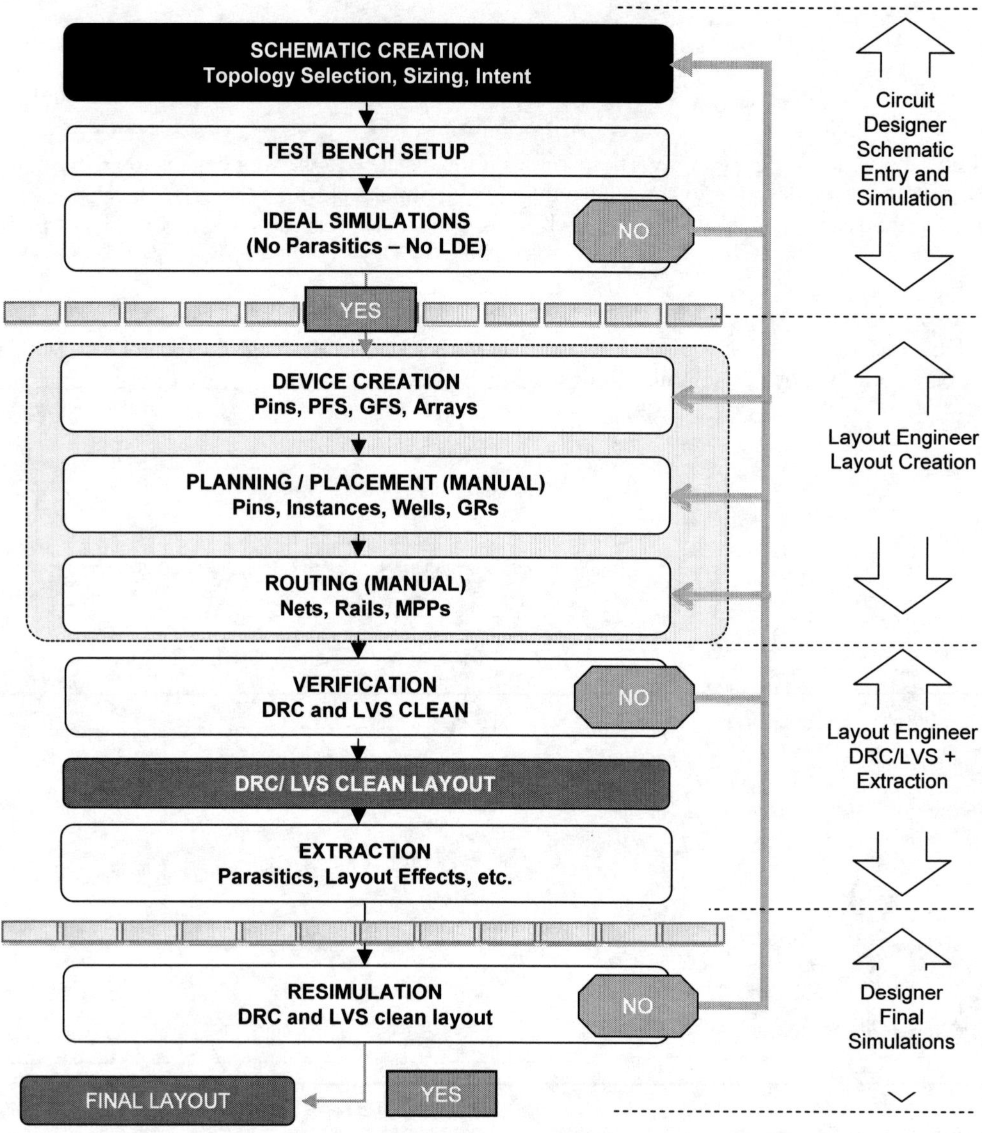

Figure 2: Traditional Design Flow

There're some limitations of the traditional flow at advanced nodes; it's hard to simulate early with accurate layout information to consider LDE impacts. That results in increased TAT and iterations; designers have to wait the full layout design to simulate the performance. And since LDE have a strong impact on performance, post-layout simulations often fail and require layout change and re-simulation, until it converges.

To analyze LDE effects in early design stage, we collaborated to develop a method to embed several LDE sources into a LDE-KIT for advanced process. This kit enables a smooth back annotation flow for even on a partial layout to predict the LDE variations on electrical device characteristics and re-simulate with parasitic and accurate LDE.

In this LDE-KIT, we involve the equations of LDE parameters calculation for several effects, and qualify the results matches with standard spice model. As shown in Figure 3, this flow is working in Virtuoso platform for custom design.

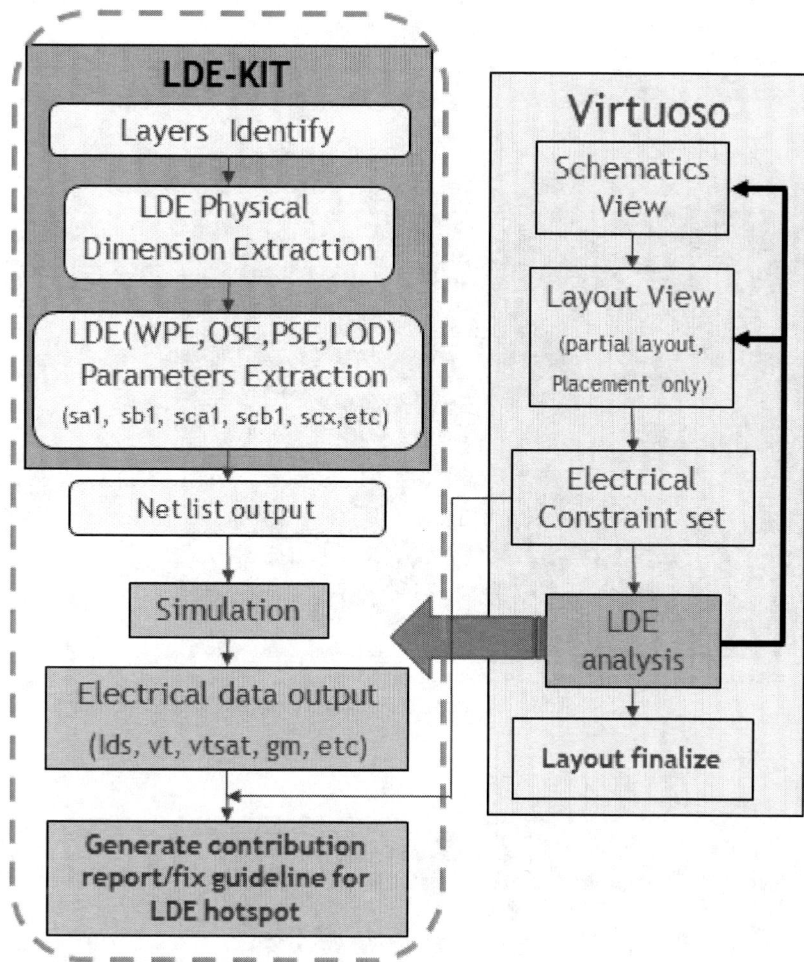

Figure 3: Custom Virtuoso LDE-KIT Flow

This LDE flow is integrated into Virtuoso and enables to the analysis of LDE from a partial layout which need not to be LVS clean. From Virtuoso Constraint Manager GUI, designers can identify the critical devices with electrical matching constraints. Setting electrical matching constraints (such as Vth/Idsat matching on current mirror pair), allows layout designer to evaluate and check the device matching in the design process through rapid and in-layout LDE analysis.

This flow also helps to accelerate the debugging by identifying large variations due to LDE, and also provide feedback to designers on root cause of variations with LDE contributions report and hotspot fixing guidelines.

3. EXPERIMENTS

On an experiment case, we setup this flow to analyze the LDE effects impacting on electrical variations and through the fixing guidelines to optimize the layout for mitigation.

3.1 Create Constraints for critical devices

In circuit design stage, circuit designer should understand which devices are critical and can set a constraint to keep these devices in a certain range. They can set the conditions in the Constraint Manager Window. Designer can select which electrical parameter (e.g. VTh or Idsat) to constrain and what is the tolerance.

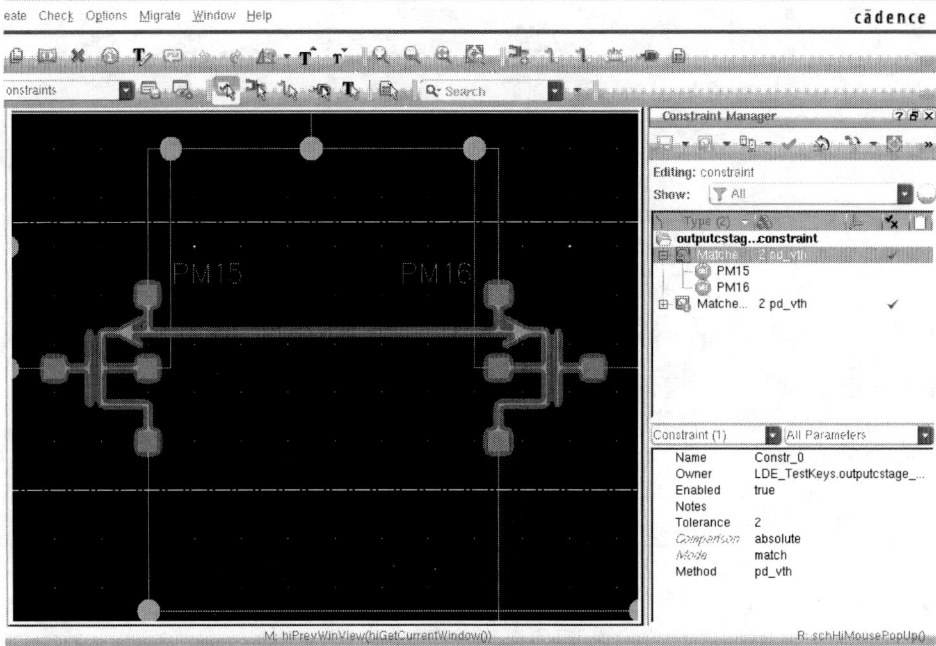

Figure 4: Add constrain to check critical devices matching

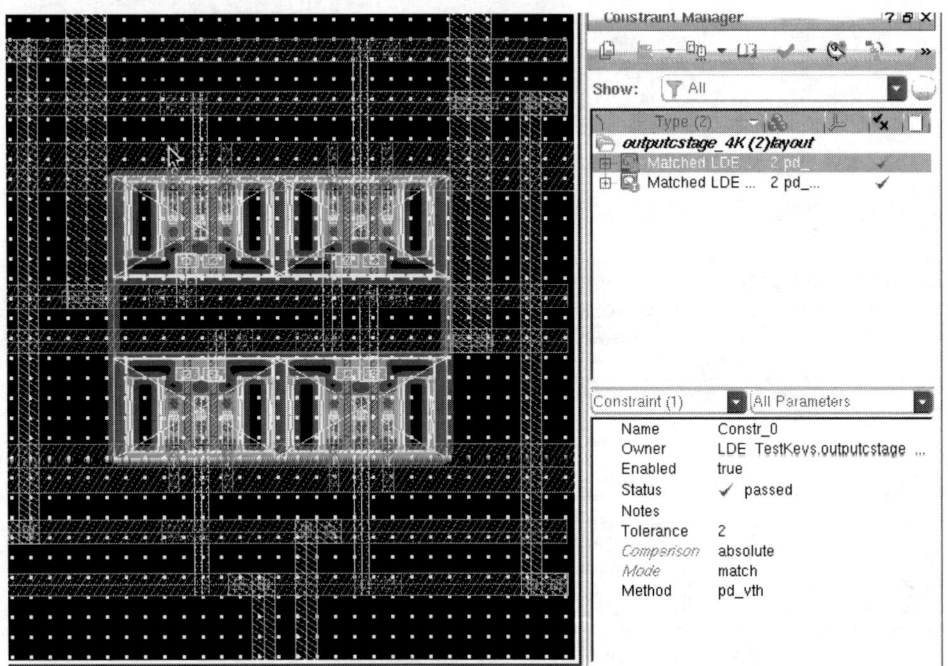

Figure 5: LDE constraint result on layout viewer considering actual LDE effects

On this circuit example, Vth variation between PM15 and PM16 is limited in 2% as a constraint to check. On layout side, by LDE-KIT, the required LDE parameters will be extracted and produce a new netlist considering the accurate location and surrounding environment. Using this new netlist LDE analysis launches automatically some SPECTRE DC simulation to evaluate the electrical parameters are and then checks if the matching constraints is met or if mismatch exceeds the threshold. This check is done automatically from the layout window, and layout designer does not need any circuit or simulation knowledge

With the LDE report, it's possible for designer to know the LDE issues early, even with just a partial placement of some critical devices.

3.2 Fixing guideline to help hotspot fixing

After getting the violation reports, designer must correct the layout to fix the hotspots. This can be quite challenging without deep knowledge of LDE or assistance. In our case, the constraints set on PM15 and PM16 (see Figure 4) are met, but the global threshold of 5% is not meet on several devices.

Refer to Figure 6, we found a total of 32 devices having deviation on Idsat exceeding the 5% threshold. As the example in Figure 6, Idsat deviation on device I8 was more than 5%. The description info shown in Annotation Browser provides a detailed violation report, providing the contribution of each LDE to this deviation. This way the layout designer understands which part of LDE contributes the most to this device electrical deviation. In our case, LOD is the main issue to cause the hotspot.

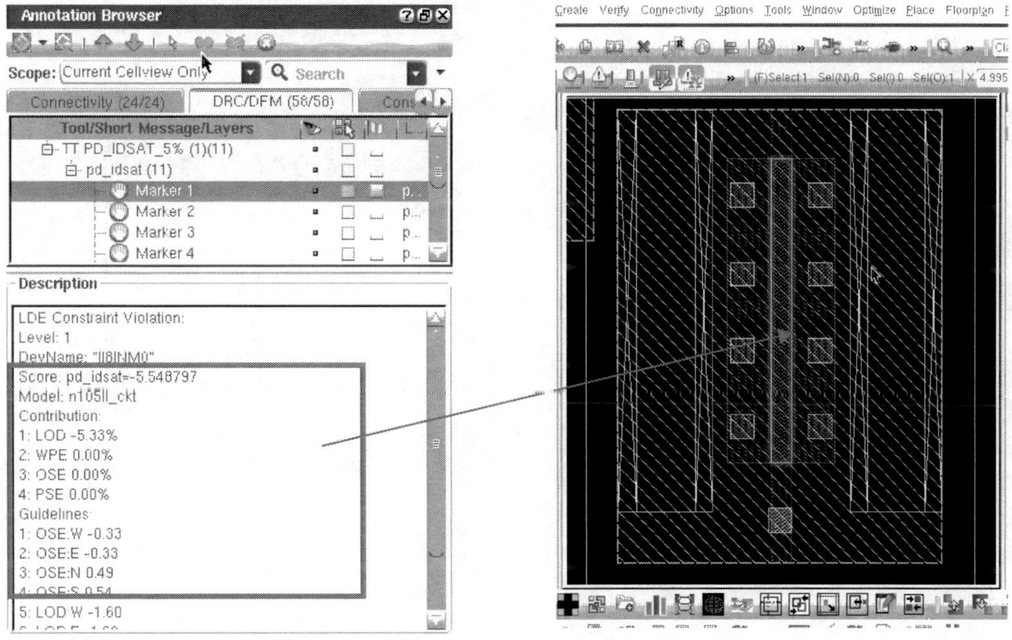

Figure 6: Check LDE hotspot report in Annotation Browser

To fix the problem, the layout designer can leverage the fixing guideline information generated during LDE checking. This info is displayed in the Annotation Browser and guides the designer on how to optimize the layout. Several fixing hints are listed and designers can choose the one that fits best their design constraints..

In our example, the main contributor for deviation of device I8 is LOD effect, so we followed Hin5 and Hint6. These 2 hints recommend that diffusion (AA) layer should be extending to blue bar location.

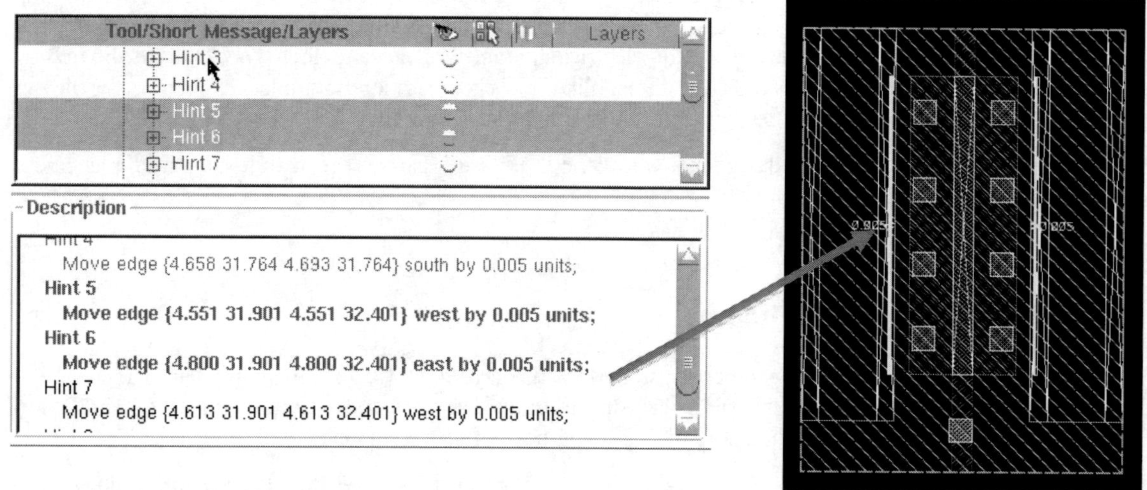

Figure 7: Fixing Guidelines for layout changing

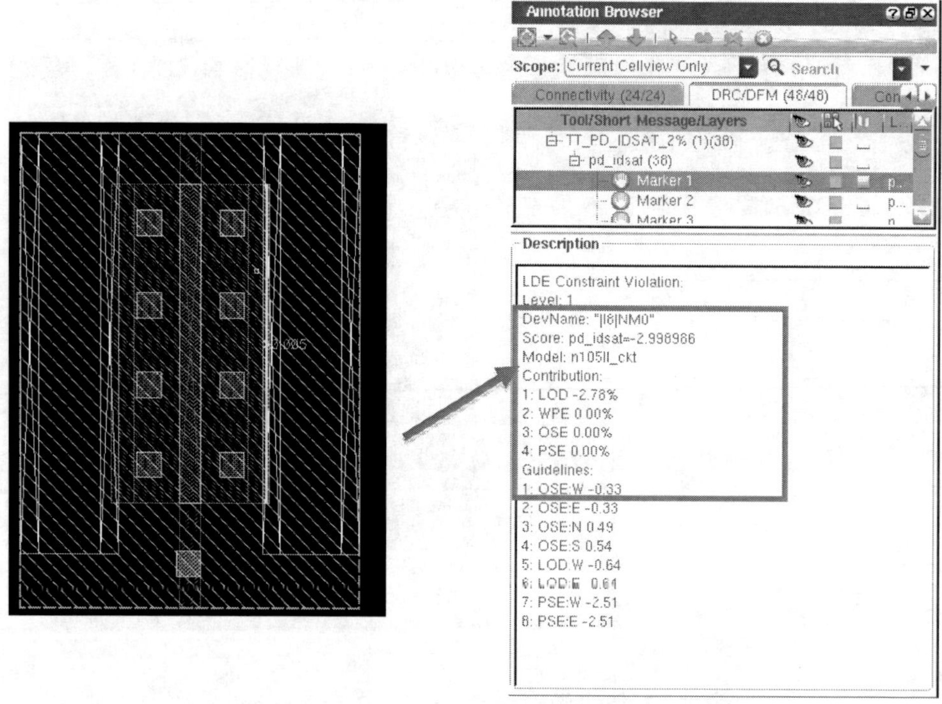

Figure 8: Modify layout on I8 to mitigate LOD effect

The Figure 8 shows the layout after the implementation of the fixing guidelines, and the subsequent LDE checking can be re-run to confirm if these hints reduced the electrical deviation due to LDE on device I8. Rerunning the LDE analysis with a global threshold to 2% shows that the Idsat deviation improved from previous 5% to current 3%.

4. CONCLUSION

In addition to post-layout simulation performed on fully laid-out design which is DRC and LVS clean, designers can detect LDE effects as early as possible during design stage, especially on 28nm and below nodes. The LDE-KIT flow developed in collaboration between us and Cadence, allows circuit designers to include LDE impact in simulation even based on a partial layout, and enables to specify electrical matching constraints on critical devices. It also empowers layout designers to check these matching constraints, and fine-grain analysis of the impact of LDE o their layout without circuit or simulation knowledge. Overall it reduces post-layout simulation iteration and accelerates convergence.

REFERENCES

[1] P. Hurat, "Layout-Dependent-Effect (LDE) Variability Design," ARM Technology Conference, ATC119, (2010).

[2] Y.-M. Sheu et al., "Modeling well edge proximity effect on highly- scaled MOSFETs," in Proc. IEEE Custom Integr. Circuits Conf.,, pp. 831–834, (2005).

Breaking Through 1-D Layout Limitations and Regaining 2-D Design Freedom Part I: 2-D Layout Decomposition and Stitching Techniques for Hybrid Optical and Self-aligned Multiple Patterning

Hongyi Liu, Jun Zhou and Yijian Chen*

School of Electronic and Computer Engineering
Shenzhen Graduate School, Peking University
Shenzhen 518055, Guangdong, China

ABSTRACT

To break through 1-D IC layout limitations, we develop computationally efficient 2-D layout decomposition and stitching techniques which combine the optical and self-aligned multiple patterning (SAMP) processes. A polynomial time algorithm is developed to decompose the target layout into two components, each containing one or multiple sets of unidirectional features that can be formed by a SAMP+cut/block process. With no need of connecting vias, the final 2-D features are formed by directly stitching two components together. This novel patterning scheme is considered as a hybrid approach as the SAMP processes offer the capability of density scaling while the stitching process creates 2-D design freedom as well as the multiple-CD/pitch capability. Its technical advantages include significant reduction of via steps and avoiding the interdigitating types of multiple patterning (for density multiplication) to improve the processing yield. The developed decomposition and synthesis algorithms are tested using 2-D layouts from NCSU open cell library. Statistical and computational characteristics of these public layout data are investigated and discussed.

Keywords: self-aligned multiple patterning (SAMP), positive-tone self-aligned sextuple patterning (pSASP), hybrid SAMP, stitching, layout decomposition and synthesis.

1. INTRODUCTION

As 193nm immersion lithography has reached its resolution limit and EUV lithography is slowly approaching its high-volume insertion point [1], alternative patterning techniques such as optical and self-aligned multiple patterning (SAMP) have been widely adopted, causing a paradigm shift of layout design methodology [2-14]. The conventional layout design is based on random 2-D pattern geometry. It enjoys extensive design flexibility due to its much less geometric limitations and scalability of optical multiple patterning beyond the resolution limit. However, optical triple/quadruple patterning faces tremendous process-window challenges and is unlikely to be scalable to sub-15nm half pitch [15]. Consequently, IC design methodology is forced to shift toward the complementary type of design scheme in which regular and dense 1-D features are cut/blocked by multiple processing steps [16-17]. For example, in 1-D gridded design, the metal layers are decomposed into two 1-D components physically located at different levels (in two different directions) while vias are used to connect them, which results in significant edge-placement yield loss [18-23]. In addition, shrinking random via/cut CD and pitch has become increasingly difficult due to its extremely small process window. Even considering EUV and DSA technologies, scaling via/cut CD and half pitch to sub-10nm still seems to be prohibitive. It remains unclear whether we finally have to adopt parallel e-beam direct writing to pattern sub-10nm vias/cuts [18]. If so, the intense use of cut/via processes must be reduced to a level such that the low throughput of e-beam direct writing can meet the high-volume manufacturing requirement. In other words, the brute-force 1-D gridded design heavily relying on cut/via patterning may not be sustainable, and our industry needs to research other scalable patterning solutions that maximize the 2-D design freedom and minimize the cut/via steps.

In this paper, we propose a hybrid optical and self-aligned multiple patterning technique (hereafter called "hybrid SAMP"), and develop computationally efficient 2-D layout decomposition and stitching algorithms. Similar to double patterning in which one critical layer is decomposed into two masks [24], the hybrid SAMP technique applies two separate SAMP processes to define the highest-density features, while a stitching process is used to form the desired 2-

*Phone: (+86) 755-21536236, Email: chenyj@pkusz.edu.cn

Design-Process-Technology Co-optimization for Manufacturability IX, edited by John L. Sturtevant, Luigi Capodieci,
Proc. of SPIE Vol. 9427, 94270G · © 2015 SPIE · CCC code: 0277-786X/15/$18 · doi: 10.1117/12.2085738

D patterns. This patterning scheme is considered as a hybrid approach as the SAMP processes offer the functionality of density scaling while the stitching process creates 2-D design freedom as well as the multiple-CD/pitch capability. It should be reminded that frequency multiplication is achieved within each single SAMP process, instead of by the interdigitating step of a multiple-patterning process, just to avoid the severe overlay challenge. Each SAMP process needs two or multiple masks: one mandrel mask and one/multiple cut/block mask(s). The concept of stitching two sets of SAMP features for 2-D design freedom is illustrated in Fig. 1 using a positive-tone (spacer is line) self-aligned quadruple patterning (SAQP) process flow. Apparently, such a stitching process does not require highly challenging via steps that can significantly degrade the patterning yield due to the edge-placement inaccuracy and small process window.

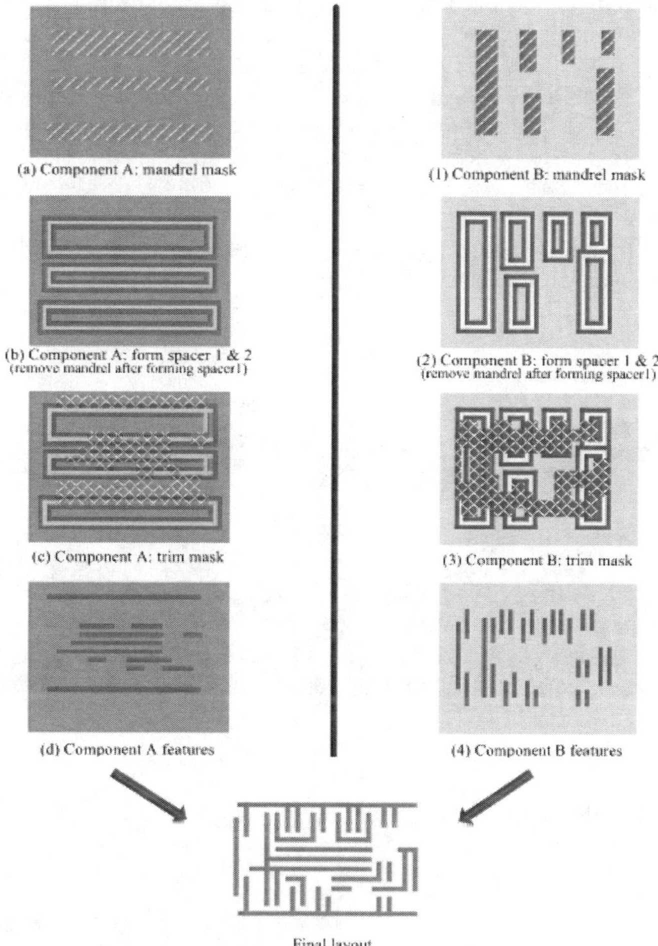

Figure 1. A schematic demonstration of using a stitching process to generate 2-D patterns from two 1-D features separately fabricated by SAQP+cut processes.

The focus of this paper is on post-SAQP patterning applications for which we naturally turn to the self-aligned sextuple patterning (SASP) scheme [7, 11, 13]. The process tone we consider is positive (pSASP, spacer is line) as the pSASP process will most likely be used for front-end layers first. However, the hybrid SAMP concept is not limited to only front-end applications and it is also applicable to back-end patterning. We develop a polynomial-time decomposition algorithm to decompose 2-D layouts directly scaled from 15-nm cell layouts designed by NCSU [25]. To accommodate more CD flexibility in a pSASP process, pattern overlapping and spacer merging techniques are proposed. Moreover, we discuss the decomposition strategy of a full-chip layout with varying pattern densities, in which double/multiple patterning can be incorporated into two/multiple cut masks to generate certain lower-density (but still sub-resolution) patterns.

2. ACCOMMODATION OF CD FLEXIBILITY

A common limitation of the positive-tone (spacer is line) SAMP processes is that they only allow limited CD ranges. For example, in a positive-tone self-aligned sextuple patterning process, there are three consecutive spacer steps that enable spatial frequency sextupling. The line features are defined by the structural spacers (e.g., spacer 1 or S1, and spacer 3 or S3) while the space features are defined by the sacrificial spacer (e.g., spacer 2) [7]. Although a few different spacer widths are possible, their arrangement follows a regular spatial distribution and may not be useful to meet the random CD requirement. In this paper, we assume that all spacers have the same width W_{spacer} , while propose two processing schemes (pattern overlapping and spacer merging) to accommodate more CD flexibility. This assumption is only for the purpose of algorithm simplicity and does not preclude the possibility that the S1 width can be different from the S3 width.

In the pattern-overlapping step for CD tuning, two features patterned by different SAMP processes overlap vertically, and a larger feature can be formed through a following pattern transfer/etch process. In a pSASP process as shown in Figure 2, there are four pattern-overlapping schemes: S1-S3 (i.e., S1-on-S3), S3-S3, S1-S1 and S3-S1. In S1-S3 and S3-S3 overlapping schemes, the space between two S3 spacers (on the first layer) is variable, which thus allows a line to be assigned into this space to overlap S3 to form a larger feature. While in S1-S1 and S3-S1overlapping schemes, the space between spacer 1 and spacer 3 (on the first layer) is the width of sacrificial spacer 2. Therefore, if a line needs to overlap S1, some neighboring features of S1 should be trimmed away to provide enough space for pattern overlapping.

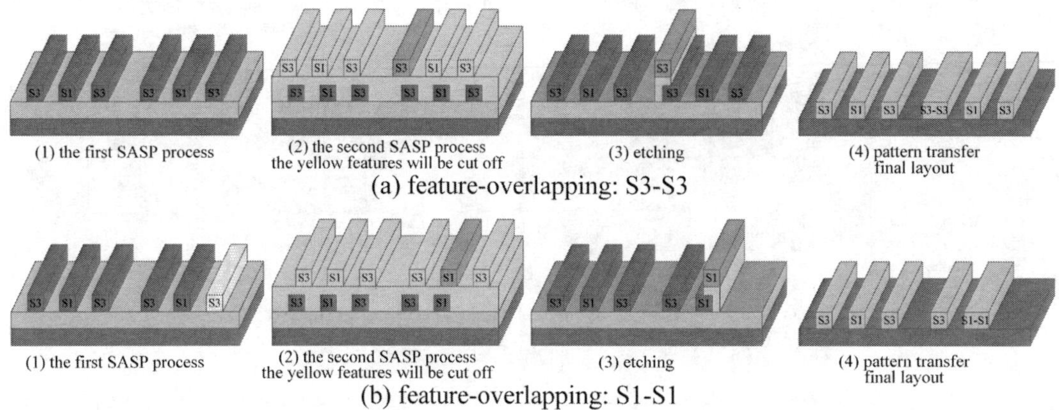

Figure 2. A conceptual demonstration of pattern overlapping to accommodate multi-CD flexibility.

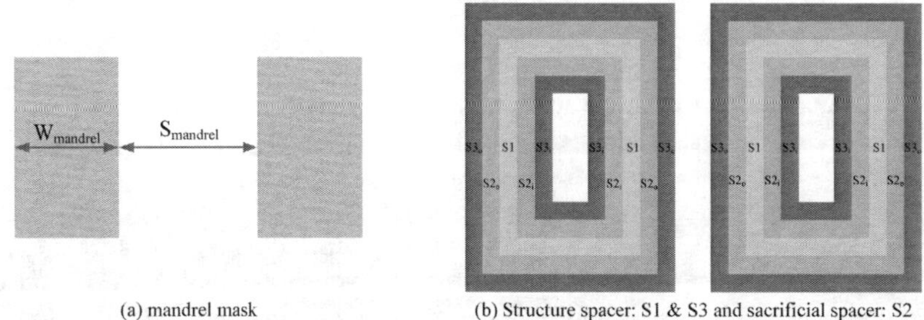

Figure 3. Feature arrangement of the positive-tone self-aligned sextuple patterning.

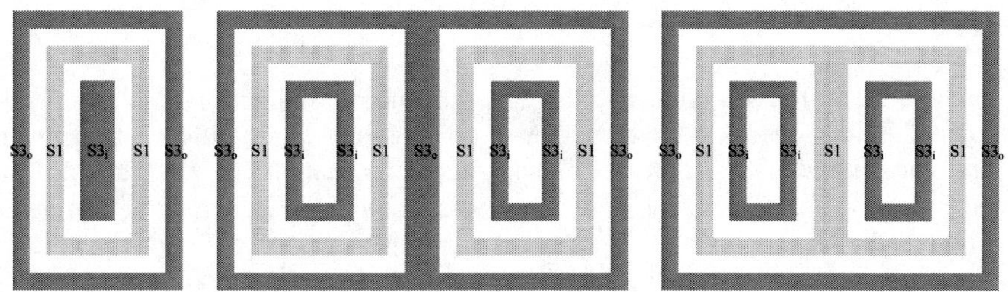

Figure 4. Some examples of structural spacer merging to form larger-CD features.

Besides the above method to form larger features by pattern overlapping, spacer merging is another possible way to form multi-CD features, provided that the desired CD is less than $2W_{spacer}$. By elaborately designing the mandrel width $W_{mandrel}$ and space $S_{mandrel}$, as illustrated in Figure 3(a), specific sidewall spacers can be merged together to form multi-CD features. Figure 4 shows the examples of structural spacer merging. As illustrated in Figure 3(b) and Table 1, we call the spacers inside the loop of spacer 1 as "inside spacers", which are marked by the subscript "$_i$" (e.g., $S3_i$ and $S2_i$); and the spacers outside the loop of spacer 1 as "outside spacers", which are marked by the subscript "$_o$" (e.g., $S3_o$ and $S2_o$). The regulations of $W_{mandrel}$ & $S_{mandrel}$ and the related processing windows in a pSASP process are shown in Table 2.

Table 1. Process symbols and their definitions.

Notation	Definition
$W_{mandrel}$	Mandrel width, the width of mandrel mask features
$S_{mandrel}$	Mandrel space, the space between two adjacent mandrel features
S1	Spacer 1
S2	Spacer 2
S3	Spacer 3
$S3_i$	The inside spacer 3 which is inside the loop of spacer 1
$S2_i$	The inside spacer 2 which is inside the loop of spacer 1
$S2_o$	The outside spacer 2 which is outside the loop of spacer 1
$S3_o$	The outside spacer 3 which is outside the loop of spacer 1

Table 2. The rules of $W_{mandrel}$& $S_{mandrel}$ during a spacer forming process.

$W_{mandrel}$	$S3_i$	$S2_i$	$S_{mandrel}$	S1	$S2_o$	$S3_o$
$W_{mandrel} \leq 2W_{spacer}$	N/A	merge	$S_{mandrel} \leq 2W_{spacer}$	merge	N/A	N/A
$2W_{spacer} < W_{mandrel} \leq 4W_{spacer}$	merge	normal	$2W_{spacer} < S_{mandrel} \leq 4W_{spacer}$	normal	merge	N/A
$W_{mandrel} > 4W_{spacer}$	normal	normal	$4W_{spacer} < S_{mandrel} \leq 6W_{spacer}$	normal	normal	merge
			$S_{mandrel} > 6W_{spacer}$	normal	normal	normal

Note 1: "N/A=not existing": there is no space to form the specific spacers.
Note 2: "merge": the spacers merge together.
Note 3: "normal": there is enough space to form separate spacers.

3. THE HYBRID SAMP DECOMPOSITION ALGORITHM

Given a 2-D target layout represented by a set of polygonal features, the goal of layout decomposition is to generate two sets of mandrel and cut/block masks. Namely, we need to decompose the target layout into two components, each of them containing one or multiple sets of unidirectional features that can be patterned by one single SAMP+cut/block process. Next, we shall demonstrate the development of a hybrid layout decomposition algorithm for the positive-tone self-aligned sextuple patterning.

3.1 Decomposition of the target layout into two components

The decomposition process starts from decomposing the target layout into multiple sets of unidirectional features. One simple treatment is to cut the 2-D patterns at the corners and separate them along different directions (e.g., X and Y) as shown in Figure 5. The features with X direction are collected into component A (Figure 5(c)), while the features with Y direction are collected into component B (Figure 5(d)). However, it should be noted that the features in each component are not limited to one single direction and may contain patterns in other directions. For instance, it may contain oblique patterns that usually account for a small portion of the layout area (e.g., in 45nm and 15nm cell layouts designed by NCSU [25, 26]) to reduce the masks and processing costs.

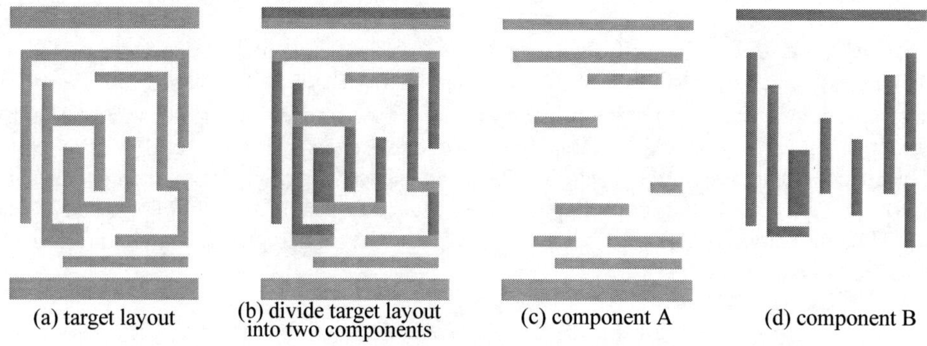

(a) target layout (b) divide target layout into two components (c) component A (d) component B

Figure 5. Decomposition of the target layout into two components.

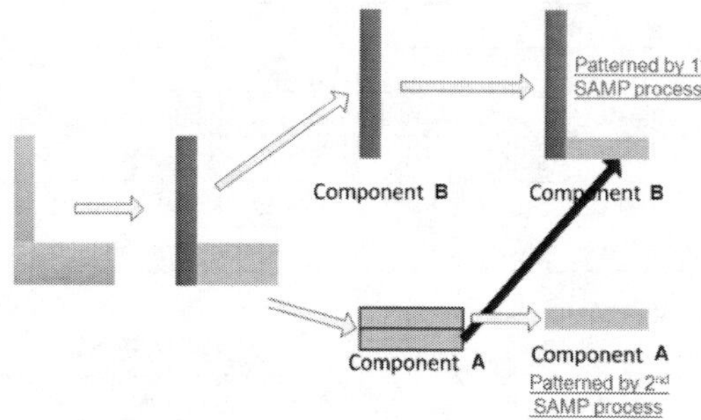

Figure 6. A cartoon to explain why 2-D features (e.g., "L" shapes) can eventually be present in each component.

The target layout usually contains large features with their CDs larger than W_{spacer}, and special attention needs to be devoted to them. For CD ranges between W_{spacer} and $2W_{spacer}$, the patterns will be decomposed into two line features (with CD equal to W_{spacer}) and recombined later by pattern overlapping. As this technique will decompose the large features into two parts (for pattern overlapping), X direction features for tuning CDs for component A need to be collected into component B which contains Y-direction features. As a result, this CD tuning capability will require 2D features (e.g., "L" shapes) to be eventually present in each decomposed component, as shown by Figure 6. On the other hand, CD tuning can also be achieved by spacer merging (see Figure 4). However, CD variations due to the overlay inaccuracy in pattern overlapping are more significant than in spacer merging.

3.2 Preliminary layout treatment

In sub-section 3.1, we show that each decomposed component mainly consists of unidirectional features in multiple directions. The resultant 2-D features (e.g., "L" shapes) should be properly treated before coloring. Taking the decomposition of component B as an example, the L-shape features can be further decomposed into two unidirectional parts (e.g., X direction and Y direction) (Figure 7(b)). In the coloring step, only Y-direction parts will be considered, while the X-direction parts can be ignored (Figure 7(c)). Apparently, the X- and Y-direction parts of any L-shape feature must have the same color (e.g., S1 or S3), and we only need to color one part in certain direction (for the purpose of developing simple coloring rules). Therefore, after layout pretreatment, only unidirectional features in each component need to be colored.

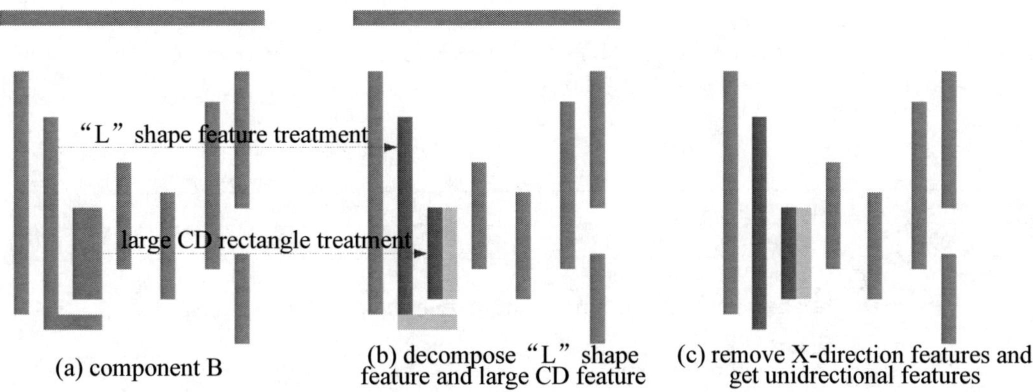

Figure 7. Preliminary layout treatment to obtain unidirectional features in each component.

3.3 Group the unidirectional features

In positive-tone SASP processes, the space between spacer S1 and spacer S3 is defined by spacer S2 (whose width is fixed), while the space between two S3 spacers are variable. The basic line arrangement in a pSASP process is S3||S1||S3 (|| means two lines are separated by the distance of a spacer CD [13]). Therefore, we can define group as: in a group, any space between two adjacent features is equal to the width of spacer 2 and the line number is not larger than three. Choosing such a maximum number of the line features in a group is based on the line number present in the basic line arrangement (S3||S1||S3) shown above. Using the above two grouping rules, we can decompose the unidirectional features into a series of groups, as illustrated in Figure 8(a).

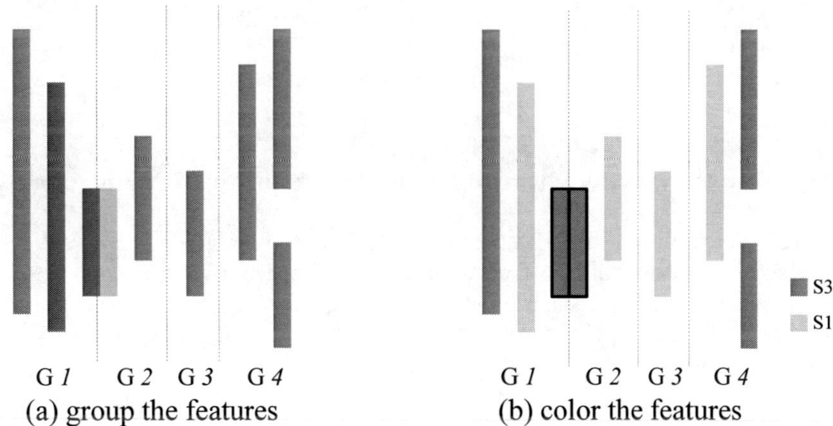

Figure 8. Grouping and coloring the unidirectional features in a component. Here, G*i* represents group *i*.

3.4 Color the unidirectional features

To develop a coloring algorithm, we set up several coloring rules:

- Assign colors to the line features in a group according to the line arrangement: S3||S1||S3.
- Only one line feature in a group should be colored as S1.

- If two 1-D features are decomposed from a large-CD feature, they should have the same color.
- Color the groups from one direction to the other (e.g., from left to right) and assume that group n is immediately on the left of group $n+1$, and group n is colored before group $n+1$, as illustrated in Figure 9.
- If a line feature has a side defined by a cut mask, this side is called a "critical edge" [27].

The above rules are based on the requirements that there should be a line feature colored as S1 which will be used to define the mandrel feature, and the line features colored as S3 should be on both sides of the S1 feature. Figure 8(b) shows an example of the coloring rules. Next, we shall discuss how to color two arbitrary groups: Group n and Group $n+1$. The related symbols used in our discussion are shown in Table 3.

Table 3. Definitions of some symbols used in our discussion.

Notation	Definition
F_{RF}	The right line feature in group n.
F_{LC}	The left line feature in group $n+1$.
d_{FC}	The distance between group n and group $n+1$, which is also the distance between F_{RF} and F_{LC}.
n_{CG}	The number of line features in group $n+1$ and its value can be 1, 2 or 3.
W_{dummy}	The width of a "dummy" feature located between two groups.
d_{FD}	The distance between F_{RF} and dummy feature, which is also the distance between group n and dummy feature.
d_{Dc}	The distance between dummy feature and F_{LC}, which is also the distance between dummy feature and group $n+1$.

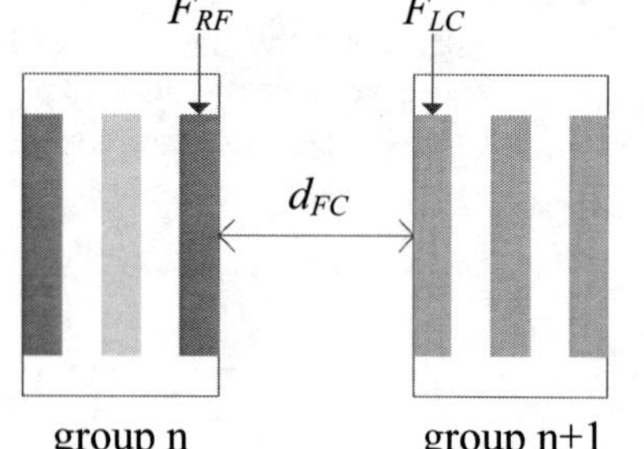

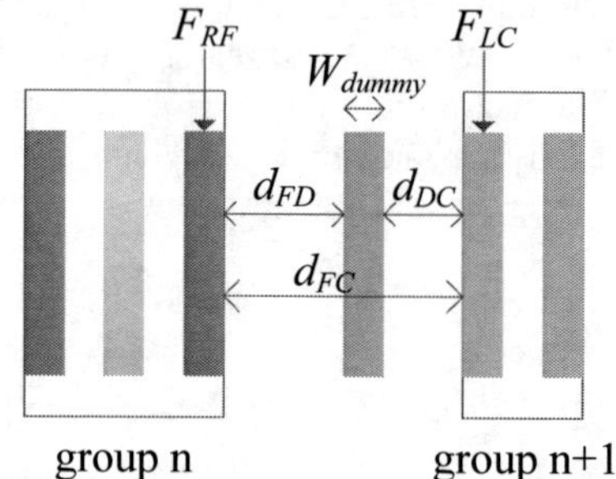

Figure 9. The geometrical relations between the features in two neighboring groups.

Note that "dummy features" are certain assisting features we insert between group n and group $n+1$ to meet the process requirement to be described later. According to the coloring rules, the dummy features can only be colored as S3 because certain feature has been colored as S1 in each group. These dummy features are produced in a pSASP process, but are redundant in the final layout and can be trimmed away using the cut mask. Once group n is colored, the coloring rules of group $n+1$ will be determined according to the number of line features in group $n+1$ (n_{CG}). The coloring rules are listed below.

(1) When n_{CG}=3, these three line features should be colored as S3||S1||S3. In this case, only one coloring scheme is allowed, as illustrated in Figure 10.
  I) When F_{RF} is colored as S3, the colored line features are decomposable without inserting any dummy feature (Figure 10(a)).
  II) When F_{RF} is colored as S1, dummy features should be introduced to meet pSASP process requirements, as illustrated in Figure 10(b). Taking the case $W_{spacer} < d_{FC} \leq 2W_{spacer}$ as an example, the space between spacer 1 and spacer 3 is W_{spacer}. Nevertheless, in some conditions, the grouping rules may result in a large distance

(e.g., larger than W_{spacer}) between S1 and S3, thus violating the process requirement. To correct this, we can insert a dummy feature (e.g., aligned to S3) to shrink the space CD to W_{spacer}. The locations and widths of various dummy features are shown in cases 1 & 2 of Table 4.

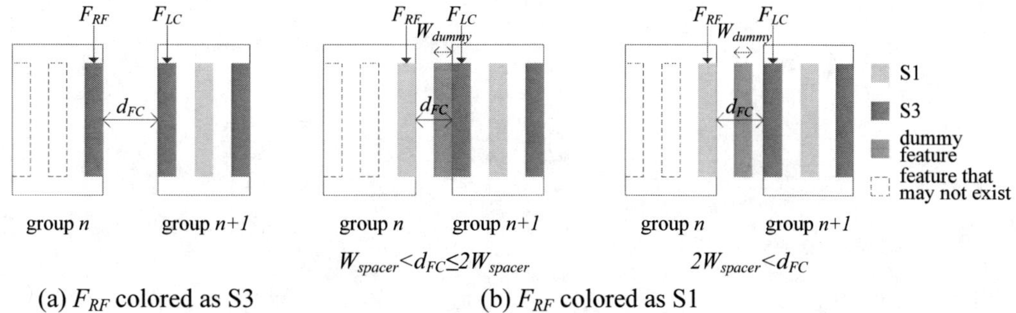

(a) F_{RF} colored as S3 (b) F_{RF} colored as S1

Figure 10. The coloring rules when n_{CG}=3.

Table 4. The dummy-feature rules to meet pSASP process requirements.

Case	n_{CG}	F_{RF}	F_{LC}	d_{FC}	W_{dummy}	d_{FD}	d_{DC}	critical edge
1	3	S1	S3	$W_{spacer} < d_{FC} \leq 2W_{spacer}$	$d_{FC} - W_{spacer}$	W_{spacer}	Merge with F_{LC}	Yes
2				$2W_{spacer} < d_{FC}$	W_{spacer}	W_{spacer}	$d_{FC} - 2W_{spacer}$	No
3	2	S3	S1	$W_{spacer} < d_{FC} \leq 2W_{spacer}$	$d_{FC} - W_{spacer}$	Merge with F_{RF}	W_{spacer}	Yes
4				$2W_{spacer} < d_{FC}$	W_{spacer}	$d_{FC} - 2W_{spacer}$	W_{spacer}	No
5		S1	S3	$W_{spacer} < d_{FC} \leq 2W_{spacer}$	$d_{FC} - W_{spacer}$	W_{spacer}	Merge with F_{LC}	Yes
6				$2W_{spacer} < d_{FC}$	W_{spacer}	W_{spacer}	$d_{FC} - 2W_{spacer}$	No
7			S1	$W_{spacer} < d_{FC} \leq 2W_{spacer}$	Not existing			No
8				$2W_{spacer} < d_{FC} \leq 4W_{spacer}$	$d_{FC} - 2W_{spacer}$	W_{spacer}	W_{spacer}	No
9				$4W_{spacer} < d_{FC}$	W_{spacer}	W_{spacer}	W_{spacer}	No
10	1	S3	S1	$W_{spacer} < d_{FC} \leq 2W_{spacer}$	$d_{FC} - W_{spacer}$	W_{spacer}	Merge with F_{LC}	Yes
11				$2W_{spacer} < d_{FC}$	W_{spacer}	W_{spacer}	$d_{FC} - 2W_{spacer}$	No
12		S1		$W_{spacer} < d_{FC} \leq 2W_{spacer}$	Not existing			No
13				$2W_{spacer} < d_{FC} \leq 4W_{spacer}$	$d_{FC} - 2W_{spacer}$	W_{spacer}	W_{spacer}	No
14				$4W_{spacer} < d_{FC}$	W_{spacer}	W_{spacer}	W_{spacer}	No

(2) When n_{CG}=2, these two line features can be colored as S3||S1, or S1||S3 (Figures 11 & 12).

III) When F_{RF} is colored as S3, the group n+1 has two coloring schemes: S3||S1 or S1||S3 (Figure 11). The coloring scheme S3||S1 does not require any dummy feature (Figure 10(a)), while the coloring scheme S1||S3 requires a dummy feature (Figure 11(b)) to meet the process requirement. The locations and widths of various dummy features are shown in cases 3 & 4 of Table 4.

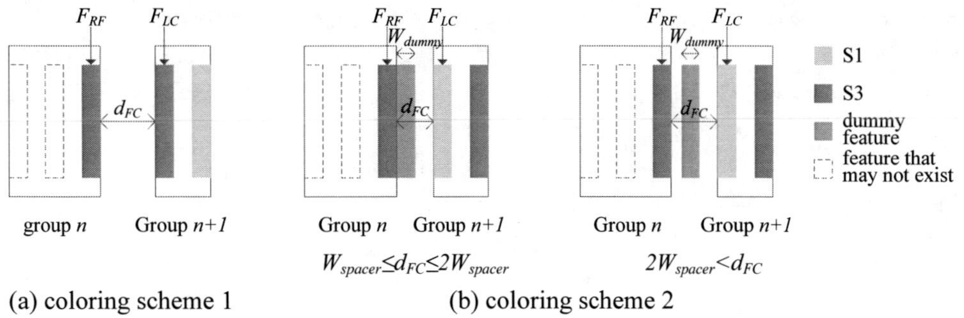

(a) coloring scheme 1 (b) coloring scheme 2

Figure 11. The coloring rules when n_{CG}=2 and F_{RF} is colored as S3.

I) When F_{RF} is colored as S1, the group n+1 also has two coloring schemes: S3||S1 or S1||S3. The coloring scheme S3||S1 requires a dummy feature to be inserted to meet the process requirement, as illustrated in Figure 12(a). The other coloring scheme, as shown in Figure 12(b), may or may not need a dummy feature, depending on d_{FC}. The dummy feature's location and width are shown in cases 5 & 6, and 7, 8 & 9 of Table 4, respectively.

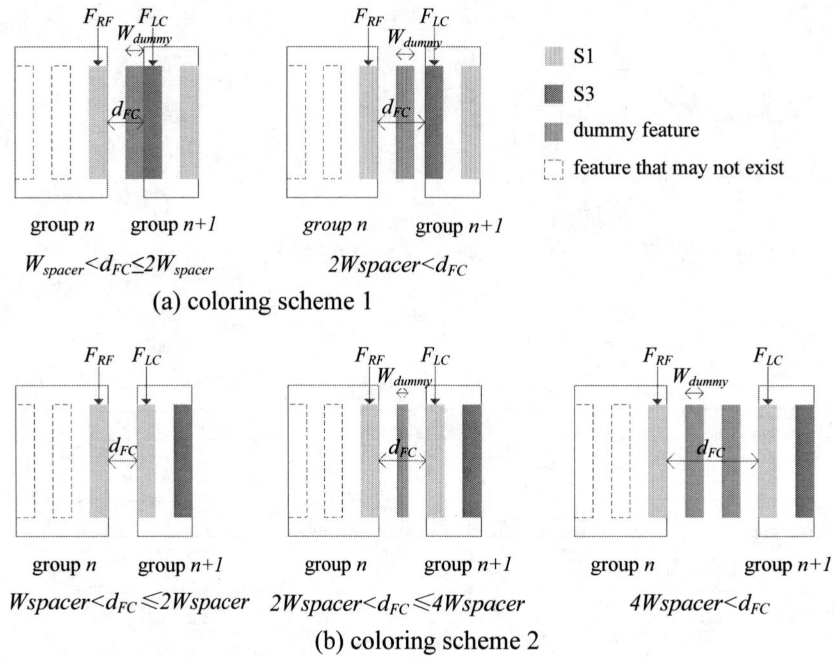

Figure 12. The coloring rules when n_{CG}=2 and F_{RF} is colored as S1.

(3) When n_{CG}=1, the group only has one coloring scheme and the line feature should be colored as S1. Figure 13(a) shows the dummy feature when F_{RF} is colored as S3 and Figure 13(b) shows the dummy feature(s) when F_{RF} are colored as S1.

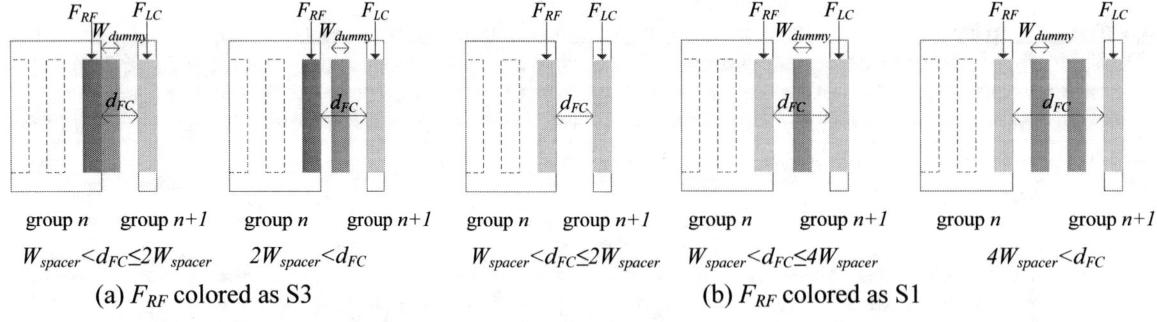

Figure 13. The coloring rules when n_{CG}=1.

3.2.4 Generate mandrel and cut masks

In our decomposition algorithm, we color the groups from one direction to the other (e.g., from left to right). The two adjacent colored groups can be seen as a cell and each cell can be treated independently (Figure 14(b)). The mandrel/cut masks can be formed by merging each cell's mandrel/cut features.

In this work, we assume the mandrel features are rectangles as only multiple sets of unidirectional patterns need to be considered in the coloring process. Namely, we just need to know the positions of its four edges to determine the shape of a mandrel feature. As exemplified by Figure 14(c). The mandrel feature's right and left edges are defined by S1, while the top and bottom edges are defined by the line ends of Y-direction features.

In the step of forming the cut features, we first search the area to see if there are dummy features (which can be accomplished by looking at Table 4). If there are dummy features, the cut mask feature will be expanded (e.g., by $d/2$ if we assume the distance between the dummy and non-dummy features is d); otherwise, the cut feature will be expanded by $W_{spacer}/2$ to tolerate overlay inaccuracy, as illustrated in Figure 14(d).

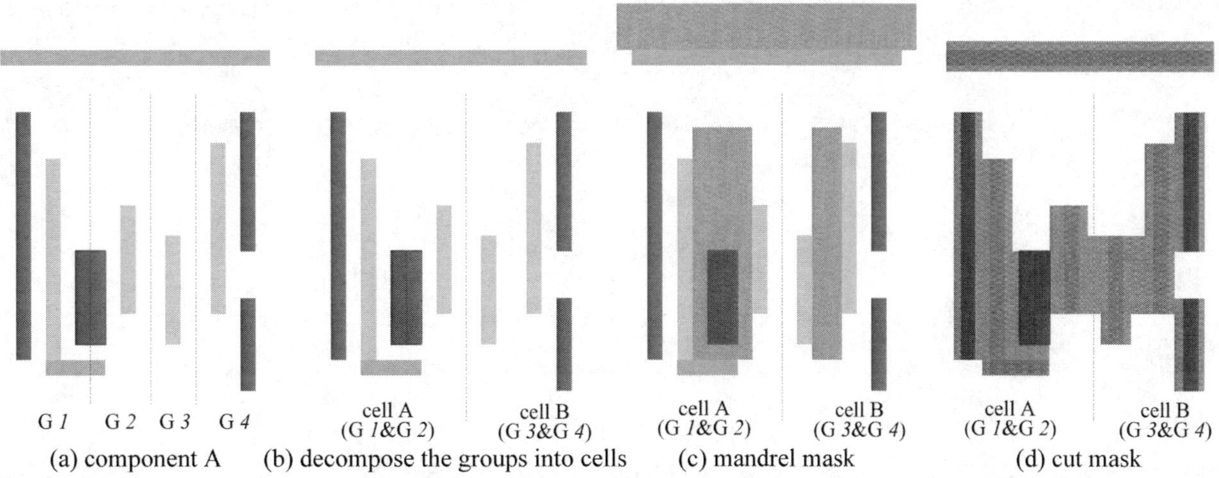

Figure 14. Forming the mandrel and cut masks. Here, G*i* represents group *i*.

3.2.5 Algorithm complexity

In Table 5, we summarize the characteristics of line feature arrangement and coloring schemes of each group. It can be seen that no matter what number of the line features in each group is, there will be at least one corresponding coloring scheme. Therefore, coloring all groups can be finished in a polynomial time even though the coloring scheme may not be necessarily optimized. When trimming away the dummy features, the critical edges are difficult to control and may negatively impact the process yield. Optimization of the coloring scheme to minimize the mandrel hotspots and critical edges can be achieved by analyzing all possible coloring schemes. Nevertheless, this needs further research effort and we shall discuss it in more detail in a follow-up paper.

Table 5. The number of line features in a group and the coloring schemes.

The number of line feature	Coloring schemes
3	S3‖S1‖S3
2	S3‖S1 or S1‖S3
1	S1

4. FULL-CHIP LAYOUT TREATMENT

In a full-chip layout, various feature densities are present in different regions and it is not necessary to decompose the whole layout using the hybrid SAMP algorithm. To accommodate more CD flexibility (e.g., large CDs), we propose incorporating optical multiple patterning into the hybrid SAMP process. If the cut masks of two decomposed components are designed as clear-field masks, they can simultaneously act in a way similar to multiple patterning to generate specific 2-D random features (whose density is beyond the optical resolution of a scanner) in the stitching process. Figure 15 shows how double patterning can be incorporated into a hybrid SAMP process.

We first divide the target layout (Figure 15(a)) into two parts: Part A (green) formed by a hybrid SAMP process, and part B (gray) formed by double patterning (see Figure 15(b)). Two sets of decomposed patterns with different colors in part B will be assigned into two "cut" masks generated in a hybrid SAMP process. If the distance between parts A and B is large enough, they can be decomposed independently. If the distance is relatively small, we need to consider the interaction between the "cut" features (of the hybrid SAMP part) and the "double-patterning" features. In other words, the mixed features share the same DP decomposition algorithm (see Figure 15(c-e)). Although certain "protection" masks may be necessary to protect these decomposed "double-patterning" features that are assigned to two cut masks (during an etch process), the "protection" patterns are often large and only need low-cost patterning processes.

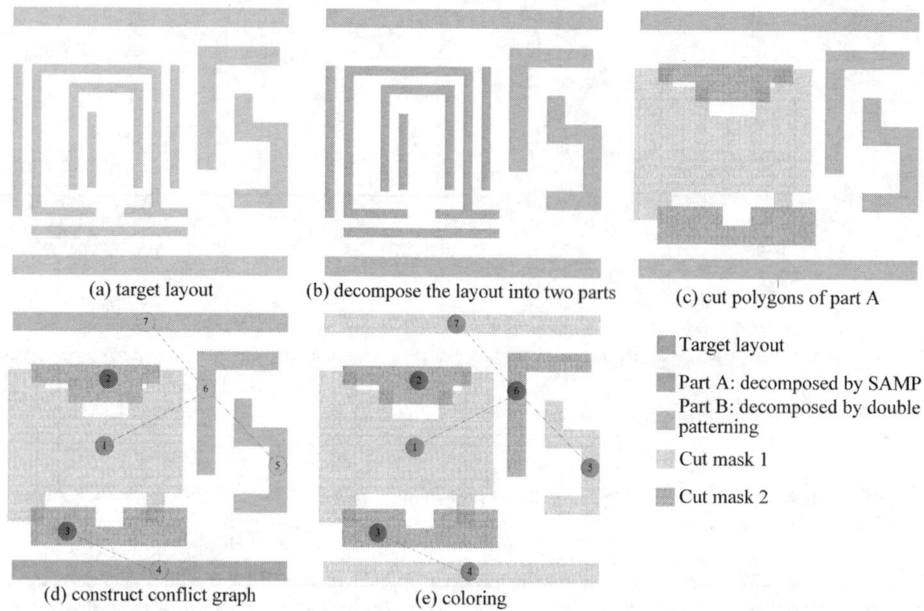

Figure 15. An example of the full-chip layout decomposition process.

5. ALGORITHM DEVELOPMENT RESULTS AND DISCUSSION

The decomposition algorithm is implemented in C++ and runs on a PC with 1G RAM and a 1GHZ CPU. NanGate FreePDK45 Generic Open Cell Library [26] and NanGate_15nm_OCL [25] are used to generate all test patterns. We randomly select the 2-D standard cells in the public library, scale them to 10 nm half-pitch, and store them in different rows of a chip to mimic a full-chip layout.

5.1 Full-chip layout decomposition

Given a full-chip layout, the first treatment is to divide the layout into two parts: A and B. The features in Part A will be formed by a hybrid SASP process, while the features in Part B will be formed by double patterning. Table 6 shows the layout decomposition results. In each table, there are five test cases, TC1 to TC5, corresponding to the increasing number of polygons. From Table 6, we can see the number of features decomposed by double patterning is far less than that formed by a hybrid SASP process.

Table 6. Layout decomposition results.

Test Case	Chip size (um)	#Feature	#Vertex	$\#Feature_{SASP}$	$\#Feature_{DPL}$
TC1	2×1	186	1253	160	26
TC2	8×3	2020	13012	1772	248
TC3	10×5	4392	27957	4014	378
TC4	100×10	80994	525089	71557	9347
TC5	200×20	321479	2078210	285670	35719

Note 1: "$\#Feature_{SASP}$": the number of features formed by a hybrid SASP process.
Note 2: "$\#Feature_{DPL}$": the number of features formed by double patterning lithography.

5.2 The decomposition results of the features in Part A

The features in Part A are decomposed by the hybrid SAMP algorithm and the results are shown in Table 7. We can see the number of Y-direction features is twice of X-direction features. Table 8 shows the numbers of different stitching types. The number of "L"-shape stitches is significantly more than that of "T"-shape and "+"-shape stitches.

Table 7. The cut results of the features in Part A.

Test Case	#Feature	$\#Feature_X$	$\#Feature_Y$	#Stitch	Runtime (s)
TC1	160	89	184	138	<1
TC2	248	938	1823	1297	5
TC3	4014	1997	4024	2711	12
TC4	9347	39075	74041	53600	232
TC5	35719	153693	293494	310139	814

Note 1: "$\#Feature_X$": the number of features in X direction.
Note 2: "$\#Feature_Y$": the number of features in Y direction.
Note 3: "#Stitch": the number of stitches.

Table 8. The numbers of different stitch types.

Test Case	#Feature	#Stitch	# "+" shape	# "L" shape	# "T" shape
TC1	160	138	6	98	34
TC2	248	1297	57	1014	226
TC3	4014	2711	92	2214	405
TC4	9347	536000	2350	41695	9555
TC5	35719	310139	8744	165208	36187

Due to the introduction of spacer merging technique, there may be some mandrel "blocks" which cannot be printed by optical lithography. As illustrated in Figure 16, there are two typical failure conditions: isolating fine lines or spaces. If the space between two adjacent mandrel blocks is smaller than the minimum space printable by optical lithography, the two mandrel blocks should be merged together and an e-beam patterning process can be applied to trim the merged feature, as shown in case 1. If the mandrel block CD is smaller than the minimum line CD printable by optical lithography, it should be expanded to be large enough and then trimmed back. Considering the low throughput of e-beam direct writer, the minimum trimming areas are preferred.

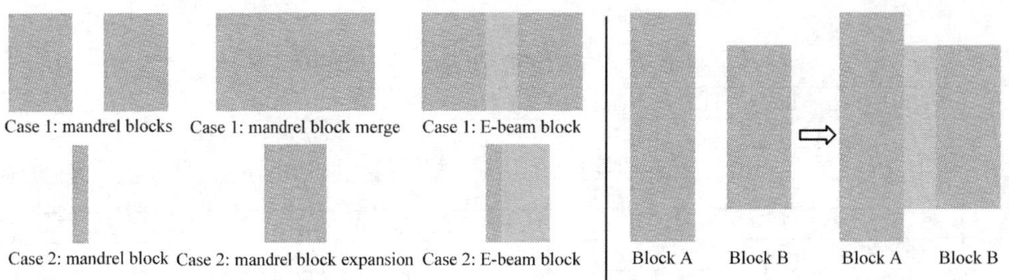

Figure 16. Fixing hotspots on the mandrel "blocks".

Table 9. The hotspot fixing results of the mandrel mask.

Test Case	Chip size (um)	#Mandrel	#Hotspot	#Merging	#Expansion	Trim area (um^2)	Percentage
TC1	2×1	58	1	0	1	0.0004	0.020%
TC2	8×3	681	29	27	2	0.0616	0.257%
TC3	10×5	1427	29	26	3	0.0516	0.103%
TC4	100×10	27144	829	672	157	1.4755	0.148%
TC5	200×20	108190	2379	1794	585	8.1210	0.203%

Table 10. The hotspot fixing results of the cut mask.

Test Case	Chip size (um)	#Cut mask	#Hotspot	#Merging	#Expansion	Trim area (um^2)	Percentage
TC1	2×1	104	46	46	0	0.0419	2.095%
TC2	8×3	1180	422	406	16	0.2615	1.090%
TC3	10×5	2455	574	546	28	0.4408	0.882%
TC4	100×10	47031	15763	15151	612	10.5931	1.059%
TC5	200×20	186902	62190	59768	2422	42.3421	1.059%

By introducing the mandrel block merging and expansion techniques, many hotspots can be fixed. From Table 9 and Table 10, we can see the number of mandrel block merging is much higher than that of mandrel block expansion. The percentage of hotspot areas that can be fixed by e-beam direct writing is about 0.2% on a mandrel mask and 1-2% on a cut mask. This confirms the potential manufacturability of introducing e-beam technology for trimming applications in hybrid SAMP processes.

6. CONCLUSION

In this paper, we propose a computationally efficient 2-D layout decomposition and stitching technique which combines optical and self-aligned multiple patterning processes. Using a positive-tone SASP processes and double patterning as an example, we demonstrate how to apply this technique to break the 1-D design limitations and accommodate more CD/pitch flexibility. Based on the various feature densities on different regions of a chip, the corresponding methods are developed to decompose the full-chip layout. The algorithm development and test results verify the functionality of our decomposition strategy, and show the statistical and computational characteristics of the public layout data under our investigation.

ACKNOWLEDGMENT

This research is supported by the Shenzhen City Fund for Fundamental Research and the start-up research fund from Shenzhen Graduate School of Peking University.

REFERENCES

[1] ITRS (International Technology Roadmap for Semiconductors) Report: Lithography, 2013.

[2] Y. Chen, P. Xu, L. Miao, Y. M. Chen, Xumou Xu, D. Mao, P. Blanco, C. Bencher, R. Hung, C. Ngai, "Self-aligned triple patterning for continuous IC scaling to half-pitch 15nm," *Proc. of SPIE*, Vol. 7973, 79731P, 2011.

[3] Y. Chen, Y. M. Chen, L. Miao, P. Xu, X. Xu, H. Chen, P. Blanco, R. Hung, C. Ngai, "Spatial frequency multiplication techniques towards half-pitch 10 nm patterning," *Proc. of SPIE*, Vol. 7973,79731T, 2011.

[4] Y. Chen, X. Xu, Y. M. Chen, L. Miao, H. Chen, P. Blanco, C. Ngai, "Recessive self-aligned double patterning with gap-fill technology," *Proc. of SPIE*, Vol. 7973,79731S, 2011.

[5] P. Xu, Y. M. Chen, Y. Chen, L. Miao, S. Sun, S-W Kim, A. Berger, D. Mao, C. Bencher, R. Hung, C. Ngai, "Sidewall spacer quadruple patterning for 15-nm half-pitch," *Proc. of SPIE*, Vol. 7973, 79731Q, 2011.

[6] C. Bencher, H. Dai, L. Miao, Y. M. Chen, P. Xu, Y. Chen, S. Oemardini, J. Sweis, V. Wiaux, J. Hermans, L.-W. Chang, X. Bao, H. Yi, H.-S. P. Wong, "Mandrel based patterning: density multiplication techniques for 15nm nodes," *Proc. of SPIE*, Vol. 7973, 79730K, 2011.

[7] Y. Chen, Q. Cheng, W. Kang, "Technological merits, process complexity, and cost analysis of self-aligned multiple patterning," *Proc. of SPIE*, 8326, 832620, 2012.

[8]Y. Chen, Q. Cheng, W. Kang, "Mandrel and spacer engineering based self-aligned triple patterning," *Proc. of SPIE*, Vol. 8328, 83280O, 2012.

[9] Y. Chen, "Process optimization and simplification of self-aligned triple patterning," *Microelectronic Engineering*, 98, 590-594, 2012.

[10]Y. Chen, Q. Cheng, W. Kang, "Analysis of process characteristics of self-aligned multiple patterning," *Microelectronic Engineering*, 98, 184-188, 2012.

[11] Y. Chen, W. Kang, P. Zhang, "A comparative study of self-aligned quadruple and sextuple patterning techniques for sub-15nm IC scaling," *Proc. of SPIE*, 8683, 86830Z, 2013.

[12] W. Kang, C. Feng and Y. Chen, "Mask strategy and layout decomposition for self-aligned quadruple patterning," *Proc. of SPIE*, 8684, 86840E, 2013.

[13] W. Kang, Y. Chen, "Process characteristics and layout decomposition of self-aligned sextuple patterning," *Proc. of SPIE*, Vol. 8684, 86840F, 2013.

[14] J. You, J. Yu, Y. Chen, "A layout decomposition algorithm for self-aligned multiple patterning," *Proc. of SPIE*, 9053, 905310, 2014.

[15] B. Yu, K. Yuan, B. Zhang, D. Ding, D. Z. Pan, "Layout decomposition for triple patterning lithography," *IEEE/ACM International Conference on Computer-Aided Design (ICCAD)*, 2011.

[16] H. Zhang, Y. Du, M. D. F. Wong, and R. O. Topaloglu, "Characterization and decomposition of self-Aligned quadruple patterning friendly layout," *Proc. of SPIE*, 8326, 83260F, 2012.

[17] K. Nakayama, C. Kodama, Toshiya Kotani,Shigeki Nojima, Shoji Mimotogi and Shinji Miyamoto, "Self-aligned double and quadruple patterning layout principle" *Proc. of SPIE*, 8327, 83270V, 2012.

[18] D. K. Lam, E. D. Liu, M. C. Smayling, T. Prescop, "E-beam to complement optical lithography for 1D layouts," *Proc. of SPIE*, 7970, 797011, 2011.

[19] H. Zhang, Y. Du, M. D. F. Wong, K. Chao, "Lithography-aware layout modification considering performance impact," *12th International Symposium on Quality Electronic Design (ISQED),* 2011.

[20] R. T. Greenway, R. Hendel, K. Jeong, A. B. Kahng, J. S. Petersen, Z. Rao, M. C. Smayling, "Interference assisted lithography for patterning of 1D gridded design," *Proc. of SPIE*, 7271, 72712U, 2009.

[21] P. Zhang, Y. Chen, "Cut-process overlay yield model for self-aligned multiple patterning and a misalignment correction technique based on dry etching," *Proc. of SPIE*, Vol. 8685, 86850K, 2013.

[22] P. Zhang, Y. Chen, "Modeling the edge-placement yield of a cut process for self-aligned multiple patterning," *Microelectronic Engineering*, 123, 73-79, 2014.

[23] P. Zhang, C. Hong, Y. Chen, "A generalized edge-placement yield model for the cut-hole patterning process," *Proc. of SPIE*, Vol. 9052, 90521Q, 2014.

[24] Y. Ban, K. Lucas and D. Pan, "Flexible 2D layout decomposition framework for spacer-type double patterning lithography," *DAC Conference*, 2011.

[25] NCSU-FreePDK-1.0, http://www.ncsu.edu.

[26] http://www.si2.org/openeda.si2.org/projects/nangatelib.

[27] Z. Xiao, Y. Du, H. Tian, M. D. F. Wong, "Optimally minimizing overlay violation in self-aligned double patterning decomposition for row-based standard cell layout in polynomial time," *ICCAD*, 2013.

Full chip two-layer CD and overlay process window analysis

Rachit Gupta[a], Shumay Shang[b], John Sturtevant[a]
[a]Mentor Graphics Corp, 8005 SW Boeckman Rd, Wilsonville, OR 97070
[b]Mentor Graphics Corp, 46871 Bayside Parkway, Fremont, CA 94538

ABSTRACT

In-line CD and overlay metrology specifications are typically established by starting with design rules and making certain assumptions about error distributions which might be encountered in manufacturing. Lot disposition criteria in photo metrology (rework or pass to etch) are set assuming worst case assumptions for CD and overlay respectively. For example poly to active overlay specs start with poly endcap design rules and make assumptions about active and poly lot average and across lot CDs, and incorporate general knowledge about poly line end rounding to ensure that leakage current is maintained within specification.

There is an opportunity to go beyond generalized guard band design rules to full-chip, design-specific, model-based exploration of worst case layout locations. Such an approach can leverage not only the above mentioned coupling of CD and overlay errors, but can interrogate all layout configurations for both layers to help determine lot-specific, design-specific CD and overlay dispositioning criteria for the fab. Such an approach can elucidate whether for a specific design layout there exist asymmetries in the response to misalignment which might be exploited in manufacturing.

This paper will investigate an example of two-layer model-based analysis of CD and overlay errors. It is shown, somewhat non-intuitively, that there can be small preferred misalignment asymmetries which should be respected to protect yield. We will show this relationship for via-metal overlap. We additionally present a new method of displaying edge placement process window variability, akin to traditional CD process window analysis.

Keywords: Simulation, overlay, process window, OPC

1. INTRODUCTION

Layout design rules are comprised of both intra-layer and inter-layer guidelines intended to guard band the layout against manufacturing variability. These rules are established or subsequently revised by considering the desired electrical characteristics in relation to the physical properties of the patterns of interest. The patterning process can be very accurately simulated and this enables investigation of test pattern printing through a variety of different process conditions.

Full-chip simulation to verify post-OPC layout correction has been a vital component of mask data preparation flows for several generations of technology manufacturing[1-3]. Full-chip contours can be simulated at multiple process window (PW) conditions, and a wide variety of automated checks such as pinch or bridge, linewidth variation (ACLV), line end pullback, and many more can provide quantitative analysis of the entire contour shoreline. The typically explored PW conditions include dose, focus, and mask CD to enable single layer CD analysis. By adding misalignment simulation, interlayer edge placement analysis is possible, including checks such as metal-via overlap, poly endcap past active, and more. This is especially useful for multi-patterning processes, where edge placement errors can lead to particular pathologies within-chip. In this work, we investigate the use of full-chip two layer simulation to study the intersection of CD and overlay variability. CD changes are induced by mask CD, dose and focus excursions, and global misalignment can be simulated at any x/y vector to characterize the combined CD-misalignment process window which ensures that pertinent two-layer edge placement metrics are maintained above specification.

Design-Process-Technology Co-optimization for Manufacturability IX, edited by John L. Sturtevant, Luigi Capodieci,
Proc. of SPIE Vol. 9427, 94270H · © 2015 SPIE · CCC code: 0277-786X/15/$18 · doi: 10.1117/12.2086368

Inter-layer design rules and in-fab CD and overlay disposition criteria are generally established in such a manner so as to mitigate worst case edge placement errors (EPE) between the layers. To first order, the EPE is given by the equation below:

$$\boldsymbol{EPE(3\sigma)} = overlay\ mean\ error + CD\ mean\ error_a + CD\ mean\ error_b + \sqrt{((CD_a\ 3\sigma)/2)^2 + (CD_b\ 3\sigma)/2)^2 + (overlay\ 3\sigma)^2)}$$

This equation distinguishes between mean CD and overlay errors and within lot variability. In this work, we only examine global misalignment vectors which translate all features in the layout the same magnitude. Mean CD is changed globally through the variables of dose, focus and mask dimension. But given that different features have different NILS, DOF, and MEEF, the result will be a distribution of CD errors within the chip. The relationship between die yield and misalignment depends upon the nature of the residual misalignment errors, specifically the distribution within die and across wafer / lot. An excellent treatment of this topic was offered by Ghaida et al[4].

In practice, on-board systems and advanced process control (APC) schemes are employed in-fab to dynamically adjust litho inputs to keep lot level mean CD and overlay errors to a very low level, but there must nonetheless be some maximum values set for rework dispositioning. The disposition criteria are typically absolute and independent of previous wafer state information. But in reality, for instance, when a given lot is measured for poly to active overlay, mean active CD and mean poly CD values are already known. This knowledge should in theory guide the allowed misalignment error which can protect yield sufficiently for that specific lot. So there is an opportunity for improved fab productivity by using such information. Some fabs have adopted such an approach for dynamic disposition criteria, but in general this information is underutilized in manufacturing.

What is further underutilized is the capability to do full-chip simulation to determine a priori what specific in die locations will be most susceptible to certain failure mechanisms through the combined CD / overlay process window, and that is the focus of this work.

2. EXPERIMENTAL

The simulation methodology utilized in this work employed Calibre OPCVerify. The process models for metal and via were for positive tone develop (PTD) of darkfield mask images to generate resist spaces on wafer. The sample layout was from a 2X node random logic, and was comprised of a metal and a via layer. The chiplet contained approximately 200,000 vias. The vias in the design have negative enclosure with respect to the metal, indicating a self-aligned via (SAV) process. The negative enclosed vias are always aligned along the X axis and the vias are always positively enclosed along the Y axis. The design rules are such that the maximum negative enclosure for SAV was 24nm and the minimum distance of the SAV to the neighboring metal line was 10nm. The minimum positive enclosure along the Y axis was 24nm. There were three different process window conditions used to generate three contours for each of the two layers, and these contours were overlayed for a total matrix of 9 different process conditions, with the center of the matrix corresponding to nominal CD for both via and metal. The small via contour was generated with 58 nm defocus, 3.4% underexposure, and a global mask bias of -.38 nm (1x). The large via contour was generated with best focus, 3.4% overexposure, and a global mask bias of + 0.38 nm (1x). The small metal contour was generated with 50 nm defocus, 3.6% underexposure, and a global mask bias of -0.25 nm. The large metal contour was generated with best focus, 4.1% overexposure, and +0.25 nm global mask bias.

Misalignment was simulated for 1 nm increments from 0 to +/-20 nm misalignment along the x and y principle axes as well as along the diagonals. The misalignment was done for the via layer only, while keeping the metal misalignment fixed. Figure 1 shows the eight different vectors in which the misalignment was introduced with each marker representing the 1nm increment. Metal-via overlap area checks were performed for each of the vias in the layout, and the overlapping area for each was calculated. The via with the minimum area overlap will presumably represent the first yield failure path, and this via's overlap area is reported as the minimum value. The average for all vias can also be reported, but is of less interest for this analysis. Figure 2 illustrates the area overlap check. In addition, an interact check was performed for

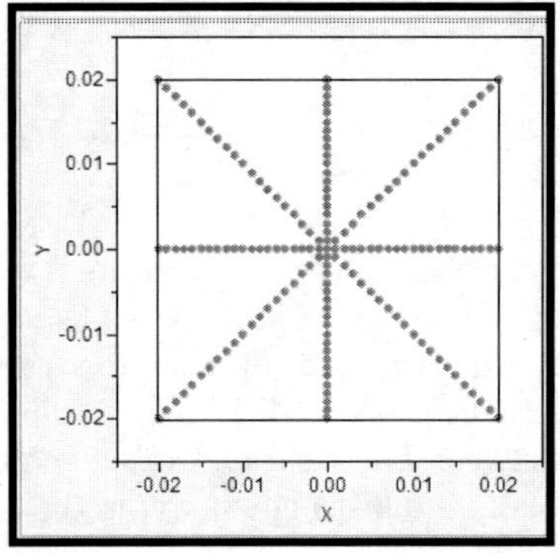

Figure 1: Eight different misalignment vectors, in 1nm increment

bridging of a via to an adjacent metal, as shown in Figure 3. This was necessary since bridging of a via to a neighboring metal line due to misalignment is a catastrophic error and puts a hard limit on the overlay window.

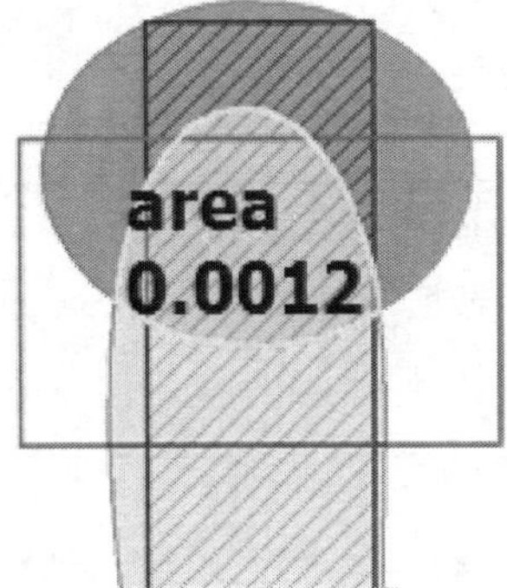

Figure2: Metal-Via area overlap check

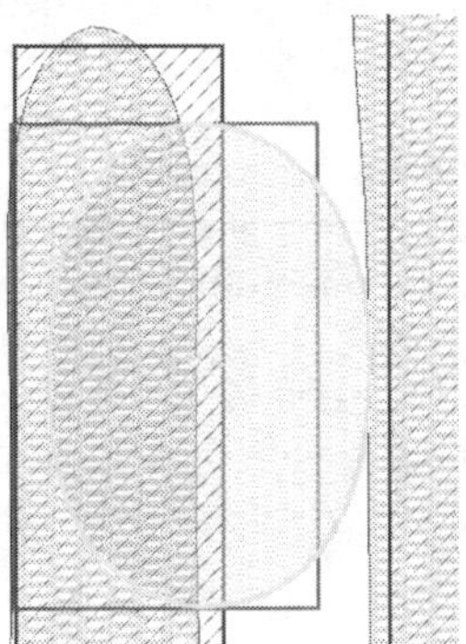

Figure 2: Metal-Via interact check, with 11nm misalignment

3. RESULTS

The EPE process window, as a function of CD variability and overlay errors, is encapsulated in the 3x3 matrix seen in Figure 4. Each individual graph in the matrix plots the minimum area overlap between the metal and via as a function of misalignment in eight different vectors. As you move from left to right, you see the effects of CD variability for the metal layer and the CD variation of the via layer is captured in the vertical direction. Additionally, the overlay error window is captured in each individual graph. The octagonal polygon seen in each graph denotes the maximum allowed overlay for that particular process condition before you get a hard bridge between the via and the adjacent metal line. This defines a clear yield limiting condition, and inside that PW exists a potentially smaller PW governed by the minimum allowable via to metal area overlap.

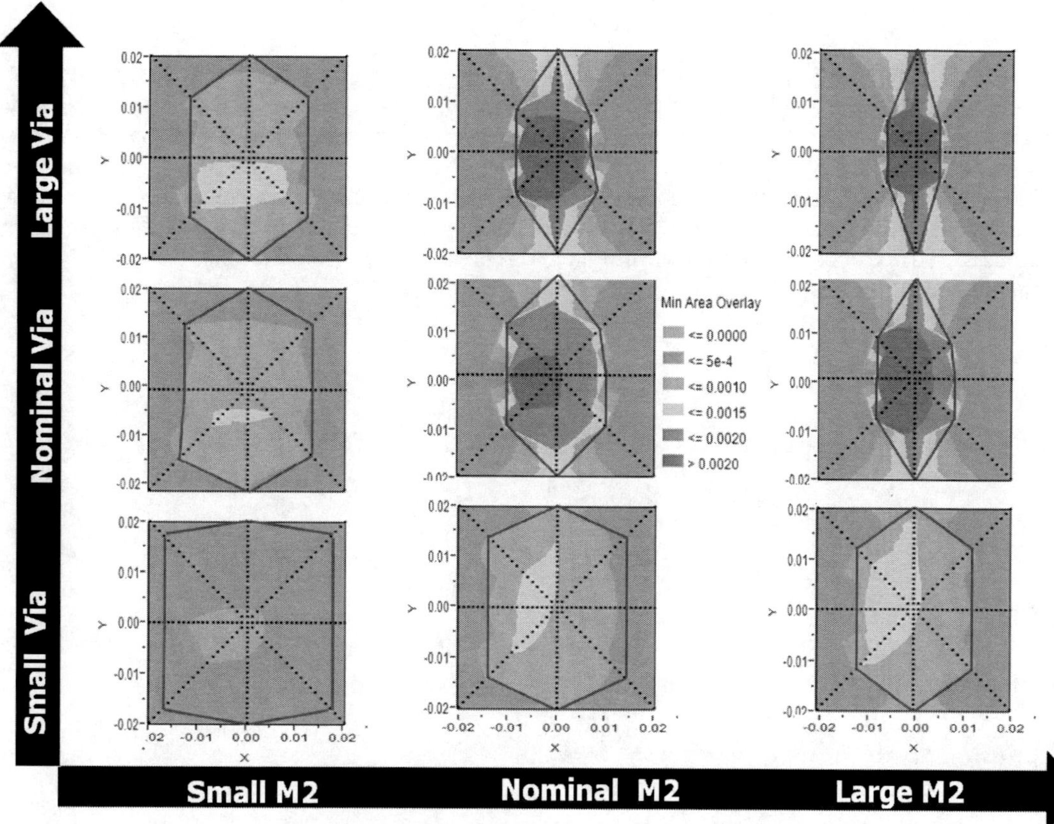

Figure 4: EPE process window, as a function of CD variability and overlay error

3.1 Nominal metal / nominal via CD

The raw data for the minimum metal-via area overlap for nominal metal – nominal via CD is plotted in Figure 5, as a function of misalignment. The X and Y axis represent the induced misalignment error, in increments of 1nm, and the contours show the minimum area overlap. The data points at which the minimum area overlap measurements were taken are shown in black. Looking at the plot, you can immediately tell that the layout is symmetric along the Y axis and asymmetric along the X axis. This result is slightly intuitive since the design has the metal layer running vertically, thereby providing little to no limitations for overlay in the Y direction. Figure 6, further shows that doing a shift of +/- 20nm along the Y direction results in identical values for the minimum metal-via area overlap. The X direction indicates a definite asymmetry, skewed towards the negative X axis. Figure 7 shows a shift of +/- 10nm along the X axis and the minimum area overlap being 35% greater in the negative X axis as compared to the positive X axis.

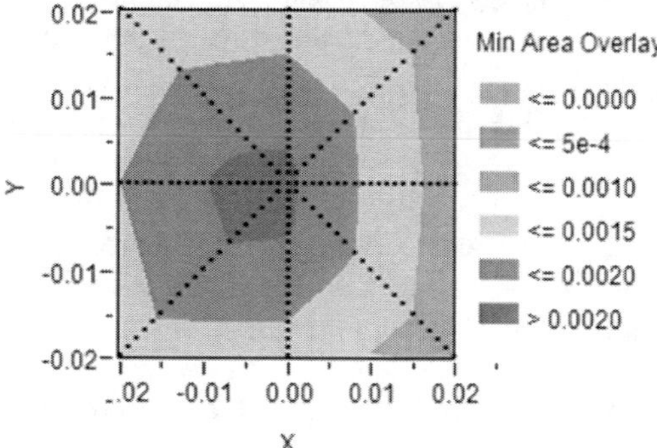

Figure 5: Minimum metal-via area overlap at nominal condition, raw data

Although Figure 5 shows us the asymmetry in the design as a function of overlay error, it does not take into account the bridging of vias with neighboring metal lines due to misalignment. This phenomenon can be seen in Figure 3 which shows that introducing a misalignment of 11nm in the positive X direction results in the bridging of via to the adjacent metal line. This bridging will result in catastrophic failure and is an immediate yield limiter. Therefore, to more accurately represent the data and to reflect the bridging limitations of overlay, the data from Figure 5 is filtered and shown in Figure 8 to give a true representation of the process. The octagonal polygon seen in Figure 8 represents the hard outer limit of overlay in each directional vector.

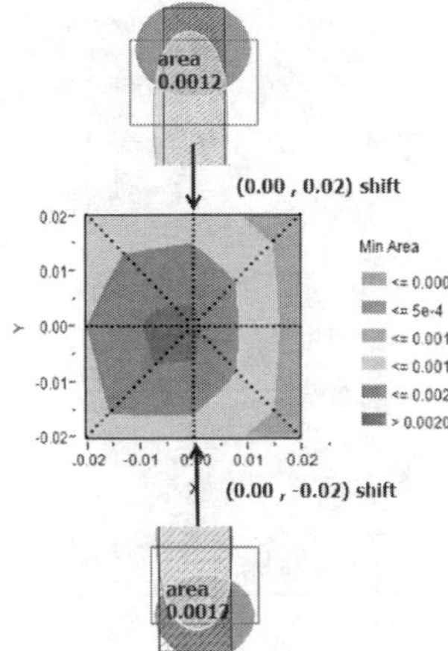

Figure 6: Symmetrical response along Y axis with +/- 20nm misalignment

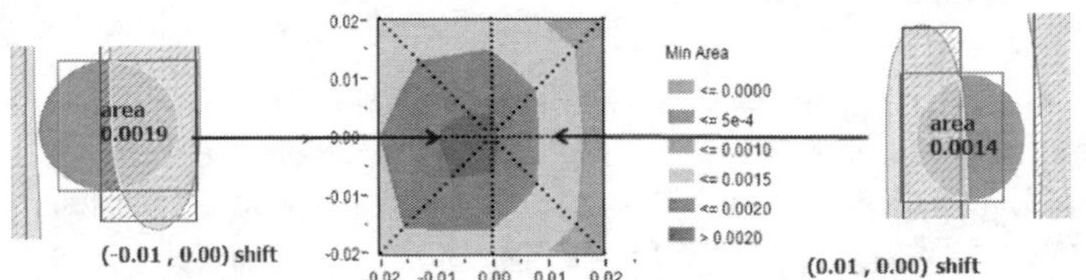

Figure 7: Asymmetric response along the X axis with +/- 10nm misalignment

3.2 Through PW metal, via CD

Now that we have established the asymmetry in the design and captured the effects of misalignment on the process, we can use this information to improve the yield. Since this is a two layer process, we can use the lot CD information of the first layer to dictate how to process the second layer and whether intentionally introducing an overlay in the correct direction can help to improve the yield and the process. Figure 4 shows the 3x3 matrix which encapsulates the CD variability and the overlay process. Let's assume that the via layer was processed first and the via CD for the lot was measured at the lower end of the process window. If we were to look at the last row of the matrix, which corresponds to "Small via" and CD variability of metal, we could do two things to help improve the yield of the metal-via overlap when processing the metal layer. The first thing would be to process the metal at the high end of the process window to get a larger metal CD, which would help to increase the metal-via overlap. The second thing would be to introduce a slight misalignment in the negative X and positive Y directions. Coupling these steps can help provide a better yield.

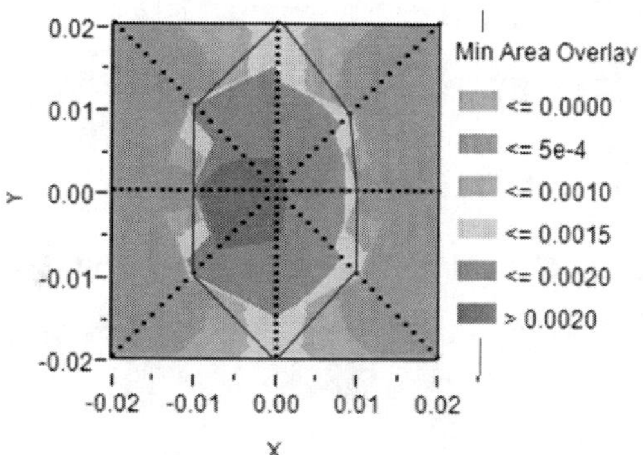

Figure 8: Minimum metal-via area overlap at nominal condition, filtered data

DISCUSSION

Edge placement control between two or more layers is critical to yield and historically the allowable CD and overlay tolerance between the constituent layers was constructed in conjunction with design rules to allow sufficient guard band of all possible cases encountered in manufacturing. In the era of computational lithography, however, it is possible to make layout-specific analysis of the impact of CD and overlay errors over the whole chip. As shown above, such an analysis can reveal interesting facets of the interaction, such as different responses along the two axes of misalignment, or asymmetric responses to positive versus negative direction misalignment. Such information can guide judicious in-fab disposition criteria, and when used in conjunction with lot level CD state analysis, further refinements can be made which are useful at protecting yield while minimizing in-fab cycle time.

CONCLUSION

Inter-layer design rules provide a guard band for EPE, which are comprised of overlay and CD variability errors. Both of these parameters are critical for yield at advanced nodes. We are able to capture the relationship between CD variation and misalignment and are able to show that there are small preferred misalignment asymmetries which should be respected to protect yield. This relationship was shown for the via-metal overlap. The coupling of PW and misalignment can also help to guide in-fab specifications such as creating design rule specific specs and different overlay criteria for X/Y directions which could help decide whether to rework the lot or pass it on to etch. For future work, we would like to extend our analysis for more complex inter-layer design rules, such as those found for multi-patterning, where the overlay error between different masks is a huge yield limiter.

REFERENCES

1. John L. Sturtevant, Srividya Jayaram, Le Hong, "Process variability band analysis for quantitative optimization of exposure conditions", Proc. SPIE. 7275, 72751Q, doi: 10.1117/12.816501 (2009).
2. John Sturtevant, Srividya Jayaram, Le Hong, Alexandre Drozdov, "Novel method for optimizing lithography exposure conditions using full-chip post-OPC simulation", Proc. SPIE 6924, 69243P (2008).
3. John Sturtevant, Srividya Jayaram; Le Hong, "Exposure tool specific post-OPC verification", Proc. SPIE 6925, 69250T (2008).
4. Ghaida, R., Gupta, M., Gupta, P., "A Framework for exploring the Interaction Between Design Rules and Overlay Control"; Proc. SPIE 8681, 86810C-1, (2013).

Quantitative evaluation of manufacturability and performance for ILT produced mask shapes using a single objective function

Heon Choi, Wei-long Wang, Chidam Kallingal

GLOBALFOUNDRIES, 400 Stone Break Rd. Ext. Malta, NY 12020

ABSTRACT

The continuous scaling of semiconductor devices is quickly outpacing the resolution improvements of lithographic exposure tools and processes. This one-sided progression has pushed optical lithography to its limits, resulting in the use of well-known techniques such as Sub-Resolution Assist Features (SRAF's), Source-Mask Optimization (SMO), and double-patterning, to name a few. These techniques, belonging to a larger category of Resolution Enhancement Techniques (RET), have extended the resolution capabilities of optical lithography at the cost of increasing mask complexity, and therefore cost. One such technique, called Inverse Lithography Technique (ILT), has attracted much attention for its ability to produce the best possible theoretical mask design. ILT treats the mask design process as an *inverse* problem, where the known transformation from mask to wafer is carried out backwards using a rigorous mathematical approach. One practical problem in the application of ILT is the resulting *contour-like* mask shapes that must be "Manhattanized" (composed of straight edges and 90-deg corners) in order to produce a manufacturable mask. This conversion process inherently degrades the mask quality as it is a departure from the "optimal mask" represented by the continuously curved shapes produced by ILT. However, simpler masks composed of longer straight edges reduce the mask cost as it lowers the shot count and saves mask writing time during mask fabrication, resulting in a conflict between manufacturability and performance for ILT produced masks[1,2]. In this study, various commonly used metrics will be combined into an *objective function* to produce a single number to quantitatively measure a particular ILT solution's ability to balance mask manufacturability and RET performance. Several metrics that relate to mask manufacturing costs (i.e. mask vertex count, ILT computation runtime) are appropriately weighted against metrics that represent RET capability (i.e. process-variation band, edge-placement-error) in order to reflect the desired practical balance. This well-defined scoring system allows direct comparison of several masks with varying degrees of complexities. Using this method, ILT masks produced with increasing mask constraints will be compared, and it will be demonstrated that using the smallest minimum width for mask shapes does not always produce the optimal solution.

1. INTRODUCTION

Resolution Enhancement Techniques (RET) refer to a category of methods applied to photomasks to extend the capabilities of photolithography processes. RETs have been instrumental in enabling the continuous scaling of semiconductor devices despite slow improvements in the resolution capabilities of exposure tools. These techniques have allowed manufacturing processes optimized for one generation to deliver devices for subsequent generations, pushing the resolution limits of optical lithography to dimensions well below the wavelength. In the past decade, RET methods have become progressively sophisticated and diverse. They range anywhere from the addition of assist features to the decomposition of the pattern into multiple exposures. Techniques such as Sub-Resolution Assist Features (SRAF's) and double-patterning are commonly used in integrated circuit (IC) manufacturing processes at 20nm technology node and below.

Design-Process-Technology Co-optimization for Manufacturability IX, edited by John L. Sturtevant, Luigi Capodieci,
Proc. of SPIE Vol. 9427, 94270I · © 2015 SPIE · CCC code: 0277-786X/15/$18 · doi: 10.1117/12.2087111

A large subset of RET deals with the distortions in the wafer images by appropriately manipulating the mask shapes to counter them. Techniques that compensate for the distortions in the optical exposure process belong to a class of Optical Proximity Corrections (OPC). As the exposure process is pushed to its physical limits, distortions become more significant and difficult to correct. To meet the resolution demands of the ever constant device scaling, OPC techniques have become increasingly complex. Initially, OPC followed a simple "rules-based" approach, whereby the target edges are fragmented and moved according to a set of pre-characterized rules that are designed to compensate for the distortions. From there it evolved into a model-based approach that uses compact models to simulate the expected distortions on wafer, which are then used to calculate the appropriate movements of the edge fragments on mask.

The latest evolution of the OPC technique is Inverse Lithography Technology (ILT). Unlike the previous OPC approaches, ILT is a more general solution as it is free from the pre-defined fragmentation of the target shape. Instead, it solves the inverse of the known transformation of mask-to-wafer on a pixel-by-pixel basis, thereby resulting in a "free-form" mask shape that will, in theory, produce the best possible print image[3]. This method is able to search over a larger solution space since it is not restricted to movements of pre-existing fragments, which presents a significant advantage for challenging patterns that require complex fragmentation rules. Over the last several years, performance advantages of ILT over "traditional" OPC techniques have been well documented[2,4].

Despite the advantages, practical application of ILT remains difficult due to the low manufacturability of ILT-produced mask shapes. The "contour-like" ILT output that represent an ideal mask are not manufacturable due to the complex shapes and the associated mask fabrication costs[5]. Therefore, ILT must be accompanied by a "Manhattanization" step to simplify the ILT outputs into shapes that conform to Mask-Rule Checks (MRC), with dimensions larger than the minimum width/space constraints of the mask fabrication process. This reduction in complexity produces manufacturable solutions at the cost of reduction in RET performance, since the MRC-compliant mask is a sub-optimal solution. Therefore, a balance much be reached between the desired RET performance and the manufacturability constraints. A sensible and optimal balance will allow for more practical and cost effective use of ILT.

The trade-off between performance and mask manufacturability is a well-known problem within ILT[1,6]. To mitigate these issues, it would be ideal to have quantitative measures available to guide in finding the ideal balance. A systematic quantitative approach can allow for a better understanding of what factors affect performance and manufacturability, and to what degree. With this idea in mind, this paper will explore the use of an *objective function* (Φ) that combines basic OPC metrics into a single score to quantitatively measure an ILT solution's ability to balance RET performance and mask manufacturability. This information will be used to determine the optimal mask constraint that produces a manufacturable mask with the desired RET performance.

2. THEORY

The goal of Φ is to combine several metrics that measure performance and manufacturability into one single "score". The metrics can be combined in a simple manner as in equation 1, where each metric is used to define a "penalty" function ϕ_n and then summed with the weight w_n for a total of N metrics. Since each metric is constructed as a penalty function to generate ϕ_n, lower values of Φ indicate a better "score". Additionally, each ϕ_n is linearly scaled and normalized ($0 \leq \phi_n \leq 1$). Given that $\sum_{n=1}^{N} w_n = 1$, values of Φ are limited to $0 \leq \Phi \leq 1$. The metrics used include measures of penalties in both performance and manufacturability. Therefore, high penalties in either categories result in high values of Φ.

$$\Phi(\eta_i) = \sum_{n=1}^{N} w_n * \phi_n(\eta_i) \quad (1)$$

The objective function here is described as a function of η_i, the mask vertex count for an ILT-based solution after Manhattanization using a particular set of MRC constraints (denoted by the index *i*). Although the most significant MRC constraint is the minimum width/space constraint (d_{mrc}), several other secondary values/limits are also used to constrain the mask shapes in various ways. This results in masks with different η_i for the same value of d_{mrc}. Therefore, η_i will be used as the independent variable throughout this study rather than d_{mrc} since mask vertex count is a good proxy for the overall mask complexity and also correlates well with the MRC minimum width/space constraint.

Figure 1 describes the metrics that comprise the objective function, along with the associated w_n. The weighting system follows a *hierarchical* structure, where the contributions to the final score are divided into categories in a tree-like fashion. At the highest level of classification, equal emphases are placed on mask manufacturability and RET performance. Manufacturability is then divided into Mask-related and OPC-related penalties. Only the computational *run-time* (typically much higher for ILT solutions) has been used as OPC penalty, while the mask-related penalties include the *mask vertex count* and the *SRAF root-mean-squared (RMS) width* and *space*. Of the manufacturability penalties, the greatest weight is placed on the mask vertex count as it most significantly affects the mask cost. The SRAF size distributions are included as they are typically the smallest and more challenging (therefore, more costly) features on a mask. Furthermore, SRAF sizes reflect the increased difficulty due to the complexities of the SRAF's generated with ILT.

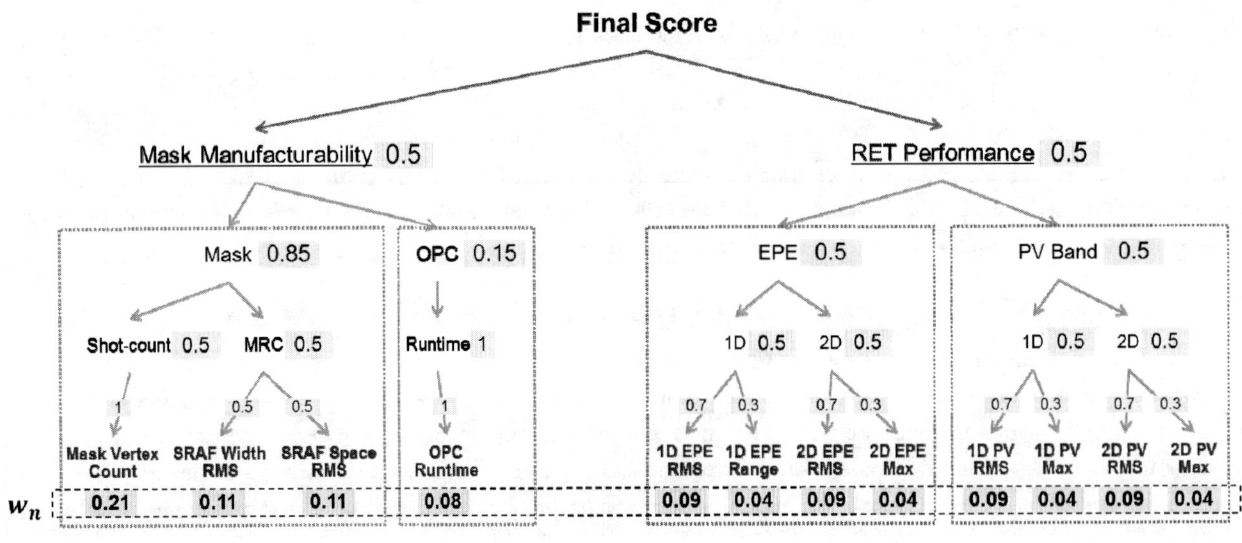

Figure 1: Description of the *hierarchical weighting* system. The contributions to the overall score Φ are categorized in a tree-like format where broad categories of metrics are further classified into narrower groupings with appropriate weights being applied at each level. The weight w_n of any individual metric is determined by the product of the weights of the parent categories.

Similarly, penalties related to RET performance are also further classified into narrower categories. The uppermost categories are *edge-placement-error* (EPE) and *process-variation* (PV) band, which are the two most basic quantitative measures of RET performance. These metrics are separated by *one-dimensional* (1D) and *two-dimensional* (2D) features so as to easily extend to a different process layer's emphasis by adjusting the corresponding weights. This separation is also used to highlight one of the advantages of ILT's pixel-based calculation, which handles 2D features quite well as it is unrestricted by rigid fragmentation rules. The quantities used as metrics are the *RMS values* and the *range*, representing the RET's overall ability to print on target and the worst case, respectively.

All the weighting values in each hierarchical node are derived and estimated from empirical observations with considerations of the cost of ownership. The w_n of an individual metric is determined by the product of the weights of the parent categories, which results in weights that accurately reflect the considerations of broader categories. It also prevents any one particular type of metrics dominating the overall score. For instance, giving equal weights to manufacturability and performance ensures that any penalties involved in mask fabrication is appropriately balanced by the associated increase in RET performance, and vise-versa. Furthermore, this hierarchical system provides the convenience of adjusting the impacts of broader categories on the overall score by adjusting the weights of high level nodes rather than the individual child nodes.

3. METHODS

The values of ϕ_n are determined from ILT simulations of a test layout, performed using a commercially available standard ILT software package. The test layout consists of two 5 x 5 μm clips taken from a typical metal layer (see Figure 2). This particular metal layer pattern is chosen as an example only, and the overall method described here can be applied to any other layer. However, 1x metal layer serves as a good example as it relies heavily on RET methods due to the challenging size and shape of the target patterns.

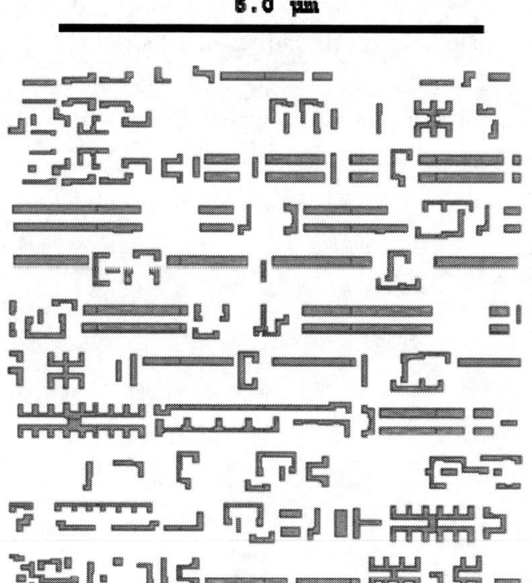

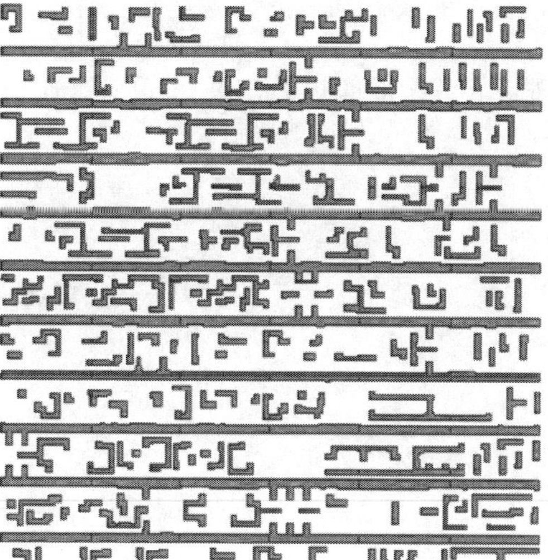

Figure 2: Test layout clips used for this study, taken from a typical metal layer. Values of the metrics are determined by simulation results of the test clips.

Mask vertex count η (scaled to the vertex count of the original target shapes) is used as the independent variable in the following analysis. As mentioned previously, η correlates well with d_{mrc}, the value of which is more physically significant as it relates directly to the dimensions on the mask. However, mask complexity is also affected by several other limits and constraints used in addition to the d_{mrc}, which produces masks with varying η for the same value of d_{mrc}. This is illustrated in Figure 3, which plots η as a function of d_{mrc} for 4 different sets of MRC settings used during Manhattanization. It demonstrates that d_{mrc} does not uniquely define η, and that the overall complexity of a mask is significantly dependent on secondary MRC limits that are imposed in addition to d_{mrc}. Therefore, η is used as the independent variable to ensure consistent analysis across all types of MRC settings. This removes differences in overall performance due to the Manhattanization approaches and reduces the analysis to be based on the resulting overall mask complexity, represented by η.

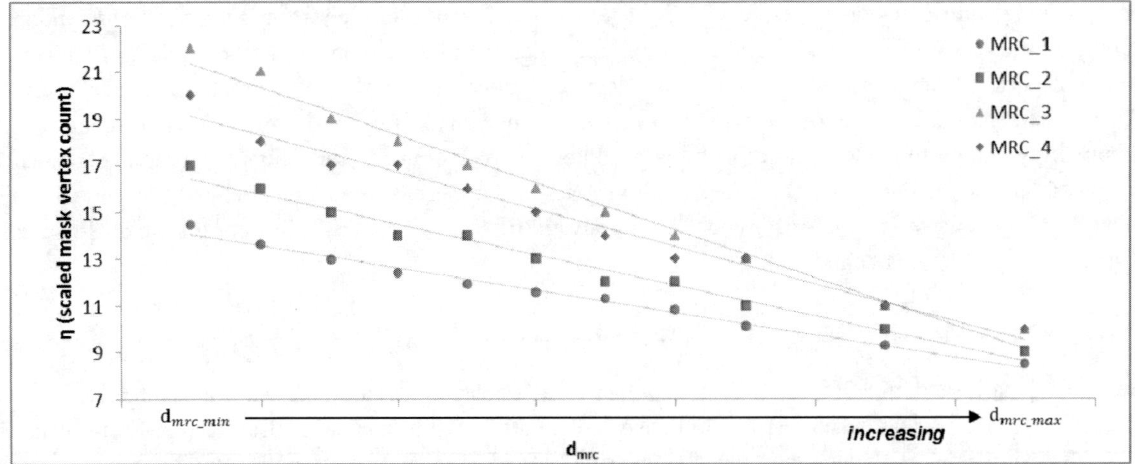

Figure 3: Relationship between mask complexity η and MRC constraint limit d_{mrc} for 4 different sets of MRC settings during Mahattanization of the ILT output. Here, η is scaled to the original target vertex count.

4. RESULTS

The main result of the study is plotted in Figure 4, which plots Φ as a function of η. It shows the score of the ILT produced masks for a wide range of η, corresponding to $d_{mrc\_min} \leq d_{mrc} \leq d_{mrc\_max}$ ($d_{mrc\_max} - d_{mrc\_min} = 12$ nm). The value of Φ varies greatly within this range, and the overall behavior illustrates the opposing impacts of the gains/losses in mask manufacturability and RET performance. The plot has three distinct regions:

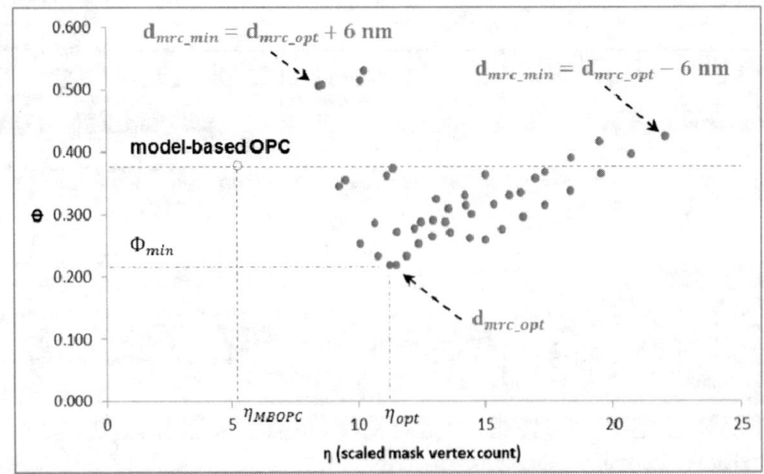

Figure 4: Plot of Φ vs η. The plot exhibits three distinct regions – high η/high Φ, low η/high Φ, optimal η/low Φ. The plot has a minimum at some moderate value of η, indicating the desired balance of mask manufacturability and RET performance.

1. High η / high Φ
   - This region represents ILT solutions with highly complex masks that stay closer to the original ILT output. Masks with $\mathrm{d}_{mrc} = \mathbf{d}_{mrc\_min}$ belong to this group. They provide higher RET performance, but the score is outweighed by mask manufacturing penalties. This results in a linear increase in Φ with increasing η.
2. Low η / high Φ
   - Masks within this group have large d_{mrc} ($= \mathbf{d}_{mrc\_max}$), and the Mahattanization is done coarsely. This produces simpler (and less costly) masks, but the loss in RET performance is too significant. The increase in Φ occurs quickly, with an exponential rise in Φ with decrease in η.
3. Optimal η / high Φ
   - At an optimal value of $\eta = \eta_{opt}$, there exists a minimum $\Phi = \Phi_{min}$ where mask manufacturing penalties are better balanced by the RET performance gains. For the test layout in Figure 2, this occurs at a minimum mask feature of $\mathrm{d}_{mrc} = \mathbf{d}_{mrc\_opt}$.

The value obtained from a mask produced by a comparable model-based OPC solution is also plotted for comparison (dashed line). Many ILT-produced masks score better (lower Φ) than the traditional OPC mask despite much higher values of η. This indicates that at an optimal value of η, use of ILT can be beneficial since the RET performance gained outweighs the additional complexities and strains added to the mask manufacturing process.

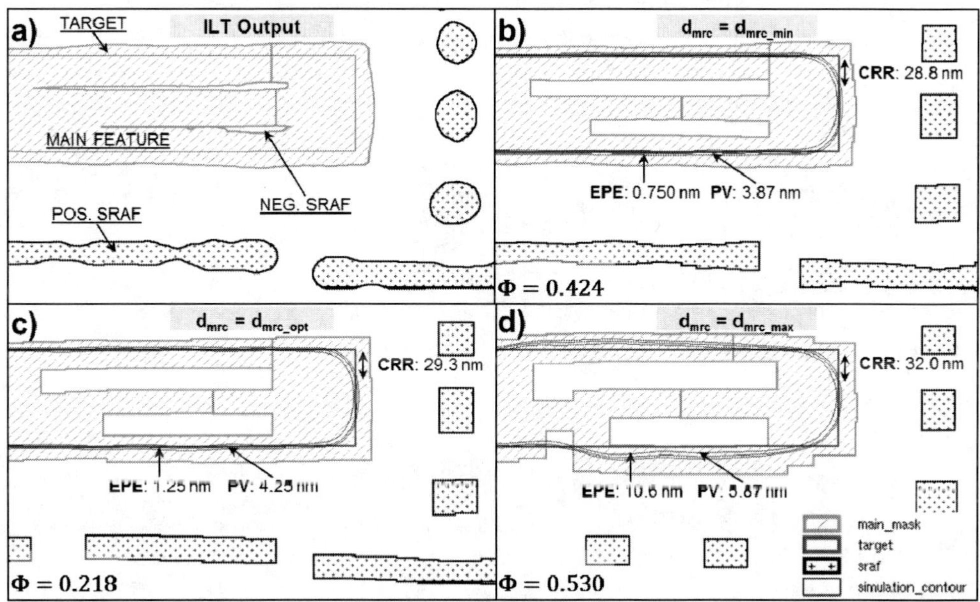

Figure 5: (a) Original ILT output. (b-c) Results of Manhattanization with varying MRC limits, corresponding to the lowest, optimal and highest η. (**EPE** = edge-placement-error, **PV** = process-variation band width, **CRR** = corner-rounding radius)

Simulation contours from masks with three different MRC constraints are shown in Figure 5, along with the original ILT output (5-a). The mask produced from the ILT output with the smallest $\mathrm{d}_{mrc} = \mathrm{d}_{mrc\_min}$ and the highest η (5-b) offers the best pattern fidelity, with the smallest EPE and PV band values. Despite the best performance, it also has a high Φ due to the mask complexity. It maintains smaller features from the original ILT output as compared to the other masks with larger MRC constraints. This results in mask shapes with more jogs and higher number of fragments, which in turn increases the shot counts (and therefore the mask writing time) required to produce the mask in an electron-beam (e-beam) writer. Additionally, the SRAF shapes are smaller and also contain more fragmented edges. These factors contribute to higher mask manufacturing costs, which is reflected in the high value of Φ.

At the other end of the spectrum (5-d), mask Manhattanized with the largest MRC constraint (d_{mrc} = $\mathbf{d}_{mrc\_max}$ and the lowest η) performs the worst, having EPE and PV band values that are significantly worse than the others. The mask shape is extremely simplified from the original ILT output, with many of the finer features being obscured by the coarse representation. The negative SRAF's inside the polygon are not sufficiently represented due to the large MRC constraint, which causes large EPE errors where the negative SRAF protrudes out towards the polygon edge.

The result with the Φ_{min} (at η_{opt}) is also shown in Figure 5-c. Although the MRC constraint is noticeably larger compared to Figure 5-b, the EPE and PV band values are not significantly deteriorated. The loss in RET performance is much less significant as compared to the mask with the largest MRC constraint. The mask shape maintains the essential features of the original ILT output while discarding the smallest features that are the most challenging to manufacture on the mask.

5. DISCUSSION

More detailed information regarding RET performance can be obtained by plotting EPE and PV band RMS values against η (Figure 6). Both plots exhibit a sharp exponential increase as η is decreased below η_{opt}. At values $> \eta_{opt}$, all curves quickly reach a limit at some minimum value. These behaviors are shared between EPE and PV band, and occur in both one-dimensional and two-dimensional features in the layout. These results indicate that beyond η_{opt}, added mask complexity does not contribute significantly to the overall RET performance. Despite the increase in mask complexity, this minimal performance improvement implies that ILT output contains finer features that only improve the pattern fidelity negligibly, and can be removed during Manhattanization. Any further attempt to replicate the original ILT output only contributes to greater mask manufacturing costs.

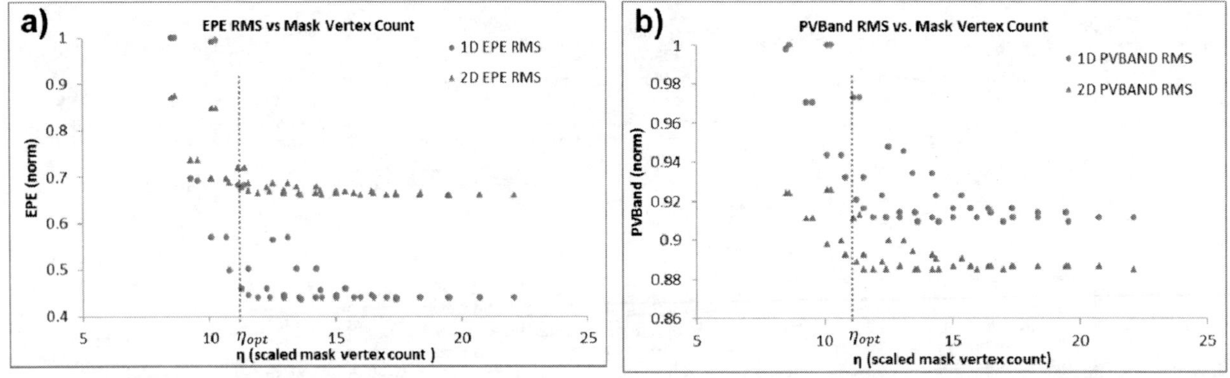

Figure 6: Plot of EPE (a) and PV band (b) RMS values vs η. In all cases, there is sharp exponential increase for $\eta < \eta_{opt}$, and constant for $\eta > \eta_{opt}$.

The overall behavior can be better understood by separating Φ into two distinct parts corresponding to mask manufacturing (Φ_m) and RET performance (Φ_p), as in equation 2-4 where *a*, *b*, *c* and *d* are all constants.

$$\Phi = \Phi_m + \Phi_p \tag{2}$$

$$\Phi_m = a * \eta + b \tag{3}$$

$$\Phi_p = c * e^{-d*\eta} \tag{4}$$

Both functions are plotted in Figure 7. As expected, the two functions have opposite behaviors with increasing η. Performance penalty Φ_p follows the same behavior as the EPE and PV band values in Figure 6, increasing exponentially with decrease in η for $\eta < \eta_{opt}$ (equation 3), and remaining nearly constant for $\eta > \eta_{opt}$. In contrast, mask manufacturing penalty Φ_m increases linearly for all η (equation 4), largely dominated by the linear increase in shot count with increasing η. When combined, it explains high values of Φ observed at the extremes of η. For high values of η, mask complexity continues to increase (linear increase in Φ_m) without improvements in performance (constant Φ_p), leading to high values of Φ. For low values of η, rapid loss in RET performance (exponential increase in Φ_p) is faster than the decrease in manufacturing difficulties (linear decrease in Φ_m), again leading to high values of Φ.

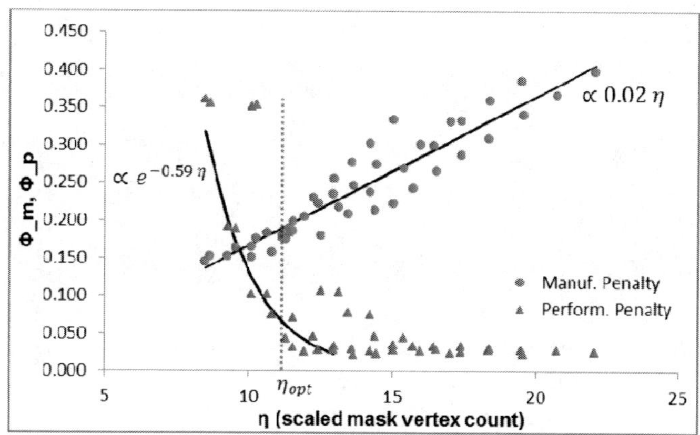

Figure 7: Plot of Φ_m and Φ_p, parts of Φ that correspond to mask manufacturability and RET performance, respectively.

The SRAF size distributions for different values of η appear to explain the sharp loss in RET performance for $\eta < \eta_{opt}$. This is expected since SRAF's are the most affected by the change in minimum width and space due to their small size. Additionally, the ability to resolve the smallest features on the mask relies heavily on SRAF's. Figure 8-a shows the SRAF size distributions for three different values of $d_{mrc} = \mathbf{d}_{mrc\_min}, \mathbf{d}_{mrc\_opt}$, and $\mathbf{d}_{mrc\_max}$, corresponding to the high, optimal, and low values of η, respectively. The smallest MRC constraint ($\mathbf{d}_{mrc\_min}$) allows the most variety of SRAF widths (d_{sraf}) with two major modes near different sizes of SRAFs. The two modes are graphically illustrated in Figure 8-(b-d) for each value of d_{mrc}. The 1st mode is largely comprised of 1st and 2nd order SRAF's, nearest to the edges of the polygons. These SRAF's likely play a big role in maintaining low EPE and PV band near those edges. For the largest MRC constraint ($\mathbf{d}_{mrc\_max}$), these modes are below the MRC limit and therefore not allowed. This is illustrated in Figure 8-d. The absence of the 1st mode SRAF's causes a sharp increase in EPE and PV band, which is reflected in the exponential increase of Φ_p for low values of η. Figure8-a also includes the SRAF size distribution for $\eta = \eta_{opt}$, with $d_{mrc} = \mathbf{d}_{mrc\_opt}$ (graphically illustrated in figure 8-c). At this optimal value, the SRAF distribution is able to cover a large portion of the 1st mode SRAF's, allowing the RET performance to be maintained without significant losses.

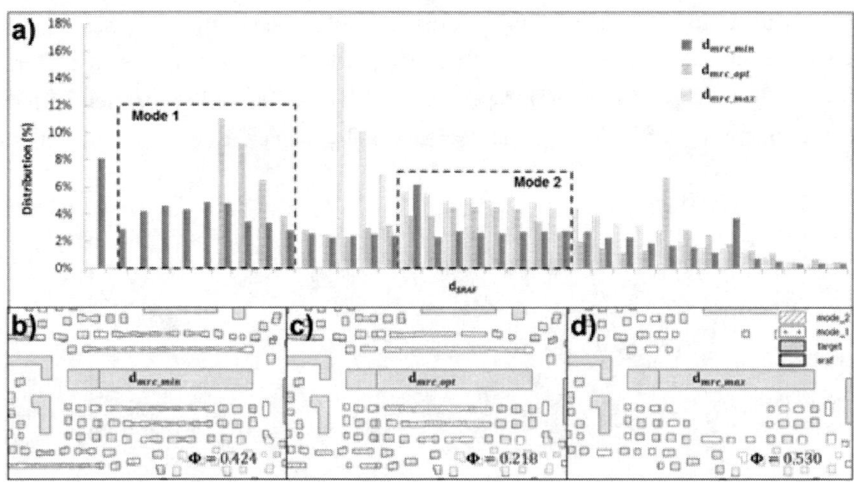

Figure 8: (a) SRAF width distributions for three separate cases that represent the high, low, and optimal values of η. (b-d) SRAF's remaining after Manhattanization for the high, moderate, and low values of η, corresponding to $d_{mrc} = \mathbf{d}_{mrc\_min}, \mathbf{d}_{mrc\_opt}$, and $\mathbf{d}_{mrc\_max}$, respectively.

6. CONCLUSION

A quantitative objective function Φ comprised of several metrics (EPE, PV-band, OPC runtime, mask vertex count, SRAF width/space) has been employed to analyze the relationship between mask manufacturability and RET performance for masks produced by Manhattanization of ILT outputs. Applying larger MRC limits during the Manhattanization reduces the complexities inherent to ILT outputs to increase manufacturability at the cost of RET performance degradation. Using values from simulations for a test layout clip, it is demonstrated here that these opposing interests and the balance between the two are well represented by the behavior of Φ, with lower values of Φ representing masks that better balance mask manufacturability with RET performance. By analyzing different components of Φ, the optimal mask complexity η_{opt} is determined as the critical point. Mask complexity beyond this value does not improve performance significantly while manufacturability penalties increase linearly. Masks with less complexity suffers exponential performance degradation while manufacturability penalties decrease linearly and unable to balance the sharp loss in performance. A closer look at the SRAF size distribution for the masks show that there are two main modes of SRAF's generated during ILT computation in the case of the tested layout clips. It is observed that the optimal mask complexity coincides with the larger MRC constraint that still allows the smallest SRAF modes, thus maintaining RET performance with lower mask complexity. These results demonstrate the usefulness of a quantitative method, whereby the manufacturing challenges and the associated performance gains in ILT produced masks are measured and compared in a consistent manner. This method is able to distinguish the conditions that provide the optimal balance of mask manufacturability and performance, allowing a more efficient and effective use of ILT in a practical manner.

References

[1] Choi, J., Kang, I.Y., Park, J. S., Shin, I. K., and Jeon, C.U., "Manufacturability of computation lithography mask: current limit and requirements for sub-20nm node," Proc. SPIE 8683, 86830L (2013).

[2] Granik, Y., "Fast pixel-based mask optimization for inverse lithography," J. Micro/Nanolith., 5(4) 043002 (2006).

[3] Kim, B. G. *et al*., "Trade-off between inverse lithography mask complexity and lithographic performance," Proc. SPIE 7379, 73791M (2009).

[4] Pang, L., Liu, Y., and Abrams, D., "Inverse Lithography Technology (ILT): what is the impact to the photomask industry?" Proc. SPIE 6283, 62830X (2006).

[5] Villaret, A., Tritchkov, A., Entradas, J., and Yesilada, E., "Inverse lithography technique for advanced CMOS nodes," Proc. SPIE 8683, 86830E (2013).

[6] Word, J. *et al*., "Inverse vs. traditional OPC for the 22nm node," Proc. SPIE 7274, 72743A (2009).

Akaike information criterion to select well-fit resist models

Andrew Burbine[*ab], David Fryer[c], John Sturtevant[c]

[a]Mentor Graphics Corporation, Leuven, Belgium
[b]Rochester Institute of Technology, 82 Lomb Memorial Drive, Rochester NY 14623, USA
[c]Mentor Graphics Corporation, 8005 S.W. Boeckman Road, Wilsonville, OR 97070

ABSTRACT

In the field of model design and selection, there is always a risk that a model is over-fit to the data used to train the model. A model is well suited when it describes the physical system and not the stochastic behavior of the particular data collected. K-fold cross validation is a method to check this potential over-fitting to the data by calibrating with k-number of folds in the data, typically between 4 and 10. Model training is a computationally expensive operation, however, and given a wide choice of candidate models, calibrating each one repeatedly becomes prohibitively time consuming. Akaike information criterion (AIC) is an information-theoretic approach to model selection based on the maximized log-likelihood for a given model that only needs a single calibration per model. It is used in this study to demonstrate model ranking and selection among compact resist modelforms that have various numbers and types of terms to describe photoresist behavior. It is shown that there is a good correspondence of AIC to K-fold cross validation in selecting the best modelform, and it is further shown that over-fitting is, in most cases, not indicated. In modelforms with more than 40 fitting parameters, the size of the calibration data set benefits from additional parameters, statistically validating the model complexity.

Keywords: Akaike information criterion, model selection, OPC, photoresist models

1. INTRODUCTION

Process modeling has been used in full-chip optical proximity correction (OPC) for more than ten years, and has evolved from an incremental process window and yield expander to required vital element in the data preparation flow[1]. Today, OPC and post-OPC verification simulation systems are a mission-critical element in multi-billion dollar fab operations. They must be stable, highly reliable, and cost-effective. The total time from a design release to yielding silicon is critical to overall manufacturing cost, and key for enabling rapid time to market for new designs. The training and validation of the best process model can be time consuming and, at times, non-definitive, especially when multiple candidate models appear to be equally valid with respect to the metrics of interest. The models which drive these solutions, while calibrated on a relatively small set of test patterns (O10^3), must faithfully predict the patterning of all design features that the factory will manufacture (O10^8), and therefore it is imperative that the user has high confidence in the chosen model.

Patterning process models for full-chip OPC and verification have utilized semi-rigorous mask and optics representation coupled to a photoresist model. These photoresist models are variously referred to as "semi-empirical", "black box", "compact", "lumped", or "behavioral", but are characterized by a mathematical formulation that provides a transfer function between inputs and measured outputs of interest. For OPC models, the input is an aerial image and the output is a predicted resist contour. An important facet of these models is that the user does not need access to sophisticated physicochemical characterization methods. Rather, all inputs required to calibrate the model are readily available in the fab: specifically the critical dimension (CD) for specific layout locations, termed "gauges". The optical exposure model can be formulated using readily available tool design information, but the equivalent information for the photoresist system typically is not known. Mathematical aspects of these process models can be related to physical phenomena which are operative in the chemical system[2], but a detailed mechanistic chemical and kinetic understanding is not necessary to yield very useful and accurate simulation results.

In practice, a full calibration flow includes calibration gauges and distinct verification or validation gauges. One common approach is to start with a master set of patterns, and use approximately one half to train the model, and the other half for verification. The goal in such a case would be for the root mean square error

Design-Process-Technology Co-optimization for Manufacturability IX, edited by John L. Sturtevant, Luigi Capodieci, Proc. of SPIE Vol. 9427, 94270J · © 2015 SPIE · CCC code: 0277-786X/15/$18 · doi: 10.1117/12.2085770

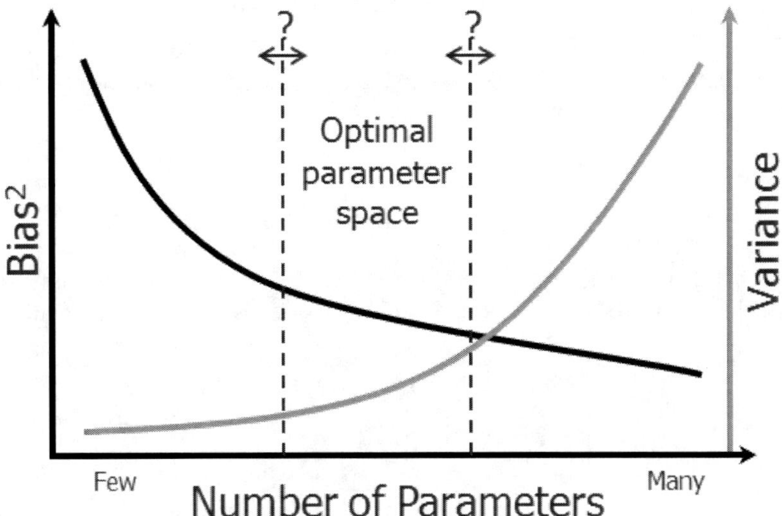

Figure 1. An illustration of the bias vs. variance trade-off based on the number of parameters in a model. The optimal parameter space expresses a good fit while not modeling stochastic or minimal effects in the system

(RMSE) fitness on the verification set to be similar to that of the calibration test suite. If verification fitness is significantly worse than calibration fitness, it may indicate a need to expand the range of design topologies incorporated into the training set.

Alternatively, such an observation may indicate that the model is over-fitting the calibration data, and the model should be simplified. Identification of the proper level of resist model complexity has been a difficult challenge facing the modeler, and there have been, to date, relatively few reports on this subject[3]. The goal is to find a well-fit model, which properly balances between bias (goodness of fit), and variance (sensitivity to training data)[4], as shown in Figure 1. This paper reports on the application of known methods in this field to the specific problem of empirical resist models for OPC.

2. K-FOLD CROSS VALIDATION

K-fold cross validation (CV) is a well-established method to assist in establishing the appropriate level of complexity in a model, by breaking up the total available pool of data into various sets or "folds" of calibration and validation groupings, and successively calibrating the model[5]. For a given data set, each sample appears in validation a single time. As shown in Figure 2 for an sample 4-fold CV for 3 modelforms, the optimum model is then the one with the lowest average K-fold validation error.

After a modelform is determined, a new model must then be calibrated for use in OPC, with either the entire data set, or some portion left out for use in final validation.

	Validation				
Modelform	I	II	III	IV	Average
Modelform A	2.56	2.13	1.98	2.42	*2.27*
Modelform B	1.81	1.55	1.63	1.81	*1.70*
Modelform C	1.22	1.20	1.32	1.17	*1.23*

Figure 2. Example 4-fold cross validation data. Shown are the RMS error values for each fold's validation set. The lowest average, modelform C, would be chosen as the champion modelform.

3. AKAIKE INFORMATION CRITERION

The Akaike information criterion (AIC) is an information-theoretic approach to model selection, penalizing models according to the number of free fitting parameters, thus requiring substantially improved calibration fitness be realized with each successive parameter added to the model[7]. Equation 1 is the formula used to calculate AIC for a single candidate model.

$$AIC = 2K - \log\left(\mathcal{L}(\hat{\theta}|y)\right) \tag{1}$$

Where K is the number of parameters (plus one for $\hat{\sigma}^2$) in the model, n is the number of samples and $\mathcal{L}$ is the maximized value of the log-likelihood function, for which we use the case for normally distributed errors:

$$\log\left(\mathcal{L}(\hat{\theta}|y)\right) = -\frac{n}{2}\log\hat{\sigma}^2 - \frac{n}{2}\log(2\pi) - \frac{n}{2} \tag{2}$$

$$\hat{\sigma}^2 = \frac{\sum \epsilon_i^2}{n} \tag{3}$$

AIC is calculated for each model, $y_i = f(\hat{\boldsymbol{\theta}}) + \epsilon_i$, and the model with the lowest value is the champion model.

It is important to note that AIC criteria compares candidate models with differing quantity of optimized parameters to determine the model that achieves the best fit relative to the amount of information (unique CD measurements) available. It provides quantitative differentiation in the spirit of the Occam's Razor postulate; it rewards parsimony in the model complexity and for two candidates of equal fitting power will always choose the simplest[6].

There is a correction for AIC, known as AIC_c, that is for use in situations where there are a high number of parameters or a small sample size; $K > n$ situations should be avoided, however. Because AIC_c approaches AIC as sample size becomes large or the number of parameters is small, it is always correct to use AIC_c over AIC, and this paper does so. As a guideline, the ratio of n/K (using the model with the highest number of parameters) should be above 40 for strict AIC use. Equation 4 shows the corrective term.

$$AIC_c = AIC + \frac{2K(K+1)}{n-K-1} \tag{4}$$

Akaike information criterion is a *comparison* metric, and as such the absolute values of each model's AIC are irrelevant - the model with the lowest AIC is the champion model, and the rest of the models are then ranked as their distance from this champion[8]. It is possible for AIC to suggest more than one model, if the distances are close enough. Equations 5 and 6 show how to attain each model's weight, which will always sum to 1 for a suite of models; the model with the lowest AIC value will have the largest weight.

$$\Delta_i = AIC_i - AIC_{min} \tag{5}$$

$$Weight = \frac{e^{-0.5\Delta_i}}{\sum e^{-0.5\Delta_i}} \tag{6}$$

As in K-fold CV, it is desired to not do statistical inference on the same data that is used to calibrate a model, so each dataset is randomly split into two halves, with one for calibration and the other is used to calculate AIC. The data used to calculate AIC must be identical across all models that are compared. The mean and variance of the modeled data are checked to be less than 1% different from those of the measured data, and residuals are checked for normality. These qualifications are to ensure the assumptions used to justify our maximum likelihood estimator are valid.

4. CM1 MODELS

The advent of dense full-chip simulation was accompanied by a new type of resist model which applied a constant threshold to a two dimensional resist surface. The resist surface is generated by applying a variety of different fast mathematical operators to the aerial image surface. These operators include neutralization, differentiation of order k, power n, kernel convolution, and n order root (equations 7 & 8). There are additional terms recently added for negative tone develop systems as well[9]. The user can specify a modelform which selects which operators and k, n, and p values are desired, thus as with the below variable threshold model, a huge number of different forms are possible. The linear coefficients C_i and continuous parameters b and s are found by minimizing the objective function during calibration.

$$R(x,y) = T = \sum_{i=0}^{N-1} C_i F_i(x,y) \tag{7}$$

$$F(x,y) = [(\nabla^k I_{\pm b}(x,y))^n \otimes G_{s,p}(x,y)]^{1/n} \tag{8}$$

The CM1 model is extremely flexible, and thus many different modelforms can be formulated with as few as one to as many as 50 fitting parameters. In this study, we use compact modelforms commonly employed in industry, as well as more recently derived modelforms for negative tone development. Table 1 shows the nested nature of these modelforms, ranging from a baseline with a few diffusion terms, to including specialized NTD terms or expanded ranges of parameters in the models. Because an exhaustive full parameter search is computationally impossible, we must instead choose from among these modelforms, using the methods outlined above.

Table 1. A selection of the modelforms available in Calibre Workbench used for testing *AIC*. The models are largely nested, as additional terms have been added to the pool over time.

Modelform	Parameters	Added Terms
A	13	Baseline
B	18	$A + I_{max}$ & I_{min}
C	20	B + Curvature
D	39	B + NTD terms & expanded b range
E	29	D + NTD terms
F	46	G + 2nd generation NTD terms
G	36	C + Additional diffusion terms
H	45	G + NTD terms

The classical minimum necessary condition required to avoid over-fitting is to ensure that the number of data points exceeds the number of degrees of freedom in the model. For OPC models, the number of data points is typically several orders of magnitude higher than the degrees of freedom, so this is not necessarily a concern. With a large number of fitting parameters, the optimization process can become very time consuming, therefore sophisticated optimization methods such as gradient descent or quasi-newton search allow for efficient exploration of the RMS response surface to locate local minima[10]. This is important in maintaining rapid turnaround time for the model calibration process, and can efficiently search vast parameter space surfaces to find globally optimum solutions.

The degrees of freedom in the model will interact with the metrology noise such that it is possible to "over-fit" the physical phenomena and start fitting the experimental noise, or fit features that appear in few samples. In this work we do not explicitly treat the experimental uncertainty associated with the experimental inputs, but in future work we will investigate the impact of measurement uncertainty which is typically dependent upon feature type. The goal in such a case would be to fit at close to 95% of the gauges within their respective 95% uncertainty values, thus constituting an alternative approach to avoiding over-fitting.

5. EXPERIMENTAL

Eight test cases were evaluated, spanning various technologies and layers, using 4-fold cross validation and Akaike information criterion. Photoresist model calibrations were performed using 40,000 iterations for each of a wide variety of CM1 modelforms. For AIC, the Δ_{min} value, weight and RMS error for the validation half is reported for each model. For 4-fold cross validation, data was randomly distributed into 4 separate groupings of 75% / 25%, where the 75% data was used for calibration and the remaining 25% was used for validation, such that each gauge appears in a validation set exactly once. The average RMS error across the validations is reported. Bolded values in each table indicate the champion value in that column.

5.1 Test case I

Test case I is from a 20 nm node data set that used an NTD resist. The gauge set was large, with 2840 gauges available for model calibration.

Table 2. Test case I with 1420 gauges used in AIC analysis. In this case, the model with the lowest RMSE is chosen by AIC despite being the most complex model.

Modelform	Parameters	Δ_{min}	Weight	RMSE	4-Fold Validation
A	13	750	0.00	1.72	1.75
B	18	470	0.00	1.52	1.53
C	20	444	0.00	1.42	1.39
D	39	175	0.00	1.28	1.33
E	29	233	0.00	1.26	1.21
F	46	0	**1.00**	**1.15**	**1.12**
G	36	427	0.00	1.28	1.26
H	45	297	0.00	1.19	1.15

For this data set, AIC chose the model with the lowest RMSE, but by doing so proclaims that there is enough data to warrant the complexity of modelform F (the most complex modelform), over any simpler, worse performing models. 4-fold CV also finds modelform F to be the champion modelform, further expressing that the model has a low over-fitting risk.

5.2 Test case II

Test case II is from a 20 nm node data set that used a PTD resist. The gauge set had 1630 gauges available for model calibration.

Table 3. Test case II with 815 gauges used in AIC analysis. In this case, there is not sufficient enough motivation to wholly choose one model, and the AIC weights reflect this.

Modelform	Parameters	Δ_{min}	Weight	RMSE	4-Fold Validation
A	13	881	0.00	3.73	3.69
B	18	279	0.00	2.56	2.57
C	20	255	0.00	2.51	2.55
D	39	198	0.00	2.37	2.21
E	29	270	0.00	2.51	2.33
F	46	106	0.00	2.22	2.21
G	36	5	0.07	2.11	2.20
H	45	0	**0.93**	**2.08**	**2.08**

In this data set, as compared to the last, there are fewer gauges available. Again, AIC picks the model with the lowest RMSE, but is not steadfast in the choice. The weight value can be thought of as the likelihood that the model is the best one, and such that modelform H is "the best" 93% of the time for data of this profile (model complexity and RMSE pairings). If there were more gauges, one could speculate that modelform H's complexity would be fully justified, however, we must evaluate the data as it stands.

The 4-fold CV results confirm the AIC notion quite well; modelform H has a much lower average validation than the rest of the candidate models, indicating it as the champion modelform.

5.3 Test case III

Test case III is from a 45 nm node data set that used a PTD resist. The gauge set had only 478 gauges available for model calibration.

Table 4. Test case III with 239 gauges used in *AIC* analysis. This case exemplifies when over-fit risk will steer AIC towards simpler models; though the fit for modelform H is better, there are not enough samples to justify a much more complex modelform.

Modelform	Parameters	Δ_{min}	Weight	RMSE	4-Fold Validation
A	13	146	0.00	1.84	1.68
B	18	22	0.00	1.32	1.35
C	20	0	**1.00**	1.26	1.27
D	39	50	0.00	1.16	**1.09**
E	29	44	0.00	1.30	1.34
F	46	87	0.00	1.17	**1.09**
G	36	46	0.00	1.24	1.23
H	45	17	0.00	**1.11**	1.14

In general, *AIC* will attempt to answer the question "Do I have enough data to justify my complex models?", and in this circumstance, the answer is: "No". This test case has a very limited gauge set, and as such is not able to justify the more complex modelforms. Despite having a worse RMSE, modelform C, with 25 fewer parameters than modelform H, is the definitive choice. Modelform H displays over-fit risk for this data, and should not be used further.

4-fold CV fails under these circumstances, and shows inconclusive results. Though modelforms D and F show the lowest validation RMSE, the data has become too fractured, with the standard deviation between fold validations is higher than for the other data sets, as well. In general, this is a tough data set because of the extremely limited gauge set, and a possible recommendation would be to collect more data before proceeding. However, given the information available to us at this juncture, modelform C represents the least risk.

5.4 Other test cases

In the remaining five test cases, model selection results fell into mostly the above form. Three additional testcases were of the form of test case I, where a single model was chosen by *AIC*, and 4-fold CV validated the choice. In some cases this was not the most complex modelform, but was always the model with the lowest RMSE. With sufficient sample size, complex modelforms (D-H) performed quite well in this study.

The remaining two cases were in the form of test case III, where a simpler model was chosen for lack of sample size (488 and 161 gauges), and 4-fold CV offered different results. For these cases, it is observed that AIC_c is the difference in the recommendation, and without the small sample size term, a more complex model would be champion. K-fold CV lacks intuition of this type, and is perhaps cause for critique of its results.

6. DISCUSSION

The methods outlined in this paper have provided quantitative rigor on the question of "over-fitting" in complex modelforms. We have demonstrated that there is risk for small sample sizes, and that *AIC* is capable of identifying such cases, and steering the modeler to the best choice among candidates. K-fold CV validates models with an abundance of gauges, but is not equipped to truly address the issue of adequate information when sample sizes are prohibitive. Because of this and runtime improvements, it appears that *AIC* is a better choice.

AIC does not provide a methodology to select feature types, an important facet of photoresist modeling which influences the physical nature of effects the modeler wishes to capture. By design, *AIC* will recommend simpler modelforms for calibration gauges with little parameter space coverage (because more complex modelforms will not improve fit substantially enough). Thus, *AIC* recommends the modelform with the best prediction accuracy based on the input calibration feature types, but can only select from modelforms and input calibration gauges which are analyzed.

7. CONCLUSIONS

Akaike information criterion is a holistic approach to model selection that provides consistent results via a simple execution of a single calibration and validation pair per candidate model. This is an improvement over K-fold CV, which revealed its weakness for small sample sizes and lack of actual knowledge on the complexity of models. AIC is capable of assessing when it is correct to use simpler models, and provides a ranking of models in the case of indecision. AIC has shown its analytical prowess to address the concern of "over-fitting" and allows for the confidence of modelers to use the necessarily complex models that drive modern photoresist OPC needs.

REFERENCES

1. J. Sturtevant, "The evolution of patterning process models in computational lithography," *Proc. SPIE* **7639**, 2010.
2. Y. Granik, D. Medvedev, and N. Cobb, "Towards standard process models for OPC," *Proc. SPIE* **6250**, 2007.
3. A. Abdo, R. Fathy, A. Seoud, J. Oberschmidt, S. Mansfield, and M. Talbi, "The effect of opc optical and resist model parameters on the model accuracy, runtime and stability," *Proc. SPIE* **6349**, 2006.
4. C. Sammut and G. I. Webb, *Encyclopedia of Machine Learning*, Springer Science & Business Media, 2011. Bias-variance decomposition.
5. S. Geisser, *Predictive Inference*, Chapman and Hall, 1993.
6. T. Hastie, R. Tibshirani, and J. Friedman, *The Elements of Statistical Learning*, Springer, 2009.
7. H. Akaike, "Information theory and an extension of the maximum likelihood principle," *International Symposium on Information Theory*, pp. 267–281, 1973.
8. K. Burnham and D. Anderson, *Model Selection and Multimodel Inference: A Practical Information-theoretic Approach*, Springer, 2002.
9. C. Ao, Y. Foong, D. Zhang, H. Zhang, D. Fryer, Y. Deng, D. Medvedev, and Y. Granik, "Evaluation of compact models for negative tone development at 20/14nm nodes," *Proc. SPIE* **TBD**, 2015.
10. J. Snyman, *Practical Mathematical Optimization: An Introduction to Basic Optimization Theory and Classical and New Gradient-Based Algorithms*, Springer, 2005.

Fast source optimization by clustering algorithm based on lithography properties

Masashi Tawada[†], Takaki Hashimoto[‡], Keishi Sakanushi[‡], Shigeki Nojima[‡], Toshiya Kotani[‡], Masao Yanagisawa[†], and Nozomu Togawa[†]

[†]Waseda University, Tokyo, Japan;
[‡]Toshiba Corporation, Yokohama, Japan

ABSTRACT

Lithography is a technology to make circuit patterns on a wafer. UV light diffracted by a photomask forms optical images on a photoresist. Then, a photoresist is melt by an amount of exposed UV light exceeding the threshold. The UV light diffracted by a photomask through lens exposes the photoresist on the wafer. Its lightness and darkness generate patterns on the photoresist. As the technology node advances, the feature sizes on photoresist becomes much smaller. Diffracted UV light is dispersed on the wafer, and then exposing photoresists has become more difficult. Exposure source optimization, SO in short, techniques for optimizing illumination shape have been studied. Although exposure source has hundreds of grid-points, all of previous works deal with them one by one. Then they consume too much running time and that increases design time extremely. How to reduce the parameters to be optimized in SO is the key to decrease source optimization time. In this paper, we propose a variation-resilient and high-speed cluster-based exposure source optimization algorithm. We focus on image log slope (ILS) and use it for generating clusters. When an optical image formed by a source shape has a small ILS value at an EPE (Edge placement error) evaluation point, dose/focus variation much affects the EPE values. When an optical image formed by a source shape has a large ILS value at an evaluation point, dose/focus variation less affects the EPE value. In our algorithm, we cluster several grid-points with similar ILS values and reduce the number of parameters to be simultaneously optimized in SO. Our clustering algorithm is composed of two STEPs: In STEP 1, we cluster grid-points into four groups based on ILS values of grid-points at each evaluation point. In STEP 2, we generate super clusters from the clusters generated in STEP 1. We consider a set of grid-points in each cluster to be a single light source element. As a result, we can optimize the SO problem very fast. Experimental results demonstrate that our algorithm runs speed-up compared to a conventional algorithm with keeping the EPE values.

Keywords: Source Optimization (SO), Image Log Slope (ILS), Lithography

1. INTRODUCTION

Lithography is a technology to make circuit patterns on a wafer. UV light diffracted by a photomask forms optical images on a photoresist. Then, a photoresist is melt by an amount of exposed UV light exceeding the threshold. Fig. 1 demonstrates the mechanism of lithography. The UV light diffracted by the photomask through the lens exposes the photoresist on the wafer. Its lightness and darkness generate patterns on the photoresist. As the technology node advances, the interval between lines on photoresist becomes too much narrower. Diffracted UV light is dispersed on the wafer, and then exposing photoresists has become more difficult.

Exposure source optimization, SO in short, techniques for optimizing illumination shape have been studied. There have been proposed several SO algorithms.[1–10] Although exposure source has hundreds of grid-points, all of previous works deal with them one by one. Then they consume too much running time and that increases design time extremely. How to reduce the parameters to be optimized in SO is the key to decrease source optimization time.

In this paper, we propose a variation-resilient and high-speed cluster-based exposure source optimization algorithm. In our algorithm, we cluster several grid-points with similar image log slope (ILS) values and reduce the number of parameters to be optimized in SO. We consider a set of grid-points in each cluster to be a single light source element. As a result, we can optimize the SO problem very fast. Experimental results demonstrate

Design-Process-Technology Co-optimization for Manufacturability IX, edited by John L. Sturtevant, Luigi Capodieci,
Proc. of SPIE Vol. 9427, 94270K · © 2015 SPIE · CCC code: 0277-786X/15/$18 · doi: 10.1117/12.2087007

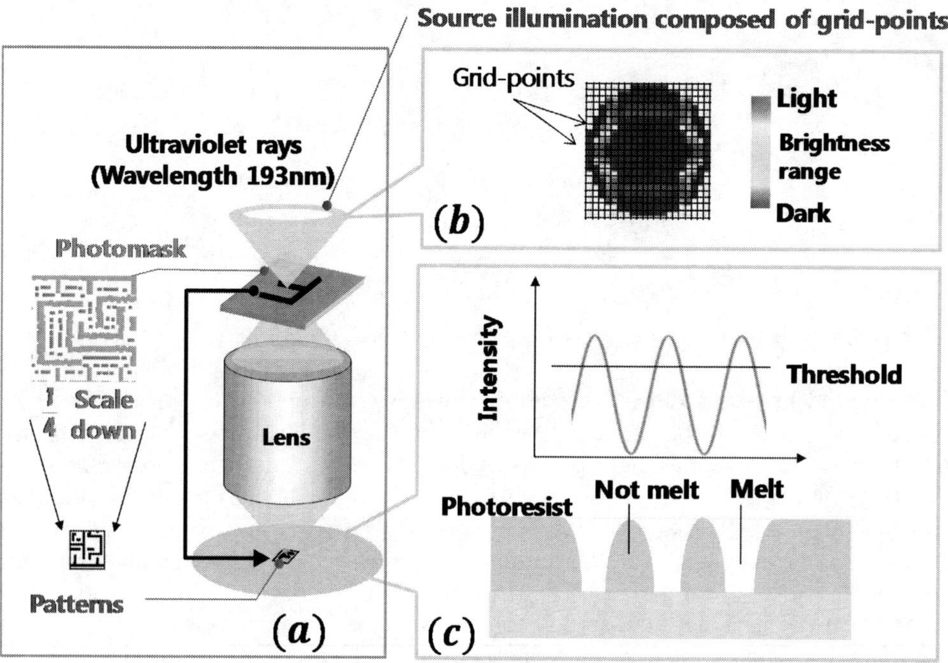

Figure 1. The mechanism of lithography.

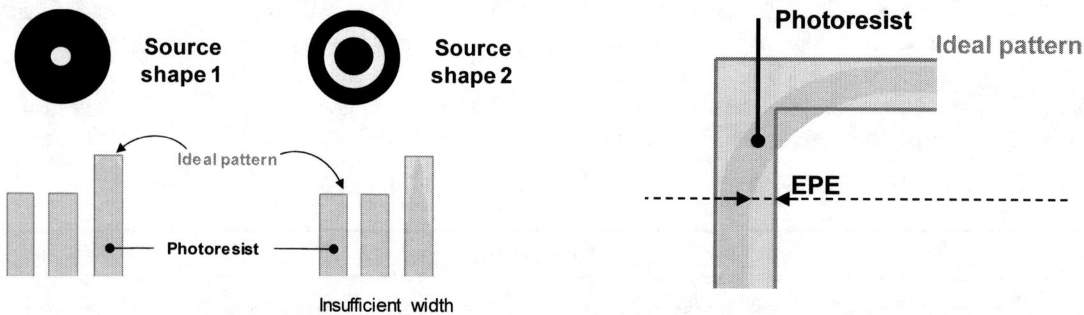

Figure 2. Source shapes and formed patterns.

Figure 3. EPE: the difference between the ideal pattern and the formed pattern.

that our algorithm achieves 8X speed-up compared to a conventional algorithm with keeping dose-variation resilience.

The main contributions in this paper are that:

- We formulate exposure source optimization as a combinatorial optimization problem. This paper opens up a new field to EDA.
- We propose an effective clustering algorithm based on one of lithographic properties to solve the SO problem and experimental results show 8X speed-up compared to a conventional approach.

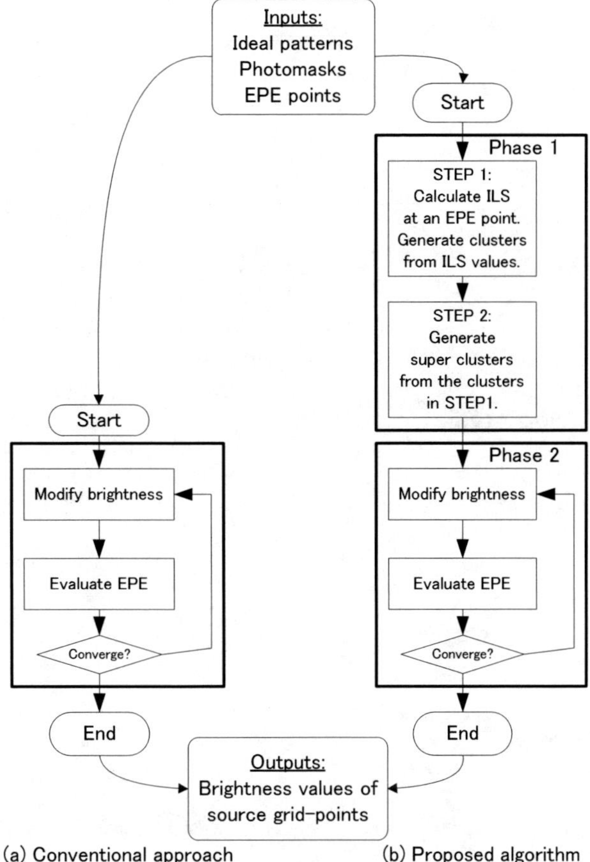

Figure 4. Flows of (a) conventional algorithm and (b) proposed algorithm for source optimization.

2. SOURCE OPTIMIZATION

2.1 Source optimization problem

Let us first consider a source optimization problem (see Fig. 4). Its inputs are ideal patterns, photomasks, and *EPE points* on ideal patterns. EPE points are the points on ideal patterns where we have to check if a sufficient circuit pattern is formed. Its outputs are brightness values of source grid-points. The objective function is shown below:

$$objective = \frac{\sqrt{\Sigma(EPE/tolerance)^2}}{n}$$

where *tolerance* shows a given rate of error allowed, and n is the number of EPE points, and EPE shows *edge placement error*. EPE is calculated on every EPE point in the optical images generated from the output source shape.

Source shape is determined by illumination brightness values. Fig. 2 shows two source shapes. Source shape 1 in Fig. 2 forms a sufficient circuit pattern but Source shape 2 in Fig. 2 forms an insufficient one. As depicted in Fig. 3, EPE represents the difference between the ideal pattern and the pattern formed by the source shape at each EPE point.

2.2 Source optimization by clustering

As we mentioned earlier, the existing works[1–10] deal with all of the grid-points and try to optimize them (Fig. 4(a)). How to reduce the number of parameters to be optimized in SO is the key to speed-up source

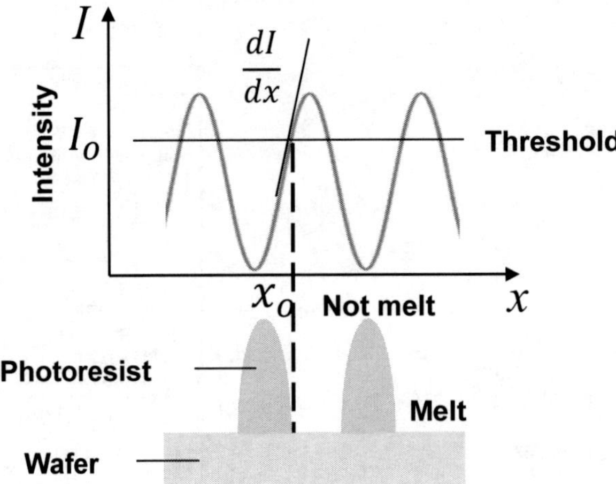

Figure 5. I_0 value and dI/dx value for calculating ILS (Image Log Slope).

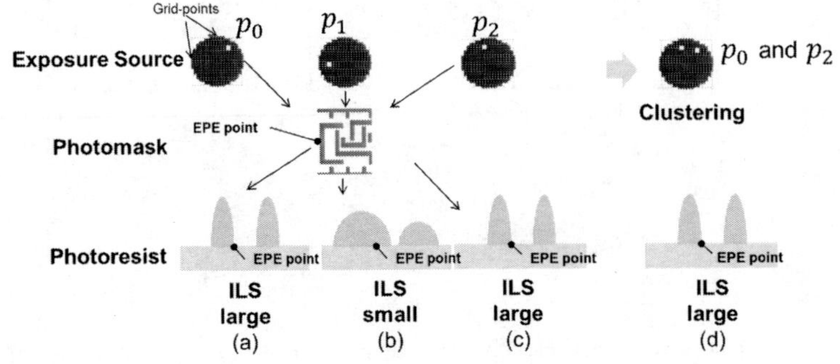

Figure 6. ILS value at an EPE point and clustering.

optimization. Clustering several grid-points based on some *lithographic properties* can lead to one of the solutions to this problem, i.e, the approach shown in Fig. 4(b) must be one of the solutions. In Fig. 4(b), we have two phases. Phase 1 is *clustering* of grid-pints on exposure source. After that, we optimize clustered grid-points as a single light source element using any combinational optimization algorithm in Phase 2.

Now we discuss how to cluster grid-points on exposure source.

2.2.1 Image Log Slope (ILS)

Image log slope (ILS) at an EPE point in the optical image is defined by the following equation:

$$ILS = \frac{1}{I_0}\frac{dI}{dx} \tag{1}$$

As in Fig. 5, ILS shows the slope of the intensity (dI/dx) of the optical image divided by the intensity value itself (I_0) at an EPE point. We consider here how much brightness values of grid-points affect ILS.

In Fig. 6(a), the grid-point p_0 shines with the brightness of some value and the other grid-points does not shine. Then the source shape exposes the photoresist through the photomask and we can have the optical image for that. When we look at the intensity value on the EPE point of the optical image, we can calculate the ILS value. In Fig. 6(a), its value is assumed to be large. In the same way, we have small ILS value in Fig. 6(b) and large ILS value in Fig. 6(c).

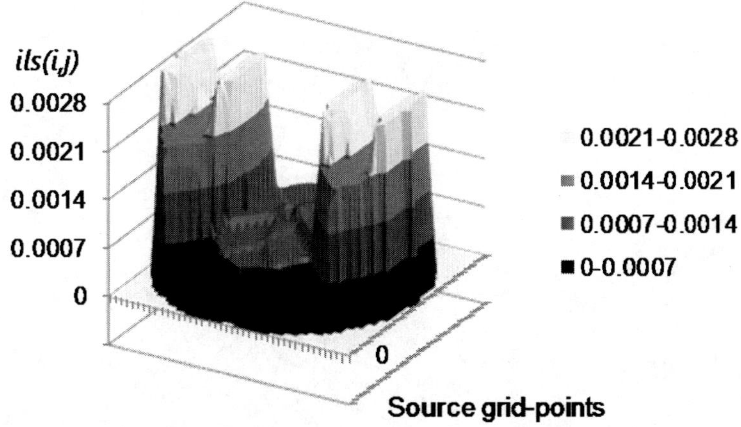

Figure 7. ILS values on each grid-point and clustering.

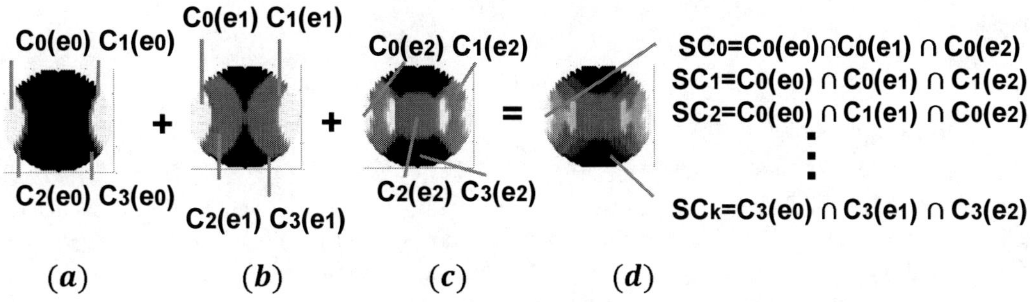

Figure 8. Generating super clusters from ILS-clusters (a) Clusters for the EPE point e_0. (b) Clusters for the EPE point e_1. (c) Clusters for the EPE point e_2. (d) Super clusters generated from (a)–(c).

When an optical image formed by a source shape has small ILS at an EPE point, dose/focus variation much affects the image. When an optical image formed by a source shape has large ILS at an EPE point, dose/focus variation less affects the image. This is because of the following reason: When dose/focus variation occurs, the threshold value shown in Fig. 5 is changed. In this case, if the ILS value on an EPE point is large enough, its optical image do not much change. In the same way, if the ILS value on an EPE point is small, its optical image may be changed much. ILS is one of lithographic properties which shows how much variation-resilient a source shape is. This means that, grid-points with similar ILS values can be optimized at the same time by clustering in SO, i.e, even if we cluster grid-points with similar ILS values and optimize them as a single light source element, we expect that it leads to a dose/focus-variation-resilient result.

Now Let P be an overall set of grid-points on exposure source. Let E be a set of EPE points. Firstly, we generate an optical image by the source shape with each grid-point $p_i \in P$ only shining and calculate its ILS value $ils(i, j)$ for every EPE point $e_j \in E$. We cluster grid-points whose ILS value is similar to each other as in Fig. 6(d) and optimize them as a single light source element.

2.2.2 Algorithm

Now we propose our variation-resilient cluster-based source optimization algorithm, which can run very fast by reducing the parameters to be optimized. Our algorithm is composed of two steps. In STEP 1, we cluster grid-points into four groups based on ILS values of grid-points at each EPE point.* In STEP 2, we generate super clusters from the clusters generated in STEP 1.

For every EPE point $e_i \in E$, let $C_0(e_i)$, $C_1(e_i)$, $C_2(e_i)$, and $C_3(e_i)$ be four clusters generated in STEP 1. $C_\ell(e_i)$ includes clustered grid-points. STEP 1 of our algorithm is described as follows:

*We have tried to cluster grid-points into 2–10 groups based on ILS values and found out that clustering grid-points into *four groups* leads to the best one.

STEP 1

```
for every EPE point e_i ∈ E do
  Calculate the maximum value max(e_i) and the minimum value min(e_i) of ils(i,j) at an EPE point e_i;
end for
for every EPE point e_i ∈ E do
  C_0(e_i)←φ;
  C_1(e_i)←φ;
  C_2(e_i)←φ;
  C_3(e_i)←φ;
  for every grid-point p_j ∈ P do
    if ils(i,j) > (3/4)max(e_i) + (1/4)min(e_i) then
      C_0(e_i)←C_0(e_i)∪{p_j};
    else if ils(i,j) > (1/2)max(e_i) + (1/2)min(e_i) then
      C_1(e_i)←C_1(e_i)∪{p_j};
    else if ils(i,j) > (1/4)max(e_i) + (3/4)min(e_i) then
      C_2(e_i)←C_2(e_i)∪{p_j};
    else
      C_3(e_i)←C_3(e_i)∪{p_j};
    end if
  end for
end for
```

Fig. 7 shows an example of STEP 1. In Fig. 7, $ils(i,j)$ values for all the grid-points at an EPE point e_i are depicted. Its maximum value is 0.0028 and its minimum value is 0.0000. Then, we cluster grid-points into four groups, $C_0(e_i)$, $C_1(e_i)$, $C_2(e_i)$, and $C_3(e_i)$. $C_0(e_i)$ includes grid-points whose ILS value is in the range from 0.0021 to 0.0028. $C_1(e_i)$ includes grid-points whose ILS value is in the range from 0.0014 to 0.0021. $C_2(e_i)$ includes grid-points whose ILS value is in the range from 0.0007 to 0.0014. $C_3(e_i)$ includes grid-points whose ILS value is in the range from 0.0000 to 0.0007.

After generating clusters for each EPE point, we generate super clusters in STEP 2. Let SC_i be the i-th super cluster generated in STEP 2. STEP 2 of our algorithm is described as follows:

STEP 2

```
i←0;
for every grid-point p_j ∈ P do
  if (p_j does not belong to any super cluster) then
    SC_i←{p_j};
    for every grid-point p_k ∈ P (j≠k) do
      if (p_k does not belong to any super cluster) then
        if (p_j and p_k belong to completely the same cluster at every EPE point) then
          SC_i←SC_i∪{p_k};
        end if
      end if
    end for
    i←i+1;
  end if
end for
```

For example, assume that we have three EPE points e_0, e_1, and e_2. We have four clusters for each of them (Figs. 8(a)–(c)). Then we generate super clusters as in Fig. 8(d).

All the grid-points in each super cluster can be considered to be a single light source element, i.e, they must have the same brightness value in SO. Based on this clustering, we can perform any combinational optimization solver such as Simulated Annealing (SA) and Genetic algorithm (GA), as Phase 2 of Fig. 4(b).

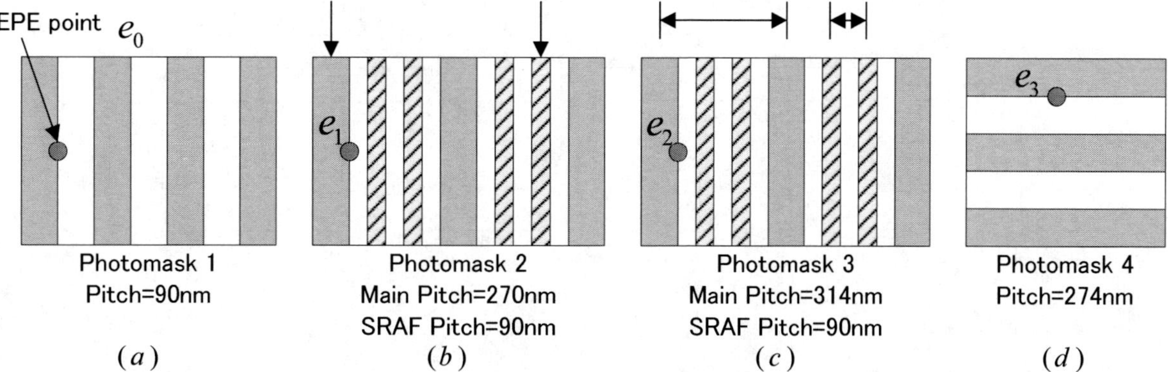

Figure 9. Input photomasks (a)–(d) and the EPE points e_0,e_1,e_2, and e_3 for the experiment.

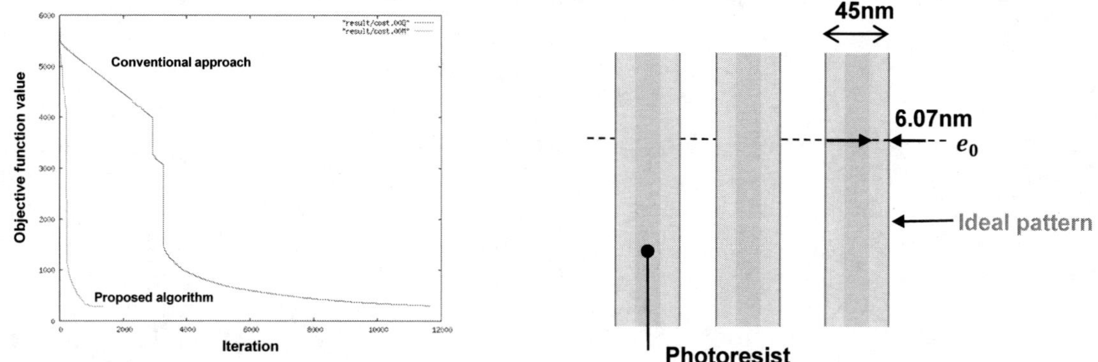

Figure 10. How the objective function value decreases as SA iterations proceed.

Figure 11. EPE value at the EPE point e_0 with source shape made by our algorithm.

3. EXPERIMENTAL EVALUATIONS

We have implemented our proposed cluster-based source optimization algorithm in C on Quad-Core AMD Opteron Processor 2360 SE PC with 1.3GHz CPU and 16GB memory.

Fig. 9 shows input photomasks on which four EPE points e_0,e_1,e_2, and e_3 are shown. We have 333 grid-points, each of which has the brightness value from 0.00 to 1.00.

We have compared our algorithm to the conventional approach, where all 333 grid-points are optimized by Simulated Annealing. In our algorithm, we generated four clusters for each EPE point in STEP 1 and then generated 26 super clusters in STEP 2. After that we applied Simulated Annealing and optimized these 26 parameters.

Table 1 summarizes the experimental results. Our algorithm optimizes exposure source in 291 seconds while the conventional approach optimizes exposure source in 2585 seconds. Our algorithm achieves 8X speed-up. The table also shows EPE values at each EPE point. Both in the conventional algorithm and our algorithm, these values are allowable ones.

Fig. 10 shows how the objective function value decreases as SA iterations proceed. As in this figure, our algorithm quickly reaches the optimal value. Fig. 11 shows the result of EPE value at the EPE point e_0.

4. CONCLUSION

In this paper, we proposed a high-speed exposure source optimization algorithm with grid-point clustering. Experimental results demonstrate that our algorithm achieves 8X speed-up compared to a conventional algorithm

Table 1. Experimental Results.

	Conventional algorithm	Proposed algorithm
CPU Time	2585 sec	291 sec
The number of iterations	25058	3122
EPE point e_0	6.41 nm	6.07 nm
e_1	4.81 nm	4.98 nm
e_2	5.96 nm	5.91 nm
e_3	7.12 nm	8.19 nm

with keeping dose-variation resilience. We will further speed-up our source optimization algorithm by utilizing another lithographic property in the future.

ACKNOWLEDGMENTS

This work was supported by Grant-in-Aid for JSPS Fellows.

REFERENCES

[1] T. Fuhner, P. Evanschizky, and A. Erdmann, "Mutual source, mask and projector pupil optimization," in *Proc. SPIE 8326,* 83260I, 2012.

[2] S. Hsu, L. Chen, Z. Li, S. Park, K. Gronlund, H. Liu, N. Callan, R. Socha, and S. Hansen, "An Innovative Source-Mask co-Optimization (SMO) method for extending low k1 imaging," in *Proc. SPIE 7140,* 75200D, 2008.

[3] S. Hsu, Z. Li, L. Chen, K. Gronlund, H. Liu, and R. Socha, "Source-mask co-optimization: optimize design for imaging and impact of source complexity on lithography performance," in *Proc. SPIE 7520,* 75200D, 2009.

[4] K. Kodera, S. Tanaka, M. Yamaji, T. Kotani, S. Nojima, K. Hashimoto, S.Mimotogi and S. Inoue, "Novel fine-tuned model-based SRAF generation method using coherence map", in *Proc. SPIE 7640,* 764017, 2010.

[5] L. Liebmann, S. Mansfield, J. Bruce, M. Cross, I. Graur, A. McGuire, J. Krueger, and D. Sunderling, "Optimizing style options for sub-resolution assist features," in *Proc. SPIE 4346,* 2001.

[6] R. Matsui, T. Noda, H. Aoyama, N. Kita, T. Matsuyama, and D. Flagello, "Global source optimization for MEEF and OPE," in *Proc. SPIE 8683,* 86830O, 2013.

[7] M. Mulder, A. Engelen, O. Noordman, G. Streutker, B. van Drieenhuizen, C. van Nuenen, W. Endendijk, J. Verbeeck, W. Bouman, A. Bouma, R. Kazinczi, and R. Socha, "Performance of flexray, a fully programmable illumination system for generation of freeform sources on high NA immersion systems," in *Proc. SPIE 7460,* 76401P, 2010.

[8] A. E. Rosenbluth, S. Bukofsky, M. Hibbs, K. Lai, A. Molless, R. N. Singh, and A. Wong, "Optimum mask and source patterns to print a given shape," in *Proc. SPIE 4346,* 2001.

[9] A. E. Rosenbluth and N. Seong, "Global optimization of the illumination distribution to maximize integrated process window," in *Proc. SPIE 6154,* 61540H, 2006.

[10] R. Socha, X. Shi, and D. Lehoty, "Simultaneous source mask optimization (SMO)," in *Proc. SPIE 5853,* 2005.

Statistical Modeling of SRAM Yield Performance and Circuit Variability

Qi Cheng, Yijian Chen*

School of Electronic and Computer Engineering
Shenzhen Graduate School, Peking University
Shenzhen 518055, Guangdong, China

ABSTRACT

In this paper, we develop statistical models to investigate SRAM yield performance and circuit variability in the presence of self-aligned multiple patterning (SAMP) process. It is assumed that SRAM fins are fabricated by a positive-tone (spacer is line) self-aligned sextuple patterning (SASP) process which accommodates two types of spacers, while gates are fabricated by a more pitch-relaxed self-aligned quadruple patterning (SAQP) process which only allows one type of spacer. A number of possible inverter and SRAM structures are identified and the related circuit multi-modality is studied using the developed failure-probability and yield models. It is shown that SRAM circuit yield is significantly impacted by the multi-modality of fins' spatial variations in a SRAM cell. The sensitivity of 6-transistor SRAM read/write failure probability to SASP process variations is calculated and the specific circuit type with the highest probability to fail in the reading/writing operation is identified. Our study suggests that the 6-transistor SRAM configuration may not be scalable to 7-nm half pitch and more robust SRAM circuit design needs to be researched.

Keywords: SRAM, SASP, SAQP, circuit variability, statistical modeling, yield.

1. INTRODUCTION

As EUV lithography is slowly approaching its high-volume manufacturing insertion[1], alternative patterning techniques such as self-aligned multiple patterning (SAMP [2-5]) have been widely adopted for critical-layer patterning of logic devices. The conventional logic circuits are designed based on random 2-D geometry, while optical multiple patterning (MP) faces tremendous process-window challenges and is unlikely to be scalable to sub-15nm half pitch. As a result, IC design has shifted to the complementary type of layout scheme in which regular and dense 1-D features fabricated by a SAMP process are cut/blocked to form the desired patterns. Among the various types of SAMP processes, self-aligned sextuple patterning (SASP) is the technique most likely to be scalable to half-pitch 7 nm. As shown in Fig. 1, SASP process complexity is comparable to that of a more pitch-relaxed self-aligned quadruple patterning (SAQP) process, while it allows IC devices to be scaled down for one more node [4]. The SRAM circuit variability in the presence of self-aligned triple patterning (SATP) has been investigated in a previous paper [6]. Understanding and modeling the statistical circuit performance and the yield loss in more advanced SAMP processes has become an urgent issue in developing the next-generation logic IC technology. Unfortunately, our industry has not published any manufacturing data on a SASP process and the only possible way to estimate them is based on other similar type of patterning scheme such as a SAQP process [1]. As illuminated in Fig. 1, the final/structural spacers in a SASP process are spacer 1 (S1) and spacer 3 (S3) and the gap between them is defined by the sacrificial spacer 2 (S2). As shown in Table 1, CDU and LWR (line-width roughness) performance of SASP spacer 1 are estimated from the data of SAQP space (due to SAQP spacer 1) [7-8], while SASP spacer 3 CDU & LWR are estimated from the data of SAQP line (due to SAQP spacer 2). Here, the terms of "spacer 1 and spacer 3" simply indicate the origin of those fine features, and the enhancement/degradation of processing performance when the spacer patterns are etched and transferred to the underneath thin film(s) has been taken into account. The estimated CDU and LWR of SASP spacer 3 are better than that of SASP spacer 1 (by a factor of about 0.7). Such a difference can cause variations of critical device parameters if the device channels (e.g., fins) are fabricated by different types of spacers, as shown in Table 2. The impact of process

*Phone: (+86) 755-21536236, Email: chenyj@pkusz.edu.cn

Design-Process-Technology Co-optimization for Manufacturability IX, edited by John L. Sturtevant, Luigi Capodieci, Proc. of SPIE Vol. 9427, 94270M · © 2015 SPIE · CCC code: 0277-786X/15/$18 · doi: 10.1117/12.2085844 variations on IC performance has been extensively studied recently [9]-[13]. Completely random process variations are often assumed in their models, thus incapable of accurately describing the specific (spatial) variation of spacer/fin structures expected in a SASP process. In this paper, we report a statistical modeling approach to predict the circuit variability and yield performance of the inverters and SRAMs fabricated by a SASP process.

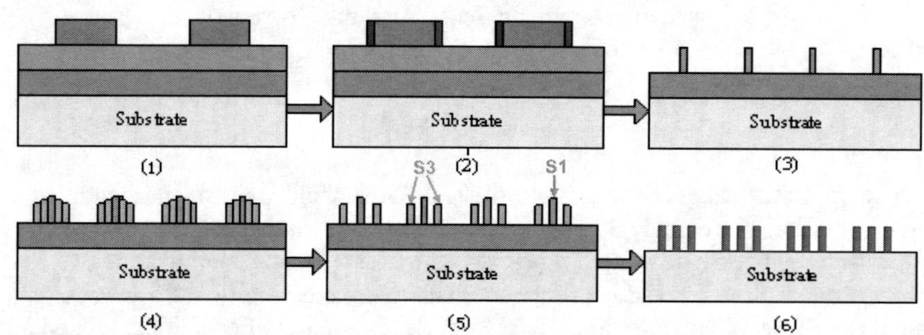

Fig. 1. A schematic description of the SASP process wherein the final/structural spacers are spacer 1 (S1) and spacer 3 (S3) while the gap between them is defined by the sacrificial spacer 2 (S2).

Table 1. Estimated CDU and LWR of the structural spacers in a SASP process [7-8].

Half Pitch (nm)	SASP Spacer1 CDU (nm)	SASP Spacer3 CDU (nm)
10	1.6	1.1
7	1.4	1.0
5	1.2	0.8

Half Pitch (nm)	SASP Spacer1 LWR (nm)	SASP Spacer3 LWR (nm)
10	1.8	1.3
7	1.6	1.1
5	1.4	1.0

Table 2. TCAD simulations of the FinFET device variability.

Half Pitch (nm)	SASP Spacer1 σVth.sat (mV)	SASP Spacer3 CDU σVth.sat (mV)
10	19.5	16.8
7	21.1	17.5

Half Pitch (nm)	SASP Spacer1 σIon (uA/um)	SASP Spacer3 CDU σIon (uA/um)
10	80	62
7	81	67

2. INVERTER VARIABILITY ANALYSIS

Multi-modality variations in CDU and LWR are often observed in a SAMP process. As shown in Fig. 2, when the fins of an inverter are fabricated by a self-aligned sextuple patterning (SASP) process, fins can consist of either spacer 1 or spacer 3. This will result in two types of inverter circuits with different statistical performance. However, it should be noted that exchanging the positions of two spacers/fins within one single transistor will not produce a new type of inverter/SRAM circuit. The load dependent inverter delay is given by the "alpha-power law" [14]:

$$D(v) = \frac{KC_v V_{dd}}{(V_{dd} - V_{th})^{\alpha}} \tag{1}$$

where C_v is the load capacitance of the transistor, V_{dd} is the supply voltage and V_{th} is the threshold voltage, K and α are the fitting parameters. We examine the dependence of inverter delay on inverter cell type. The upper and lower lines of the delay time are calculated by $V_{th} = \overline{V_{th}} + 3\sigma V_{th}$ and $V_{th} = \overline{V_{th}} - 3\sigma V_{th}$, respectively. Where $\overline{V_{th}}$ and σV_{th} are the mean value and standard deviation (considering the CD variation only) of the threshold voltage calculated by the model developed in [15]. As indicated by Fig. 3, the delay of the inverter shown in Fig. 2(b) is more sensitive to $\sigma V_{th.spacer1}$, while the delay of the inverter shown in Fig. 2(a) increases more rapidly with $\sigma V_{th.spacer3}$. The calculation results clearly indicate that significant difference of the inverter delay exists between these two types of circuits.

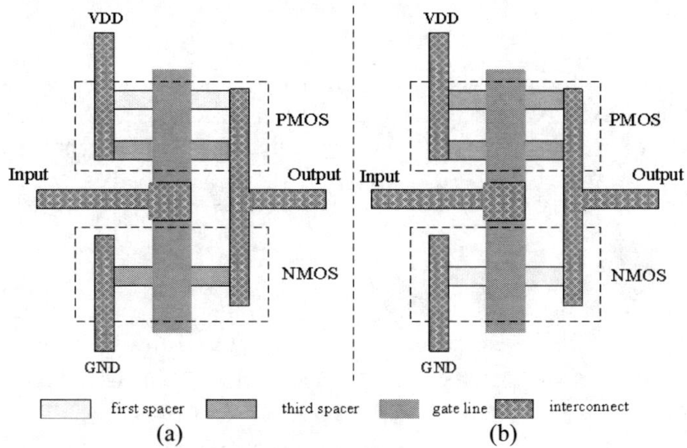

Fig. 2. Two possible inverter structures fabricated by a SASP process.
We assume one fin in NMOS and two fins in PMOS of an inverter (line/space is not to scale).

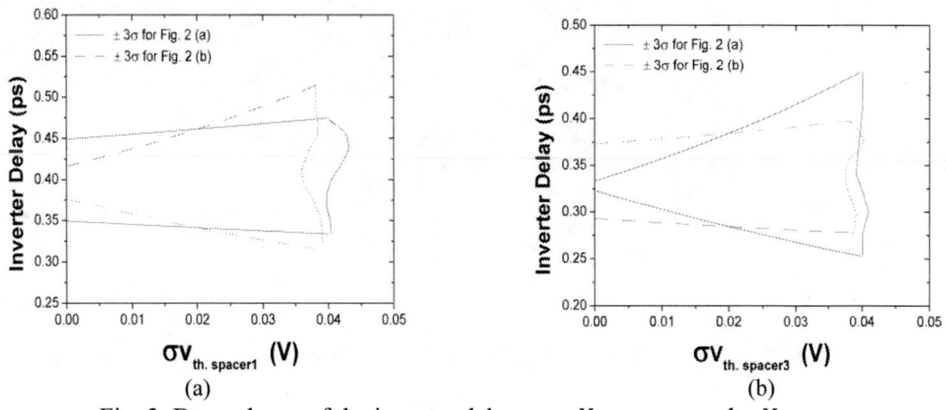

Fig. 3. Dependence of the inverter delay on $\sigma V_{th.spacer1}$ and $\sigma V_{th.spacer3}$.

3. SRAM VARIABILITY ANALYSIS

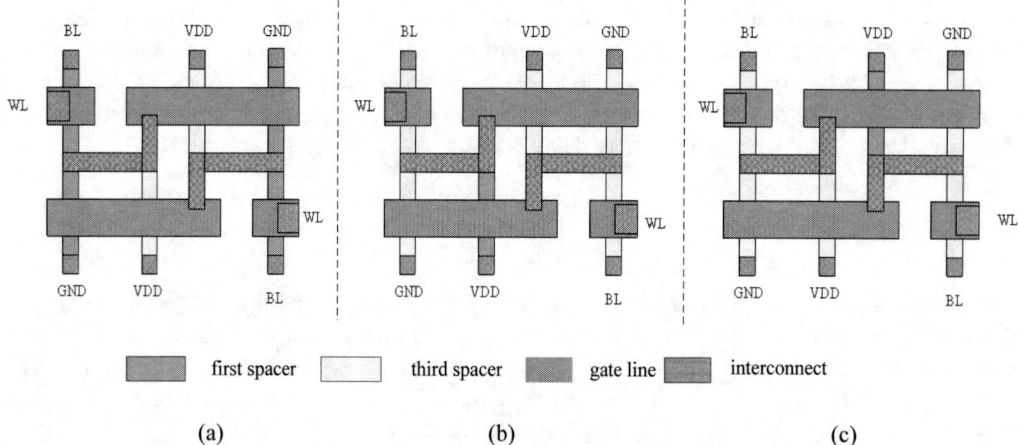

Fig. 4. Three possible SRAM structures with fins fabricated by a positive-tone (spacer is line) SASP process which accommodates two types of spacers (i.e., spacer 1 and spacer 3, [4]). However, gates can be fabricated by a more pitch-relaxed patterning technique such as positive-tone SAQP process which only allows on type of spacer [1], [4].

As shown in Fig. 4, due to the multi-modality of fins and related FinFET device variability, three typical SRAM structures can be fabricated in a SASP process. To enable a fast and accurate SRAM performance analysis, a simple "square law" is used to calculate the drain current. In the linear region, the drain current is calculated by:

$$I_{ds} = \mu C_{ox}\frac{mW}{L}\left[(V_{gs} - V_{th})V_{ds} - \frac{V_{ds}^2}{2}\right] \tag{2}$$

In the saturation region,

$$I_{ds} = \mu C_{ox}\frac{mW}{2L}(V_{gs} - V_{th})^2 \tag{3}$$

where μ is the carrier mobility (u_e for the electron and u_p for the hole), C_{ox} is the gate oxide capacitance, L is the channel length, m is a parameter that can be tuned to reflect the number of fins within a FinFET, and W is twice of the fin height. A first-order approximation to describe the DIBL effect can be obtained by [9]:

$$V_{th} = V_{th.sat} + \eta(V_{th.lin} - V_{th.sat})$$

where $V_{th.sat}$ and $V_{th.lin}$ are the threshold voltages extracted from the saturation and linear regions, respectively. η reflects the impact of DIBL and is calculated by $(V_{th.lin} - V_{th.sat})/(V_{dd} - 0.05)$ [16]. All these device parameters such as fin thickness (t_{si}) , supply voltage (V_{dd}) and doping concentration (N_a) are extracted from 2013 ITRS and listed as below:

Table 3. FinFET parameters used in our simulations.

Half Pitch (nm)	t_{si} (nm)	V_{dd} (V)	N_a (cm^{-3})
10 nm	3.7	0.77	1e17
7 nm	3.1	0.74	1e17

In the following sections, the device parameters at HP=10 nm will be used unless specified otherwise.

3.1 Read Failure Analyses

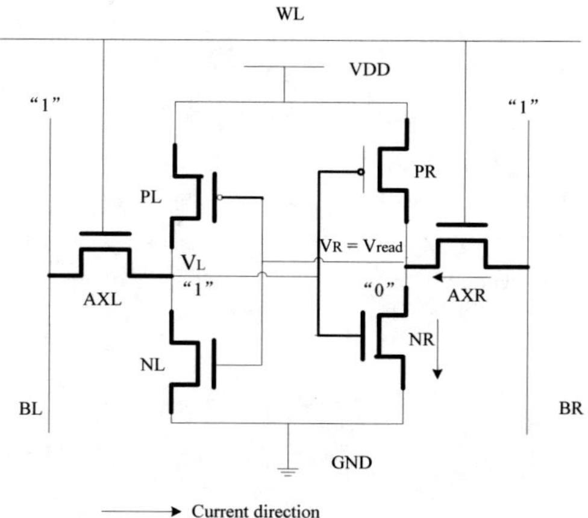

Fig. 5. A 6-transistor SRAM circuit in the reading operation.

While reading a SRAM shown in Fig. 5 ($V_L = 1$ V and $V_R = 0$ V), the voltage at "R" is increased to V_{read}. If V_{read} is higher than the trip-point voltage (V_{trip}), the stored data will be changed and a read failure occurs. V_{trip} can be obtained by solving Kirchhoff's law at "L" as:

$$I_{NL.sat}(V_g = V_{trip}, V_d = V_{trip}, V_s = gnd) = \\ I_{AXL.sat}(V_g = V_{dd}, V_d = V_{dd}, V_s = V_{trip}) + I_{PL.sat}(V_g = V_{trip}, V_d = V_{trip}, V_s = V_{DD}) \quad (4)$$

where $I_{NL.sat}$, $I_{AXL.sat}$ and $I_{PL.sat}$ are the saturation currents of NL, AXL and PL parts. Inserting (3) into (4), (4) can be reformulated as:

$$\frac{\mu_e W}{2L}(V_{trip} - V_{th.NL})^2 = \frac{\mu_e W}{2L}(V_{dd} - V_{trip} - V_{th.AXL})^2 + \frac{\mu_P W}{2L}(V_{trip} - V_{dd} - V_{th.PL})^2 \quad (5)$$

V_{trip} can be approximated as a linear function of $\Delta V_{th.NLsat}$, $\Delta V_{th.AXLsat}$ and $\Delta V_{th.PLsat}$ around the operating point:

$$V_{trip} \approx \overline{V_{trip}} + A_{trip}\Delta V_{th.NL\,sat} + B_{trip}\Delta V_{th.AXL.sat} + C_{trip}\Delta V_{th.PLsat} \quad (6)$$

where $\overline{V_{trip}} = 0.415$ V is the value of V_{trip} without any threshold-voltage variation, $A_{trip} = 0.4365$, $B_{trip} = -0.335$, and $C_{trip} = 0.2234$. In Fig. 6, we plot the relationship between V_{trip} and several threshold-voltage variations. V_{trip} linearly increases with $V_{th.NL}$ and $V_{th.PL}$, while decreases with the enhancement of $V_{th.AXL}$. The comparison with TCAD simulations indicates that the linearized model accurately describes the fluctuation tendency of V_{trip} near the operating point.

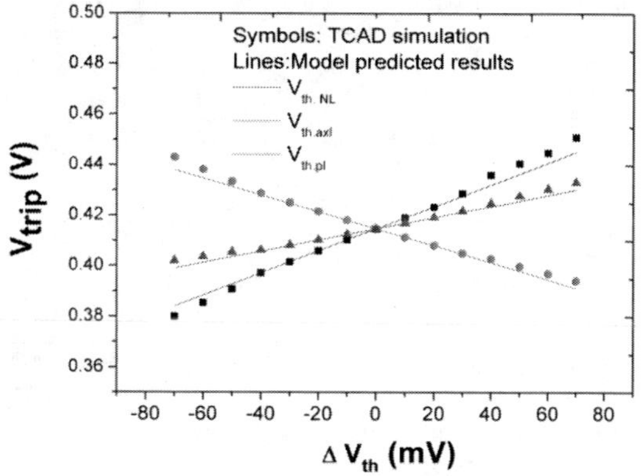

Fig. 6. V_{trip} vs. FinFET threshold-voltage variations.

During the reading operation, V_{read} can be obtained by solving Kirchhoff's law at "R" as:

$$I_{NR.lin}(V_g = V_{dd}, V_d = V_{read}, V_s = gnd) = I_{AXR.sat}(V_g = V_{dd}, V_d = V_{dd}, V_s = V_{read}) \quad (7)$$

Substituting (2) and (3) into (7) yields:

$$\frac{u_e W}{L}\left(V_{dd} - V_{th,NR} - \frac{V_{read}}{2}\right)V_{read} = \frac{u_e W}{2L}(V_{dd} - V_{read} - V_{th.AXR})^2$$

where $V_{th,NR}$ and $V_{th.AXR}$ are the threshold voltages of NR and AXR parts, respectively. The relationship between V_{read} and $V_{th,NR}$, $V_{th.AXR}$ is shown in Fig. 7. Similarly, V_{read} can be approximated as:

$$V_{read} \approx \overline{V_{read}} + A_{read}\Delta V_{th.NR} + B_{read}\Delta V_{th.AXR} \quad (8)$$

where $\overline{V_{read}} = 0.187$ V $A_{read1} = 0.264$, $B_{read1} = -0.548$. The model's result is compared with TCAD simulation in Fig. 7. It shows that higher $V_{th.AXL}$ and lower $V_{th.NL}$ will result in a lower V_{read}, which will reduce the probability of read failure.

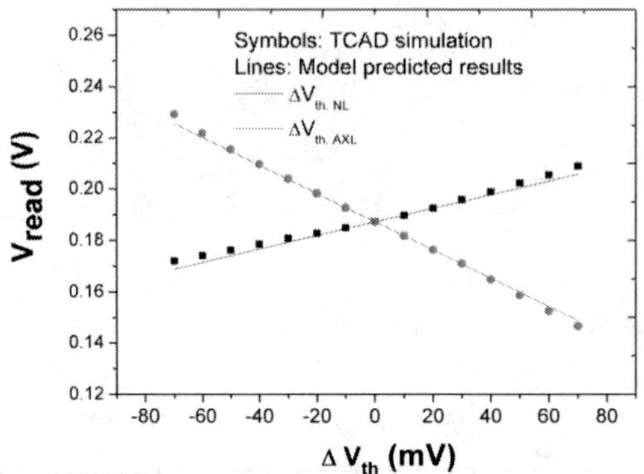

Fig. 7. V_{read} vs. FinFET threshold-voltage variations.

The standard deviations of V_{trip} and V_{read} as a function of the threshold-voltage variations can be obtained from (6) and (8), and the calculation results are shown in Fig. 8. Read failure probability is calculated as:

$$P_{read} = P(V_{read} > V_{trip}) = P(R = (V_{read} - V_{trip}) > 0)$$

where $\bar{R} = \overline{V_{read}} - \overline{V_{trip}}$, $\sigma_R{}^2 = \sigma_{trip}{}^2 + \sigma_{read}{}^2$, the probability of read failure for various devices with fins fabricated by a SASP process is shown in Fig. 9. The result indicates that the read failure probability of a SRAM cell is closely related to the fin type and the related threshold voltage fluctuation. For example, when $\sigma V_{th.spacer3}$ is fixed at 0.03 V, the cell type of Fig. 4 (c) has the highest probability to fail in the reading operation.

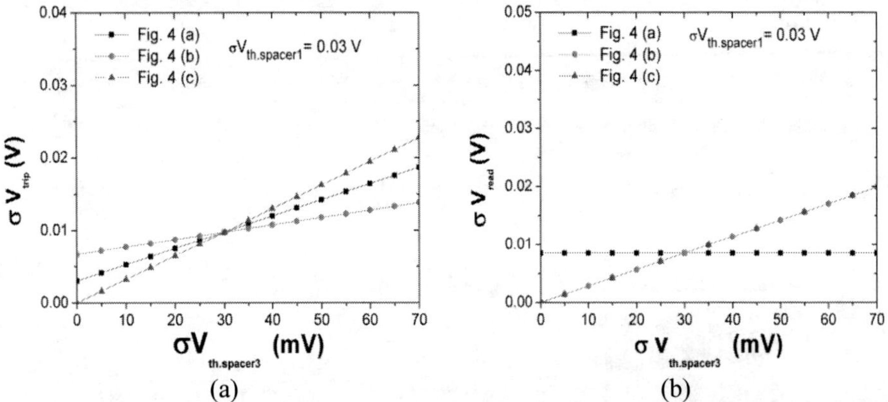

(a) (b)

Fig. 8. The standard deviations of (a) V_{trip} and (b) V_{read} vs. $\sigma V_{th.spacer3}$.

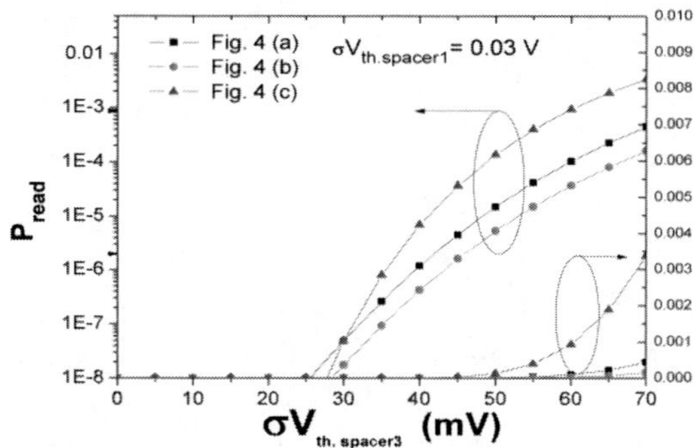

Fig. 9. Read failure probability vs. $\sigma V_{th.spacer3}$.

3.2 Write Failure Analyses

During the writing operation (Fig. 10), write failure occurs when the voltage of node "L" is not reduced below the trip-point voltage of the right inverter (PR-NR) within the time when the word line remains high (T_{WL}). The time required to change the data stored in a SRAM (T_{write}) can be obtained by solving Kirchhoff's law at "L" as [10]:

$$T_{write} = \frac{\int_{V_{dd}}^{V_{trip}} C_L dV}{I_{PL} - I_{AXL}}$$

where I_{PL} and I_{AXL} are the currents passing through PL and AXL parts. T_{write} can be further approximated as:

$$T_{write} = \overline{T_{write}} + A_{write}\Delta V_{th.NR} + B_{write}\Delta V_{th.AXR}$$

where $A_{write} = -0.048$, $B_{write} = 0.108$, $\overline{T_{write}}$ is the mean value of T_{write}. The calculated results are shown in Fig. 11. The write failure probability (P_{write}) is calculated as: $P_{write} = P(T_{write} > T_{WL})$, and the results are shown in Fig. 12. It indicates that the failure probability of the SRAM cell in Fig. 4 (c) increases rapidly with $\sigma V_{th.spacer3}$ when $\sigma V_{th.spacer1} = 0.03\ V$, the IC designers should pay more attention to this type of SRAM when $\sigma V_{th.spacer3}$ is high.

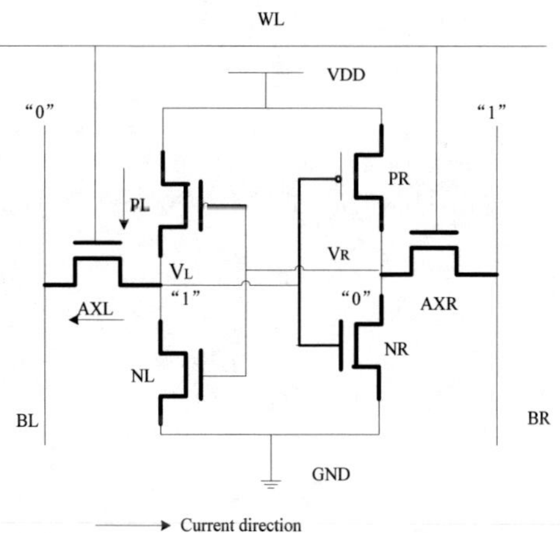

Fig. 10. A 6-transistor SRAM circuit in the writing operation.

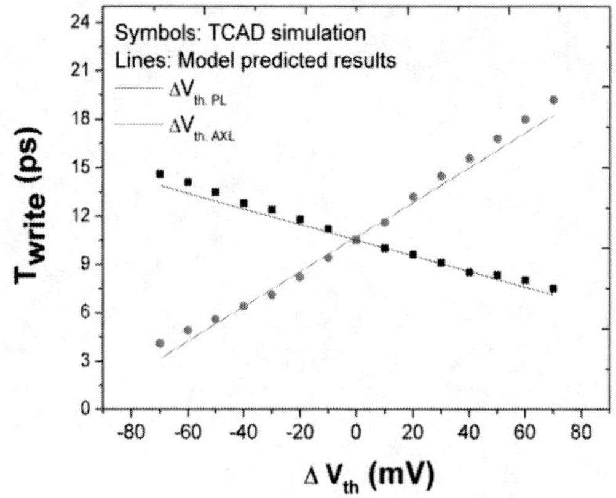

Fig. 11. T_{write} vs. FinFET threshold-voltage variations.

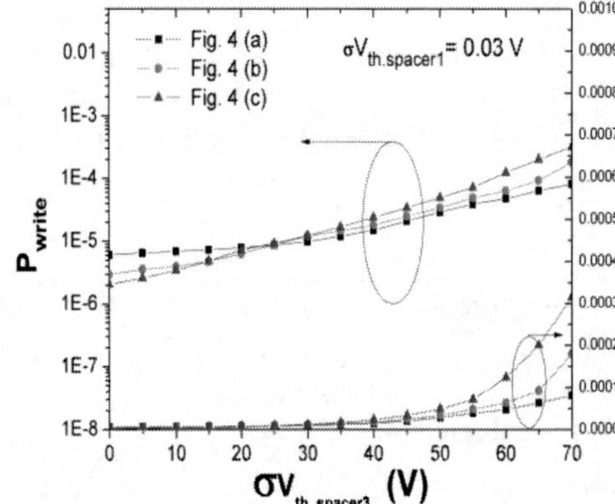

Fig. 12. Write failure probability vs. $\sigma V_{th.spacer3}$.

3.3 SNM and Yield Performance

HP=10 nm

SRAM Type	μ(SNM) (mV)	σ(SNM) (mV)	μ(SNM)-6σ(SNM) >0.04Vdd?
(a)	92.3	9.1	Yes
(b)	94.4	7.8	Yes
(c)	93.7	7.1	Yes

HP=7 nm

SRAM Type	μ(SNM) (mV)	σ(SNM) (mV)	μ(SNM)-6σ(SNM) >0.04Vdd?
(a)	71.3	9.2	No
(b)	72.6	8.4	No
(c)	73.7	8.2	No

Fig. 13. SNM characteristics and yield performance index (right column, [17]) of various types of SRAM circuits fabricated by a SASP process.

SNM characteristics of the above three basic types of SRAM circuits obtained from statistical TCAD simulations for $HP = 10$ nm and $HP = 7$ nm are shown in Fig. 13. The LWR and CDU parameters for different technology (half-pitch) nodes are listed in Table 1. 50 SRAM cell samples considering CDU and LWR are simulated in this study. The method reported in [18] is applied to generate LER/LWR functions. Due to a good correlation between the roughness of two edges of spacers, we assume $LER = LWR/1.2$, and σ_{LER} and σ_{CDU} are set according to Table 1. The correlation length λ is assumed to be 15 nm [19]. The simulation results indicate that 6-transistor SRAM circuit design does not meet the six-sigma yield requirement [17] when the half pitch is scaled down to 7nm. Therefore, more robust SRAM circuit design incorporating the information of process variability may be required in future IC design.

4. CONCLUSION

In this paper, we analyze the statistical characteristics multi-modality variations in a SASP process and its impacts on the inverter and SRAM circuit behavior. The inverter delay time, SRAM read/write failure probability, and SNM and yield performance are investigated using both statistical modeling and TCAD simulations. It is shown that SRAM circuit yield is significantly degraded by the multi-modality variation of the fin structures in a SRAM cell, especially when the half pitch of FinFET devices is driven down to 7 nm. The specific SRAM type with the highest probability to fail in the reading/writing operation is also identified. Our study suggests that the 6-transistor SRAM configuration may not be functional at deep nano-scale and more robust SRAM circuit design needs to be explored.

ACKNOWLEDGEMENT

This research is supported by the Shenzhen City Fund for Fundamental Research and the start-up research fund from Shenzhen Graduate School of Peking University.

REFERENCES

[1] International Technology Roadmap for Semiconductors: Lithography, 2013.

[2] C. Bencher, Y. Chen, H. Dai, W. Montgomery, and L. Huli, "22 nm half-pitch patterning by CVD spacer self-alignment double patterning (SADP)," *Proc. SPIE*, vol. 6924, 69244E, 2008.

[3] Y. Chen, W. Kang, P. Zhang, "A comparative study of self-aligned quadruple and sextuple patterning techniques for sub-15nm IC scaling," *Proc. SPIE*, vol. 8683, 86830Z, 2013.

[4] Y. Chen, Q. Cheng, W. Kang, "Technological merits, process complexity, and cost analysis of self-aligned multiple patterning," *Proc. SPIE*, vol. 8326, 832620, 2012.

[5] Y. Chen, W. Kang, P. Zhang, "A comparative study of self-aligned quadruple and sextuple patterning techniques for sub-15nm IC scaling," *Proc. SPIE*, vol. 8683, 86830Z, 2013.

[6] W. Xiao, Q. Cheng and Y. Chen, "SRAM circuit performance in the presence of process variability of self-aligned multiple patterning," *Proc. SPIE*, vol. 8684, 86840K, 2013.

[7] P. Xu, Y. M. Chen, Y. Chen, L. Miao, S. Sun, S-W Kim, A. Berger, D. Mao, C. Bencher, R. Hung, C. Ngai, "Sidewall spacer quadruple patterning for 15-nm half-pitch," *Proc. of SPIE*, Vol. 7973, 79731Q, 2011.

[8] K. Oyama, S. Natori, S. Yamauchi, A. Hara and H. Yaegashi, "CD error budget analysis for self-aligned multiple patterning," *Proc. SPIE*, vol. 8325, 832517, 2012.

[9] A. Carlson, Device and Circuit Techniques for Reducing Variation in Nanoscale SRAM, Ph.D. dissertation, Dept. Elect. Eng. Computer Scis., UC Berkeley, CA, 2008.

[10] S. Mukhopadhyay, H. Mahmoodi-Meimand, and K. Roy, "Modeling of failure probability and statistical design of SRAM array for yield enhancement in nanoscaled CMOS," *IEEE Trans. Comput.-Aided Design Integr. Circuits Syst.*, vol. 24, no. 12, pp. 1859–1880, Dec. 2005.

[11] B. Cheng, S. Roy, G. Roy, F. Adamu-Lema, and A. Asenov, "Impact of intrinsic parameter fluctuations in decanano MOSFETs on yield and functionality of SRAM cells," *Solid-State Electron.*, vol. 49, no. 5, pp. 740–746, 2005.

[12] J. Wu and C. H. Diaz, "Expanding role of predictive TCAD in advanced technology development," in *Proc. 18th Int. Conf. Simulation Semiconductor Processes Devices*, Sep. 2013, pp. 169–171.

[13] X. Wang *et al.*, "Interplay between process-induced and statistical variability in 14-nm CMOS technology double-

gate SOI FinFETs," *IEEE Trans. Electron Devices*, vol. 60, no. 8, pp. 2485–2492, Aug. 2013.
[14] X. Chen, Y. Wang, Y. Cao, Y. Ma, and H. Yang, "Variation-aware supply voltage assignment for simultaneous power and aging optimization," *IEEE Trans. Very Large Scale Integr. (VLSI) Syst.*, vol. 20, no. 11, pp. 2143-2147, Nov. 2012.
[15] Q. Cheng, J. You and Y. Chen, "Correlating FinFET device variability to the spatial fluctuation," *Microelectron. Eng.* vol. 119, pp. 53-60, Feb, 2014.
[16] A. Bansal, S. Mukhopadhayay and K. Roy, "Device-optimization technique for robust and low-power FinFET SRAM design in nano-scale Era," *IEEE Trans. Electron Devices*, vol. 54, no. 6. pp. 1409-1419, Jun. 2007.
[17] A. Pavlov and M. Sachdev, "CMOS SRAM Design and parametric test in nano-scaled technologies", Springer, 2008, pp. 59-62.
[18] S. Yu, Y. Zhao, L. Zeng, G. Du, J. Kang, R. Han, X. Liu, "Impact of Line-Edge Roughness on Double-Gate Schottky-Barrier Field-Effect Transistors," *IEEE Trans. Electron Devices*, vol.56, no.6, pp.1211-1219, Mar. 2009.
[19] G. Leung , L. Lai , P. Gupta and C. O. Chui "Device-and circuit-level variability caused by line edge roughness for sub-32-nm FinFET technologies," *IEEE Trans. Electron Devices*, vol. 59, pp. 2057-2062, Aug. 2012.

Layout optimization and trade-off between 193i and EUV-based patterning for SRAM cells to improve performance and process variability at 7nm technology node

Sushil Sakhare[1], Darko Trivkovic[1], Tom Mountsier[2], Min-Soo Kim[1], Dan Mocuta[1], Julien Ryckaert[1], Abdelkarim Mercha[1], Diederik Verkest[1], Aaron Thean[1], and Mircea Dusa[3]
[1]IMEC, Kapeldreef 75, B-3001 Leuven, Belgium; [2]Lam Research, Leuven, Belgium, [3]ASML, Leuven, Belgium.

ABSTRACT

The Fin-FET Technology scaling to sub 7nm node, using 193 immersion scanner is restricted due to reduced margins for process. The cost of the process and complexity of designs is increasing due to multi-patterning to achieve area scaling using 193i scanner. In this paper, we propose a two Fin-cut mask design for Fin-pattering of 112 SRAM (two Fins for pull-down and one Fin for pull-up and pass-gate device) cell using 193i lithography and its comparison with EUVL single print. We also propose two keep masks for middle of line patterning ,with increased height of the SRAM cell using 193i, that results in area of a uniform-Fin SRAM cell area at 7nm technology; whereas EUVL can enable non-uniform SRAM cell at reduced area. Due to unidirectional patterning, margins for VIA0 landing over MOL are drastically reduced at 42nm gate pitch and hence to improve margins, the orientation for 1st metal is proposed to be orthogonal to the gate. This results in improved performance for SRAM and reliability of the technology.

Keywords: EUVL, 193i lithography, Design Technology co-optimization, 7nm technology, SRAM cell design, multi-patterning, Fin-CUT optimization, middle of line scheme for SRAM design. Orientation of 1st metal, Mint.

1. INTRODUCTION

Area of the SRAM cell is always used as metric to measure scaling of technology. At 7 nm technology node, aggressive scaling of gate and metal pitch results in aggressive scaling of SRAM cell area using multiple patterning. Self-aligned quadruple patterning (SAQP) [1][2] is required to realize Fin at 24nm pitch as well as 32nm metal pitch. The desired performance is achieved by scaling gate length (Lg) at 42nm pitch, realized using LELE/SADP. Using EUVL, most of multi-patterns can be realized in single print at reduced effective variability observed for multi-patterning. Due to uncertainty about introduction of EUVL for production at 7nm node, this paper proposes and compares solutions for low cost patterning & design using 193i and EUV. Thus, we compare schemes for patterning of Fins and Middle of Line as they define effective height of the SRAM cell in Fin-FET technology. Under the 193i assumption, all patterns are forced to have unidirectional patterning where reliability and electrical performance of cell is improved by having first metal orthogonal to gate.

2. FEOL FIN-CUT PATTERNING

Patterning Fins using SAQP is extremely challenging due to increasing aspect ratio of Fins at 24nm pitch. For the SRAM cell, the Fin design requires etching of those Fins in a very narrow window of 42nm; Hence patterning requires an optimum solution to lower the cost. One of patterning option selected is SAQP for Fin grating with EUV fin-cut. Minimum patterning margins are observed for Fins in the 112 SRAM cell. In Fig. 1, SAQP mandrel and Fin-cut patterning for 2x2 array of 112 SRAM array is shown. The single print contour for Fin-cut is used for cutting generated fins as shown in Fig.

Design-Process-Technology Co-optimization for Manufacturability IX, edited by John L. Sturtevant, Luigi Capodieci, Proc. of SPIE Vol. 9427, 94270O · © 2015 SPIE · CCC code: 0277-786X/15/$18 · doi: 10.1117/12.2086100

1 (d). In a variability prone environment, Fin mandrel and Fin cut collectively define fin structure. In Table 2, the impact of CD variation and overlay over the resultant Fins is captured using process simulation (Coventor). We observed that the optimized Fin-cut contour pattern gives sufficient margin for Fins in the 112 SRAM cell. One of the worst case scenarios is shown in Fig. 2, where a 6nm CD reduction of the mandrel along with ±3 nm Fin-cut overlay results in etching of the sidewall of an isolated single Fin as well as sharp extensions of Fins that are intended to tuck under the gate. From Table 2, it is seen that a target overlay of 3nm and CD variation of 3nm can give sufficient margins for designing 112 SRAM using EUV single print Fin-cut at 45% area scaling in comparison with 10nm node.

Table 1: Gate and metal pitches assumed for N10 and N7 technology

Technology node	N16	N10	N7
Gate pitch	90nm	64nm	42nm
M1/Mx pitch	64nm	48nm	32nm

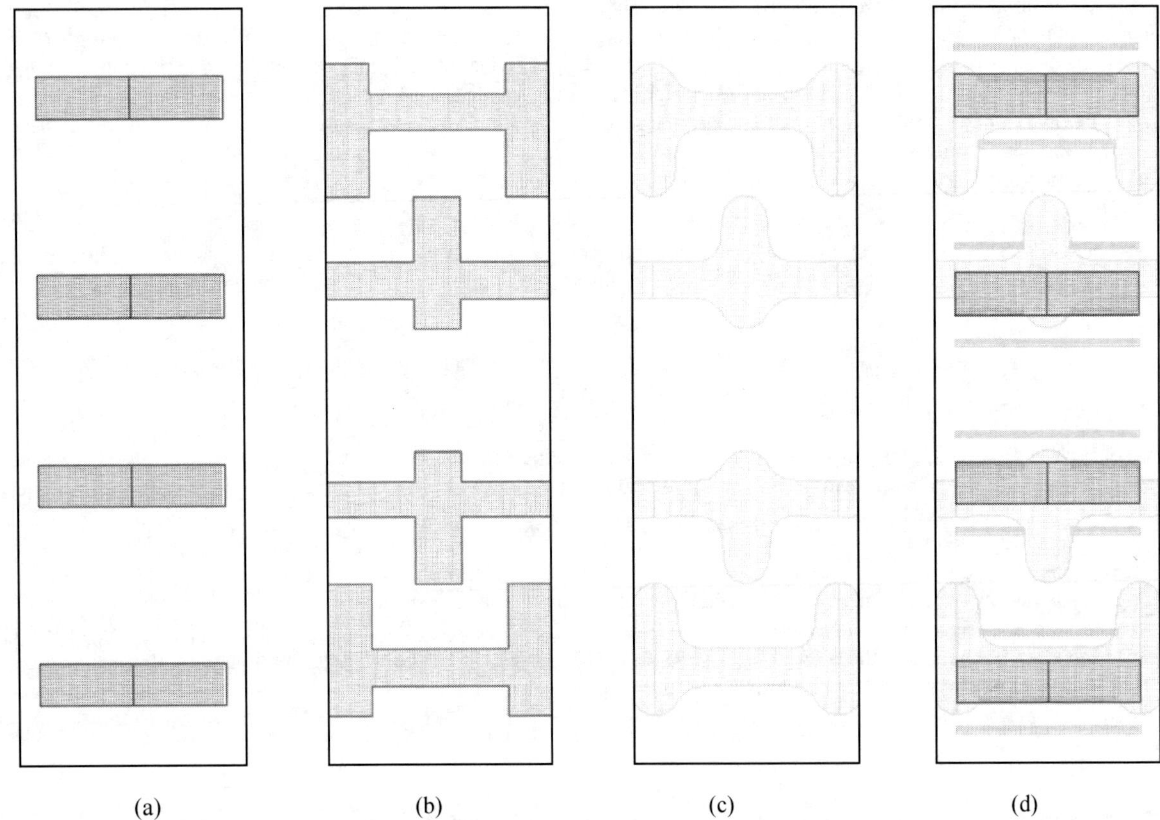

Fig. 1: Fins are realized using Self- align quadruple patterning (SAQP). As shown in (a) mandrels are processed using 193i whereas (b) shows drawn EUV fin-cut. (c) Captures contour analyzed for Fin-cut patterns whereas (d) captures resultant fins realized using SAQP for Non-uniform Fin 112 SRAM cell.

Table 2: Due to overlay and CD variation is analyzed for Fin-core and Fin-cut using Coventor process simulator where worst cases are seen for -6nm core CD and overlay variation of ±3nm for Fin-cut. The minimum margins required are met by EUV fin-cut patterning.

Experiment	FINCORE, nm	Spacer 1, nm	Spacer 2, nm	FINCUT Overlay, nm	FINCUT Bias, nm	Result
Grating variation	±2, ±4, ±6, Nom	±1.25, Nom	±0.75, Nom	Nom	Nom	No Failures
Add FINCUT overlay	±4, ±6, Nom	±1.25, Nom	±0.75, Nom	±3, ±5, Nom	Nom	All -4 and -6 core runs failed as did nom grating at ±5 overlay
Scale back grating bias and overlay	±2, ±4, Nom	±1.25, Nom	±0.75, Nom	±2, ±3, Nom	Nom	All conditions pass except -4 core at ±2 overlay
FINCUT bias with nominal grating	Nom	Nom	Nom	±2, Nom	±1, -2, -3, -4, Nom	No Failures
All factors considered	±2, Nom	±1.25, Nom	±0.75, Nom	±2, Nom	±1, -2, -3, -4, Nom	No Failures
Reduce Spacer 1 bias, Reduce FINCUT CD	±2, Nom	±0.75, Nom	±0.75, Nom	±3, Nom	-1 thru -5, Nom	No Failures
Test at greater FINCORE bias	±4, Nom	±0.75, Nom	±0.75, Nom	±3, Nom	-1 thru -5, Nom	No failures except -4 core at -1 FINCUT bias

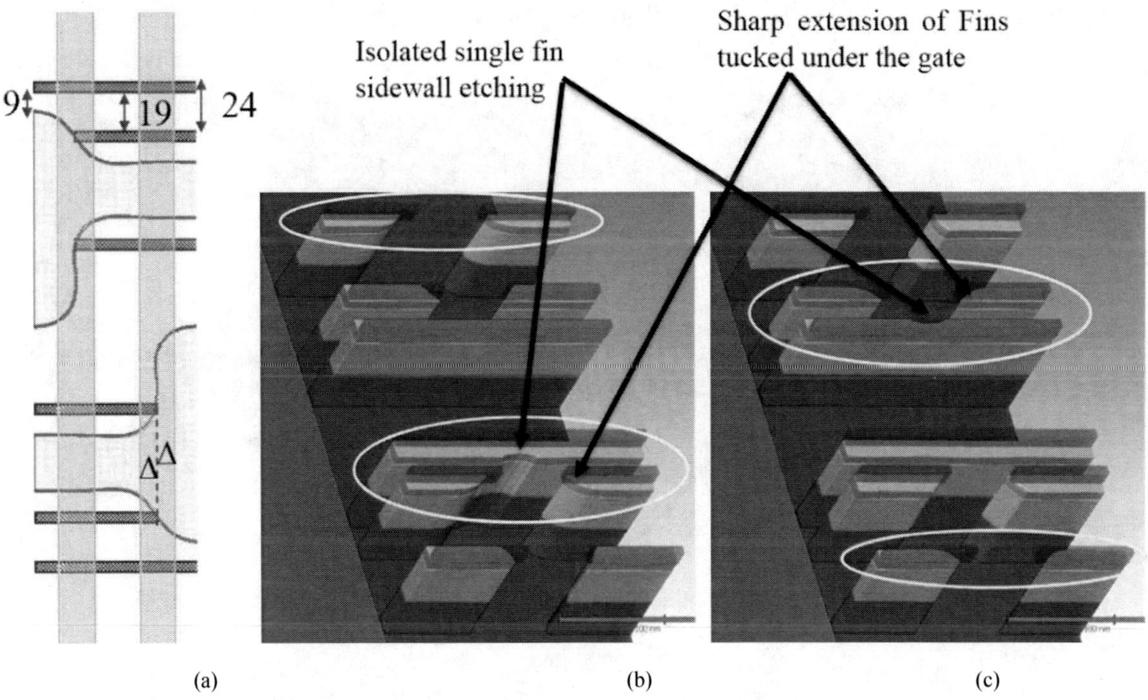

Fig. 2: -6nm Fin-mandrel CD variation and ±3nm EUV Fin-cut shows worst case patterning margins in (b) and (c) respectively. The contour of Fin-cut for single SRAM cell is captured in (a) where 9nm margin for Fin-cut is set along the gate pattern. The Δ margin ensures that Fin edge is tucked under the gate.

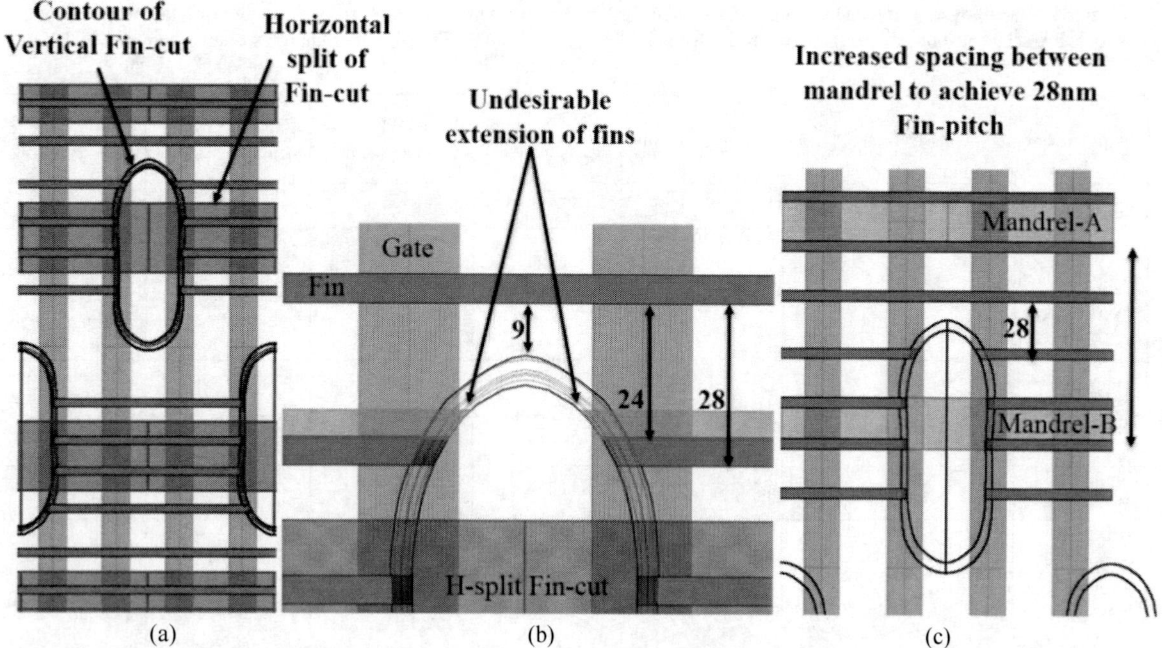

Fig. 3: (a) Contour assessment of Fin-cut pattern split along the gate, as well as Fin-cut split orthogonal to gate. (b) Under 3nm overlay margins, Fin-pitch should be increased more than 28nm as Fin extensions resulting at 24nm pitch are undesirable to ensure Fin tucking under the gate.(c) 28nm Fin-pitch can be achieved by increasing spacing between Fin mandrels(A &B).

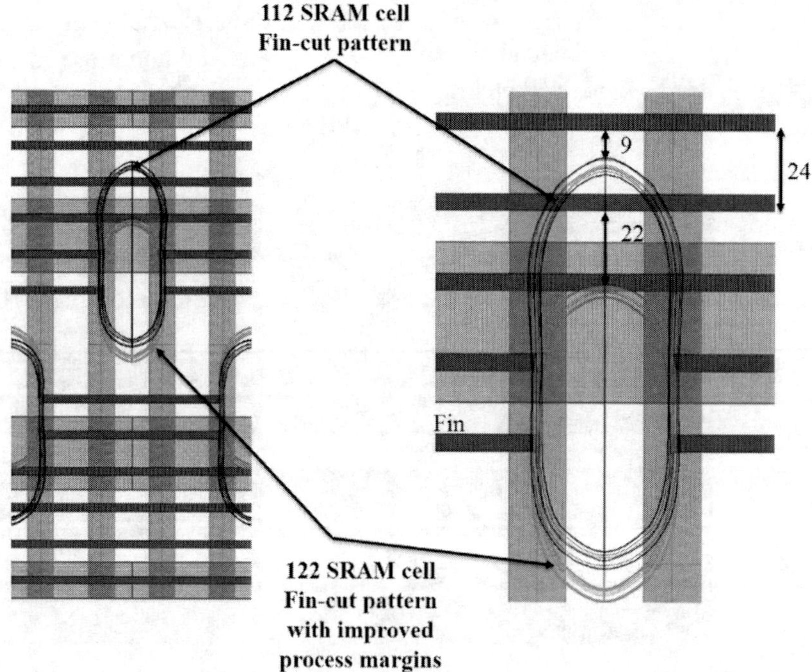

Fig. 4: 122 SRAM cell Fin-cut patterns simplifies SRAM cell design with improved process margins from 9nm to 22nm, which will definitely improve yield.

Under 193i assumption, the Fin-cut patterning requires at least two splits. One split is optimized for cutting the fin in a narrow 42nm window insuring the same margins as EUV. The contour assessment in Fig. 3 shows two critical patterning margins: fin sidewall and fin tip. At 24nm Fin-pitch, the 193i solution does not provide the desired margins for fins tucking under the gate. An undesirable extension of Fin can be pulled back by increasing Fin-pitch to more than 28nm. Thus to ensure same margins for process and overlay, the 112 SRAM cell must be designed with a higher Fin-pitch of 28nm, which is achieved by increasing spacing between mandrels as shown in Fig. 3(c). At 10nm technology node, the 122 SRAM cell (One Fin for pull up device, 2 Fins for pull-down and pass-gate device) is an optimum electrical and cost effective solution [1] for the technology. Fig. 4 compares Fin-cut contours for the 112 and 122 SRAM cells where the cut pattern for 122 shows 22nm improved spacing from active isolated fin, which was 9nm in the case of 112 at 28nm fin-pitch. Designing the 122 SRAM cell at 7nm node simplifies patterning by retaining a Fin-pitch of 24nm for both SRAM and logic. Thus with two splits of Fin-cut pattern, it is possible to realize High Density SRAM cell of technology using a 193i scanner whereas single Fin-cut patterning requires EUV.

3. MIDDLE OF LINE (M0A)

SRAM cell layout with non-uniform fins gives the maximum area scaling [2] for Fin-FET technology. Using 193i lithography for realizing active local interconnect (M0A) at 42nm pitch requires a 5 layer block mask design to create a non-uniformed Fin 112/122 SRAM cell (Fig. 5 (a)). Five block masks for M0A will see relative overlay and CD variation resulting in reduced process margins at increased cost. M0A can be realized with single lithography using EUV scanner as shown in Fig. 5 (b). To realize MOA with 193i, we propose a solution for M0A patterning where MOA is realized with two splits of keep masks instead of five blocks.

Fig. 6 captures the contour of two keep masks designed for 193i lithography assuming that the process can allow M0A fully self-aligned along the gate. To enable such design, it is essential to increase spacing between two patterns of the same split. Increasing spacing to 45nm results in increased area for the SRAM cell. A margin of 30nm between the two splits of keep mask provides the required overlay and CD variation tolerance. Simplification of M0A patterning shifts complexity of routing to higher metals where bit-lines are routed in metal-3 instead of metal-1 to ensure sufficient margins for VIA0. An increased height of the cell reduces SRAM area scaling from 45% to 53% as shown in Fig. 5 (c).

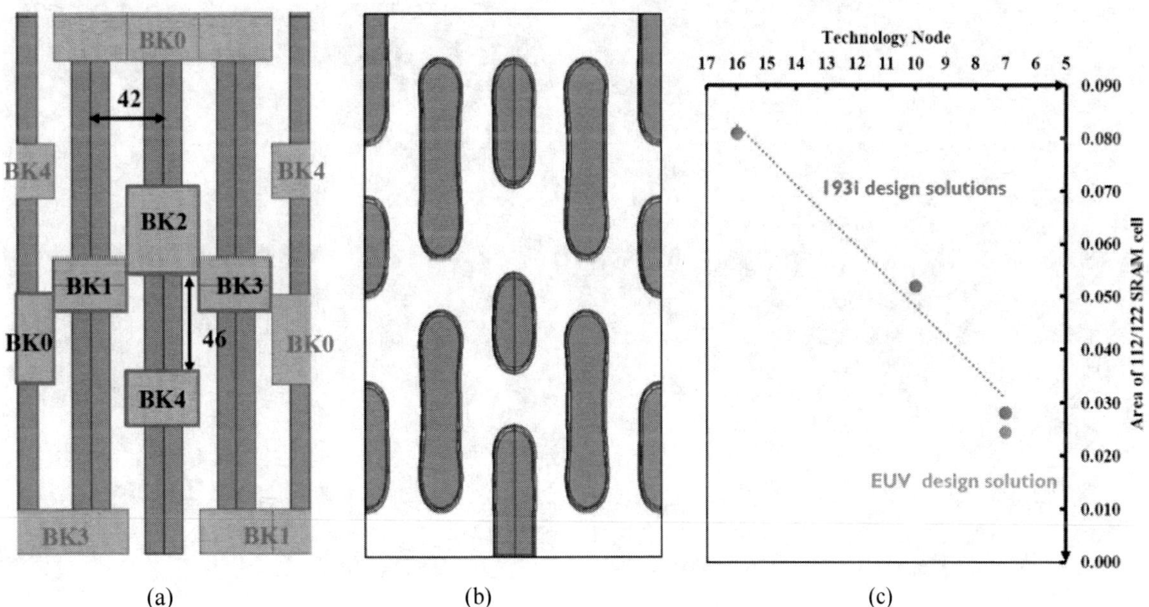

Fig. 5: (a) Five block mask requirement to pattern M0A using 193i scanner. (b) M0A contour replacing five block mask by single print EUV. (c) Area trend for 112/122 SRAM cell shows maximum scaling of 45% for EUVL design whereas 54% for 193i design.

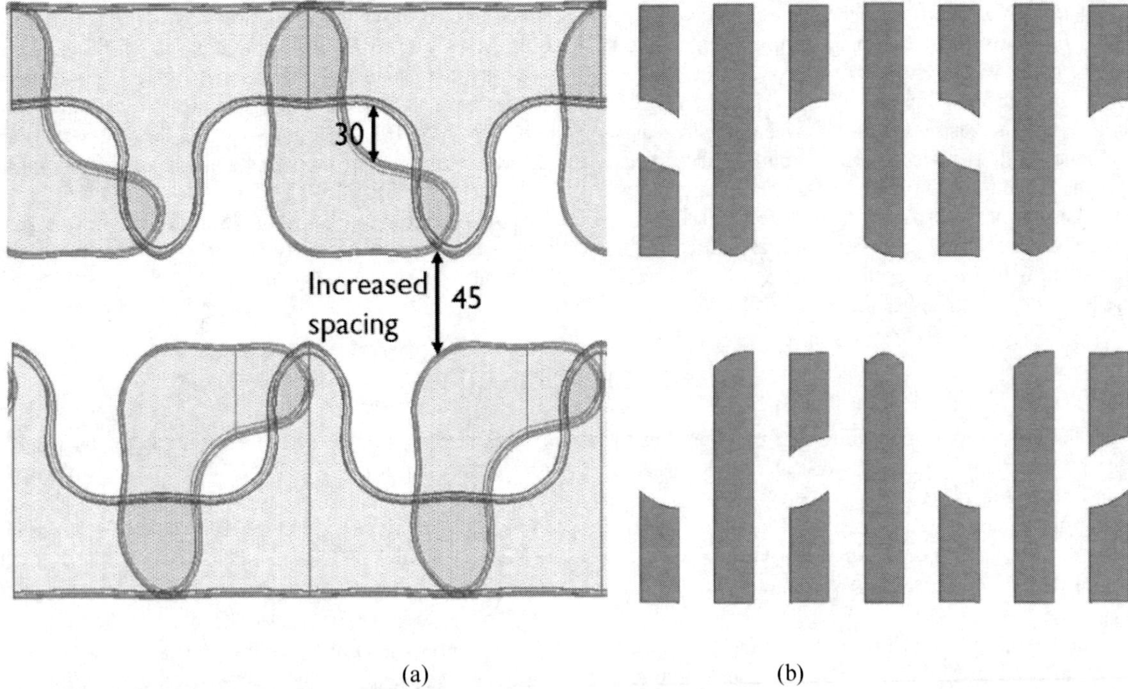

Fig. 6: (a) Contour of two split keep mask for M0A patterning, To enable design the spacing between two keep mask is increased resulting in increased height (b) Resultant M0A pattern for 2 keep solution, simplification of M0A shifts complexity to higher metal

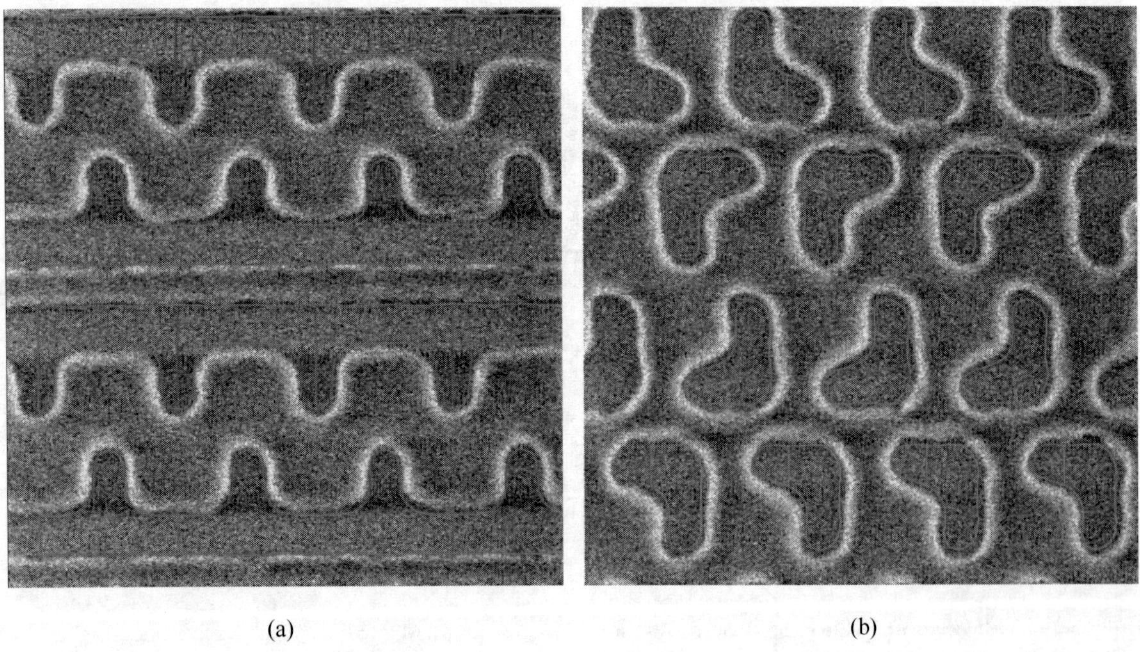

Fig. 7: Wafer realization of two splits of M0A patterns in resist.

4. FIRST METAL

It has been shown that for an aggressive scaling of gate pitch, effective margins for Via landing over gate middle of line (M0G) are not sufficient and reduce SRAM cell performance by 30% at the 10nm technology node [2]. MOG and M0A provides access of device terminals to BEOL, where VIA0 landing from Metal-1 establishes connection with M0A and M0G. Traditionally metal-2 is always orthogonal to gate to enable standard-cell designs where Metal-1 can be bidirectional in order to access M0G and M0A. At 42nm pitch, Metal-1 patterning is restricted by process to have unidirectional metal at 7nm node whereas EUVL allows freedom to design bidirectional metal-1. Dual damascene metal-1 drops VIA0 over M0A and M0G as shown in Fig. 8. For the 193i solution, Metal-1 should be at 42nm pitch to intercept M0A, where 50% spacing between pattern results in 21nm lines spaced 21nm apart, like M1-B and M1-C as shown in Fig. 8. The effective VIA0 landing from M1-B has no margin for CD and overlay; thus M1-B pattern width should be reduced by 10nm, resulting in 11nm wide M1-A pattern to achieve 5nm margin on both edges. MOG pattern is used for opening the gate cap and spacer surrounding the metal gate. Patterning of M0G-C must ensure margin for the spacer around Gate-D to avoid etching. Due to placement of MOG away from the spacer, the effective contacting area for Via0 landing from M1-C is reduced substantially, which is undesirable.

Additional margin is required to compensate CD and overlay of Metal-1 patterning and demands M1-C to shrink by 20nm resulting in an extremely narrow wire M1-D. Thus unidirectional Metal-1 oriented parallel to gate does not show sufficient margin for Via0 landing. It has been shown for 10nm technology [2] that an SRAM cell with the word-line routed in M1 shows 30% performance degradation compared to routing in Metal-2. The drastically reduced width of M1 word-line to satisfy margins for Via0 has elevated the problem of resistance at 7nm technology node. If the first metal is routed orthogonal to gate then it improves contact area and margin by placement of Via0 as shown for M1-D connection in Fig. 9. In this case Via0 as dropped from M1-D is allowed to shift left in order to improve margins from 5nm to 10nm to align it to center of M0G-D. Thus having first metal orthogonal to gate effectively increases margins for Via0 and process reliability. The orthogonal Metal-1 decouples via0 landing over M0G margins from the width of Metal-1 resulting in improved process reliability .

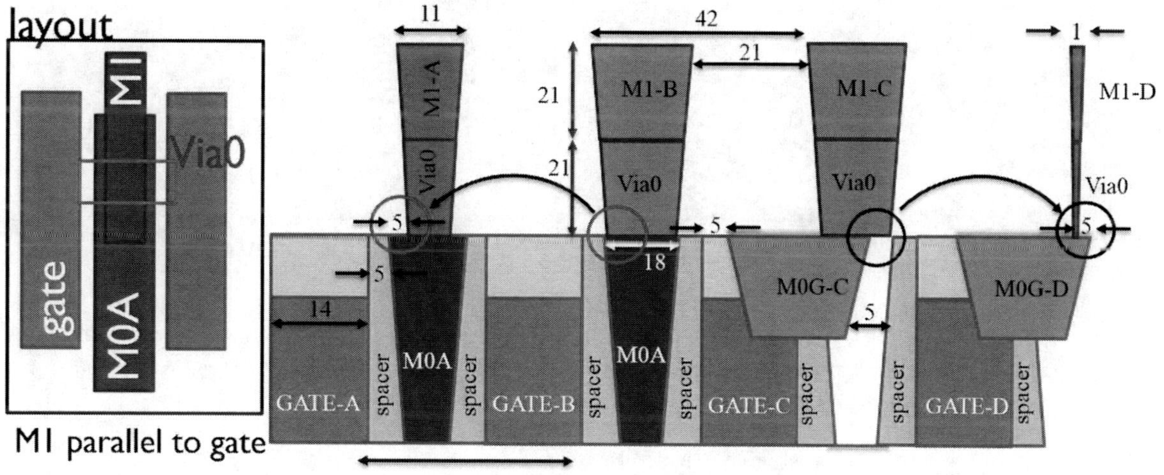

Fig. 8: M1 at 42nm pitch results into no margins for Via0 landing over MOA present between Gate-B and Gate C. To buy CD & overlay margins of 5nm, the M1-B width should be reduced to 11nm resulting in M1-A. The Via0 is landing outside MOG-C by 5nm, to buy CD and overlay margin M1-C should be reduced to 1nm resulting in M1-D which is undesirable.

At 7nm node, orthogonal M1 can be used for routing the critical signals BL/BLB, as shown in Fig. 10 (a) - realized using SAQP at 32nm pitch. In Fig. 10 (b) the resultant patterns are simplified by single print EUVL or LELE+SADP process to improve resistivity of metal lines. To maximize yield of SRAM cell, the area of smallest metal pattern can be increased by increasing the height of the SRAM cell. The M1 horizontal lines depicted in Fig. 10 (c) show maximum increased in metal area compared with M1 vertical pattern in Fig. 10 (d), for the same increased in height of SRAM cell. Thus keeping first metal helps in resolving the minimum area issue.

A 193i-based solution for SRAM would use a simplified, two keep-mask, M0A patterning scheme that transfers complexity to higher metal, hence M1 horizontal is used as routing layer instead of signal layer as shown in Fig. 11 (a). As first metal is used for improving margins of Via0 landing, it will see the maximum number of minimum area islands. To improve minimum area, M1 pitch is more relaxed than M3. It is possible to have M1 pitch > 40nm to enable SADP [3][4] as well as having aggressive M1 pitch realized using SAQP to enable standard cell designs. The SAQP patterning process developed for M3 can be used, as it is at the same pitch for M1 to save the cost of the technology [5]. The comparison between M2 patterns in Fig. 11 (b) and M1 vertical patterning in Fig. 10 (d) represents similar design for metal patterning. As M1 horizontal can have relaxed pitch and it is the layer which is used for improving contact between FEOL and BEOL, we renamed it as **Mint** which changes all layer naming for the technology stack as shown in .

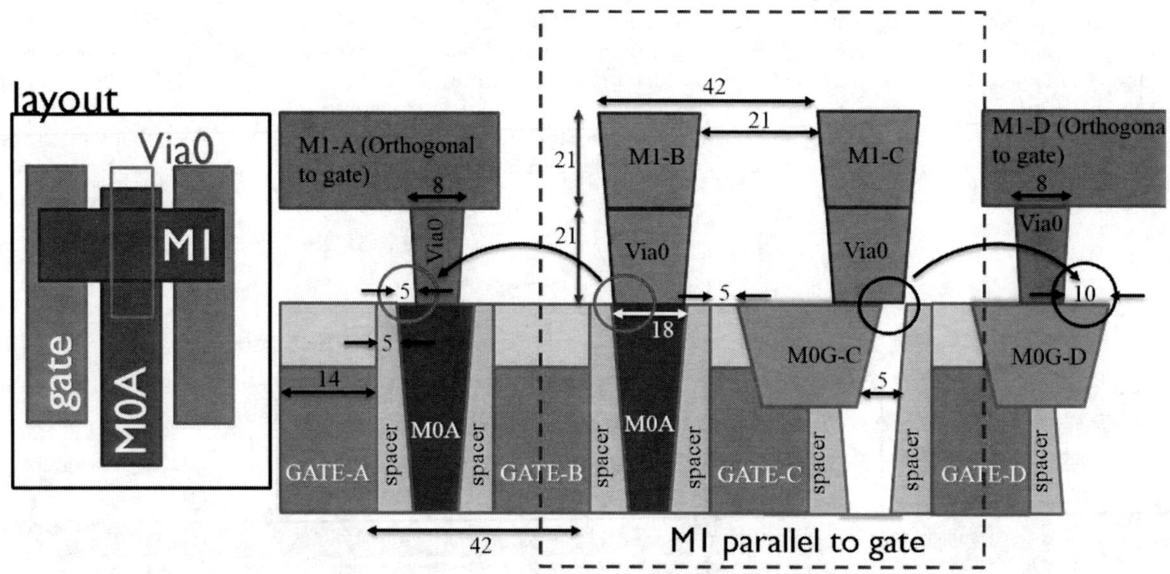

Fig. 9 : Changing M1 orientation orthogonal to gate results into excellent improvement in Via0 landing over M0G and M0A, The drawn Via0 is aligned with M0A and hence improved contact area & margins as shown for M1D and M1A pattern.

Table 3: Metal orientation and renaming of M1 horizontal as Mint.

Naming used	Orientation	Pitch	New names
M1	Orthogonal to gate (Horizontal)	>40nm /SADP or 32nm/SAQP	**Mint**
M2	Parallel to gate	42nm/SADP	**M1**
M3	Orthogonal to gate (Horizontal)	32nm/SAQP	**M2**

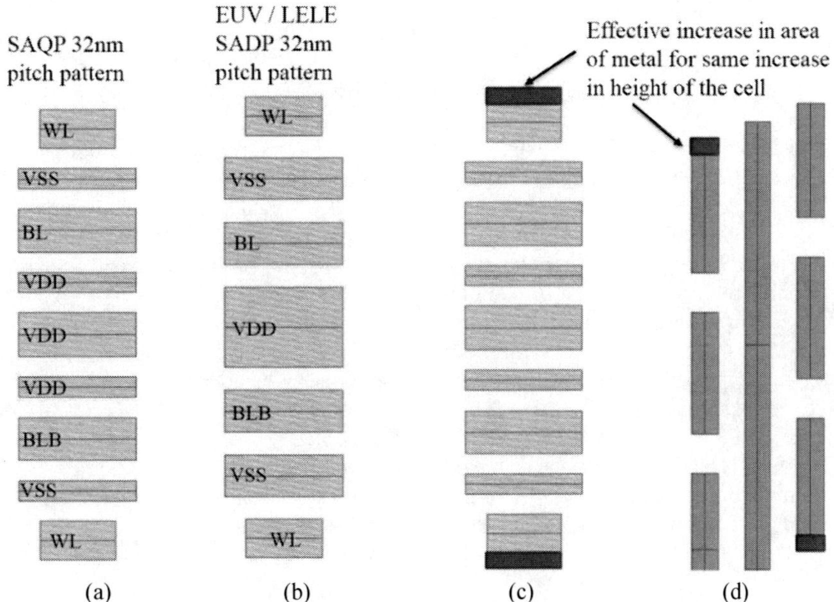

Fig. 10: (a) M1 orthogonal to gate realized using 32nm SAQP patterning. (b) Single print EUV or LELE+SADP can simplify patterning to improve resistivity of signals.(c) The minimum area of smallest metal piece is fixed by increasing height of the cell, M1 horizontal in (c) compared with M1 vertical (d) in SRAM layout shows maximum benifit when M1 is horizontal (orthogonal to gate).

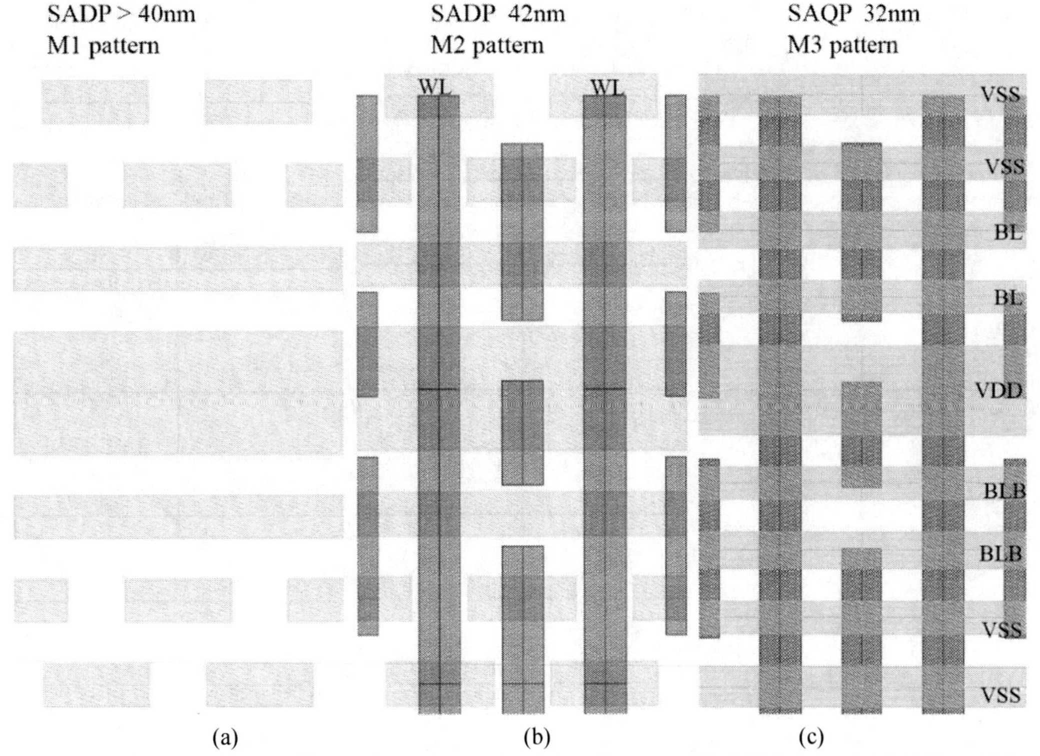

Fig. 11: (a) M1 Horizontal is simplified further to have relax pitch > 40nm to enable SADP patterning (b) M2 parallel to gate (vertical) improves WL resistance and hence performance [2]. (c) M3 is parallel to M1, Critical signals like BL/BLB & supply are routed in M3 at 32nm pitch using SAQP.

5. CONCLUSIONS

At 7nm technology, under EUVL and 193i assumptions, the first metal (Mint) should be orthogonal to gate in order to compensate for overlay and CD variation margins at aggressive pitches. The first metal is used for establishing connection between FEOL and BEOL at more relaxed pitch than the desired M2 horizontal pitch; hence first metal is named as **Mint**. In the case of an EUVL solution, the patterning of first metal can enable routing of BL/BLB signals at variable widths, making design simple at reduced resistance for signals as well as finishing design at second metal layer. Saving of an additional layer is possible due to enablement of complex M0A patterning using EUVL. The area of the SRAM cell is defined by the M0A and M0G patterning used. Under the EUVL assumption for M0G and M0A, we can achieve 45% scaling compared with 10nm technology, whereas for the 193i solution with a two keep-mask solution, the effective scaling reached 53%. To enable design using a two keep scheme instead of a 5-level block mask requires increased spacing between the two keep mask splits, which increases height of the cell resulting in increased area of the SRAM cell using 193i patterning. For the same SRAM cell area, 112 SRAM can be enabled only if the Fin-pitch is increased beyond 28nm from 24nm. For 24nm SAQP Fin patterning, the spacing between two mandrels can be increased to enable designing of 112 SRAM cell at 28nm fin pitch. The electrical performance of 122 SRAM cell is better than 112 SRAM cell at an added advantage of simplified Fin-patterning at 24nm pitch that also improves process margins.

6. REFERENCES

[1] Sakhare, S.S.; Miyaguchi, K.; Raghavan, P.; Mercha, A., " Simplistic Simulation-Based Device-VT-Targeting. Technique to Determine Technology High-Density. LELE-Gate-Patterned FinFET SRAM in Sub-10 nm," *Electron Devices, IEEE Transactions on* , vol.PP, no.99, pp.1,1.

[2] Ryckaert, J.; Raghavan, P.; Baert, R.; Bardon, M.G.; Dusa, M.; Mallik, A.; Sakhare, S.; Vandewalle, B.; Wambacq, P.; Chava, B.; Croes, K.; Dehan, M.; Jang, D.; Leray, P.; Liu, T.-T.; Miyaguchi, K.; Parvais, B.; Schuddinck, P.; Weemaes, P.; Mercha, A.; Bommels, J.; Horiguchi, N.; McIntyre, G.; Thean, A.; Tokei, Z.; Cheng, S., "Design Technology co-optimization for N10," *Custom Integrated Circuits Conference, 2014 IEEE Proceedings of the* , vol., no., pp.1,8, 15-17 Sept. 2014 doi: 10.1109/CICC.2014.6946037.

[3] Chris Bencher ; Yongmei Chen ; Huixiong Dai ; Warren Montgomery ; Lior Huli; 22nm half-pitch patterning by CVD spacer self-alignment double patterning (SADP). Proc. SPIE 6924, Optical Microlithography XXI, 69244E (March 07, 2008); doi:10.1117/12.772953.

[4] Yong Kong Siew; Versluijs, J.; Kunnen, E.; Ciofi, I.; Alaerts, W.; Dekkers, H.; Volders, H.; Suhard, S.; Cockburn, A.; Sleeckx, E.; Van Besien, E.; Struyf, Herbert; Maenhoudt, M.; Noori, A.; Padhi, D.; Shah, K.; Gravey, V.; Beyer, G., "Integration of 20nm half pitch single damascene copper trenches by spacer-defined double patterning (SDDP) on metal hard mask (MHM),"*Interconnect Technology Conference (IITC), 2010 International*, vol., no., pp.1,3, 6-9 June 2010; doi: 10.1109/IITC.2010.5510743

[5] Arindam Mallik, Julien Ryckaert, Abdelkarim Mercha, Diederik Verkest, Kurt G. Ronse, Aaron Thean "No more of Moore's Law: the high cost for dimensional scaling" IMEC (Belgium) [9422-58] SPIE 2015.

Incorporating DSA in multipatterning semiconductor manufacturing technologies

Yasmine Badr[a], J. Andres Torres[b] , Yuansheng Ma[c], Joydeep Mitra[d], Puneet Gupta[a]
[a]University of California, Los Angeles. Engineering IV, Westwood Blvd, Los Angeles, CA 90095
[b]Mentor Graphics Corporation. 8005 S.W. Boeckman Road, Wilsonville, OR 97070
[c]Mentor Graphics Corporation. 46885 Bayside Parkway, Fremont, CA 94538
[d]Mentor Graphics Corporation, 5000 Plaza on the Lake, Austin, TX 78746

ABSTRACT

Multi-patterning (MP) is the process of record for many sub-10nm process technologies. The drive to higher densities has required the use of double and triple patterning for several layers; but this increases the cost of the new processes especially for low volume products in which the mask set is a large percentage of the total cost. For that reason there has been a strong incentive to develop technologies like Directed Self Assembly (DSA), EUV or E-beam direct write to reduce the total number of masks needed in a new technology node.

Because of the nature of the technology, DSA cylinder graphoepitaxy only allows single-size holes in a single patterning approach. However, by integrating DSA and MP into a hybrid DSA-MP process, it is possible to come up with decomposition approaches that increase the design flexibility, allowing different size holes or bar structures by independently changing the process for every patterning step.

A simple approach to integrate multi-patterning with DSA is to perform DSA grouping and MP decomposition in sequence whether it is: grouping-then-decomposition or decomposition-then-grouping; and each of the two sequences has its pros and cons. However, this paper describes why these intuitive approaches do not produce results of acceptable quality from the point of view of design compliance and we highlight the need for custom DSA-aware MP algorithms.

Keywords: Directed Self Assembly (DSA), DFM, Lithography checks, layout verification, multi patterning, physical verifications, grapho epitaxy

1. Introduction.

1.1. Introduction to Directed Self-Assembly (DSA)

Self-Assembly is the phenomenon that occurs when block co-polymers composed of immiscible blocks phase-separate into organized structures [1]. For example, a diblock co-polymer can self-assemble into periodic structures of one type of block into a matrix of the other. Lithographically-printed patterns (in a scheme called Graphoepitaxy) or chemically-treated surfaces (in a scheme called Chemoepitaxy) are used to *direct* the self-assembly process [2].

The realizable assembled pitch depends on the characteristics of the used block co-polymer. The graphoepitaxy process for contact holes is shown in Figure 1, where trenches are lithographically printed first, and then the surface is spin-coated with the block co-polymer (BCP). Upon thermal annealing, the phase separation occurs, and with a particular BCP and surface treatment of substrate [3], cylinders are obtained within the guiding pattern.

Design-Process-Technology Co-optimization for Manufacturability IX, edited by John L. Sturtevant, Luigi Capodieci,
Proc. of SPIE Vol. 9427, 94270P · © 2015 SPIE · CCC code: 0277-786X/15/$18 · doi: 10.1117/12.2084776

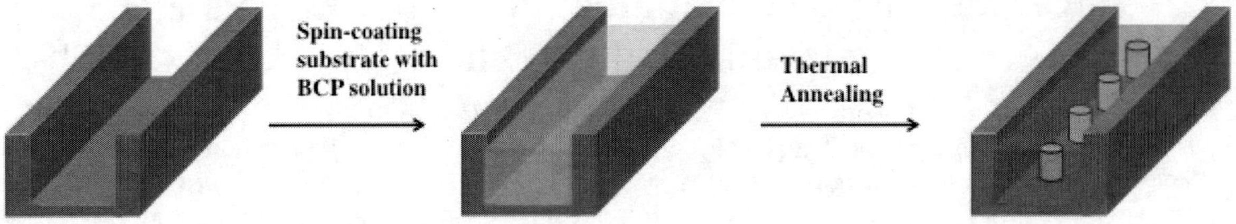

Figure 1. An example directed self- assembly process of a diblock co-polymer using Graphoepitaxy

In Graphoepitaxy, there are two styles of guiding templates: trenches [4] and posts [5]. Currently, the former option is more mature and understood, and more attractive to the industry since the posts templates need to be manufactured by e-beam in turn limiting the throughput of the process.

DSA has been successfully demonstrated for contact holes [4] and lamellae [6]. Since DSA is capable of printing dense nano-features of roughly uniform dimensions [7], it is a very good fit for contact and via layers. For that reason, this work is concerned with graphoepitaxy for contact/via hole structures.

1.2. Implications of DSA on Contact/Via holes

The natural multiplicative nature of DSA makes it very attractive as a low-cost resolution enhancement technology. However, it imposes some restrictions on the design. The manufacturable pitches depend on the natural pitch L_0 of the BCP, creating polygons of uniform width. Only a certain range of pitches greater than L_0 is manufacturable in DSA. Moreover, the BCP cannot be locally modified on a per guiding-pattern basis. Thus, if the process uses DSA along with a single mask to print the templates, only single-size contact/via holes can be manufactured. In addition, some layout configurations are forbidden because they require guiding templates that cannot be manufactured by optical lithography.

In the contact/via holes DSA scheme where the templates are manufactured using 193nm wavelength, the easiest configurations are one-dimensional arrays of contacts, with pitch equal to L_0, with higher reliability achieved if smaller number of holes are to be assembled per guiding template [8].

1.3. DSA in Hybrid Lithography Schemes with Multiple Patterning

In order to achieve the sub-5nm nodes, DSA will need to be complemented with other technologies in hybrid lithography schemes like DSA with Multiple Patterning (MP), DSA with EUVL or DSA with E-beam [5].

By using DSA along with MP, earlier research [8] advocated that it is possible to reduce the number of exposures used in the process, for example a Triple Patterning (TP) process coupled with DSA, could replace a traditional Quadruple Patterning Process (QP), in order to have a less costly technology.

In a hybrid process, where DSA is applied altogether with MP, it is required to perform the DSA-grouping of the polygons as well as the mask assignment. DSA-grouping is the process of selecting which contacts are to be formed together by one template while mask assignment is the process of determining the mask that is to print each polygon.

There are a couple of different options for such a hybrid DSA-MP process. The substrate can be spin-coated with the BCP only after all the exposures have been done, thus self assembly takes place after all the guiding templates have already been printed. In this approach, all contact holes are defined through the self-assembly process and none is directly printed by the lithography process.

Alternatively, one can apply self assembly on each mask separately. In this latter approach, self-assembly can be bypassed for a subset of the masks making it possible to use conventional lithography to print some of the contact holes directly, which gives some flexibility in the size of the printed holes allowing contact/via bar shapes which cannot be printed with a pure DSA process. In this work however, we are assuming none of the masks can bypass the self-assembly, and hence the lithography step only creates the guiding templates, and does not create contact/via holes directly.

There are several important parameters in this problem:

a) **Minimum Grouping distance** (*min_dsa*): Minimum distance that can exist between two contacts in a DSA group. This distance is usually derived from the natural pitch (L_0) of the block copolymer as follows: $min\_dsa = L_0 - contact\_width$.

b) **Maximum grouping distance** (*max_dsa*): Maximum distance that can exist between two neighboring contacts in one DSA group. This is derived from the properties of the block copolymer, because its self-assembly pitch cannot be stretched beyond a certain threshold.

c) **Minimum Lithography Distance** (*litho_dist*): Minimum space that can occur on a single mask.

d) **Maximum DSA Group Size** (*max_g*): Maximum number of contacts that can be grouped together.

e) **Number of masks** (*N*): number of exposures in the process.

The DSA groups are determined by constraints from the self assembly process itself as well as the constraints of the photolithography which is used to print the guiding templates. While stronger confinement can lead to less placement error for the holes [8], some templates are not optically manufacturable especially in the case of 193i lithography, while they may be available with higher resolution processes like EUVL. Since the grouping of the contacts determines the template shape, some contact grouping configurations are not allowed.

In this work, we assume that only collinear contact holes that are aligned on same horizontal or vertical axis can be grouped together, forming a manhattan one-dimensional array. We make this assumption to guarantee that there is a 193nm process available to print the necessary guiding patterns. In addition, there is an upper limit on the number of contacts that can lie within one group (*max_g*).

In this work we study the different flows that can be used in enabling the hybrid DSA-MP process, highlight their pros and cons and conclude whether sequentially combining Multiple Patterning decomposition algorithms and pure DSA grouping algorithms can work well to deliver a hybrid DSA-MP technology.

2. Alternative Flows for solving the DSA-MP problem

As introduced in section 1.3, having a hybrid DSA-MP process requires a solution which performs the DSA grouping as well as the mask assignment. This problem can be handled by multiple flows as shown in Figure 2.

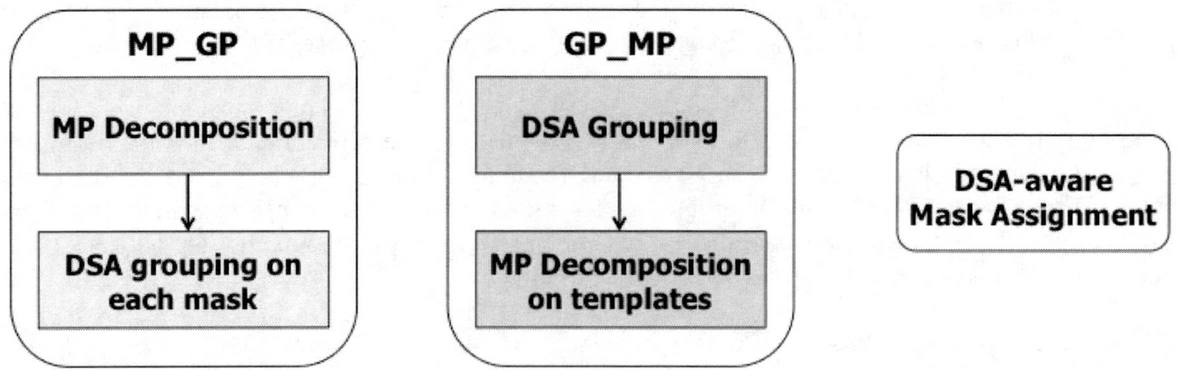

Figure 2. Different flows to do DSA grouping and MP decomposition for hybrid DSA-MP processes.

The first two flows are the decomposition-then-grouping (**MP_GP**) and the sequential grouping-then-decomposition (**GP_MP**) approaches. In **MP_GP**, mask decomposition is performed first, then DSA grouping is performed on each resultant mask separately. Alternatively, in **GP_MP**, DSA grouping is attempted on the complete layer, and then mask assignment is done on the resulting groups. The third alternative is a DSA-aware mask decomposition algorithm. Each of these flows has its pros and cons.

The **MP_GP** approach is friendly to DSA because it inherently favors smaller groups which produce smaller placement errors and a lower defect rates [8], [9], [10], [11]. This is because the mask decomposition essentially scatters the neighboring contacts onto different masks. However it tries to resolve conflicts for a larger number of entities (in comparison to doing the decomposition on the grouped contacts). This can make it harder to get to a conflict-free decomposition result, especially that Triple Patterning and Quadruple Patterning problems are NP-hard problems [12] and thus MP solvers are usually approximate and sub-optimal.

The **GP_MP** approach on the other hand, leads to fewer decomposition conflicts since the decomposition is performed on the DSA groups and not on the individual contacts; however, this flow is not friendly to DSA because it does not give higher priority to the smaller groups which are more reliable from a DSA perspective. In addition, the DSA-incompliance that stems from DSA topology constraints may not be resolved by MP decomposition which is distance-driven.

Finally, a DSA-aware coloring should produce better results, if it considers DSA challenges within the formulation of the problem, but handling the two problems (grouping and decomposition) simultaneously is expected to be a harder problem to solve. In this paper we investigate the results of the first two flows, and we leave the study of the third flow in length to subsequent work.

3. Experiments

In this section we show results of the **MP_GP** and **GP_MP** approaches. Both **MP_GP** and **GP_MP** have been implemented by using existing Multiple Patterning and Directed Self Assembly tools, which are not aware of the process being hybrid DSA-MP.

We ran our experiments on the Via1 layer of AES and MIPS from Open Cores [13], as well as an ARM Cortex M0 processor and a Leon3 Sparc V8 processor. These layouts that have been synthesized, placed and routed using commercial 45nm SOI libraries then sized and scaled.

After modification of the layouts, the via-width is 14nm and the minimum spacing is 21nm. The size of each of the test cases, in number of vias is shown in Table 1.

Test case	Number of vias
AES	48123
CortexM0	35255
LEON3	93474
MIPS	34784

Table 1. Number of vias in test cases.

The assumed parameters of the used DSA process are shown in Table 2. The flows were executed on every test case, for 2 masks [Double Patterning (DP)] and three masks [Triple Patterning (TP)].

min_dsa	20
max_dsa	42
$litho\_dist$	66
max_g	4
$contact\ width$	14
L_0	34
N	2 (DP) and 3(TP)

Table 2. Parameter Values used in experiments [nm].

The results of running **MP_GP** and **GP_MP** approaches are shown in Table 3 where the total numbers of spacing violations between the resulting DSA groups on the same mask are shown. It is important to point out that in some cases a DSA group can be composed by a single contact.

On all test cases except one, **MP_GP** outperforms **GP_MP**, from the point of view of producing less number of violations. This is contrary to the expectation discussed in section 2. The reason for that is that the DSA grouping algorithm is rule-based and in many cases when there are several pairs of contacts within the DSA grouping distance, it does not selectively determine groups that would prevent violations and groups them in a sub-optimal fashion. Thus for that type of grouping algorithm, having the contact scattered onto different masks before grouping produces a fewer violations.

Test case	N	MP_GP Violations	GP_MP Violations
AES	2	641	696
CortexM0	2	488	487
LEON3	2	642	680
MIPS	2	315	324
AES	3	6	29
CortexM0	3	7	28
LEON3	3	1	7
MIPS	3	5	13

Table 3. Results of **MP_GP** and **GP_MP** decomposition approaches.

In order to assess the quality of the **MP_GP** and **GP_MP** approaches, we present some layout snippets, along with the results of the two approaches.

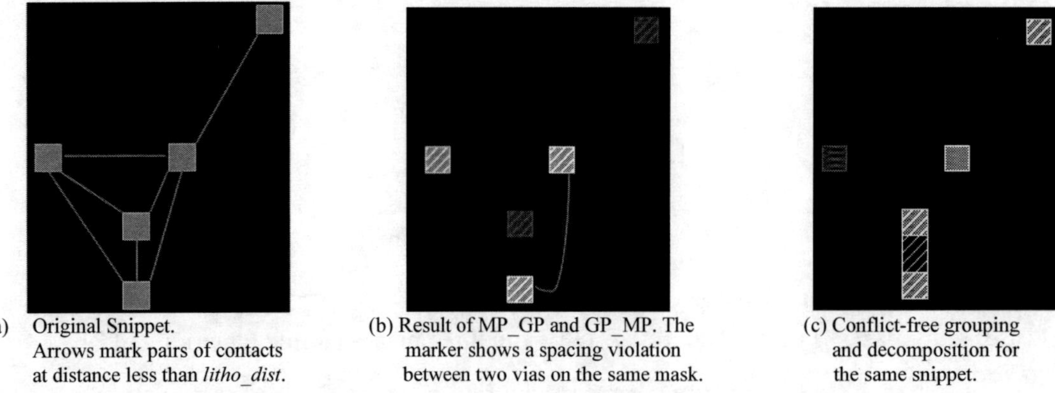

(a) Original Snippet. Arrows mark pairs of contacts at distance less than *litho_dist*.

(b) Result of MP_GP and GP_MP. The marker shows a spacing violation between two vias on the same mask.

(c) Conflict-free grouping and decomposition for the same snippet.

Figure 3. Sample of **MP_GP** and **GP_MP** results on snippet #1, with TP. The output shows that the sequential **MP_GP** and **GP_MP** approaches can fail to produce a good solution even for simple configurations.

In Figure 3a, we show a simple layout snippet of five vias. The two approaches **MP_GP** and **GP_MP** got to the same solution, which is shown in Figure 3b. For this snippet, **MP_GP** and **GP_DP** failed to use the grouping to resolve conflicts, resulting in all DSA groups with one contact each. A possible conflict-free solution for the same snippet is shown in Figure 3c.

Two other examples are shown in Figure 4 and Figure 5. It is clear that the sequential approaches **MP_GP** and **GP_MP** failed even on these simple snippets. **MP_GP** failed because the MP decomposer gave equal priority to all pairs of polygons having an intra-distance less than *litho_dist*, without considering that the pairs of contacts that are aligned on the same vertical or horizontal axis could have been DSA-grouped and accordingly assigned to the same mask. The **GP_MP** approach failed because a lot of contacts were within *max_dsa* distance, which led to very complex groups that are disallowed by DSA compliance, due to the lithography and self-assembly constraints. Accordingly, many DSA groups were disqualified, leaving the spacing violations to be handled mostly by the MP decomposition step, which in turn led to a large number of violations. Thus, a DSA-aware coloring algorithm is required to handle the hybrid DSA-MP process since the sequential *MP_GP* and *GP_MP* performed poorly.

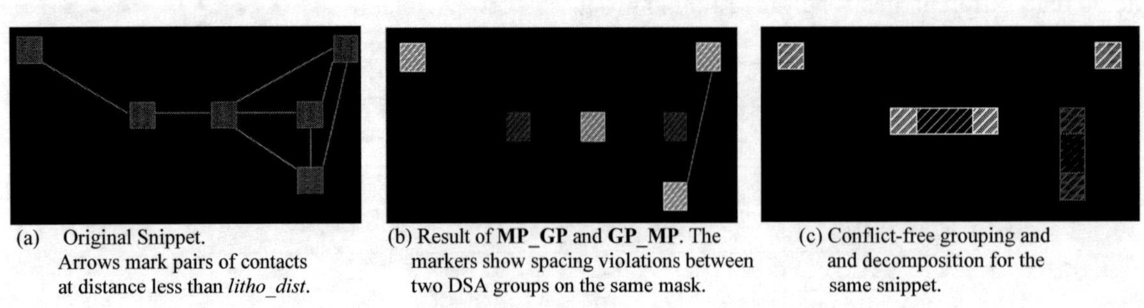

(a) Original Snippet. Arrows mark pairs of contacts at distance less than *litho_dist*.

(b) Result of **MP_GP** and **GP_MP**. The markers show spacing violations between two DSA groups on the same mask.

(c) Conflict-free grouping and and decomposition for the same snippet.

Figure 4. Sample of **MP_GP** and **GP_MP** results on snippet #2 with TP.

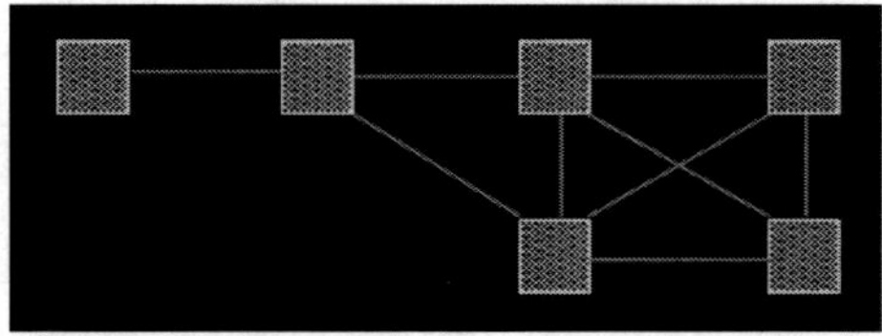

(a) Original Snippet.
The arrows mark pairs of contacts at a distance less than *litho_dist*.

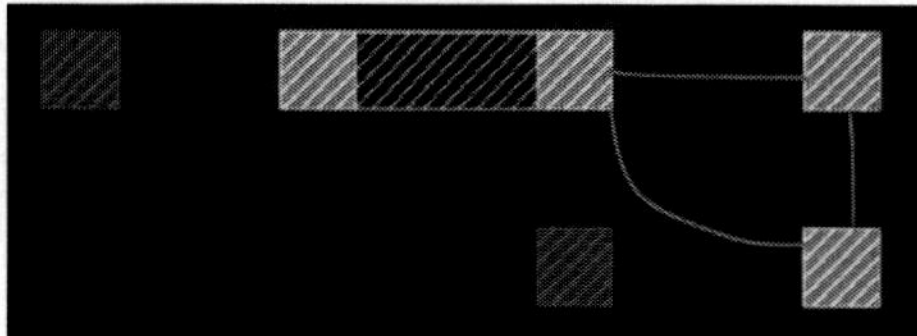

(b) Result of **MP_GP**. The red marker shows a spacing violation between two DSA groups.

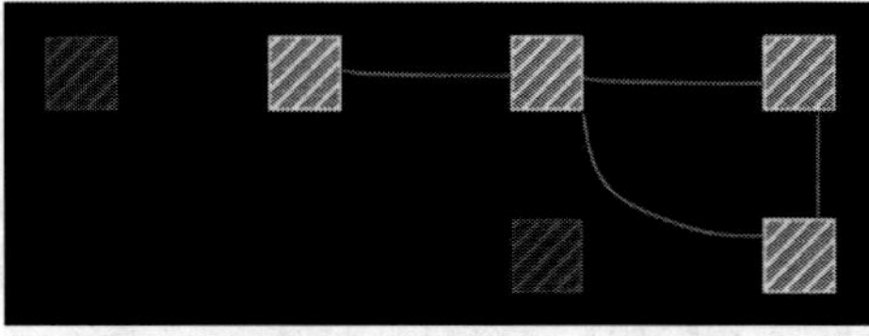

(c) Result of **GP_MP**. The red marker shows a spacing violation between two DSA groups.

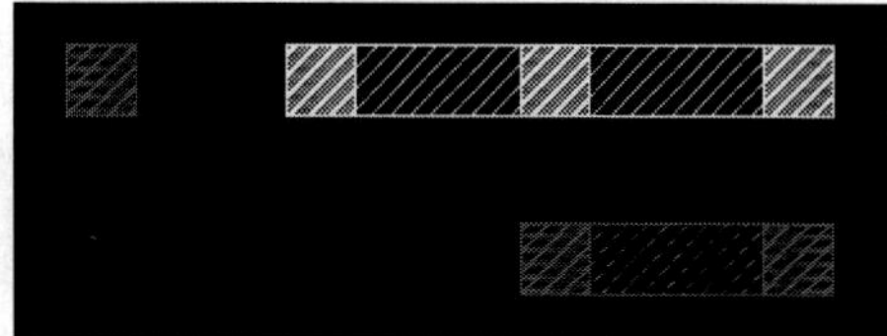

(d) Conflict-free grouping and decomposition for the same snippet.

Figure 5. Sample of **MP_GP** and **GP_MP** results on snippet #3 with DP.

4. Conclusions

In this paper, we have considered the integration of DSA into MP technologies with the objective of saving one mask for a less costly process. We studied two sequential approaches considering MP decomposition and DSA grouping as two independent steps. Results indicate that these two sequential approaches can fail to find a solution to the problem even under very simple layout configurations. Thus, a solution that simultaneously considers the constraints of DSA and those of MP, and can perform DSA-aware mask assignment, is required and will be the focus of our future work.

Acknowledgments

This work was partly supported by IMPACT+ center (http://impact.ee.ucla.edu).

Bibliography

[1] B. Xu, R. Piñol, M. Nono-Djamen, S. Pensec, P. Keller, P. Albouy, D. Lévy, and M.-H. Li, "Self-assembly of liquid crystal block copolymer peg-b-smectic polymer in pure state and in dilute aqueous solution," *Faraday discussions* 143 (2009).

[2] N. Jarnagin, High X Block Copolymers For Sub 20 Nm Pitch Patterning: Synthesis, Solvent Annealing, Directed Self Assembly, And Selective Block Removal. *PhD thesis*, Georgia Institute of Technology (2013).

[3] M. Kim, E. Han, D. P. Sweat, and P. Gopalan, "Interplay of surface chemical composition and film thickness on graphoepitaxial assembly of asymmetric block copolymers," *Soft Matter* 9(26), 6135–6141 (2013).

[4] L. Chang, X. Bao, C. Bencher, and H.-S. Wong, "Experimental demonstration of aperiodic patterns of directed self-assembly by block copolymer lithography for random logic circuit layout," in *IEEE International Electron Devices Meeting (IEDM)* (2010).

[5] J. Chang, H. K. Choi, A. F. Hannon, A. Alexander-Katz, C. A. Ross, and K. K. Berggren, "Design rules for self-assembled block copolymer patterns using tiled templates," *Nature Communications* 5 (2014).

[6] L. S. Wan, P. A. Rincon Delgadillo, R. Gronheid, and P. F. Nealey, "Directed self-assembly of ternary blends of block copolymer and homopolymers on chemical patterns," *Journal of Vacuum Science Technology B: Microelectronics and Nanometer Structures* 31, 06F301–06F301–6 (2013).

[7] R. Ruiz, H. Kang, F. A. Detcheverry, E. Dobisz, D. S. Kercher, T. R. Albrecht, J. J. de Pablo, and P. F. Nealey,

"Density multiplication and improved lithography by directed block copolymer assembly," *Science* 321(5891), 936–939 (2008).

[8] Y. Ma, J. A. Torres, G. Fenger, Y. Granik, J. Ryckaert, G. Vanderberghe, J. Bekaert, and J. Word, "Challenges and opportunities in applying grapho-epitaxy DSA lithography to metal cut and contact/via applications," in *30th European Mask and Lithography Conference*, 92310T–92310T, International Society for Optics and Photonics (2014).

[9] H. Kang, G. S. Craig, E. Han, P. Gopalan, and P. F. Nealey, "Degree of perfection and pattern uniformity in the directed assembly of cylinder-forming block copolymer on chemically patterned surfaces," *Macromolecules* 45(1), 159–164 (2011).

[10] C. T. Black, R. Ruiz, G. Breyta, J. Y. Cheng, M. E. Colburn, K. W. Guarini, H. Kim, and Y. Zhang, "Polymer self assembly in semiconductor microelectronics," *IBM Journal of Research and Development* 51(5), 605–633 (2007).

[11] D. Sundrani, S. Darling, and S. Sibener, "Hierarchical assembly and compliance of aligned nanoscale polymer cylinders in confinement," *Langmuir* 20(12), 5091–5099 (2004).

[12] B. Yu, K. Yuan, B. Zhang, D. Ding, and D. Pan, "Layout decomposition for triple patterning lithography," in *International Conference on Computer-Aided Design (ICCAD)* (2011).

[13] "http://www.opencores.org."

Design Layout Analysis and DFM Optimization using Topological Patterns

Ji Xu[a], Karthik N. Krishnamoorthy[a], Edward Teoh[a], Vito Dai[a], Luigi Capodieci[a], Jason Sweis[b], Ya-Chieh Lai[b],
[a]GLOBALFOUNDRIES Inc., 2600 Great America Way, Santa Clara, CA 95054;
[b]Cadence Design Systems Inc., 2655 Seely Avenue San Jose, CA 95134

ABSTRACT

During the yield ramp of semi-conductor manufacturing, data is gathered on specific design-related process window limiters, or yield detractors, through a combination of test structures, failure analysis, and model-based printability simulations. Case-by-case, this data is translated into design for manufacturability (DFM) checks to restrict design usage of problematic constructs. This case-by-case approach is inherently reactive: DFM solutions are created in response to known manufacturing marginalities as they are identified.

In this paper, we propose an alternative, yet complementary approach. Using design-only topological pattern analysis, all possible layout constructs of a particular type appearing in a design are categorized. For example, all possible ways via forms a connection with the metal above it may be categorized. The frequency of occurrence of each category indicates the importance of that category for yield. Categories may be split into sub-categories to align to specific manufacturing defect mechanisms. Frequency of categories can be compared from product to product, and unexpectedly high frequencies can be highlighted for further monitoring. Each category can be weighted for yield impact, once manufacturing data is available.

This methodology is demonstrated on representative layout designs from the 28 nm node. We fully analyze all possible categories and sub-categories of via enclosure such that 100% of all vias are covered. The frequency of specific categories is compared across multiple designs. The 10 most frequent via enclosure categories cover ≥90% of all the vias in all designs. KL divergence is used to compare the frequency distribution of categories between products. Outlier categories with unexpected high frequency are found in some designs, indicating the need to monitor such categories for potential impact on yield.

Keywords: topology, pattern, design rule, layout, analysis, manufacturability

1. INTRODUCTION

As the semiconductor manufacturing industry moves into advanced nodes, more complex integration schemes emerge to enable reduction of the size of the physical structure of electronic components. This trend in turn drives design rules to be more complicated in order to avoid as much non-manufacturable design as possible. On the other hand, pattern-based approaches such as DRC plus[1] augment DRC (Design Rule Checking) by recognizing problematic designs through a library of patterns. Such patterns can be identified by gathering data on specific design-related process window limiters, or yield detractors, through a combination of test structures, failure analysis, and model-based printability simulations. Case-by-base, DFM solutions are created in response to known manufacturing marginalities as they are identified.

In this work, we propose a complementary approach by utilizing design-only topological pattern analysis to profile and categorize all possible layout constructs of a particular type appearing in a design. In Section 2, we describe the methodology and flow to build a complete library of topological patterns that quantifies the frequency of occurrence of a category of layout constructs. The methodology is demonstrated through a study on the enclosure of vias by the metal above the via across multiple designs. In Section 3 we compute the KL divergence as a statistical measure to compare the frequency distribution of different categories of via enclosure between designs. In Section 4, we conclude with an outlook on the possible future work.

Design-Process-Technology Co-optimization for Manufacturability IX, edited by John L. Sturtevant, Luigi Capodieci, Proc. of SPIE Vol. 9427, 94270Q · © 2015 SPIE · CCC code: 0277-786X/15/$18 · doi: 10.1117/12.2086904

2. DESIGN LAYOUT ANALYSIS USING TOPOLOGICAL PATTERNS

A topological pattern consists of a bitmap representing the placement and alignment of polygon edges and a vector of dimensional constraints. An example of 3×3 line-end to side bitmap is shown in Figure 1a and the corresponding constraints on row heights and column widths are shown in Figure 1b. The bitmap shown in Figure 1a may also be written as "1 1 1 / 0 0 0 / 0 1 0" where the " / " represents a new row in the bitmap. Previous work has shown that any rectangular region of Manhattan layout has a 1:1 mapping to a topological pattern.[2,3] Therefore, it is possible to represent the space of all finite rectangular regions of Manhattan layout through enumeration of the topological pattern bitmap. Conversely, it is possible to classify all finite rectangular regions of Manhattan layout as topological patterns. In this section we describe basic methodologies of enumeration and classification for assembling topological patterns to represent a full design. We use the example of assembling the topological pattern library for all metal-via enclosure configurations to demonstrate our methodologies. Section 2.1 describes the method of enumeration in detail. Section 2.2 describes the application of enumeration to construct all 2×2 patterns to capture all the via enclosure configurations at each corner of every via. Section 2.3 extends the application of enumeration to construct 3×3 and 4×3 patterns. Section 2.4 describes the method of classification in detail. Section 2.5 describes how to apply classification to find the patterns of via enclosure with higher dimensionalities.

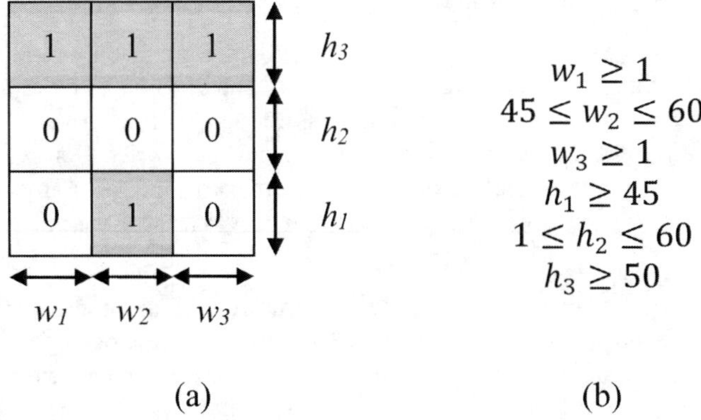

Figure 1. A topological pattern consists of (a) a bitmap and (b) a vector of constraints

2.1 Enumeration

Enumeration is a process of orderly counting of a finite number of unique elements. Specifically, the finite set of all possible values for each variable is identified and then the rest are combined through a Cartesian product. For example, if we want to enumerate all the possible two digit binary numbers, we first list all the options for first digit, and then list all the options for the second digit, and finally list all the possible combinations of the first and the second digit. As there are only two options for each binary digit, "0" and "1", the result should have a list of 2×2=4 elements, namely "00", "01", "10", and "11". In general, N variables with M values result in M^N unique combinations. Therefore enumeration is feasible only to a small number of variables with limited number of variations per variable. When applying enumeration to the construction of topological patterns, patterns which are redundant after reflection or rotation are excluded.

2.2 The construction of 2×2 patterns for via enclosures

The smallest 2D topological patterns to build have a dimension of 2×2. Although a 2×2 pattern is not large enough to define a via and its vicinity, it defines the via corners. Specifically, a via corner is defined by the topological pattern "0 0 / 1 0". The metal and space above or around the via corner can be defined by another 2×2 pattern "x_1 x_2 / x_3 x_4", where "x_i" can have the value of either "0" for space or "2" for metal. By overlapping "0 0 / 1 0" with "x_1 x_2 / x_3 x_4" we obtain the 2×2 pattern that defines the via corner and its overlaying metal in vicinity, *i.e.* "x_1 x_2 / x_3+1 x_4". Enumeration produces 2^4 = 16 possible patterns for "x_1 x_2 / x_3+1 x_4", which can be further reduced to 12 unique ones after eliminating the redundant patterns through rotation and reflection. This process is illustrated a flow chart in Figure 2a, and the resulting list of 2×2 patterns is shown in Figure 2b.

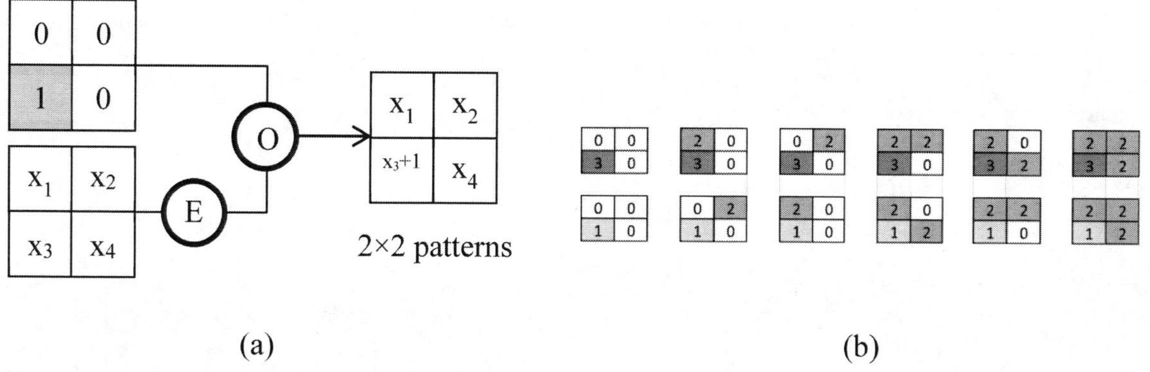

Figure 2. The flow chart of assembling 2×2 topological patterns for metal-via enclosure configurations using enumeration methodology (a) and the resultant patterns (b). In the flow, E stands for enumeration and O stands for overlaying.

Implementing this library of patterns with appropriate dimension constraints ($w_1 \geq 1$, $w_2 \geq 1$, $h_1 \geq 1$, $h_2 \geq 1$) into a pattern-matching deck, we are able to match all via corners for 4 different 28nm node designs. In this paper we exclude SRAM from the analysis, since the SRAM is composed of a single repeated pattern. A plot of log-scale count of pattern matches versus the pattern bitmap is shown below in Figure 3. Some clearly DRC violating patterns have zero count as expected, such as "0 0 / 3 0" and "0 2 / 3 0", while the zero count for all the "free floating" via patterns such as "0 0 / 1 0" implies the non-existence of such configurations. This result also indicates the internal similarity among such 4 designs: they all share approximately the same order of magnitude in counts for each of the seen patterns. This sense of similarity/dissimilarity by category leads us to compare designs from a particular standpoint, which will prove to be a very useful analysis methodology in the following sections of this paper. Another very important learning from this study is that, all the to-be-constructed topological patterns with higher dimensions, such as 3×3 patterns and 4×3 patterns, can be ignored or placed in an "unlikely to occur" deck if they are composed of any of the zero-counting 2×2's.

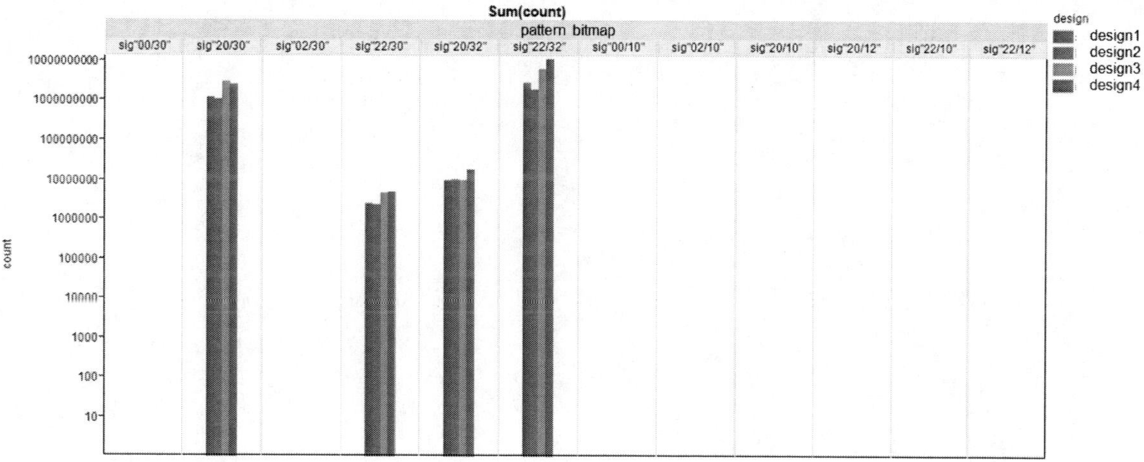

Figure 3. A log-scaled bar chart plot of raw count in matched patterns versus the pattern bitmap.

2.3 The construction of 3×3 and 4×3 patterns for via enclosures

As previously described, the smallest 2D topological pattern to define a via and its vicinity has a dimension of 3×3, *i.e.* overlaying a matrix of "x_1 x_2 x_3 / x_4 x_5 x_6 / x_7 x_8 x_9" upon a matrix of "0 0 0 / 0 1 0 / 0 0 0", where "x_i" can be either "0" or "2". The enumeration step produces $2^9 = 512$ possible overlaps of metal over the via. Next, we filtered out all patterns containing 2×2 sub-patterns with zero counts, *i.e.* patterns that are design rule violations, as well as patterns which are redundant under rotations and reflections. The resulting 23 unique patterns are shown in Figure 4b. This collection of 3×3 patterns can represent any via where any metal edge in the vicinity of the via is aligned with the via edges. As noted before, if any metal edge is not aligned yet projecting onto the via edges, the 3×3 patterns are no longer applicable and new patterns with higher dimensionalities such as 4×3 patterns need to be constructed to represent such configurations. Similar to the construction of 3×3 patterns, we overlay a matrix of "x_1 x_2 x_3 x_4 / x_5 x_6 x_7 x_8 / x_9 x_{10} x_{11} x_{12}" upon a matrix of "0 0 0 0 / 0 1 1 0 / 0 0 0 0", where "x_i" can be either "0" or "2". After enumeration and filtration, the final list of 79 unique 4×3 patterns can be obtained, as shown in Figure 5b.

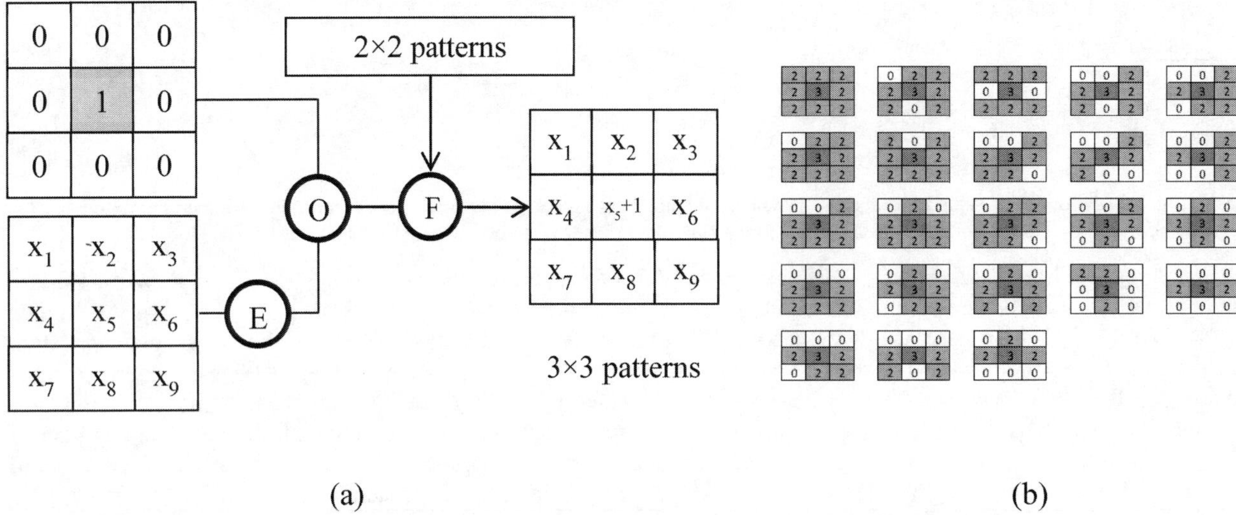

Figure 4. The flow chart of assembling 3×3 topological patterns for metal-via enclosure configurations using enumeration methodology (a) and the collection of 23 unique 3×3 bitmaps after filtration (b).

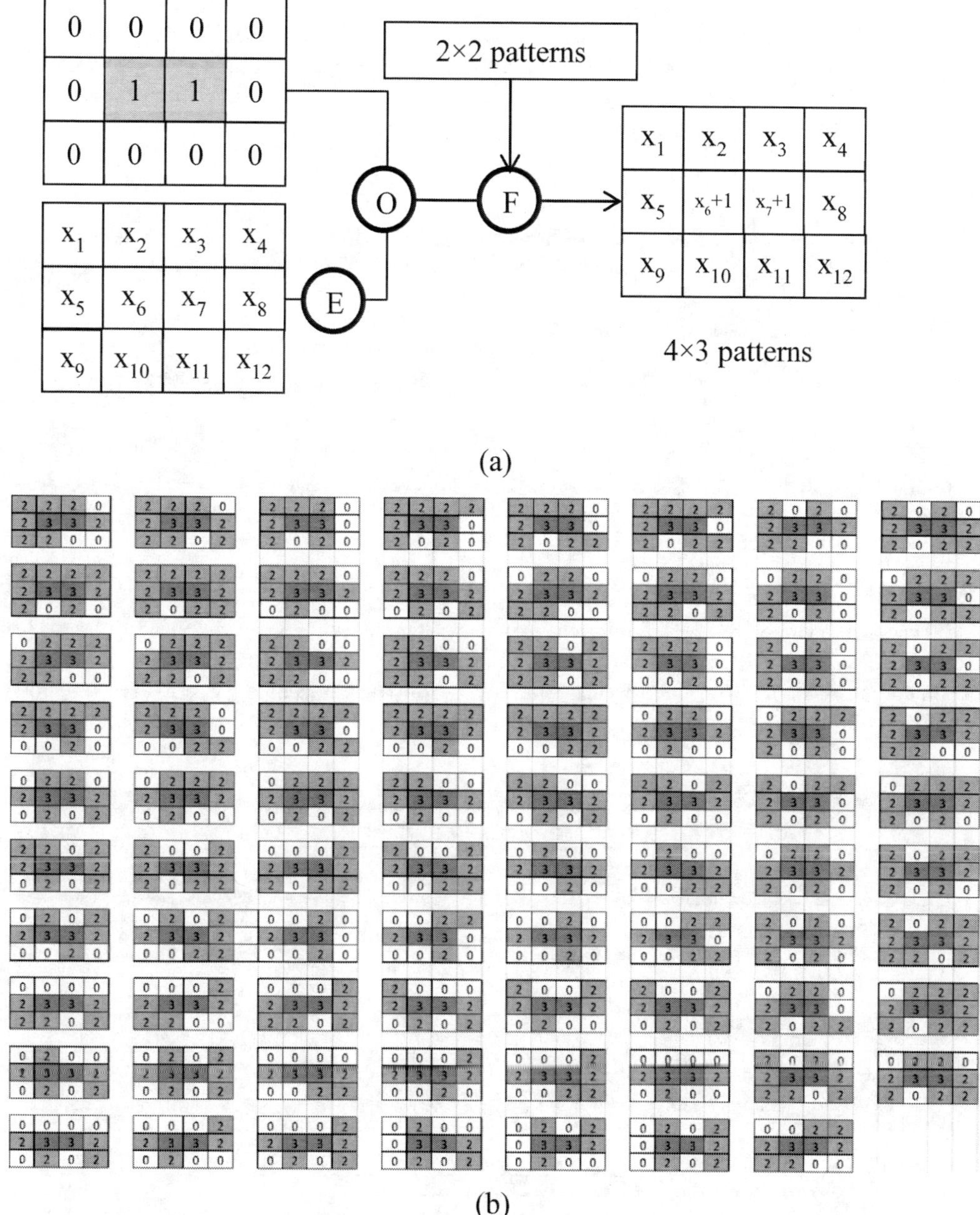

Figure 5. The flow chart of assembling 4×3 topological patterns for metal-via enclosure configurations using enumeration methodology (a) and the collection of 79 unique 4×3 bitmaps after filtration (b). In the flow, E stands for enumeration, O stands for overlaying, and F stands for filtration.

2.4 Classification

Classification is the process of grouping existing objects into categories according to certain criteria. Different criteria results in different classification categories. For example, if a list of numbers "1, 2, 4, 5" is classified based on numeric value, then each of the numbers in the list would be a separate category. However, if the same list of numbers is classified based on an even/odd criteria, then "1" and "5" will be grouped as odd while "2" and "4" will be grouped as even. A key difference between classification and enumeration is that classification builds just enough categories such that every existing object is accounted for, whereas enumeration defines all possible categories, many of which maybe empty. Therefore, enumeration is efficient where the number of objects is much greater than the number of categories, such that every category contains many objects. Conversely, classification is efficient where the number of potential categories is much greater than the number of objects, such that most categories are empty. In the example of assembling topological patterns for metal-via enclosure configurations, classification is used as a catch-all for vias not found by enumerated 3×3 and 4×3 patterns. The resulting topological patterns must have higher dimensions (5×3 patterns, 6×3 patterns, etc.). This work flow is shown in Figure 6.

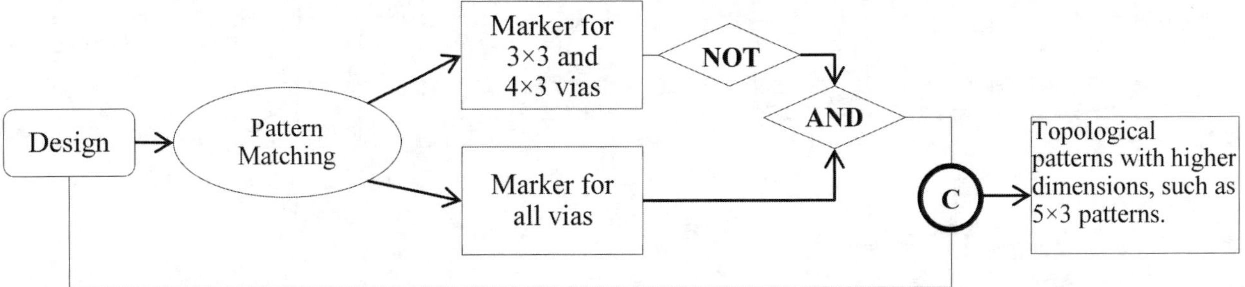

Figure 6. The flow chart of assemble 5×3 and higher-dimensional topological patterns for metal-via enclosure configurations using classification methodology. In the flow, NOT and AND stands for the Boolean operation NOT and AND, and C stands for classification.

2.5 The construction of topological patterns with higher dimensions for via enclosure using pattern classification

On a design, we use a Boolean NOT operation to remove the markers of all the analyzed vias from the markers of all the vias, so that only the markers of the non-analyzed vias are left. Using this new marker layer, we apply pattern classification specifically with a radius that reaches 1 nm away from the edges of the via. As a result shown in Figure 7, we obtain 47 new topological patterns, 28 of which are 5×3 patterns, 15 are 6×3 patterns, and 4 are 7×3 patterns.

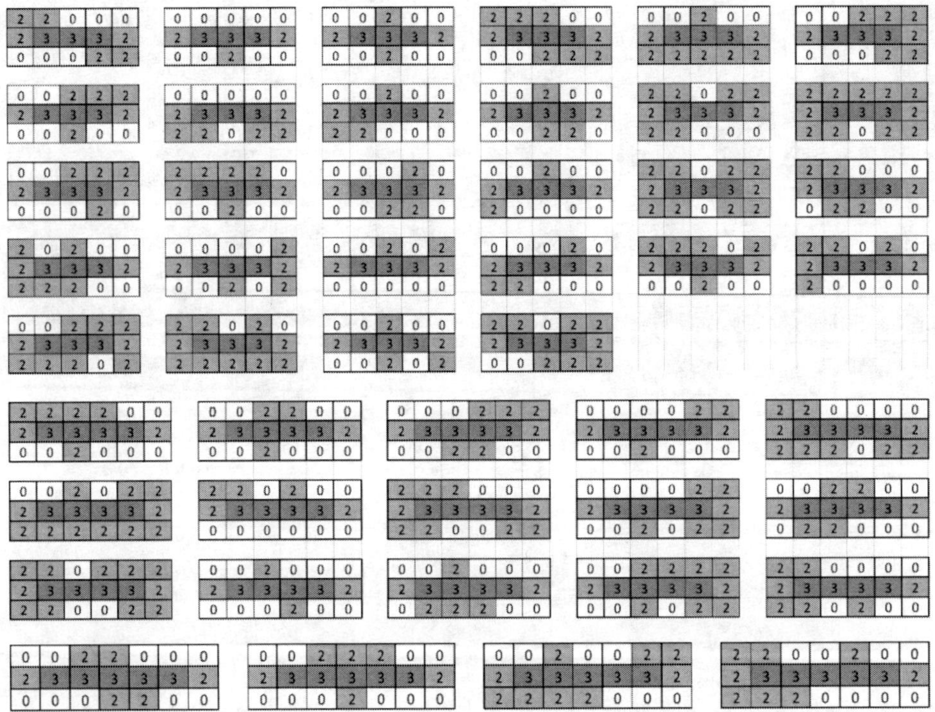

Figure 7. The 5×3, 6×3, and 7×3 topological patterns found through pattern classification.

3. STATISTICAL LAYOUT ANALYSIS AND DFM OPTIMIZATION

3.1 The Coverage

One of the goals of layout analysis is to make sure we have a complete coverage of all cases. To define this metric we establish a definition of what accounts towards all cases. For the example of via enclosure, all cases mean all vias, of which the defining topological pattern is shown in Figure 8a. The dimension of vias also needs to be defined, and as an example the dimension constraints of vias in 28 nm technology are shown in Figure 8b, c, and d.

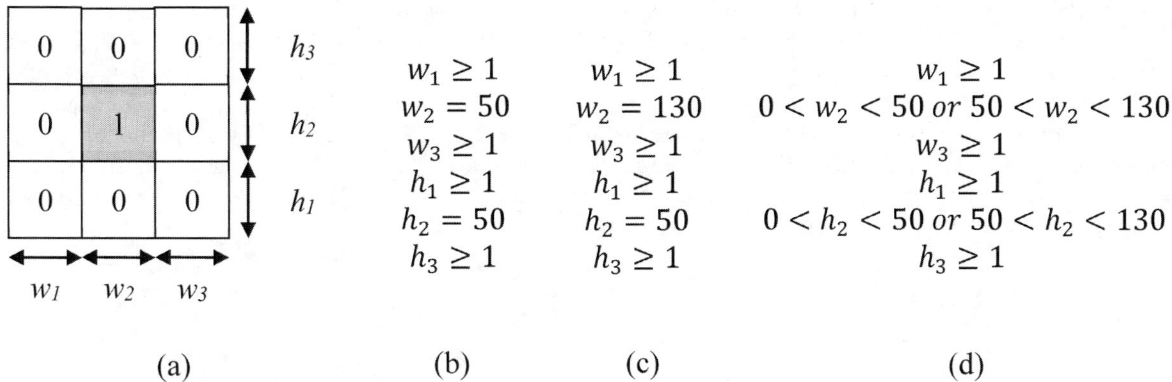

Figure 8. the topological pattern which defines all vias (a) and the corresponding 28nm node dimensional constraints for (b) square via (50 nm by 50 nm), (c) via bar (50 nm by 130 nm), and (d) other via with irregular dimensions.

The space of all vias differ from design to design. This point can be proved by examining the four 28nm designs that we utilize in section 2. As seen in the Figure 9 bar chart, design 3 and 4 have significantly larger amounts of via than design 1 and 2, which implies that design 3 and 4 are physically larger in size. The disproportionally larger count of vias in V1 layer against other via layers indicates a heavy M1-M2 inter-layer routing in design 3 and 4, which is much less prominent in design 1 and 2. Proportion-wise as seen in the pie charts inserted in Figure 9, a 3:2 ratio in count of total square via vs. via bar is universally seen across the four designs under study, but if we break the data down by layers, a clear signature of heavier reliance on square via at layer V1 can be seen in design 3 and 4 if compared with design 1 and 2. In contrast, a heavier reliance on via bar at layer V2 through layer V7 can be observed in design 3 and 4.

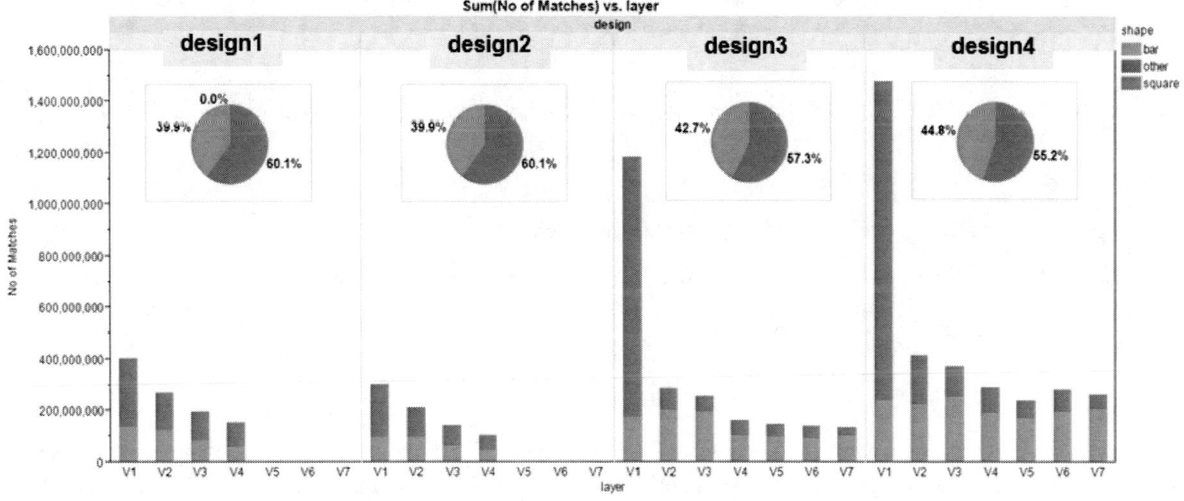

Figure 9. (a) stacked bar charts showing the number of vias vs. the layer for 4 different designs, with the colored legend indicating the shape of the via and the total number of via on the top-right corner of each chart. (b) a matrix of pie charts showing the composition of vias in terms of shape at each layer for the 4 designs.

By incorporating these topological patterns into a pattern matching deck, we are able to cover 100% of all the vias for metal enclosure configurations, as shown in Figure 10. In this chart, every matched pattern is identified as a category and assigned with a unique color. If any matched pattern is representing less than 0.1% of the vias of a particular via shape at certain layer, then it is categorized as the "Other" category. The length of the colored portion of the bar represents how much in percentage of the vias have metal enclosure configuration of the corresponding pattern. For example, if we look at the red portion of the bar chart for via bars of design 1 at V1.M2 level, we can clearly tell that about 12% of the via bars in V1 layer of design 1 have a M2 enclosure characterized by the "0 0 0 / 2 3 2 / 0 0 0" topological pattern. As indicated by the legend of the chart, we see that there are 18 most popular topological pattern that represent >0.1% of the square vias or the via bars in any via layer of the 4 designs, out of which 8 patterns are 3×3 patterns, 4 are 4×3 patterns and 6 are 5×3 patterns. If we compare the bar charts of design 1 and design 2, we can observe that these two designs are very similar in via enclosure configurations as the proportions and distribution of colored bars are quite close. More interestingly, although the blue ("2 2 2 / 2 3 2 / 2 2 2") and red ("0 0 0 / 2 3 2 / 0 0 0") bars seem to dominate all 4 designs, the yellow ("0 0 0 0 / 2 3 3 2 / 0 0 2 2") and green ("0 0 0 / 2 3 2 / 2 2 2") portion of bars stands out in design 3 with significant length, which are much less observable in design 1, 2, and 4. While it is good to have an overview of the via enclosure configurations with the help of stacked bar charts, a deeper and more quantitative analysis is needed to compare the design styles among different design.

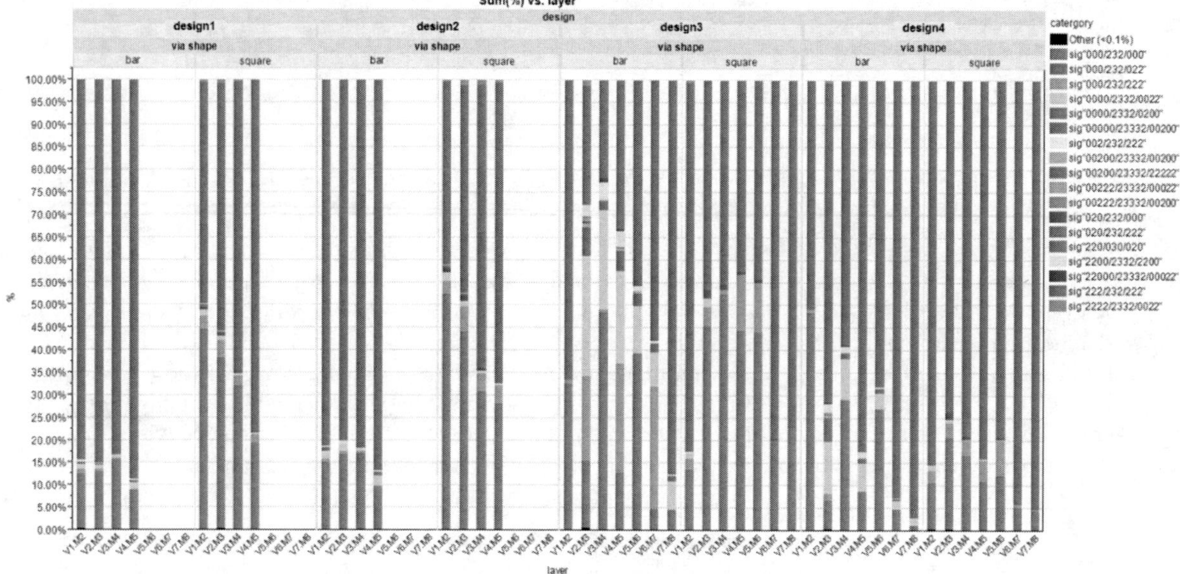

Figure 10. Stacked bar charts of profiled vias in metal enclosure configurations at every via layer for 4 different designs (100% of all the vias are profiled)

3.2 Statistical layout analysis using Kullback Leibler divergence

Kullback Leibler (KL) divergence, is a metric used in information theory to measure the dissimilarity between two probability distributions.[4] If P and Q are two distributions with the same set of outcome i with associated probabilities $P(i)$ and $Q(i)$, the KL divergence $D_{KL}(P||Q)$ is defined as the following:

$$D_{KL}(P||Q) = \sum_i ln\left(\frac{P(i)}{Q(i)}\right)P(i) \tag{1}$$

KL divergence is always non-negative and it is zero if and only if the distributions P and Q are completely identical. KL divergence is also non-symmetrical, which means that $D_{KL}(P||Q) \neq D_{KL}(Q||P)$ as long as P and Q are not identical. Each term summed in Equation (1) measures the relative likelihood of seeing outcome i in P when compared with Q. A large positive value indicates that outcome i is much more likely to happen in P than Q.

To apply KL divergence to layout analysis, we treat the percentage coverage of each pattern on a design as a distribution outcome i. We pick a design to use as the baseline distribution Q to compare against other designs as P. For example, if we pick design 1 as the baseline distribution Q and treat the distribution in design 2, 3, or 4 as P, we can calculate $ln\left(\frac{P(i)}{Q(i)}\right)P(i)$ for each via shape and obtain the sum as the KL divergence, as shown in Figure 11. Based on the KL divergence in Figure 11a, it is clear that design 2 is most similar to design 1, whereas design 3 and 4 are less similar. This corroborates the fact that design 1 & 2 come from company A and design 3 & 4 come from company B. Examining each term summed in KL divergence in Figure 11b, we observe that the largest positive term is the via bars of design 4. This implies that, if we are experiencing some manufacturing difficulties in design 1 that is generally associated with via bars, we are very likely to encounter more of such difficulties in design 4. For another example, we can calculate $ln\left(\frac{P(i)}{Q(i)}\right)P(i)$ for each via enclosure configurations and obtain the sum as the KL divergence, as shown in Figure 12. We pick any pattern as an outcome i if it matches more than 0.1% of the vias in any design. Therefore the number of outcomes narrows down to 19, which are composed of 18 unique patterns plus a group of all the other patterns. Based on the KL divergence in Figure 12a, design 2 is the one that resembles design 1 the most, followed by design 4, and finally design 3. Closely inspecting each term summed in KL divergence in Figure 12b, we discover that the majority of the differences reside in 4 patterns, namely "2 2 2 / 2 3 2 / 2 2 2" (fully via enclosure), "0 0 0 / 2 3 2 / 0 0 0" (zero side enclosure on both sides of the via), "0 0 0 / 2 3 2 / 2 2 2" (zero side enclosure on only 1side of the via), and "0 0 0 0 / 2 3 3 2 / 0 0 2 2" (zero side enclosure on both sides of the via, with an overlaying metal edge projecting into the via). Design 3 has such a large percentage of vias with the "0 0 0 0 / 2 3 3 2 / 0 0 2 2" configuration that this pattern becomes its most observable outlier. This indicates that more attention needs to be paid to the manufacturing of "0 0 0 0 / 2 3 3 2 / 0 0 2 2" vias in design 3, because it may very likely become a major yield detractor for design 3, even if it wasn't one in design 1. Also, compared to other designs, design 4 has the largest positive term in full via enclosure and the largest negative term in zero side enclosure for vias. This trend is friendlier to manufacturing since better via enclosure leads to better yield in general. However, sometimes zero side enclosure is not necessarily a bad practice for design, if self-aligned via process is involved. What's more, we can also observe that design 4 has a larger KL divergence than design 3 in Figure 11a, while the reverse is true in Figure 12a. This indicates that the result of KL divergence calculation may change if the perspective changes. For example, if there is a manufacturing problem associated with the via bars specifically in the "0 0 0 0 / 2 3 3 2 / 0 0 2 2" configuration in design 1, then design 3 is more likely to be affected than design 4.

A DFM optimization proposal is: if there is a "golden" basis design (Q) to which the fab is aligning/optimizing its processes, then any new design (P) needs to be compared to this basis design through statistical analysis such as KL divergence to find the outliers that may become manufacturing challenges. DFM solutions may be implemented to suppress these outliers and change the new designs to resemble the "golden" design as much as possible, paving the way for a smooth ramp of the manufacturing yield for new designs.

Design	KL divergence
2	0.000000996
3	0.001563892
4	0.004975463

(a)

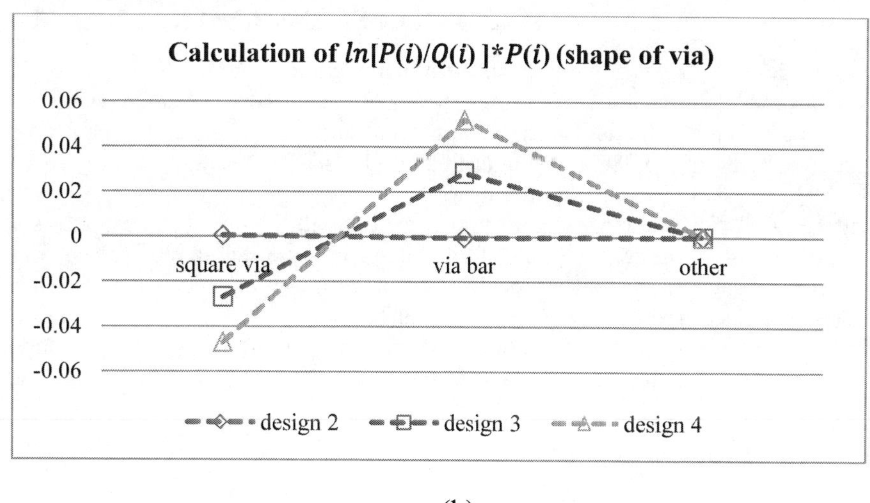

(b)

Figure 11. The table of KL divergence in via shape for design 2, 3, and 4 (a) and the calculation of $\ln\left(\frac{P(i)}{Q(i)}\right)P(i)$ for each via shape (b).

Design	KL divergence
2	0.009779801
3	0.099940352
4	0.050565141

(a)

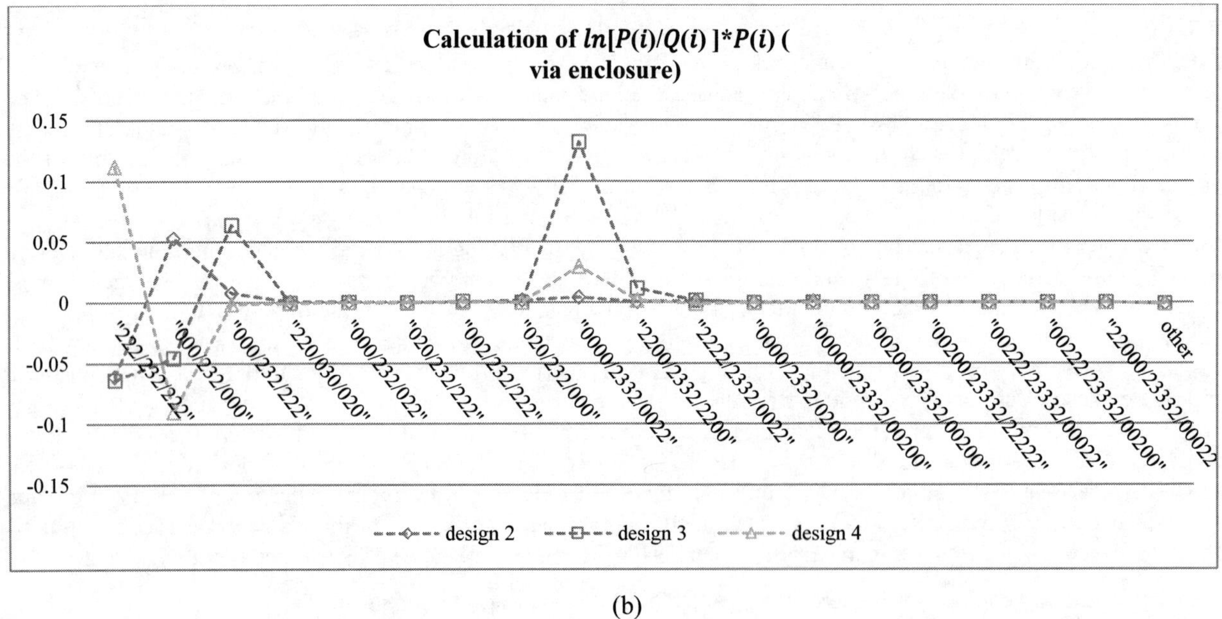

(b)

Figure 12. The table of KL divergence in via enclosure for design 2, 3, and 4 (a) and the calculation of $\ln\left(\frac{P(i)}{Q(i)}\right)P(i)$ for each pattern of via enclosure (b).

4. CONCLUSIONS AND FUTURE WORK

In this paper we have described the methodology of design layout analysis using topological patterns. We have demonstrated this methodology on the analysis of via enclosure of 28nm designs. But this methodology is not limited to only one technology node. On the contrary, our approach is generally applicable to any current or future nodes. For example, by using the 28nm topological pattern library and changing the dimension constraints to values corresponding to the 14 nm node, we can plot the distribution of via-metal enclosure configurations (3×3 patterns, 4×3 patterns, 5×3 patterns, 6×3 patterns and 7×3 patterns) as we did with 28 nm node designs, as shown in Figure 13. Interestingly, the majority of 14 nm designs in terms of via-metal enclosure can be represented with just 6 3×3 topological patterns. This observation suggests that 14 nm designs are much more regular, as the reduction in dimension of structures requires more regular design styles for tighter manufacturing process control. Further analysis requires the expansion of the radius for topological patterns, and corresponding increase in the dimension of topological pattern bitmaps. But the same methodologies, enumeration and classification, can be generally applied to solve such problems.

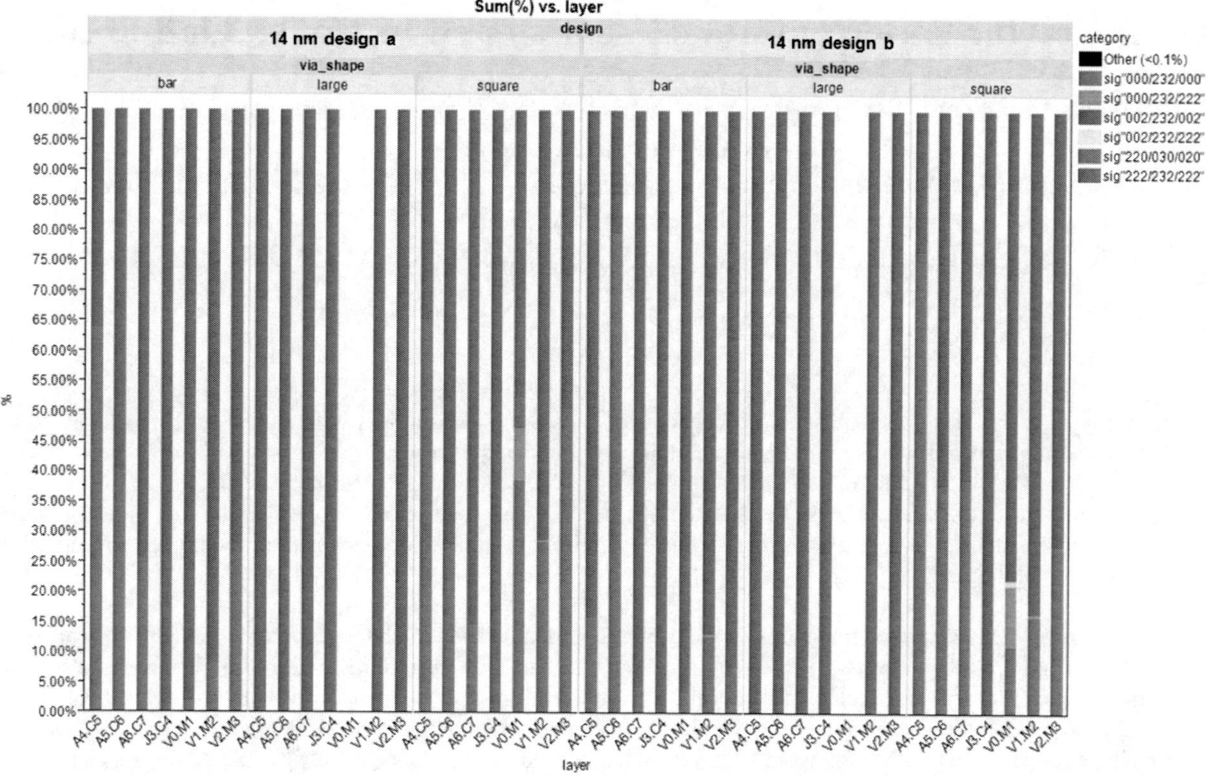

Figure 13. Stacked bar charts of via enclosure configurations at every via layer for 2 different 14nm designs (100% of all the vias are profiled).

REFERENCES

[1] Dai, V., et al., "DRC Plus: augmenting standard DRC with pattern matching on 2D geometries," Proc. SPIE 6521, 65210A (2007).

[2] Dai, V., et al., "Systematic Physical Verification with Topological Patterns," Proc. SPIE 9053, (2014).

[3] Teoh, E., et al., "Systematic Data Mining Using a Pattern Database to Accelerate Yield Ramp," Proc. SPIE 9503, (2014).

[4] Cover, T. and Thomas, J., [Elements of Information Theory], 2nd edition, Wiley, Hoboken (2006).

Automation for Pattern Library Creation and In-Design Optimization

Rock Deng[1], Elain Zou[1], Sid Hong[1], Jinyan Wang[1], Yifan Zhang*[2], Jason Sweis[2], Ya-Chieh Lai[2], Hua Ding[2], Jason Huang[2]

1 Semiconductor Manufacturing International Corporation, No. 18 Zhangjiang Road, Pudong New Area, Shanghai, China

2 Cadence Design Systems, Inc., 2655 Seely Ave., San Jose, CA 95134, USA

ABSTRACT

Semiconductor manufacturing technologies are becoming increasingly complex with every passing node. Newer technology nodes are pushing the limits of optical lithography and requiring multiple exposures with exotic material stacks for each critical layer. All of this added complexity usually amounts to further restrictions in what can be designed. Furthermore, the designs must be checked against all these restrictions in verification and sign-off stages. Design rules are intended to capture all the manufacturing limitations such that yield can be maximized for any given design adhering to all the rules.

Most manufacturing steps employ some sort of model based simulation which characterizes the behavior of each step. The lithography models play a very big part of the overall yield and design restrictions in patterning. However, lithography models are not practical to run during design creation due to their slow and prohibitive run times. Furthermore, the models are not usually given to foundry customers because of the confidential and sensitive nature of every foundry's processes. The design layout locations where a model flags unacceptable simulated results can be used to define pattern rules which can be shared with customers.

With advanced technology nodes we see a large growth of pattern based rules. This is due to the fact that pattern matching is very fast and the rules themselves can be very complex to describe in a standard DRC language. Therefore, the patterns are left as either pattern layout clips or abstracted into pattern-like syntax which a pattern matcher can use directly. The patterns themselves can be multi-layered with "fuzzy" designations such that groups of similar patterns can be found using one description. The pattern matcher is often integrated with a DRC tool such that verification and sign-off can be done in one step. The patterns can be layout constructs that are "forbidden", "waived", or simply low-yielding in nature. The patterns can also contain remedies built in so that fixing happens either automatically or in a guided manner.

Building a comprehensive library of patterns is a very difficult task especially when a new technology node is being developed or the process keeps changing. The main dilemma is not having enough representative layouts to use for model simulation where pattern locations can be marked and extracted. This paper will present an automatic pattern library creation flow by using a few known yield detractor patterns to systematically expand the pattern library and generate optimized patterns. We will also look at the specific fixing hints in terms of edge movements, additive, or subtractive changes needed during optimization. Optimization will be shown for both the digital physical implementation and custom design methods.

Keywords: DFM, pattern matching, auto fixing

1. INTRODUCTION

Through 28nm process development, the manufacture process steps become more complex, and the design chip size grows exponentially large. Using lithography model to run printability simulation to check the limits of optical lithography becomes prohibitive for designers in terms of runtime, and readiness of the lithography model.

*yifanzh@cadence.com

Design-Process-Technology Co-optimization for Manufacturability IX, edited by John L. Sturtevant, Luigi Capodieci, Proc. of SPIE Vol. 9427, 94270R · © 2015 SPIE · CCC code: 0277-786X/15/$18 · doi: 10.1117/12.2087100

We have developed a pattern library creation system that can collect yield detractor patterns from printability checks [1] [2] [3] [4] and systematic defects from silicon testing data. By using the pattern library we are able to track defect patterns and to monitor these patterns on new incoming designs then to detect these patterns in early process steps in order to accelerate yield ramp.

We have implemented an enumeration flow that can create a topological pattern deck for all the three value logic (TVL) patterns stored in the pattern library. This topological pattern deck has different layout context surroundings then we can define a value range of multiple metal lines' width and space of the targeted central geometry on the pattern to enumerate more systematic patterns. Furthermore we can add, subtract or move a polygon that is located on targeted location to enumerate new patterns. All these enumerated patterns could be optimized patterns or new yield detractor patterns. New yield detractor patterns will be stored into a TVL pattern library and optimized patterns can provide information on how to create fixing guidelines for this particular pattern.

We can compile a pattern-based techfile from the library that can be used in the design phase to check the critical hotspots found on the design layout and it is also compatible with the router to do auto-fixing to improve the manufacturability of the design.

The goal of this paper is to present an engineering learning process of building a yield detractor pattern library that can help designers to rule out all the bad patterns which occur in their design and have a tracking/checking system to monitor the manufacture process. Section 2 describes the methodologies used to translate a printability simulation hotspot or a silicon failure data into a pattern. Section 3 introduces the enumeration method used to expand the pattern library. Section 4 presents pattern matching technology for in-design use.

2. PATTERN LIBRARY CREATION SYSTEM

2.1 Collect patterns based on printability simulation hotspot

First run printability verification simulation on reference layouts and the lithographic hotspots are identified by a marker. The hotspots are divided into five types as spacing, width, enclosure-above, enclosure-below and line-end. The marker produced by printability simulation flags a location where the space, width or enclosure measurement of the contour for this portion of geometry at a process corner is lower than the manufacturability threshold. Figure 1 shows the overview of the pattern library creation flow.

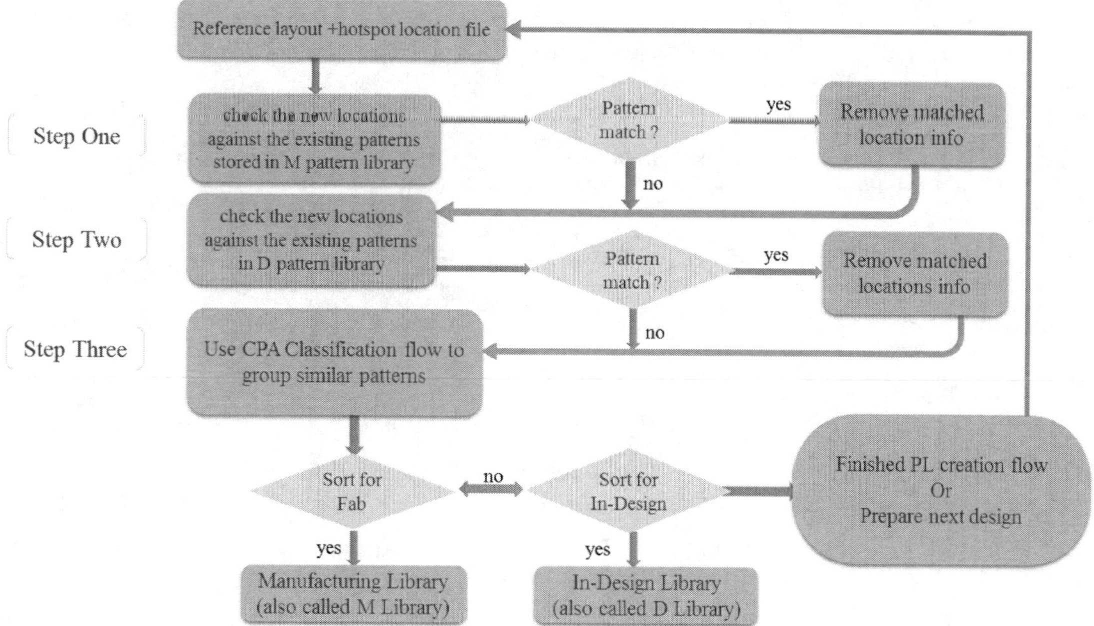

Figure 1: The overview flowchart of the pattern library creation flow.

For the very first reference layout, the library building process can start from the step three and using pattern classification methodology to systematically sort hotspot patterns. So the patterns have same failure mechanism but may have small difference in surrounding context are grouped together. Based on the information produced by step three, the patterns are stored into two pattern libraries; manufacturing library (M library) and in-design library (D library). M library contains the problematic patterns that detected from the beginning of the manufacturing process development. D library contains the problematic patterns that will be released to design customers for in-design phase hotspot clean up. For the second reference layout, the building process starts from step one, first to check all the found hotspot locations, any of these patterns are already stored in the pattern libraries.

A yield detractor pattern is clipped based on the radius should be covered from the center of the marker, but the printability hotspot markers are created based on the contour so marker sometimes has a big off-set from where the actual hotspot location is in the design layout. Figure 2 shows the marker is shifted from the actual hotspot location on design layout. In this paper we are mainly focused on the problematic patterns that has lithographic issue due to VIA/Contact placement on the metal layers.

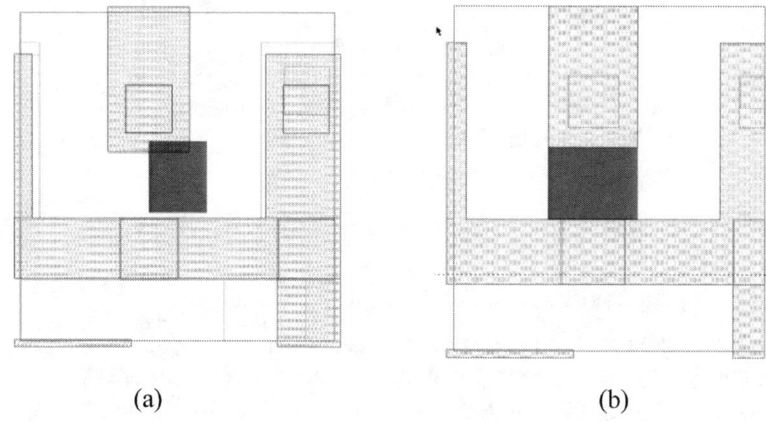

Figure 2: A yield detractor pattern clipped from reference layout (a) clipped based on the original marker, (b) clipped based on the fixed marker.

After a comprehensive study of all the hotspots that were detected by printability simulation on different reference layouts, a fixing methodology for marker placement for each yield detractor pattern has been developed. The marker fixing rules are listed in table 1.

Table 1: The marker fixing rule (Polygon in blue is the marker).

Fixing rule	Original Marker	Fixed Marker
The marker should fit into the metal spacing and the marker width should be the common running length of the two metals. The distance between fixed marker center and original marker center is within the defined off-set range.		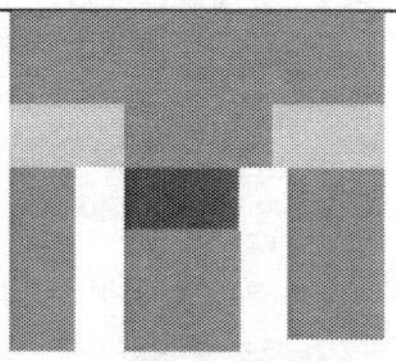
The marker should fit into the metal width and the marker width should be the running length of the two metals. The distance between fixed marker center and original marker center is within the defined off-set range.	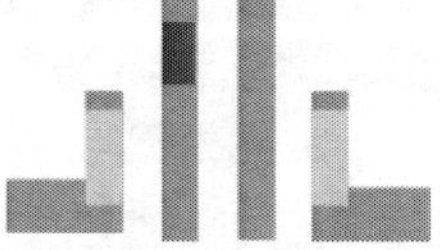	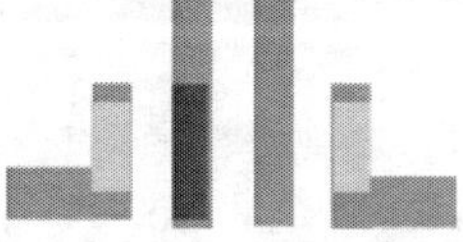

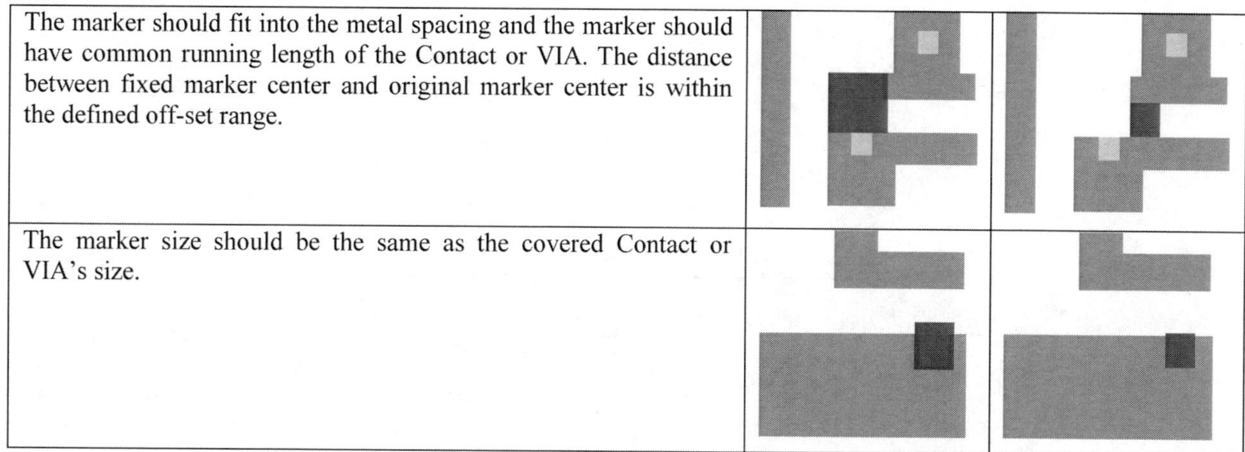

The marker should fit into the metal spacing and the marker should have common running length of the Contact or VIA. The distance between fixed marker center and original marker center is within the defined off-set range.		
The marker size should be the same as the covered Contact or VIA's size.		

The output from printability simulation can have thousands or tens of thousands flagged locations and by using a suitable classification method, the hotspots can be grouped into classes. In order to collect useful patterns, we have used topological method (squish) to analyze the metal width, metal space and enclosure coverage for the detected hotspots and layout regularity to find the suitable radius for pattern clip.

We have run pattern profiling on the collected patterns and find some of the patterns are not from the same class but have common failure mechanism so we have implemented a pattern altering process that can extend the don't care (DC) regions of all known patterns and regroup the processed patterns. The pattern altering process can extend pattern coverage which allows us to match more potential hotspots locations. After fixing the classification constraints and patterns size, we have implemented an automatic pattern library creation flow as shown in figure 3.

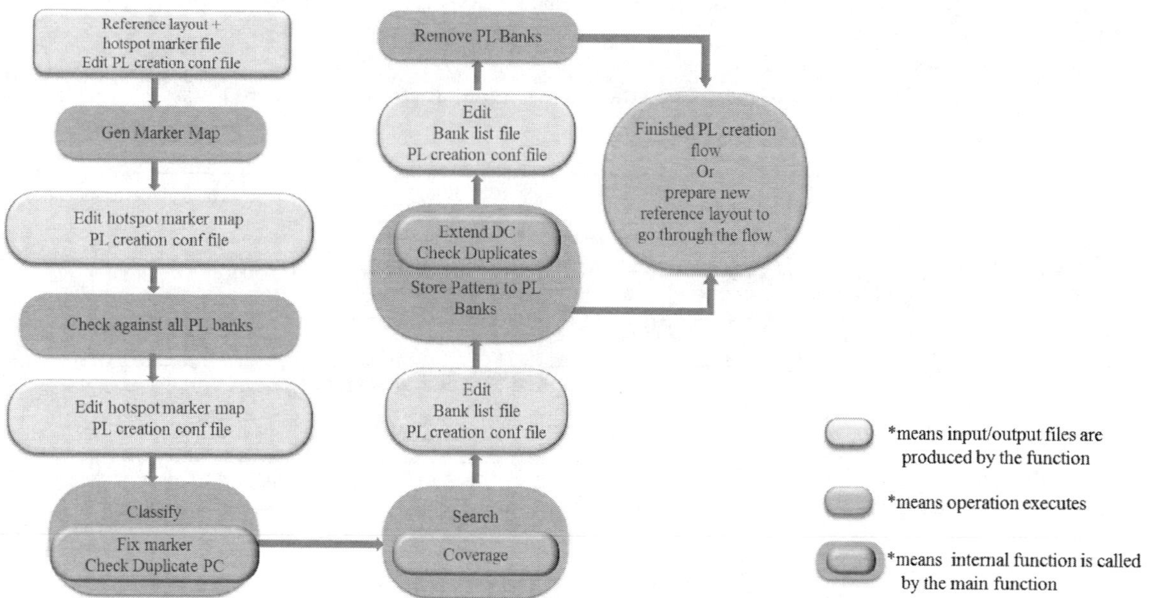

Figure 3: the flowchart for pattern creation system.

2.2 Collect patterns based on analysis silicon failure mechanism

Yield detractor patterns can be collected based on the known process weak points. The failure mechanism can be described in a topological deck with constraint.

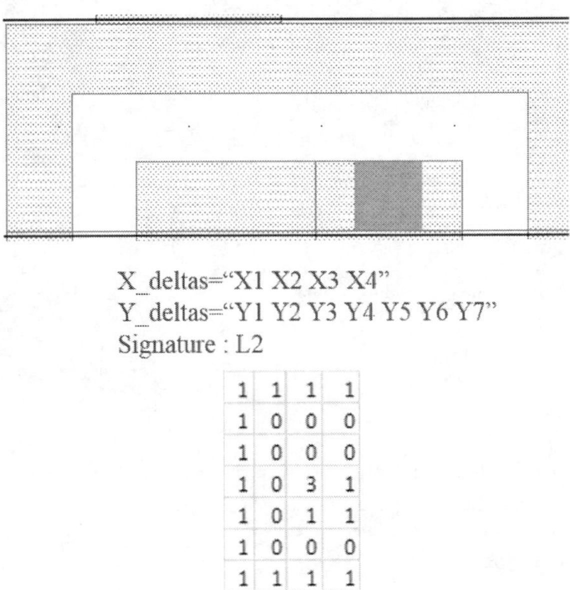

Figure 4: Example of systemtic defect found on the silicon and the topological rule deck for pattern match. (for illustrative purpose and does not represent actual data).

The topological rule deck can be used to detect all the possible problematic patterns on all the reference layouts. For consistence, we will run pattern creation flow to collect all the matched patterns.

All the yield detractor patterns are collected from the reference layout are assigned with a property and kept into different function libraries.

Table 2: The property of the collected yield detractor patterns in the pattern library

Pattern	Property
	Hotspot type
	Severity level
	Pattern tolerance type
	From which reference layout
	Should be applied to which design layers
	How many patterns are grouped into this pattern class
	The date the pattern was installed in the library.

3. PATTERN ENUMERATION FLOW

The patterns stored in the pattern library have a fixed marker to indicate where the failure locations are and using topological pattern description method we can systematically analyze these failure locations and convert into value range constraints in a topological enumeration deck. It is worthwhile to analyze the most frequently occurring and representative problematic patterns in the wider constraint range and to find the worst constraint value that is far lower than manufacturability threshold but still passes DRC. We can also determine what constraint value that the problematic pattern should have which will have good yieldable dimensions. We are using simulation to determine the pattern defect rate. The pattern defect rate is also related to the pattern surrounding each situation. So each enumerated pattern is also analyzed with different surrounding contexts.

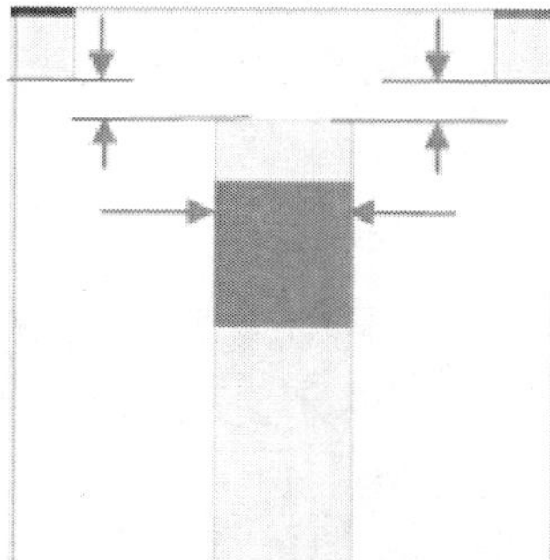

Firgure 5: Example of problematic pattern stored in the pattern library. Shows the VIA encolusre and metal spacing constraint ranges are analyzed (for illustrative purpose and does not represent actual data).

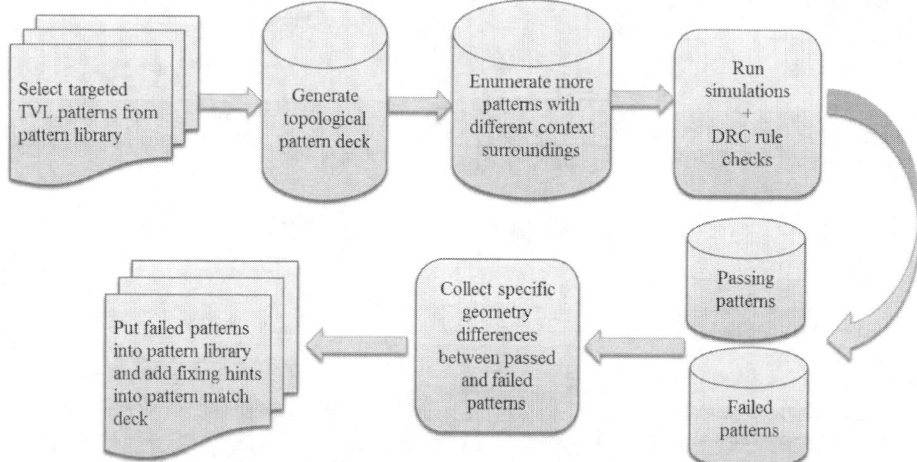

Firgure 6. The overview of the pattern enumeration flow.

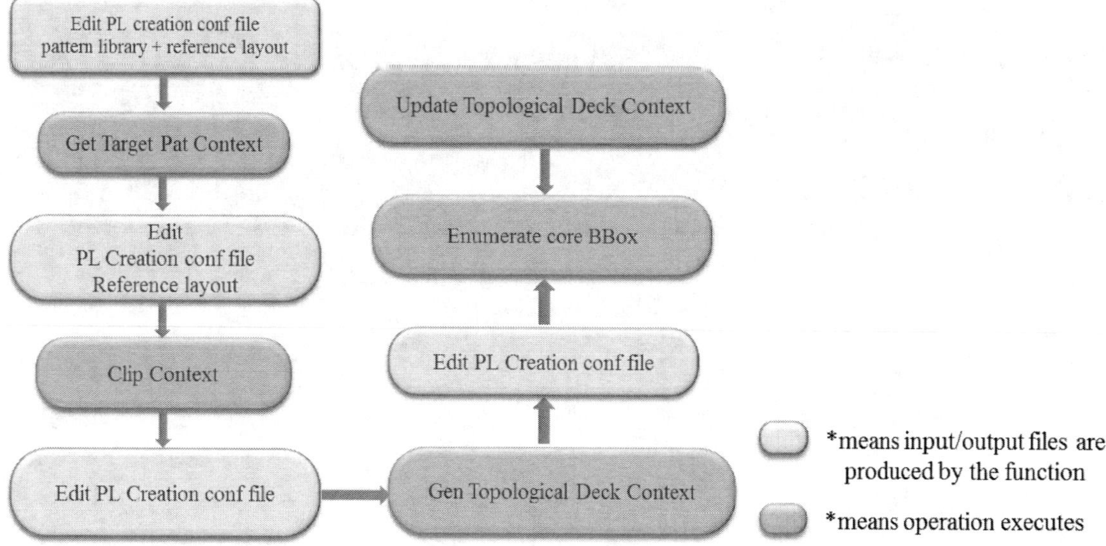

Figure 7: The flowchart of the pattern enumeration flow.

We have implemetated an automatic pattern enumeration flow as shown below. The system can collect geometry information for each failed and passed pattern. For each pattern class, the patterns have the same topological siginatures so the dimension difference between failed pattern and passed pattern can be calculated. Based on the dimesion difference complied from the system, this information can be coverted into fixing gudienlines. The flowchart of the pattern enumeration flow is shown in the figure 7.

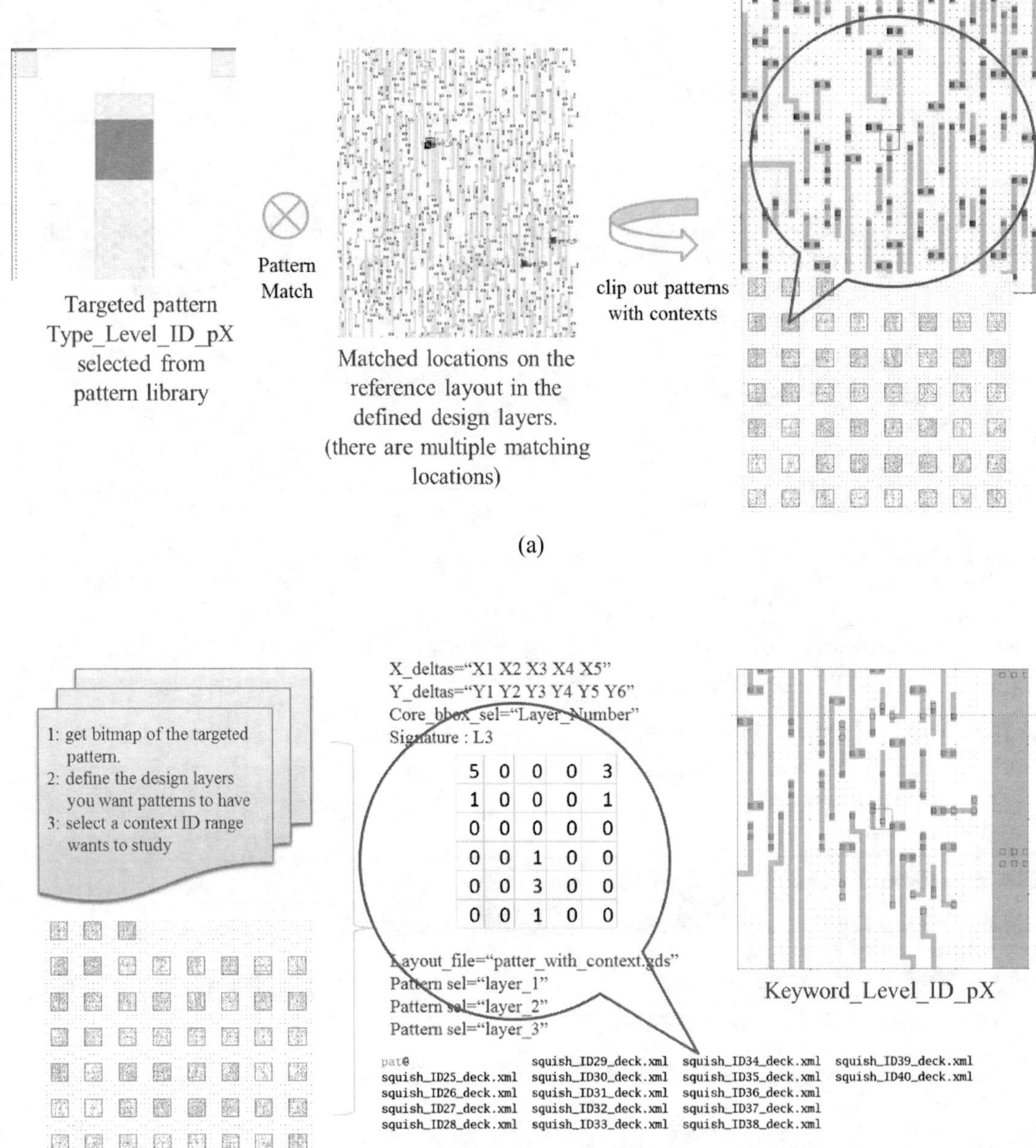

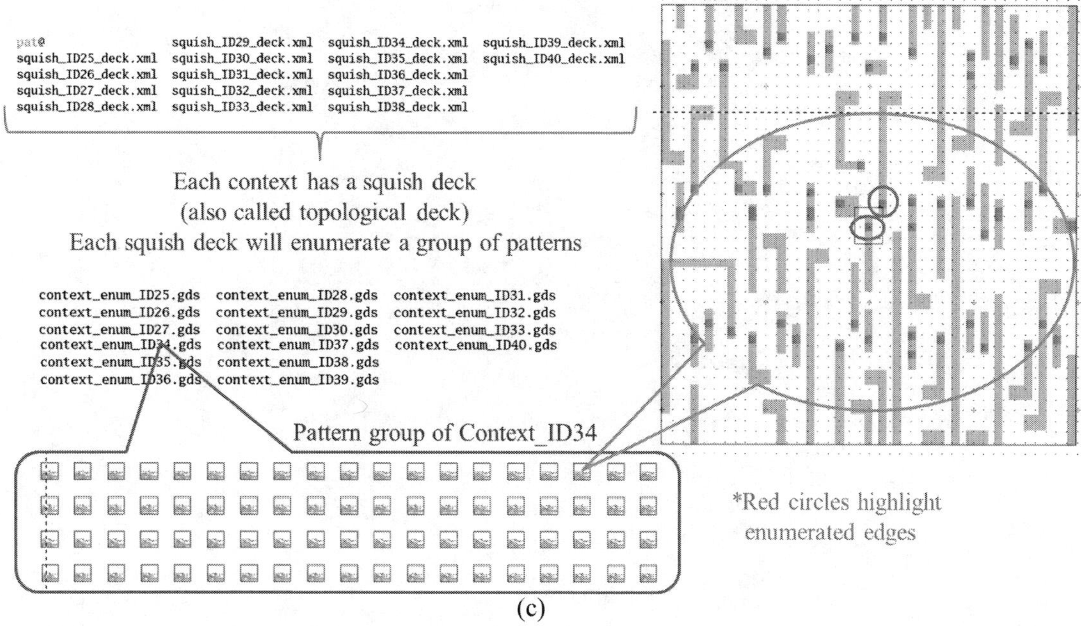

Figure 8: The illustrative presentation of the pattern enumeration flow. (a) the pattern selected from pattern library and different context situations are extracted. (b) for each pattern, the topological enumeration deck has generated. (c) the enumerated patterns with same situation context. And for each situation context, a group core patterns are enumerated.

4. YIELD DETRACTOR PATTERNS CLEAN UP FOR IN-DESIGN USE

At the 28nm technology node, the physical signoff for the design chip should be done during the design phase. Using pattern matching technology, the foundry can provide manufacturing process limit information to their design customers without revealing their sensitive process information. The early stage yield detractor removal process is shown in Figure 9. Based on the information stored in the pattern techfile, the pattern matcher can run from the placement implementation design step to prevent yield limiting patterns from happening on the cell pin nets and then re-checked at the post-route implementation step to clean up all the hotspots.

Figure 10 shows the experiment result on a 28nm test chip, pattern match check run after route implementation step, then designer can apply auto-fix to all the found hotspot locations. For the designer, the real concern for yield detractor removal flow is runtime and fixing method. Pattern matching shows much faster runtime and easy deployable fixing method.

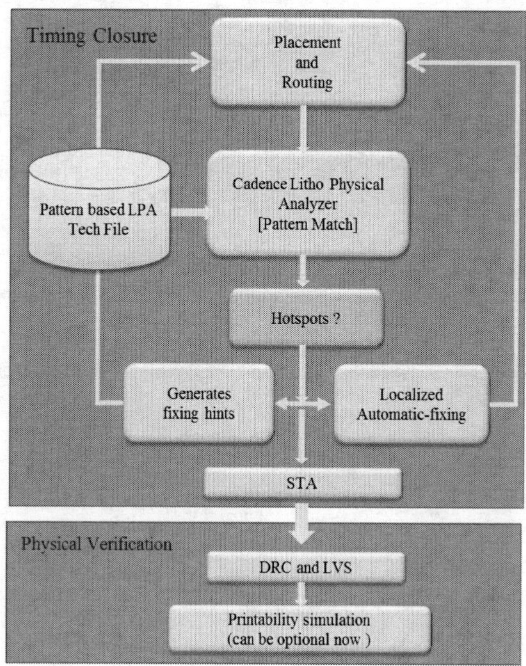

Figure 9: Digital design yield detractor removal flow.

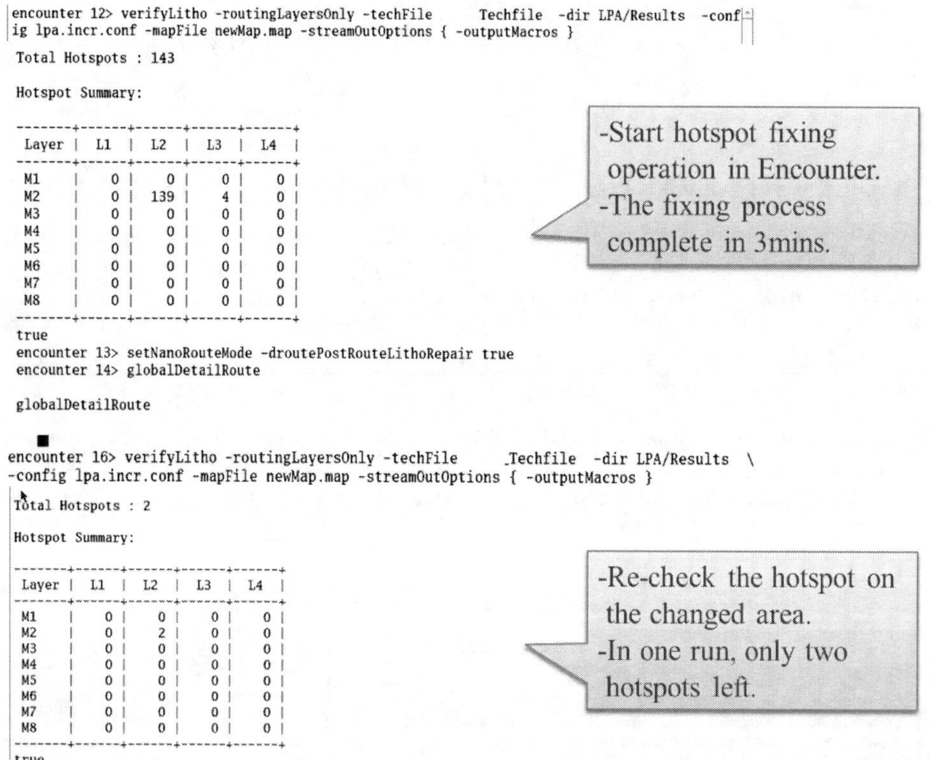

```
encounter 12> verifyLitho -routingLayersOnly -techFile      Techfile  -dir LPA/Results  -conf
ig lpa.incr.conf -mapFile newMap.map -streamOutOptions { -outputMacros }
Total Hotspots : 143

Hotspot Summary:

-------+------+------+------+------+
 Layer |  L1  |  L2  |  L3  |  L4  |
-------+------+------+------+------+
 M1    |    0 |    0 |    0 |    0 |
 M2    |    0 |  139 |    4 |    0 |
 M3    |    0 |    0 |    0 |    0 |
 M4    |    0 |    0 |    0 |    0 |
 M5    |    0 |    0 |    0 |    0 |
 M6    |    0 |    0 |    0 |    0 |
 M7    |    0 |    0 |    0 |    0 |
 M8    |    0 |    0 |    0 |    0 |
-------+------+------+------+------+
true
encounter 13> setNanoRouteMode -droutePostRouteLithoRepair true
encounter 14> globalDetailRoute

globalDetailRoute

encounter 16> verifyLitho -routingLayersOnly -techFile      _Techfile  -dir LPA/Results  \
-config lpa.incr.conf -mapFile newMap.map -streamOutOptions { -outputMacros }
Total Hotspots : 2

Hotspot Summary:

-------+------+------+------+------+
 Layer |  L1  |  L2  |  L3  |  L4  |
-------+------+------+------+------+
 M1    |    0 |    0 |    0 |    0 |
 M2    |    0 |    2 |    0 |    0 |
 M3    |    0 |    0 |    0 |    0 |
 M4    |    0 |    0 |    0 |    0 |
 M5    |    0 |    0 |    0 |    0 |
 M6    |    0 |    0 |    0 |    0 |
 M7    |    0 |    0 |    0 |    0 |
 M8    |    0 |    0 |    0 |    0 |
-------+------+------+------+------+
true
```


Figure 10: The experiment result for pattern matching after route implementation step.

Figure 11 shows the yield detractor removal flow in custom design. The clear and suggestive error marker is implemented for IP designers to understand where the hotspot locations are and what kind of problem it is. Figure 10 shows the pattern match run on an IP layout that is implemented in Virtuoso. The error marker shows where the hotspot happened and the rule name suggests that it is an enclosure issue and the enclosure coverage is not enough for the bottom layer. Also the suggestive hint can be displayed to show what is the suggest enclosure value.

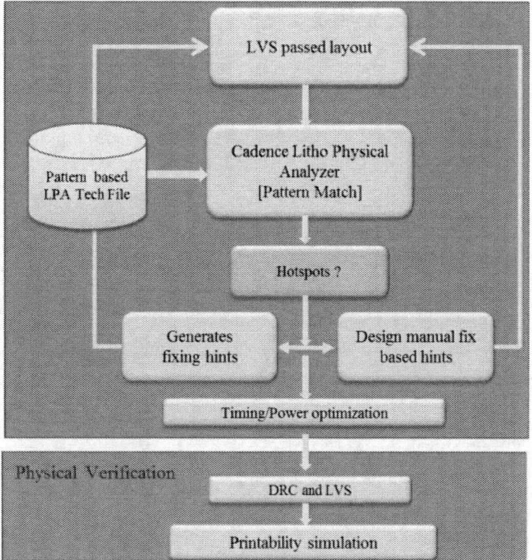

Figure 11: Custom design yield detractor removal flow.

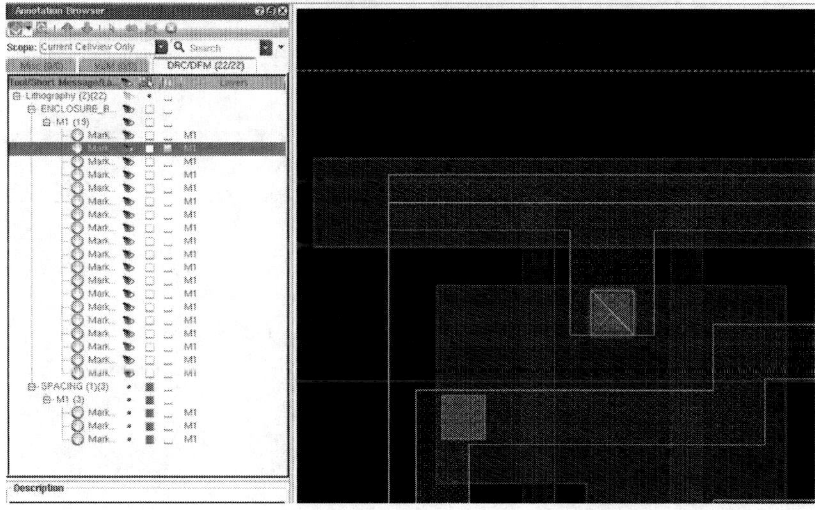

Figure 12: The experiment result for pattern matching on an IP design that implemented in virtuoso.

5. CONCLUSION

In this paper, we described the successful development of an automated pattern creation system that can help cross-functional teams collect yield detractor patterns and a system to efficiently process large amount data and output filtered/classed patterns in easy-to-review metrics. An automated pattern enumeration system that can systematically expand the pattern library and generate optimized patterns and gives dimension information between problematic patterns and optimized patterns was also implemented. Based on the dimension information, we can enumerate more potential yield detractors to monitor the improvement of the manufacturing process during yield ramp.

The pattern based techfile are ready and been used in SMIC digital design reference flow to run pattern matching and can automatically fix yield detractor patterns.

REFERENCES

[1] V. Dai, et al., "DRC Plus: augmenting standard DRC with pattern matching on 2D geometries", Proc. SPIE, Vol. 6521, 65210A (2007).

[2] V. Dai, et al., "Developing DRC Plus rules through 2D pattern extraction and clustering techniques", Proc. SPIE, Vol. 7275, 727517 (2009).

[3] J. Yang, et al., "DRC Plus in a router: automatic elimination of lithography hotspots using 2D pattern detection and correction", Proc. SPIE, Vol. 7641, 76410Q (2010).

A New Lithography Hotspot Detection Framework Based on AdaBoost Classifier and Simplified Feature Extraction

Tetsuaki Matsunawa[a], Jhih-Rong Gao[b], Bei Yu[b] and David Z. Pan[b]

[a] Center for Semiconductor Research & Development, Toshiba Corp., Kawasaki, Japan
[b] ECE Department, Univ. of Texas at Austin, Austin, TX, USA
tetsuaki.matsunawa@toshiba.co.jp, {jrgao, bei, dpan}@cerc.utexas.edu

ABSTRACT

Under the low-k1 lithography process, lithography hotspot detection and elimination in the physical verification phase have become much more important for reducing the process optimization cost and improving manufacturing yield. This paper proposes a highly accurate and low-false-alarm hotspot detection framework. To define an appropriate and simplified layout feature for classification model training, we propose a novel feature space evaluation index. Furthermore, by applying a robust classifier based on the probability distribution function of layout features, our framework can achieve very high accuracy and almost zero false alarm. The experimental results demonstrate the effectiveness of the proposed method in that our detector outperforms other works in the 2012 ICCAD contest in terms of both accuracy and false alarm.

Keywords: Design for Manufacturability, Lithography Hotspot Detection, Machine Learning, Real AdaBoost

1. INTRODUCTION

Shrinking device feature sizes is the holy grail for the semiconductor industry and it is being pursued through the use of shorter wavelength and several design for manufacturing (DFM) technologies, such as optical proximity correction (OPC) and resolution enhancement techniques (RETs). In view of the delay of next generation lithography technologies, such as extreme ultra-violet lithography (EUVL), optimization of design for manufacturing process has recently become much more important.[1] However, for 45nm node and below, *lithography hotspot*, which is lower fidelity pattern on a wafer, still exists even after application of these DFM techniques. In the physical design and verification phase, a technique to detect these lithography hotspots is essential for improving manufacturing yield.

So far, lithography simulation is the most widely used hotspot detection method.[2] Although this method is expected to achieve very high accuracy, it is also known to be extremely time-consuming.[3] Therefore, a hotspot detection method with reasonable runtime without accuracy loss is sorely needed in the physical verification phase to avoid increase of turn-around time (TAT) and manufacturing cost. To achieve a balance between runtime and accuracy, several hotspot detection methods without lithography simulation have been proposed. Popular candidates are pattern matching based methods[4–8] and machine learning based methods.[9–20] Pattern matching based methods define a pattern library consisting of known hotspots, followed by the whole layout scanning with this library. Although pattern matching based methods are very fast and accurate in detecting known hotspots, these methods have a fundamental problem in that unknown-hotspot detection is impossible.

On the other hand, machine learning based methods generate classification models based on given hotspots and non-hotspots (training data set) and can detect unknown hotspots. Machine learning based methods have been shown to be of benefit for the hotspot detection problem in terms of runtime and detection accuracy. However, the contradiction between high detection accuracy and low false alarm remains a critical issue that inhibits the practical use of machine learning based methods, for the following reasons:

(1) Layout feature selection: The optimal layout feature that represents geometrical attributes of hotspots and non-hotspots appropriately is uncertain. On the one hand, short runtime can be expected from a simple layout feature in low-dimensional feature space, but this simplicity may also lead to deterioration of detection accuracy because it is unable to describe hotspot-specific attributes exactly. On the other hand, highly accurate detection in high-dimensional space causes over-fitting (false alarm) and long runtime due to complexity of feature space.

Design-Process-Technology Co-optimization for Manufacturability IX, edited by John L. Sturtevant, Luigi Capodieci,
Proc. of SPIE Vol. 9427, 94270S · © 2015 SPIE · CCC code: 0277-786X/15/$18 · doi: 10.1117/12.2085790

(2) Classification model training: It is difficult to define an efficient classification algorithm for a complicated feature space with the relatively small number of training data sets. Non-hotspots usually significantly outnumber hotspots in a layout, and furthermore, the amount of real hotspots may be less than 50 in a later phase of process development. To improve detection accuracy with a small number of training data, a nonlinear classification algorithm is essential. However, an algorithm with strong nonlinearity that classifies patterns in high-dimensional space, such as SVM, increases the possibility of false alarms, resulting in difficulty of highly accurate and low-false alarm detection.

So far, a fragmentation-based feature[14] and a density-based feature,[13] have been proposed to represent layout patterns. In addition, several learning models, e.g., artificial neural network (ANN)[9,12] and support vector machine (SVM),[11] have been applied. Multi-level learning[13] based on multiple SVM kernels, hierarchical learning,[14] and a hybrid method combining pattern matching and machine learning have recently been proposed.[15–18,20] The basic idea of these state-of-the-art methods is the construction of multiple nonlinear classifiers in a restricted feature space through classifying training data before model calibration. Although these works have shown that higher accuracy and lower false alarm can be achieved compared to previous techniques, there is still much room for improvement in view of runtime increase and many remaining false alarms.

To achieve a good trade-off between accuracy and false alarm, it is necessary to utilize an uninvolved layout feature that is simple but sufficiently appropriate to represent geometrical attributes of hotspots. In addition, a smart nonlinear classification algorithm with lower complexity than SVM is required as well. In this paper, we propose a high-accuracy and low-false alarm hotspot detection framework. A robust classifier in conjunction with base classifiers using a probability distribution function of simplified layout features realizes high-accuracy detection without false alarm increase. Our key contributions are as follows:

- We develop a feature space evaluation index to quantify different layout extraction methods and to define a learning-algorithm-friendly layout feature.
- We propose a simple but highly effective layout feature based on the feature space index, which provides an efficient description of layout patterns in low-dimensional space.
- We develop a new Adaboost machine learning algorithm that works together with our simplified layout feature to boost detection accuracy without causing false alarm increase.
- We demonstrate high accuracy and low false alarm under industry-strength benchmarks.

The rest of the paper is organized as follows. In Section 2, we give the problem formulation and the overall hotspot detection flow. In Section 3 and Section 4, we present details of the proposed simplified features and classification models, respectively. Section 5 presents the experiment results, followed by the conclusion in Section 6.

2. PRELIMINARIES

2.1 Problem Formulation

A hotspot is defined as a low-image-fidelity pattern on a wafer. As mentioned in the introduction, detecting all hotspots and eliminating them in the early phase of physical design and verification is becoming important. Fig. 1 (a) and (b) indicate examples of non-hotspot layout and hotspot layout, respectively. A "Frame" and a "Core" show the ambit associated with its center and the computational domain, respectively. The contour of a resist shape calculated by lithography simulation is also shown in each figure. We can see that, compared with non-hotspot layout, the hotspot layout shows a clear difference between the contour and the target shape. To evaluate the effectiveness of hotspot detection, we define several terms used throughout this paper.

Definition 1 (Accuracy). The detection accuracy rate is as follows:

$$Accuracy = \frac{Hit}{\#of\ hotspots} \tag{1}$$

where Hit is the number of correctly detected hotspots.

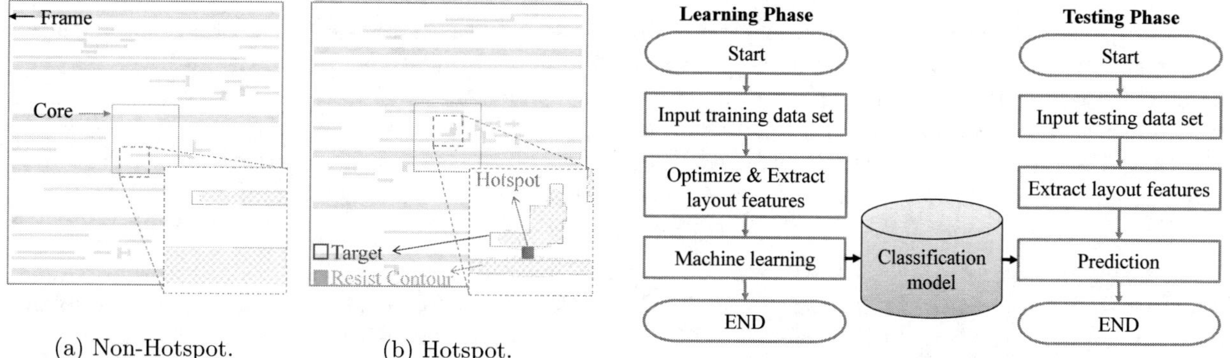

(a) Non-Hotspot.

(b) Hotspot.

Figure 1. Examples of hotspot pattern.

Figure 2. Overview of the proposed method.

Definition 2 (False alarm). The false alarm (extra) is as follows:

$$False\ Alarm = \#of\ falsely\ detected\ hotspots \tag{2}$$

Now we give the problem formulation of hotspot detection.

Problem 1 (Hotspot Detection). *Given layout data including hotspots and non-hotspots, a prediction model is calibrated to detect unknown hotspots from a verification layout. The goal of the hotspot detection is to maximize the accuracy and minimize the false alarm.*

In previous work it has been shown that optimization for either accuracy or false alarm is not difficult. However, it is really challenging to optimize accuracy and false alarm simultaneously.

2.2 Overview of Hotspot Detection Flow

Our hotspot detection method consists of two phases, "Learning phase" and "Testing phase", as shown in Fig. 2. In the learning phase, a training layout is given and a classification model is calibrated after optimization and extraction of a layout feature. The details of the layout feature and classification model making are described in Section 3 and 4. In the testing phase, a verification layout that is not the same as the training layout and includes unknown hotspots is used as the input. After the feature extraction from the verification layout, labels, which consist of -1 for non-hotspots and $+1$ for hotspots, are predicted by using the classification model trained in the learning phase.

As shown in Fig. 2, our proposed flow is simple and complicated processing such as layout classification before model learning in hierarchical or multi-level learning[13, 14] does not need to be performed. Also, optimization with complicated nonlinear classification kernels is not required. The following are two reasons why high-accuracy detection and low false alarm can be expected with our simple flow. (1) Simplified Layout Feature: Simple but highly effective layout feature with high linear separability and low-dimensional space is defined by using our feature space index. (2) Boosting-Based Classification Model: Utilization of a weakly nonlinear learning algorithm is provided by using simplified layout feature.

3. SIMPLIFIED LAYOUT FEATURE

In hotspot detection, feature extraction is a vitally important phase, in which the initial geometrical information is translated into a set of layout features. Although several layout features have been proposed, the question of how to define the optimal parameters of layout features remains open. In this section, we present a feature space index to resolve this issue.

3.1 Feature Space Index

In order to define an appropriate layout feature, we propose a feature space index capable of optimizing the parameters of layout features. The basic idea is to evaluate the performance of layout features by measuring the distances between hotspots and non-hotspots in low-dimensional feature space. Based on the index, we can select an appropriate layout feature with a small amount of data while suppressing complexity of feature space. The definition of the index is as follows. The feature vectors X are projected to form new low-dimensional feature vectors P.

$$P = XU_k \tag{3}$$

In this paper, eigenvectors U are determined by using principal component analysis (PCA) to avoid redundant layout attributes and reduce dimensions.[20] In the PCA calculation, U are calculated by solving the following eigenvalue problem:

$$R_x U = U\Lambda \tag{4}$$

where R_x is the variance-covariance matrix of feature vectors X, Λ is the diagonal matrix of the eigenvalues in $(\lambda_1, ..., \lambda_M)$, and k in equation (3) is defined by the value of contribution rate η_k, which is 1.

$$\eta_k = \frac{\sum_{i=1}^{k} \lambda_i}{\sum_{i=1}^{M} \lambda_i} \tag{5}$$

In the projected feature space P, feature space index H is given by the following equation.

$$H = \left| 1 - \frac{1}{\alpha + \exp(-Z)} \right| \tag{6}$$

where α, discussed in detail below, is a hyper-parameter in consideration of generalization capability and Z is average distance of hotspot features to non-hotspot features.

$$Z = \frac{1}{N} \sum_{i=1}^{N} d_i \tag{7}$$

$$d_i = \frac{\sqrt{(x_i - \mu)^{\mathrm{T}} V^{-1} (x_i - \mu)} - d_{NHS_{min}}}{d_{NHS_{max}} - d_{NHS_{min}}} \tag{8}$$

where N is the total number of real hotspots in the training set, d is the Mahalanobis distance[21] normalized by non-hotspot features, μ is the center of mass of non-hotspot features, V is the variance covariance matrix of non-hotspot features, $d_{NHS_{max}}$ is the maximum Mahalanobis distance of non-hotspot features, and $d_{NHS_{min}}$ is the minimum Mahalanobis distance of non-hotspot features.

With the above definitions, the linear separability and predicting performance can be estimated by H and Z in a given layout feature space. For a layout extraction feature, if its resulting Z is in the range of $1 < Z < 10$ and its resulting H is near or toward zero, it indicates that it is an appropriate layout feature. Specifically, $Z < 1$ means that it is difficult to separate hotspots and non-hotspots linearly as most hotspots are located in the non-hotspot feature space. In contrast, $Z > 1$ shows linear-separation-friendly features because of the distance of hotspots from non-hotspots. However, too large Z indicates deterioration of prediction accuracy. When the hotspot features are too far away from the non-hotspot features, the allowable range of decision boundary is broadened, and as a result, generalization capability for the testing layout cannot be ensured. The appropriate upper limit of Z depends on training data, but according to our preliminary experiments, it is up to 10.

In our framework, generalization capability is evaluated using H . In equation (6), $\alpha = 1$ is equivalent to sigmoid function, but this means the bigger Z the better. To prevent decreases in prediction accuracy, an appropriate criterion is defined by adjusting α. This parameter depends on technology node and automatic determination of the optimal value of α is a subject for future work. In this paper, we set α value as 0.9 by the prescribed preliminary experiments.

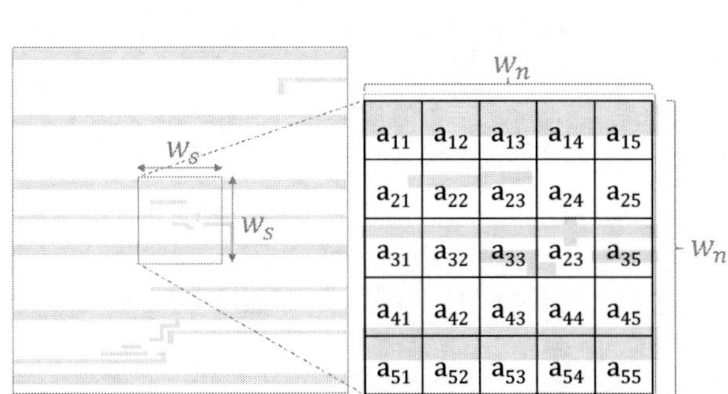

Figure 3. Density-based layout feature. Feature Vector is represented as: $X = \{a_{11}, a_{12}, ..., a_{54}, a_{55}\}$.

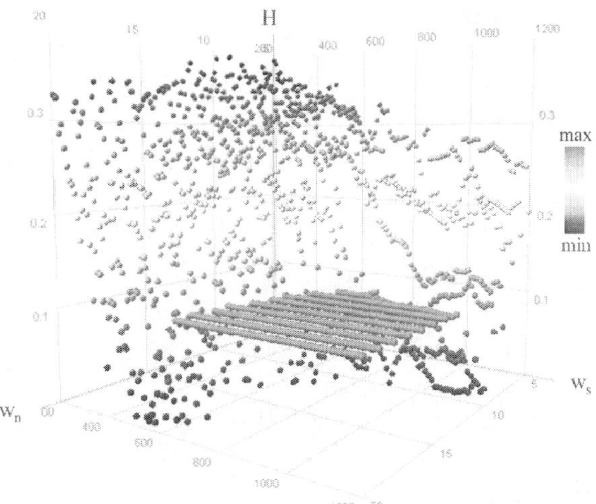

Figure 4. Solution space of density-based parameters.

3.2 Feature Selection and Optimization

We utilize a simplified density-based feature as an appropriate feature. "Simplified" means the feature with high linear separability in low-dimensional space and its parameters are optimized by using equation (6) and (7). Fig. 3 indicates the basic concept of the density-based feature. Feature vectors X show arrangement of area values of layout patterns in a given grid. Parameters of a feature w consist of the total size of the encoding area w_s and the number of grids w_n. Based on the discussion in the previous section, these parameters can be optimized by the following formula:

$$\text{minimize } H(w) \tag{9}$$

$$\text{subject to } w_s \in \{1, 2, ..., \mathcal{S}\} \tag{10}$$

$$w_n \in \{1, 2, ..., \mathcal{N}\} \tag{11}$$

Because equation (6) and (7) are nonlinear and parameters w are integer values, the above formulation is a nonlinear integer programing problem (NIPP). Fig. 4 indicates an example of the solution space of density-based parameters for training data set of Case 1 in Table 1 (See section 5). Since both integer linear programming and non-linear programming are well-known hard problems, it is not surprising that solving the combined version, NIPP, is challenging. However, as illustrated in Fig. 4, due to the limited solution space, we can find the optimal w using brute-force search method. The w_s is to be restricted within $\mathcal{S} = 1200$nm depending on core area of the layout data, and because the small grid size leads to increase in data, computational time and complexity of feature space, w_n is preferably $\mathcal{N} = 20$ or less. Fig. 4 shows a bottom (low H) region representing parameters expected to ensure high prediction accuracy with good linear separability.

The flat region indicates parameters restricted by α because the large w_n features are able to represent layout attributes in detail but it can also lead to a big difference in Z due to its complexity.

Although the density-based layout feature has been used successfully for the hotspot detection problem,[13,19] it is unclear how the parameters of the feature are defined. In this framework, our simplified feature can define the optimal parameters and represent larger area of the layout with lower-dimensional space compared to the conventional density-based method.[10] We will further discuss the advantages of the proposed layout feature compared to the conventional features in Section 5.

4. HOTSPOT DETECTION

As mentioned in the introduction, strong nonlinear classification models such as ANN and SVM are inappropriate for the hotspot detection problem in terms of false alarm reduction. Moreover, multi-nonlinear kernel models are unfit for low-false alarm detection and cause TAT increase. We therefore employ a decision tree classifier[22] to prevent increasing of false alarms. A decision tree is simple to interpret and has advantages both of fast runtime and good performance even if its assumptions are somewhat violated by the true model. Furthermore, we enhance accuracy based on the probability distribution function of layout features corresponding with the complexity of parity structure in layout feature space.

We briefly describe the formulation to build a robust classifier as follows. Given the training data $x_n : n = 1, ..., N$, label of the training data $t_n \in \{-1, +1\}$ and number of base classifiers $m = 1, ..., M$, first the sample weight D is initialized.

$$D_1(n) = 1/N \tag{12}$$

Then the base decision tree classifier y_m is fitted using D_m to define candidate split $\theta = (j, s_k)$ of a feature j and threshold s at node k.

$$\theta = \text{argmin}_\theta (G, \theta) \tag{13}$$

$$G = \frac{n_l}{N_k} F(P(Q_{left})) + \frac{n_r}{N_k} F(P(Q_{right})) \tag{14}$$

$$Q_{left}(\theta) = (x, t) | x_j \leq s_k \tag{15}$$

$$Q_{right}(\theta) = Q \setminus Q_{left}(\theta) \tag{16}$$

$$P = \frac{1}{N_k} \sum_{i \in x^k} D_m(t_i = l) \tag{17}$$

$$F = \sum_l P(1 - P) \tag{18}$$

where G is Gini index, Q is new tree, n_l, n_r are the numbers of x in new trees, N_k is the total number of x at node k, P is proportion of class $l = \pm 1$ and F is impurity function. Next, the probability distribution function W is defined and the output of the base classifier is set as follows:

$$W_l^j = \sum_{n: j \in J \wedge t_n = l} D_m(n) \tag{19}$$

$$y_m(x) = \frac{1}{2} \ln \left(\frac{W_{+1}^j + \epsilon}{W_{-1}^j + \epsilon} \right) \tag{20}$$

$$D_{m+1}(n) = D_m(n) \exp\left(-t_n y_m(x_n)\right) \tag{21}$$

where ϵ is a small positive constant. The processing above is repeated as many times as the number of M, and finally the robust classifier Y is given by the following equation.

$$Y(x) = \text{sign} \left[\sum_{m=1}^{M} y_m(x) \right] \tag{22}$$

This kind of classifier is known as "Real Adaboost", which was proposed by Friedman et al.[23] We show the details for calibration of our detection model in Algorithm 1, where "DECISIONTREE" is decision tree function to define θ in equation (13). The output of the base classifier is calculated by using the sample weight based on θ in (19) and (20).

In our framework, layout features are simplified by our feature space index but still contain a little complexity of parity structure, discussed in detail in Section 5.1, that includes several disjoint clusters in layout feature space. Thus, SVM cannot resolve the trade-off relation between accuracy and false alarm because it is difficult to select an appropriate kernel for this kind of layout feature. Although ANN might be able to learn this problem, it is very difficult to optimize its parameters while minimizing the false alarm. In contrast, a simple decision tree achieves a certain degree of accuracy without false alarm increase because this is a weakly nonlinear algorithm.[22] It should be noted that a weakly nonlinear algorithm is indicated to be a classifier which is only slightly correlated with the true classification model. However, our classifier is able to achieve a performance superior to that of other classification algorithms by adding base classifiers.

Algorithm 1 Real Adaboost with Decision Tree

Require: x, t, m

Initialize the sample weights D_1
for all base classifier **do**
    Tree depth = 0
    DECISIONTREE(x, t, D_m)
    Calculate the probability distribution function W
    Set the output of base classifier y_m
    Update the sample weights D_m
end for
return Final classifier Y

Generally, Adaboost is sensitive to the data including lots of noise and outliers. However, it can be superior with respect to the over-fitting issue compared to other classification algorithms. In the hotspot detection problem, layout features contain few or no factors corresponding to noise or outliers because the layout patterns are restricted by design rules. Although the minuscule differences among the patterns, such as small jogs, may be regarded as a noisy factor, Real Adaboost is able to construct a robust classifier considering such small differences based on the probability distribution function of feature space.

5. EXPERIMENTAL RESULTS

The proposed methodologies are implemented in Python and accelerated by Cython on a Linux machine with eight 3.4GHz CPUs and 32GB memory. The industrial benchmark suite released by[24] is applied. This benchmark suite includes five test cases, and the statistics of each case are listed in Table 1. It should be noted that Case 1 and 3 have a sufficient number of hotspots (HS) compared to the non-hotspots (NHS). In Case 2, 4 and 5, however, model learning is expected to be very difficult because they have small numbers of hotspots, which are less than 5% compared to their total numbers of non-hotspots. These cases can be seen as disadvantageous for machine learning, and the performance of the model is greatly influenced by feature selection and the classification algorithm. We conducted two experiments related to feature space evaluation and comparison for hotspot detection accuracy, respectively.

Table 1. ICCAD 2012 Benchmark Data.

Name	Tech	Training data			Testing data	
		#HS	#NHS	HS rate	#HS	Area (mm^2)
Case 1	32nm	99	340	29.12%	226	12516
Case 2	28nm	174	5285	3.29%	498	106954
Case 3	28nm	909	4643	19.58%	1808	122565
Case 4	28nm	95	4452	2.13%	177	82010
Case 5	28nm	26	2716	0.96%	41	49583

5.1 Feature Space Evaluation

In order to define an appropriate layout feature in consideration of the complexity of feature space, we compare our simplified density-based feature with fragmentation-based feature[17] and conventional density-based feature[13] by using the method described in Section 3.1. Motif data constitutes the core area in the training data of the Case 1 layout. Fig. 5 shows the histograms of d_i values for each layout feature. The parameters in our feature are set as the size of encoding area w_s = 1200 nm and the number of grids w_n = 10.We use following parameters for the conventional density-based method: w_s = 420 nm and w_n = 12.Fig. 5(a) and (b) indicate that in the fragmentation-based and conventional density-based features, linear separation is hard as most d_i values are smaller than 1. In contrast, Fig. 5(c) represents that, with our simplified feature, linear separation is less difficult than with the other two methods because all d_i are larger than 1.

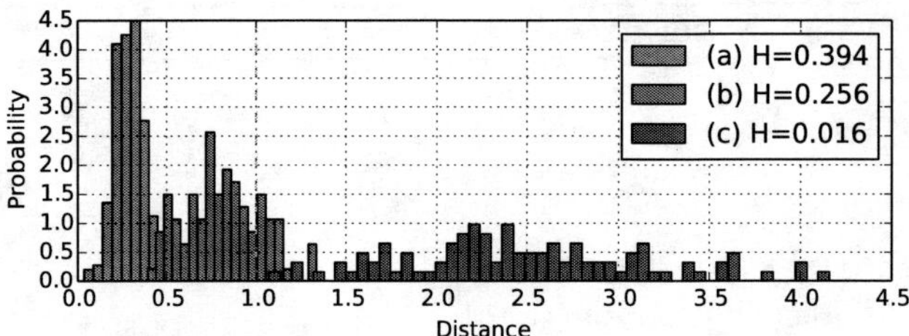

Figure 5. Histograms of distances (d_i) between hotspots and non-hotspots, (a) Fragmentation-based, (b) Conventional density-based, (c) Our simplified density-based.

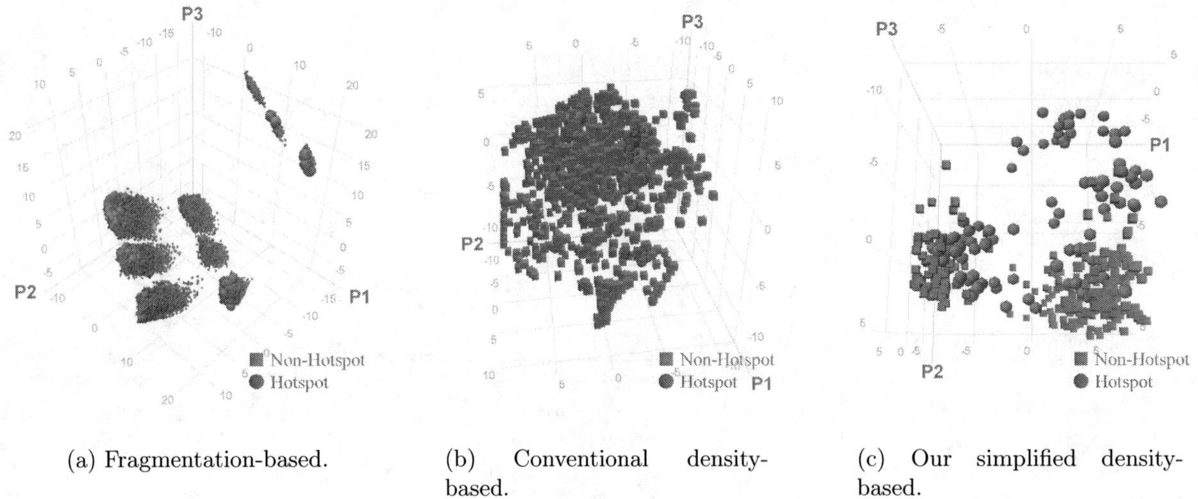

(a) Fragmentation-based.

(b) Conventional density-based.

(c) Our simplified density-based.

Figure 6. 3D visualized images of feature space.

To check the complexity of feature space, reference data to which principal component analysis was applied is shown in Fig. 6. The three axes in Fig. 6 correspond to the orthogonal basis of the three higher eigenvalues. Comparing these figures, layout features in fragmentation-based space are separated into nine clusters, and hotspots are mixed with non-hotspots in each cluster. Thus, a multi nonlinear classification kernel technique such as those used in[13] or[14] is required for this kind of complicated feature space. However, a multi kernel method may cause over-fitting.Moreover, runtime will also increase because the data clustering process after the feature extraction phase is added in the testing phase. For the same reason, the conventional density-based feature is too complicated for a single classification kernel because most hotspot features are covered with non-hotspot features. In contrast, Fig. 6(c) indicates that our simplified feature can define the decision boundary easily due to its unsophisticated feature space.

5.2 Effectiveness of Hotspot Detection

In order to confirm the hotspot detection accuracy and false alarm, we compared our results with[18] as the highest level of accuracy and[19] as the lowest level of false alarm. Since the benchmark data consists of layout clips, we limited the verification region to the core area (layout clips only) in order to avoid uncertainty effect from outside the frame area. In parameter setting for Real Adaboost classifier, we adjust them by increasing the number of base classifiers and tree depth within the following ranges until the prediction accuracy in the testing data is stabilized: #classifiers are 10 to 300 and tree depth is 1 to 4. As shown in Table 2, with the exception of the Case 4, our approach achieves more than 95% average accuracy and 0 false alarm. Though the runtime of our method is comparable even with Python implementation, further acceleration, e.g., 10 to 100 times, is possible by implementation in a statically typed language such as C++.

To evaluate the impact of our proposed feature space index, we compare our simplified density-based feature

Table 2. Comparison with [18] and [19].

Layout	Methods	Accuracy	False Alarm	Runtime
Case 1	[18]	94.69%	1,493	38.1s
	[19]	62.30%	17	14.4s
	ours	100.00%	0	7.0s
Case 2	[18]	98.20%	11,834	3m54s
	[19]	43.30%	75	3m1s
	ours	98.60%	0	5m51s
Case 3	[18]	91.88%	13,850	14m58s
	[19]	42.50%	7	4m3s
	ours	97.20%	0	4m57s
Case 4	[18]	85.94%	3,664	5m56s
	[19]	52.60%	41	1m37s
	ours	87.01%	1	2m50s
Case 5	[18]	92.86%	1,205	20s
	[19]	53.70%	4	44.6s
	ours	92.68%	0	1m9s

Table 3. Comparison on different w_n.

Layout	Methods	Accuracy	False Alarm	Runtime
Case 1	Low-w_n	92.00%	3	4.8s
	High-w_n	100.00%	4	10.4s
	ours	100.00%	0	7.0s
Case 2	Low-w_n	83.10%	2	3m47s
	High-w_n	95.80%	0	9m1s
	ours	98.60%	0	5m51s
Case 3	Low-w_n	94.70%	43	3m44s
	High-w_n	97.60%	5	6m51s
	ours	97.20%	0	4m57s
Case 4	Low-w_n	72.88%	7	2m6s
	High-w_n	79.10%	2	3m46s
	ours	87.01%	1	2m50s
Case 5	Low-w_n	65.85%	2	1m5s
	High-w_n	90.24%	0	1m28s
	ours	92.68%	0	1m9s

Table 4. Comparison of different algorithms.

Layout	Methods	Accuracy	False Alarm	Runtime
Case 1	LR	92.50%	32	10.2s
	SVM	95.10%	14	10.21s
	ours	100.00%	0	7.0s
Case 2	LR	47.20%	320	5m49s
	SVM	77.50%	142	4m55s
	ours	98.60%	0	5m51s
Case 3	LR	54.70%	1,001	4m31s
	SVM	90.80%	258	4m26s
	ours	97.20%	0	4m57s
Case 4	LR	63.84%	149	3m55s
	SVM	66.67%	74	2m42s
	ours	87.01%	1	2m50s
Case 5	LR	63.41%	82	1m9s
	SVM	65.85%	21	1m6s
	ours	92.68%	0	1m9s

with the other two parameters. In Table 3, Low-w_n and High-w_n indicate $w_n = 5$ and $w_n = 20$, and H values of Low-w_n, High-w_n and ours are 0.196, 0.111 and 0.016, respectively. All w_s are set to 1200 nm. Table 3 shows that the feature parameters are optimized by our proposed feature space index to maximize the accuracy and minimize the false alarm.

We further compare our proposed Real Adaboost method with two other classification algorithms: Logistic Regression (LR) and SVM. We use the parameters for SVM defined by the grid search method: $C = 1000$, $\gamma = 100$. Table 4 shows the comparison results and indicates that Real Adaboost has the best performance for both accuracy and false alarm. It should be noted that LR is able to achieve reasonable accuracy even though it is a simple linear classification model. This result shows that the hotspot detection problem is simplified by larger-area representation based on our feature space index. This result also shows a little complexity of layout features due to the large number of false alarms in LR. However, the result in SVM indicates that it is difficult to resolve the trade-off relation between accuracy and false alarm with a strong nonlinear classification algorithm.

6. CONCLUSIONS

This paper proposes an accuracy-boosted hotspot detection method for layout optimization. By applying our feature space evaluation index for layout representation, a density-based layout feature is defined, which realizes extraction of layout attributes with low-dimensional space. The robust Real Adaboost classifier is able to detect hotspots accurately with extremely low false alarm. The experimental results show that our method can achieve over 95% hotspot detection accuracy with almost zero false alarm and outperform the best published results for the ICCAD 2012 benchmarks. We believe this simple but effective methodology is promising to dramatically reduce the manufacturing and process optimization cost.

REFERENCES

[1] Pan, D. Z., Yu, B., and Gao, J.-R., "Design for manufacturing with emerging nanolithography," *IEEE Transactions on Computer-Aided Design of Integrated Circuits and Systems (TCAD)* **32**(10), 1453–1472 (2013).

[2] Inoue, S., Kotani, T., Nojima, S., Tanaka, S., Hashimoto, K., and Mori, I., "Total hot spot management from design rule definition to silicon fabrication," in *Electronic Design Processes Workshop EDP*, (2003).

[3] Simmons, M. C., Kang, J.-H., Kim, Y., Park, J. I., weon Paek, S., and Kim, K.-s., "A state-of-the-art hotspot recognition system for full chip verification with lithographic simulation," in *Proc. of SPIE*, **7974** (2011).

[4] Kahng, A. B., Park, C.-H., and Xu, X., "Fast dual graph based hotspot detection," in *Proc. of SPIE*, **6925** (2006).

[5] Kang, J.-H., Choi, J.-Y., Shim, Y.-A., Lee, H.-S., Su, B., Chan, W., Zhang, P., Wu, J., and Kim, K.-Y., "Combination of rule and pattern based lithography unfriendly pattern detection in opc flow," in *Proc. of SPIE*, **71221** (2008).

[6] Yu, Y.-T., Chan, Y.-C., Sinha, S., Jiang, I. H.-R., and Chiang, C., "Accurate process-hotspot detection using critical design rule extraction," in *IEEE/ACM Design Automation Conference (DAC)*, 1167–1172 (2012).

[7] H.Yao, Sinha, S., Chiang, C., Hong, X., and Cai, Y., "Efficient process-hotspot detection using range pattern matching," in *IEEE/ACM International Conference on Computer-Aided Design (ICCAD)*, 625–632 (2006).

[8] Xu, J., Sinha, S., and Chiang, C. C., "Accurate detection for process-hotspots with vias and incomplete specification," in *IEEE/ACM International Conference on Computer-Aided Design (ICCAD)*, 839–846 (2007).

[9] Nagase, N., Suzuki, K., Takahashi, K., Minemura, M., Yamauchi, S., and Okada, T., "Study of hot spot detection using neural network judgment," in *Proc. of SPIE*, **6607** (2007).

[10] Wuu, J.-Y., Pikus, F. G., Torres, A., and Marek-Sadowska, M., "Detecting context sensitive hot spots in standard cell libraries," in *Proc. of SPIE*, **7275** (2009).

[11] Drmanac, D. G., Liu, F., and Wang, L.-C., "Predicting variability in nanoscale lithography processes," in *IEEE/ACM Design Automation Conference (DAC)*, 545–550 (2009).

[12] Ding, D., Wu, X., Ghosh, J., and Pan, D. Z., "Machine learning based lithographic hotspot detection with critical-feature extraction and classification," in *IEEE International Conference on IC Design and Technology (ICICDT)*, 219–222 (2009).

[13] Wuu, J.-Y., Pikus, F. G., Torres, A., and Marek-Sadowska, M., "Rapid layout pattern classification," in *IEEE/ACM Asia and South Pacific Design Automation Conference (ASPDAC)*, 781–786 (2011).

[14] Ding, D., Torres, A. J., Pikus, F. G., and Pan, D. Z., "High performance lithographic hotspot detection using hierarchically refined machine learning," in *IEEE/ACM Asia and South Pacific Design Automation Conference (ASPDAC)*, 775–780 (2011).

[15] Wuu, J.-Y., Pikus, F. G., and Marek-Sadowska, M., "Efficient approach to early detection of lithographic hotspots using machine learning systems and pattern matching," in *Proc. of SPIE*, **7974** (2011).

[16] Mostafa, S., Torres, J. A., Rezk, P., and Madkour, K., "Multi-selection method for physical design verification applications," in *Proc. of SPIE*, **7974** (2011).

[17] Ding, D., Yu, B., Ghosh, J., and Pan, D. Z., "EPIC: Efficient prediction of ic manufacturing hotspots with a unified meta-classification formulation," in *IEEE/ACM Asia and South Pacific Design Automation Conference (ASPDAC)*, 263–270 (2012).

[18] Yu, Y.-T., Lin, G.-H., Jiang, I. H.-R., and Chiang, C., "Machine-learning-based hotspot detection using topological classification and critical feature extraction," in *IEEE/ACM Design Automation Conference (DAC)*, 671–676 (2013).

[19] Lin, S.-Y., Chen, J.-Y., Li, J.-C., Wen, W.-y., and Chang, S.-C., "A novel fuzzy matching model for lithography hotspot detection," in *IEEE/ACM Design Automation Conference (DAC)*, 681–686 (2013).

[20] Gao, J.-R., Yu, B., and Pan, D. Z., "Accurate lithography hotspot detection based on pca-svm classifier with hierarchical data clustering," in *Proc. of SPIE*, 90530E–90530E (2014).

[21] Mahalanobis, P. C., "On the generalized distance in statistics," *Proceedings of the National Institute of Sciences (Calcutta)* **2**, 49–55 (1936).

[22] Hastie, T., Tibshirani, R., Friedman, J., and Franklin, J., "The elements of statistical learning: data mining, inference and prediction," *The Mathematical Intelligencer* **27**(2), 83–85 (2005).

[23] Friedman, J., Hastie, T., and Tibshirani, R., "Special invited paper. additive logistic regression: A statistical view of boosting," *Annals of statistics* , 337–374 (2000).

[24] "ICCAD contest 2012." `http://cadcontest.cs.nctu.edu.tw/CAD-contest-at-ICCAD2012/problems/p3/p3.html`.

A methodology to optimize design pattern context size for higher sensitivity to hotspot detection using pattern association tree (PAT)

Shikha Somani[1*], **Piyush Pathak**[1], **Piyush Verma**[1], **Sriram Madhavan**[1] **and Luigi Capodieci**[1]

[1]*GLOBALFOUNDRIES Inc, 2600 Great America Way, Santa Clara, CA, USA 95054*

*shikha.somani@globalfoundries.com

Abstract

Pattern based design rule checks have emerged as an alternative to the traditional rule based design rule checks in the VLSI verification flow [1]. Typically, the design-process weak-points, also referred as design hotspots, are classified into patterns of fixed size. The size of the pattern defines the radius of influence for the process. These fixed sized patterns are used to search and detect process weak points in new designs without running computationally expensive process simulations. However, both the complexity of the pattern and different kinds of physical processes affect the radii of influence. Therefore, there is a need to determine the optimal pattern radius (size) for efficient hotspot detection. The methodology described here uses a combination of pattern classification and pattern search techniques to create a directed graph, referred to as the Pattern Association Tree (PAT). The pattern association tree is then filtered based on the relevance, sensitivity and context area of each pattern node. The critical patterns are identified by traversing the tree and ranking the patterns. This method has plausible applications in various areas such as process characterization, physical design verification and physical design optimization. Our initial experiments in the area of physical design verification confirm that a pattern deck with the radius optimized for each pattern is significantly more accurate at predicting design hotspots when compared to a conventional deck of fixed sized patterns.

1. Introduction

At advanced technology nodes, design-process interactions are very complex. Traditional design rule checks in the VLSI design verification flow are intended to detect simple geometric configurations in the design that the wafer fabrication process cannot support. These rules can be width, space, overlap and enclosure checks computed over simple geometric configuration constructs spanning single or multiple design layers. In addition to the configurations, it is important now to include more context (2D geometries surrounding the configuration) during rule based checking as context plays a key role in determining the 'root-cause' of the design hotspot. This leads to additional complexity to the rule definition. Pattern based checks have emerged

Design-Process-Technology Co-optimization for Manufacturability IX, edited by John L. Sturtevant, Luigi Capodieci, Proc. of SPIE Vol. 9427, 94270T · © 2015 SPIE · CCC code: 0277-786X/15/$18 · doi: 10.1117/12.2086954 as a successful alternative to the rule based checks as they help alleviate the problem associated with capturing context along with the configuration. It has been demonstrated that the pattern based checks are significantly more accurate at detecting design hotspots as compared to the simple rule based checks [1–3]. Figure 1 shows a tip-to-line configuration which is included inside a pattern which also includes additional 2D geometries surrounding the configuration.

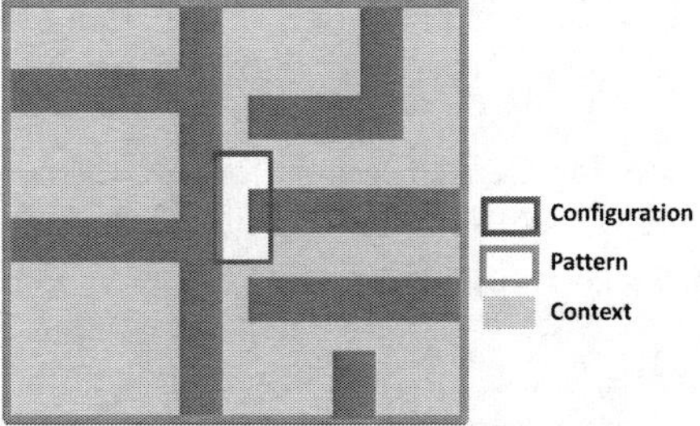

Fig. 1. The configuration is a simple geometric construct like a tip to line space shown in the center of the figure above. A pattern is shown here as the configuration in the center along with the surrounding context or situation.

Typical pattern based decks consist of a set of patterns with fixed sized context (pattern radius). The intent is to capture both the configuration and context around the design hotspot. This assumption is valid if the design-process interaction has a constant radius of influence. However, different process steps in the fabrication flow rely on different physical mechanisms, each with its own characteristic radius of influence as shown in figure 2.

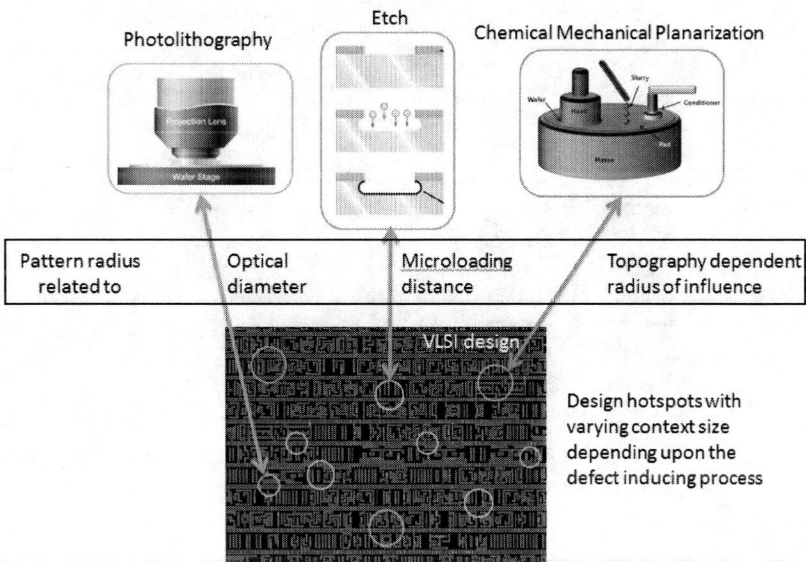

Fig. 2. Systematic design dependent defects during fabrication can occur during various process steps, each with its own physical mechanism and a corresponding characteristic radius of influence.

Furthermore, for a given process the variation in the density and geometry within a pattern results in different context sizes for different patterns. The size of the context (pattern radius) in the pattern is relevant in improving the defect detection rate of the pattern. As described in figure 2, in the process of photolithography, the radius of influence is governed by the optical diameter of the light source and the diffusivity of chemical species within the photoresist. For the etch process, the radius of influence depends upon the diffusivity of the etchant to the wafer surface and the etched byproduct away from the wafer surface (microloading effect). For Chemical Mechanical Planarization, the radius depends upon the local wafer topography and the slurry characteristics. Thus, an ideal pattern deck for hotpot detection would be one where each pattern has an optimal context size based on the defect inducing process step that interacts with it.

The key challenge in identifying optimal context for patterns in a pattern deck is the lack of a 'root cause' analysis methodology for systematic design dependent defects on the wafer. A root cause methodology in the case of patterns involves a context sensitivity analysis. Such an analysis in turn involves systematically varying the pattern radius and observing the accuracy of wafer defect detection by performing pattern matching with the said pattern. The radius at which the defect detection accuracy is the highest corresponds to the 'root cause' pattern. In this work, one such analysis is demonstrated using a novel approach to build pattern associations called the Pattern Association Tree (PAT).

The algorithm used to create the PAT using the training design data set (which vaguely represents sampling of the design-space) is presented as a part of this work. Using the generated PAT and an associated pattern ranking algorithm, a method is proposed to select optimal context for patterns in the pattern deck. The pattern deck created from PAT is compared with the traditional deck of fixed size patterns in terms of the accuracy in detecting hotspots. Figure 3 on the left shows an example of a traditional deck with fixed size pattern, and on the right it shows an example of a proposed deck containing patterns of different sizes.

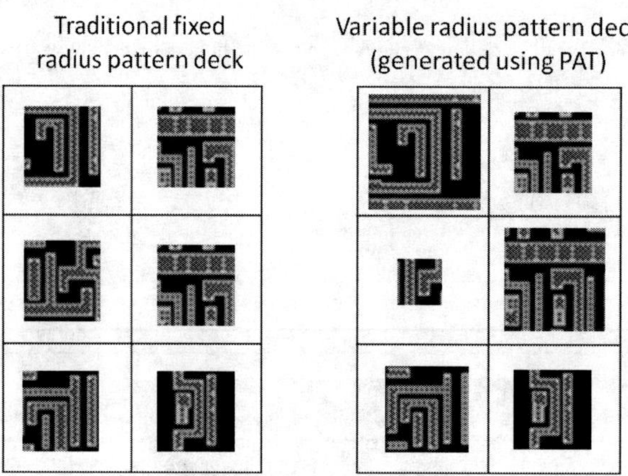

Fig. 3. Left: Traditional pattern deck with all patterns having a fixed common radius. Right: Proposed pattern deck with pattern radius optimized for each pattern separately.

2. Pattern Association Tree

The PAT is a directed graph with nodes and edges, where each node corresponds to a pattern with a specific radius and the edges represent the association between the patterns. The association between two patterns is defined on the basis of an alignment and containment criteria as shown in figure 4.

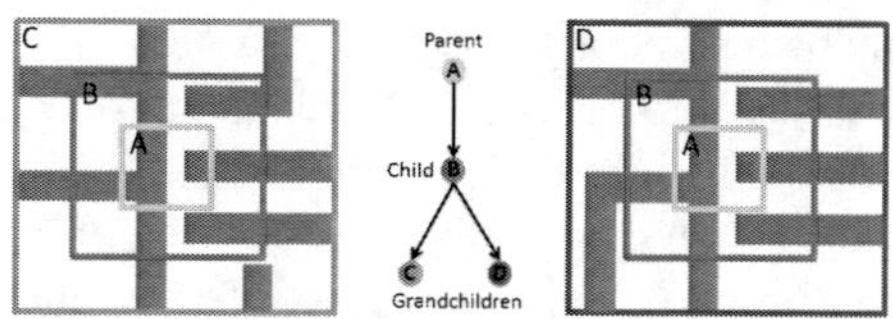

Fig. 4. Pattern Association Tree (PAT) for four patterns A, B, C and D with different radii.

Here, the pattern 'A' is considered a parent of pattern 'B' if and only if the pattern 'A' is fully contained within pattern 'B' (containment criteria) and the centers of patterns 'A' and 'B' coincide (alignment criteria). In figure 4 on the left, pattern 'A' is a parent of pattern 'B'. Also, pattern 'B' is the parent of pattern 'C'. In figure 4 on the right, Pattern 'B' is also fully contained and aligned within another pattern 'D'. Therefore, pattern 'B' is also the parent of pattern 'D'. Now, we have constructed the PAT for patterns 'A', 'B', 'C', and 'D' which are represented as its nodes shown in figure 4 (center).

2.1. Construction

In this study first we selected our training data set in the form of VLSI designs with known defect locations. For simplicity, the training set corresponds to one category of hotspots on one design layer like bridging of metal wires. A pattern association tree is created using the algorithm for PAT creation over these marked hotspot locations in the design (also shown in figures 6, 8 and 9). The overall algorithmic flow for PAT construction is shown in figure 5. Detailed steps involved in the PAT creation are described below.

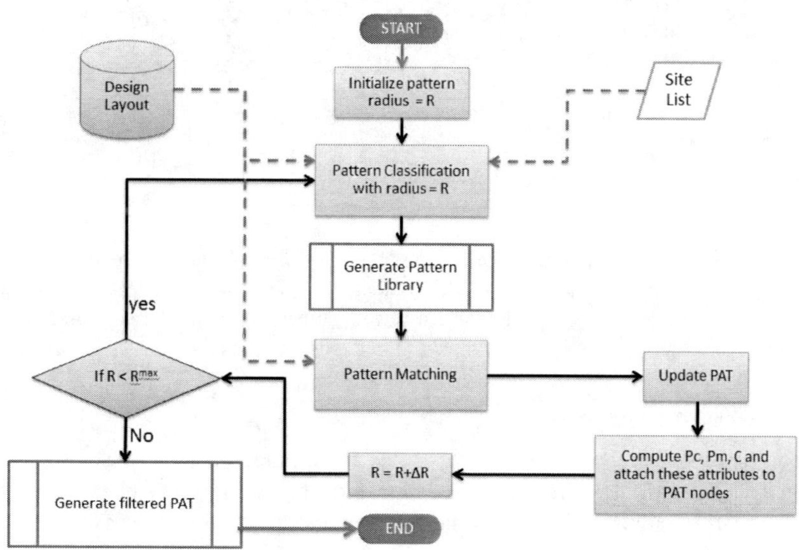

Fig. 5. Flowchart for the generation of PAT

2.1.1. Step 1 - Generation of nodes

As shown in figure 6, the first step is pattern classification which is performed on the full design at the marked locations (design hotspots). The process of pattern classification involves the clustering of various marked locations into unique patterns of a fixed given radius. Thus, the number of unique patterns obtained after the pattern classification step is less than or equal to the number of defect locations marked in the design. The pattern classification step is repeated with varying pattern radius values to cover a large range of context sizes. Each classification run produces a list of patterns at a given radius (a pattern deck at every radius), and a count of the defect locations covered by each pattern (N_c) is computed.

In the example shown in figure 6, the VLSI design has 6 known defect locations. These locations are classified using radii r_1 and r_2, where $r_1 < r_2$. At the smaller radius r_1, only two unique patterns are found. However at radius r_2, four unique patterns are found.

In the classification step, it is expected that as the classification radius is increased, the number of unique patterns found in the classification step also increases. This is due to the trivial observation that at higher radii, the pattern context is larger, which results in higher pattern complexity, ultimately resulting in a larger number of unique pattern clusters. Since the number of unique patterns goes up as the classification radius is increased while the total number of defect locations stays constant (as determined by the input to the classification step), the average value of N_c decreases with increasing classification radius.

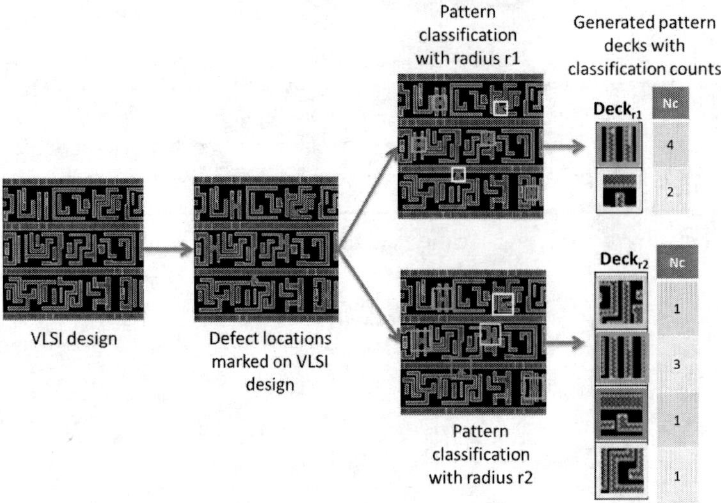

Fig. 6. Pattern classification is the first step in the creation of the PAT. Here, classification is performed on a VLSI design at marked defect locations by systematically varying the classification radius. At each radius, a list of unique patterns is generated, along with a count of the defect locations that each pattern in the list matches.

2.1.2. Step 2 Building associations

The second step involves the generation of the PAT by utilizing the pattern classification output from step 1 along with the input defect location information. For any given defect location, as the classification radius is varied, unique patterns with that radius are assigned to it. These unique patterns assigned to a specific defect site are associated, and the pattern with smaller radius is regarded as the parent of the pattern with larger radius (figure 8). Note that the patterns assigned to the same defect site on the design naturally pass the containment and alignment criteria discussed earlier. When this association rule is applied across patterns spanning all the defect sites in the design, a full PAT is created as shown in figure 7.

In the example shown in figure 8, the PAT consists of six nodes. Two of these are patterns from the classification output at radius r_1 and the remaining four are from the classification output at radius r_2. Note, how associations are built between patterns based on their assignment to defect sites on the design.

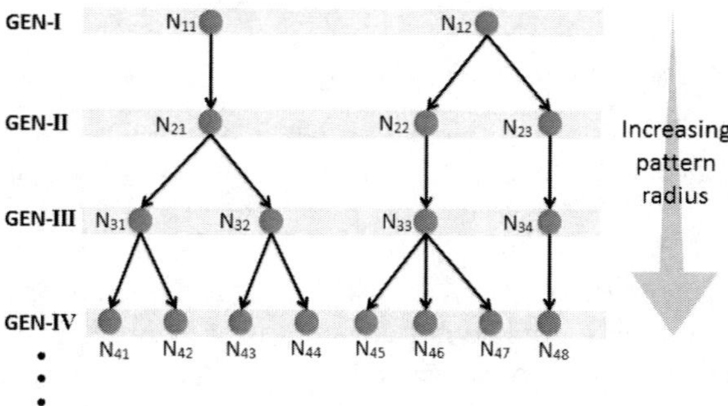

Fig. 7. Full PAT showing with several layers representing various 'generations' of patterns. Gen-I consists of patterns with the smallest radius. Pattern radius increases with every generation.

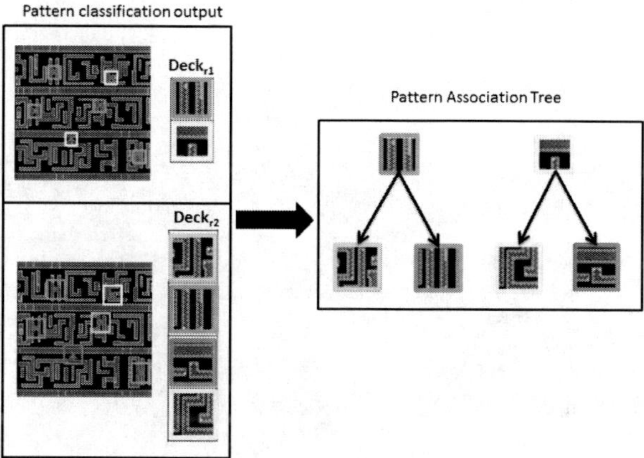

Fig. 8. PAT is created using the pattern classification output. Patterns which share a common defect site location on the design are regarded as being associated. The pattern with smaller radius is the parent of the one with a larger radius.

2.1.3. Step 3 - Annotating nodes with attributes

The third step is pattern matching which is used to identify all locations of patterns in the pattern decks (generated from the pattern classification step) that matches in the design. This step does not require the information about defect locations. After the pattern matching, each pattern in the pattern deck is associated with the count of the number of locations that are matched in the design and is referred to as N_m. Note that the number of pattern decks generated from the previous classification step is equal to the number of radii that is used to classify the given design hotspot locations. Therefore, the number of pattern matching runs equal to the number of pattern decks created at pattern classification step.

In the example shown in figure 9, pattern matching is performed twice, once with the deck at radius r_1 and once with the one at radius r_2. For the 2 patterns in the deck at radius r_2, 9 and 6 locations respectively are found to match in the entire design. In figure 9, count of hotspots covered (N_c) and the number of matched locations (N_m) are also shown.

Now, a statistical metric called 'precision' (P) is used to annotate all patterns in the pattern decks. 'Precision' is defined for as the ratio of N_c to N_m. Here, it is computed for each pattern.

$$P = \frac{N_c}{N_m} \tag{1}$$

Thus, P represents the ratio of the number of matched locations that are real defects to the total number of matched locations for a given pattern in the design. For the example shown in figure 9, the table in the right shows how the precision (P) and sensitivity (S) are computed (C) for each pattern in the deck. Please note that even though in this work we have used only 4 attributes for the pattern nodes - N_c, N_m, P and S, many more attributes can be defined for these nodes like the number of children of a pattern node (with varying weight for different generation children), the number of active children (children with P¿P_c^* etc), the number of higher generation patterns which are contained within the pattern but not aligned and hence not associated, and so on and so forth. The framework explained here is applicable for any number of such attributes.

It is important to note that the pattern matching step is not a requirement to generate the PAT. However, it is used in the study to compute the 'precision' value and annotate each pattern node in the PAT with this additional attribute. As discussed in the next section, this attribute is critical in post processing the tree to generate the pattern deck with patterns of variable context size.

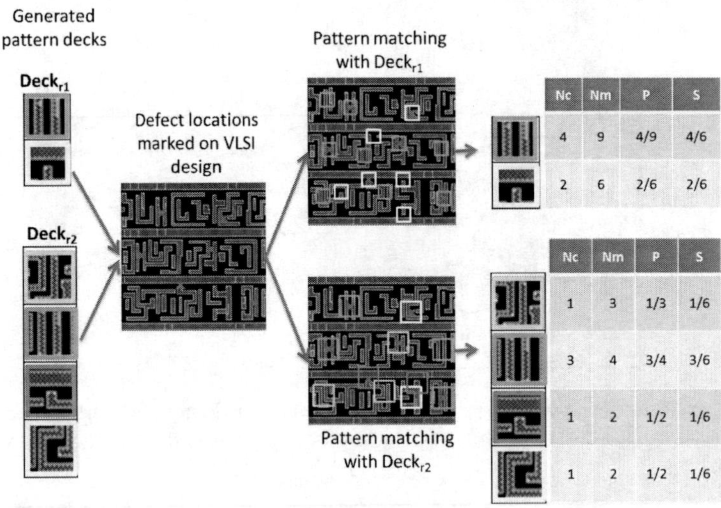

Fig. 9. Process of creating PAT and annotating each pattern (node in the PAT) with the 'precision' and sensitivity value computed for each pattern

2.2. *PAT salient features*

1. Every child node in PAT can have one and only one parent node. Since both containment and alignment criteria are used to create associations in PAT. The pattern with a larger context size can have only one pattern with a smaller context size when their centers are aligned. If the alignment criteria was waived for building pattern associations, then this will not hold true.

2. A child node can have more than one parent node. As the radius of a pattern is increased, its context may vary, resulting in multiple child patterns.

3. The number of child nodes for any given node represents the extent of variation in pattern context observed across different design hotspot locations in the given design.

4. The number of nodes at a smaller radius is always less than or equal to the number of nodes at a larger radius. This is due to the nature of pattern classification and is discussed in the previous section

5. The number of unique tree traversal paths from the first generation to the last generation in the PAT is equal to the number of nodes in the last generation. This is not specific to PAT and is true for any directed acyclic graph. The significance of these tree traversal paths is discussed in the following section.

6. The average precision of a pattern increases as the pattern radius increases. This is due to addition of specific context with the increase in pattern radii. The nodes in the last generation, on an average, have the highest precision among all nodes across all radii.

3. Pattern deck with variable context

One of the many applications of the PAT is to generate a pattern deck with patterns of variable context size or pattern radius for higher precision in the design hotspot detection. In such a deck, the radius of each pattern is optimized to reduce the rate of false positives and increase the rate of true negatives. Optimizing pattern radius is equivalent to finding the 'root cause' pattern for each defect hotspot as discussed previously. Following sections discuss how an annotated PAT can be used to identify the 'root cause' patterns.

3.1. Discussion

The PAT consists of a series of patterns at different radius for each defect known in the VLSI physical design. Multiple subset of patterns is formed by traversing the PAT from a node in the last generation to the node in the first generation. Under the assumption that the defect has a strong correlation with the design due to a undesirable process-design interaction, one of the patterns in the series truly represents the context relevant for that defect (as shown in the highlighted node for one of the traversal paths described in figure 10).

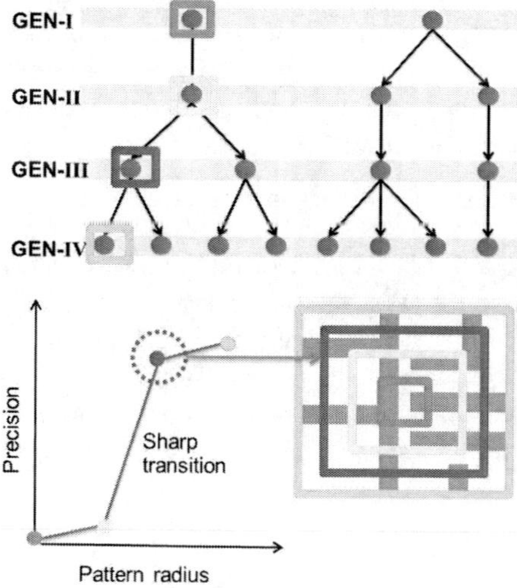

Fig. 10. Selection criterion for patterns from a traversal path spanning the largest radii pattern node to the smallest radii node

Optimization of pattern radius for each defect location in the design is equivalent to filtering the nodes in PAT based on attributes associated with the nodes. In our analysis, each node in the PAT is annotated with the corresponding values of N_c, N_m, P and S. Each of these attributes represents a vital trait of the node which can be further used to make a decision about the selection of a node in the final multi-radius pattern deck. N_c represents the number of defect locations that a certain node covers while N_m represents the usage frequency of the said pattern in the design. Precision (P) represents the accuracy of defect detection by a pattern, or equivalently, the probability that a matched location of the pattern would capture a defect on the wafer. Sensitivity (S) represents how many hotspots were covered by the pattern.

For a defect that is strongly correlated with undesirable design-process interaction, there will be a unique pattern radius optimal for defect detection. To identify this optimal pattern radius, the precision trend as a function of pattern radius can be analyzed to provide a context sensitivity measurement. It is expected that at very small values of radius, the precision will be very low as the pattern will correspond to a simple geometric configuration with widespread use in the design (very high N_m). As the radius is increased, on an average, the precision is expected to increase, with a sharp rise around the optimal pattern radius. After this sharp rise in precision, it is expected to plateau out with any further increase in pattern radius. Thus, to identify the optimal radius, is is important to look for a distinct rising signature in the precision vs radius plot for any defect location (shown in figure 10).

The sharp rise in the precision value along a traversal path in the PAT is an important condition for the identification of the optimal radius. However, there are other factors that are considered in filtering of the PAT nodes. These factors relate to the absolute values of N_c and N_m, which provide statistical weight to the measurement of P. Nodes for which N_c is very small, represent those patterns that have covered a very few defect sites and therefore have less coverage in the training set and therefore will not have a high confidence in predicting a defect on a new design.Those with small N_m correspond to those patterns which are seldom used in the design. Depending upon the end application of the generated pattern decks, these nodes may or may not be important to retain. For example, if the deck is used in the design verification flow, such patterns should be included so that the design can be free of these less frequent design constructs.

An important corollary of the sharp rise in precision vs pattern radius trend is for those cases where such a trend is missing. Such cases fall into one of the two categories. One where the wafer defect does not have a strong correlation with the fabrication process and thus the defect is ‘more’ random than systematic. The second where the design data for the precision measurement is not sufficient, thus resulting in a large error bar.

3.2. *PAT Deck generation*

The deck generation process involves the algorithm described above to filter the nodes in the PAT based on a set of criterion.

As mentioned before, the first step in the filtration process is to enumerate all the traversal paths in the tree from the node with the largest radius to the node with the smallest radius. The number of such paths is equal to the number of nodes in the last generation. As an example, for the tree in figure 7, some of the traversal path sequences are shown in the rows of the matrix below.

$$\begin{pmatrix} N_{41} & N_{31} & N_{21} & N_{11} \\ N_{42} & N_{31} & N_{21} & N_{11} \\ N_{43} & N_{32} & N_{21} & N_{11} \\ N_{44} & N_{32} & N_{21} & N_{11} \\ . & . & . & . \end{pmatrix}$$

The next step involves the analysis of the N_c, N_m, P and S trends along each of these paths. An initial set of filters that can be applied directly, involve the picking of only those nodes where each of these attributes is greater than a threshold value, denoted by N_c^*, N_m^*, P^* and S^* respectively.

$$N_m > N_m^*, N_c > N_c^*, P > P^* S > S^* \tag{2}$$

Note that any complex combination of similar filtering criteria (say can also be used depending on the intended application. So only those nodes are retained that contribute to a statistically significant defect population on wafer and are frequently used in the physical design (dependent on the training set size). A second set of filters is applied by retaining only those tree traversal paths where the precision vs radius trend shows a sharp rise as discussed previously.

$$\frac{\Delta P}{\Delta r} > \frac{\Delta P^*}{\Delta r} \tag{3}$$

To compute this numerically, a difference matrix for the precision value of the patterns for each traversal path is computed. This can be written as follows.

$$\begin{pmatrix} P_{41} & P_{31} & P_{21} & P_{11} \\ P_{42} & P_{31} & P_{21} & P_{11} \\ P_{43} & P_{32} & P_{21} & P_{11} \\ P_{44} & P_{32} & P_{21} & P_{11} \\ . & . & . & . \end{pmatrix}$$

Each row in the matrix above corresponds to a unique traversal path along the tree. The number of rows in the matrix is equal to the number of nodes in the last generation of the PAT. The number of columns is equal to the number of distinct radii values chosen for the analysis. To measure the change in precision as a function of radius, a numerical forward difference across the columns of the precision matrix can be computed as shown below.

$$\begin{pmatrix} P_{41}-P_{31} & P_{31}-P_{21} & P_{21}-P_{11} \\ P_{42}-P_{31} & P_{31}-P_{21} & P_{21}-P_{11} \\ P_{43}-P_{32} & P_{32}-P_{21} & P_{21}-P_{11} \\ P_{44}-P_{32} & P_{32}-P_{21} & P_{21}-P_{11} \\ . & . & . \end{pmatrix}$$

One node per row is chosen on the basis of the maximum value of the precision difference in the said row (shown in figure 10). Thus, the number of unique patterns selected on the basis of the difference matrix is equal to the number of rows in the matrix. Following section shows a case-study for the PAT creation and pattern deck extraction based on the ideas discussed in the current section.

4. Experiment

The case-study used for this work involved the construction of PAT to generate a deck constituting patterns of varying sizes. For this study, the training set included multiple design layouts with known bridge hotspots on the metal layer. These hotspots were obtained from lithographic printability simulations. The design data represents a VLSI design of area 2.16 mm^2 and 19913 bridge hotspot locations were used for pattern classification.

Design hotspots are classified into patterns for a range of pattern radii. If the starting radius is 'r' (minimum space of the metal in the design), classification of hotspots is done for radius 'r', '2r', '3r', '4r', '5r', and '6r'. All design hotspots were covered by the unique pattern classes found for each radius. Note that each hotspot is classified by only one pattern class at each fixed radius.

After creating pattern decks for each fixed radius, we create the PAT using the information on the location of design hotspots to check the alignment and containment criterion and identify the parent-child relationships among patterns of different fixed pattern radius.

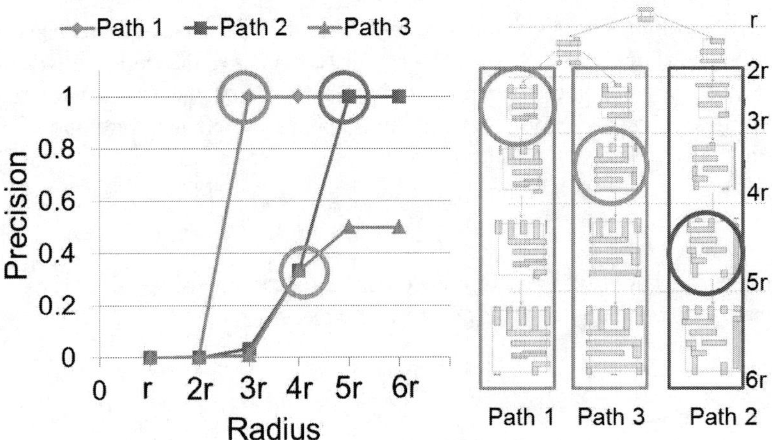

Fig. 11. Left: shows the precision value of a pattern node for three PAT paths each spanning the patterns nodes from radius *r* to 6*r*, Right: shows the corresponding pattern geometries associated with each pattern node of the three PAT paths

Now we have selected patterns and created a pattern deck. This pattern deck has patterns of sizes spanning from radius 3r to 6r. This composition is shown in figure 12. After building the PAT, we generated pattern match data on the training design to annotate each PAT node with the precision value of the pattern. Figure 11 on the left shows the precision value for each node along the three traversal paths spanning nodes from ‘r’ to ‘6r’ radius. Using the filtering algorithm described earlier, one pattern is selected from each traversal path using the criterion of the highest change in the precision value. Physically this means that moving from a smaller radius to a higher radius along a traversal path, context is added incrementally. Thus, if there is a sharp jump in the precision value, it implies that the incremental context added between the two radius increments has the highest contribution to the defect for that particular configuration (which is shared by all the patterns along the traversal path in the PAT). Additional filtering is performed on PAT to select nodes based on certain threshold values of precision.

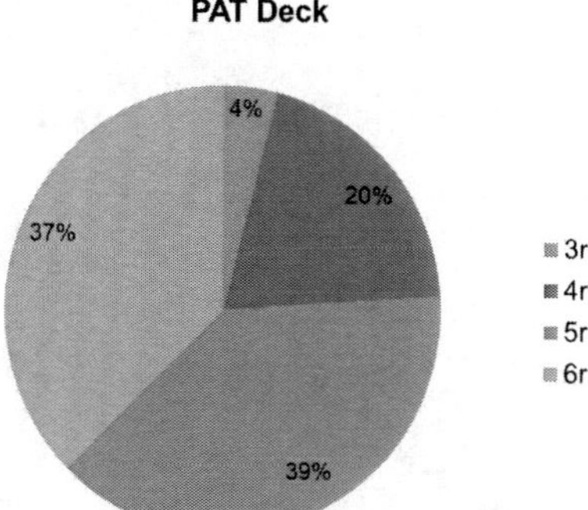

Fig. 12. Composition of pattern deck created from PAT after applying filters

5. Results

The pattern deck is formed from the training design data. In order to test the performance of the pattern deck, another test design data is used. Here, to evaluate the performance of the pattern deck, hotspots for the test design were identified using lithographic printability simulations. Here, only the 'bridge' hotspots are taken into account to evaluate the deck performance. Pattern search is performed for both the variable sized (PAT deck) and fixed sized (FR deck) patterns in the respective decks. Now, the matched locations of patterns is compared with the locations of known hotspots. This resulted in computing the number of hotspots covered by the deck (N_c). Using this data, the performance of each pattern deck is shown in terms of precision and sensitivity. Here, sensitivity of the deck is defined as number of hotspot locations that are covered by the pattern matching locations of the deck and is normalized to the total number of hotspots. As defined earlier, precision of the deck is defined as the number of matched locations that covered hotspots and is normalized to the total number of matched locations of the deck. Note that these metrics are defined for the deck and not a single pattern class.

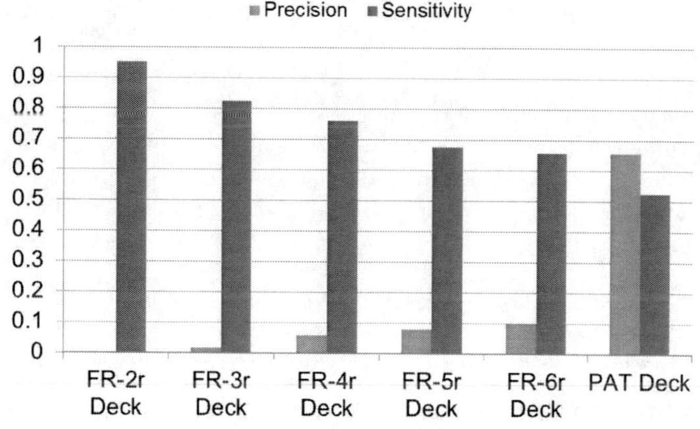

Fig. 13. shows sensitivity, precision and normalized (with number of patterns in the FR-6r deck) fraction of patterns for PAT Deck, and baseline fixed radii (FR) pattern decks for radius $2r$ (FR-2r), $3r$ (FR-3r), $4r$ (FR-4r), $5r$ (FR-5r), and $6r$ (FR-6r)

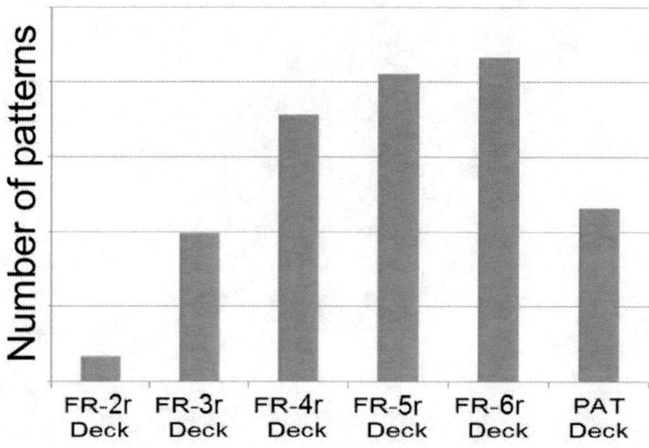

Fig. 14. Deck Size for fixed radii (FR) pattern decks and the PAT deck

The number of patterns in the decks are shown in figure 14. It can be observed that the number of patterns in the PAT deck are almost half as many in the deck at largest fixed radius (FR-6r deck). Figure 13 shows the precision and sensitivity of each of these decks for the test design data. Here, the deck with variable sized patterns (PAT deck) has the precision value of 0.65 compared to the highest value of precision at 0.1 among decks (FR decks) with fixed sized patterns. Further, the sensitivity for the PAT deck is around 80 percent of the sensitivity computed for the deck at largest fixed radius (FR-6r deck). The sensitivity of the PAT deck can be further improved by increasing the training data set.

6. Conclusions and Future work

We have demonstrated a methodology to capture the DFM learning on a given process and design with the directed graph of patterns called pattern association tree (PAT). Further, we have shown a method to filter and select critical patterns from the PAT generated from the training design data-set. This resulted in the deck with patterns of varying sizes or radius of influence. PAT generated deck was compared with the conventional deck of fixed sized patterns and it was shown that the PAT generated deck has a higher rate of hotspot identification when the pattern was found in the test design-set.

Future work would include the exploration of the analysis with more training design data to improve the training sample size and devising new methods to perform optimization of pattern selection from PAT for applications in VLSI design verification, and the identification of process weak-points.

References and links

1. Vito Dai, Jie Yang, Norma Rodriguez, and Luigi Capodieci. Drc plus: augmenting standard drc with pattern matching on 2d geometries. In *Advanced Lithography*, pages 65210A–65210A. International Society for Optics and Photonics, 2007.
2. Piyush Pathak, Sriram Madhavan, Shobhit Malik, Lynn T Wang, and Luigi Capodieci. Framework for identifying recommended rules and dfm scoring model to improve manufacturability of sub-20nm layout design. In *SPIE Advanced Lithography*, pages 83270U–83270U. International Society for Optics and Photonics, 2012.
3. Piyush Pathak, Piyush Verma, and Sarah N McGowan. Methods of modifying a physical design of an electrical circuit used in the manufacture of a semiconductor device, May 27 2014. US Patent 8,739,077.

20NM CMP MODEL CALIBRATION WITH OPTIMIZED METROLOGY DATA AND CMP MODEL APPLICATIONS

Ushasree Katakamsetty[1], Dinesh Koli[2], Yeo Sky[1], Hui Colin[1],
Ruben Ghulghazaryan[3], Burak Aytuna[4], Jeff Wilson[4]

[1] GLOBALFOUNDRIES, 60 Woodlands Ind. Park D Street 2, 738406, Singapore
[2] GLOBALFOUNDRIES, 400 Stone Break Road extension Malta, NY 12020, United States
[3] MENTOR GRAPHICS DEVELOPMENT SERVICES, Halabyan St., 16 Building, 0038, Yerevan, Armenia
[4]MENTOR GRAPHICS, 8005 SW Boeckman Road Wilsonville, OR 97070, United States

ABSTRACT

Chemical Mechanical Polishing (CMP) is the essential process for planarization of wafer surface in semiconductor manufacturing. CMP process helps to produce smaller ICs with more electronic circuits improving chip speed and performance. CMP also helps to increase throughput and yield, which results in reduction of IC manufacturer's total production costs. CMP simulation model will help to early predict CMP manufacturing hotspots and minimize the CMP and CMP induced Lithography and Etch defects [2]. In the advanced process nodes, conventional dummy fill insertion for uniform density is not able to address all the CMP short-range, long-range, multi-layer stacking and other effects like pad conditioning, slurry selectivity, etc.

In this paper, we present the flow for 20nm CMP modeling using Mentor Graphics CMP modeling tools to build a multilayer Cu-CMP model and study hotspots. We present the inputs required for good CMP model calibration, challenges faced with metrology collections and techniques to optimize the wafer cost. We showcase the CMP model validation results and the model applications to predict multilayer topography accumulation affects for hotspot detection. We provide the flow for early detection of CMP hotspots with Calibre CMPAnalyzer to improve Design-for-Manufacturability (DFM) robustness.

Keywords: CMP, Topography, Lithography, Depth of Focus, DFM, CMP modeling, CMP hotspots, Fill

1. INTRODUCTION

Chemical Mechanical Polishing/Planarization is a process of smoothing surfaces with the combination of chemical reactions and mechanical forces. The process uses chemical slurry in conjunction with a polishing pad and retaining ring, to remove material and even out irregular topography for planar wafer surface [1]. CMP is necessary to bring the entire surface within the depth of focus of a photolithography system. As the number of layers increases, the difficulty in maintaining the focus in the lithography process increases [2]. The multiple metallization layers in today's devices require extreme planarity to maintain high manufacturing yields [3]. CMP topography variation increases for the higher metal layers due to multi-layer topography accumulations resulting in larger inter and intra-level variations as shown in Figure 1.

Design-Process-Technology Co-optimization for Manufacturability IX, edited by John L. Sturtevant, Luigi Capodieci, Proc. of SPIE Vol. 9427, 94270U · © 2015 SPIE · CCC code: 0277-786X/15/$18 · doi: 10.1117/12.2085728

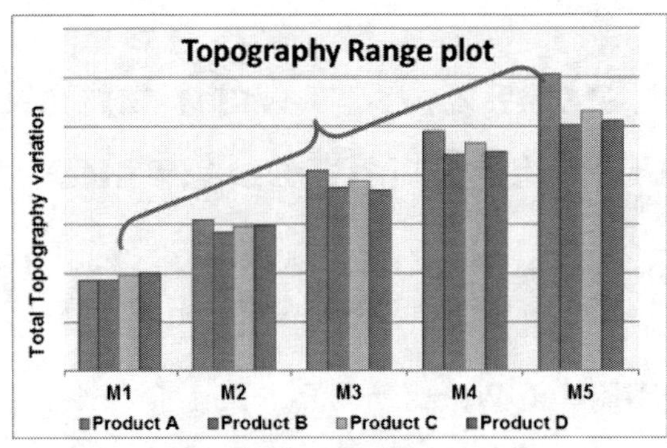

Figure 1: Graph showing accumulated topography variations

2. NEED FOR CMP MODEL BASED SOLUTION

Traditional way to avoid CMP manufacturing weak points or hotspots in the IC designs is by following the density rules defined in the foundry design manuals. With smaller feature sizes and multi-functional complex SOC chips in advanced technologies, it is hard to support all the process variations using rule-based approach. And with increase in metallization layers in current IC designs, rule based analysis is not able to capture the multi-layer accumulation of CMP dishing and erosion hotspots.

In smaller geometries, the CMP effects may extend beyond the cell boundary and create additional interactions that may not be caught by traditional design rules. These effects may also introduce additional variations on timing, power and other parameters of the IC chip. Figure 2 shows CMP hotspots appearing at nominal density region due to multi-layer topography accumulations, while these issues were not captured by rule based approach.

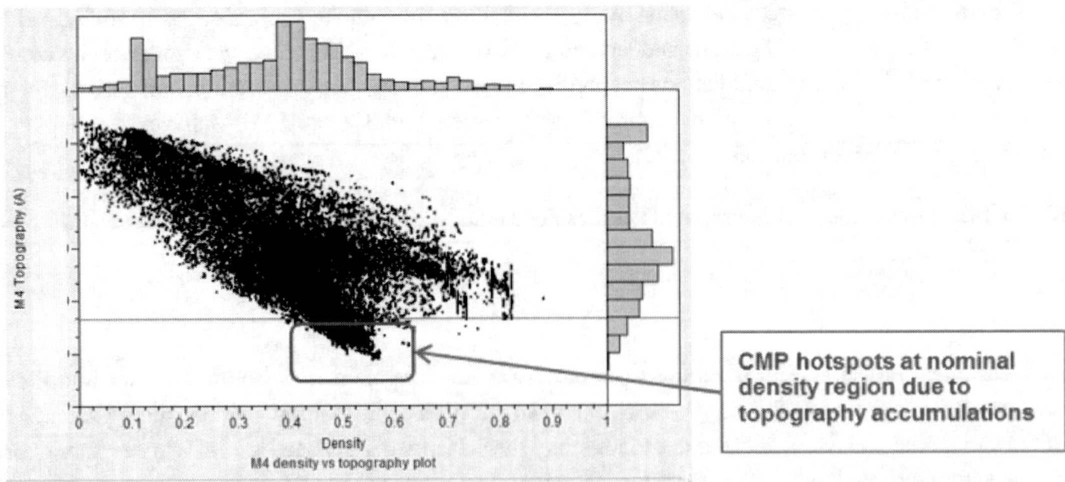

Figure 2: Density vs cumulative topography

Figure 3 shows CMP topography profile variation for the same density with different line width. CMP model based analysis will provide a better solution to handle manufacturing variations in addition to the rule based analysis.

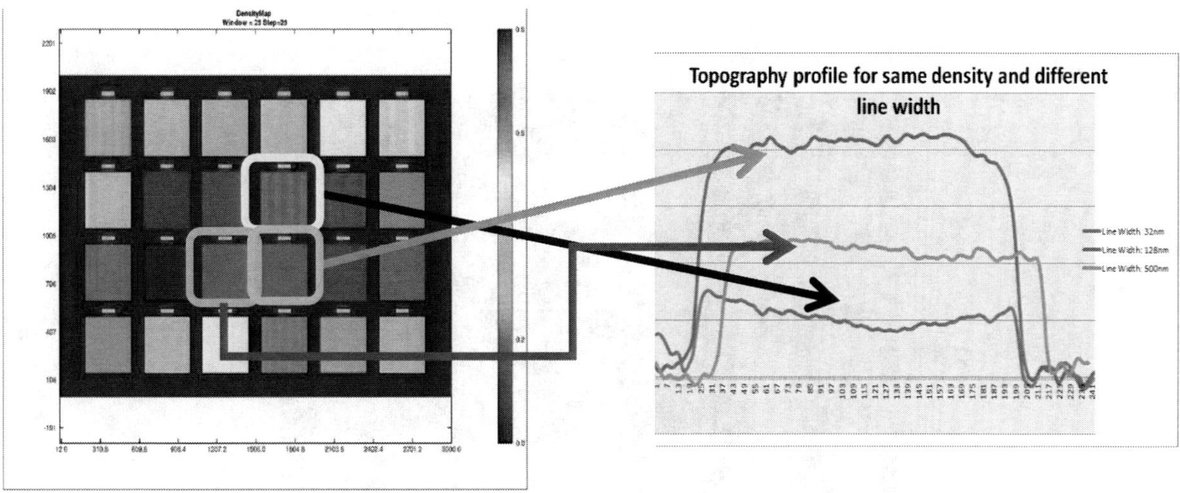

Figure 3: Same density structures topography variation against line width

3. CMP MODEL CALIBRATION FLOW

In this paper, CMP modeling technology from Mentor Graphics is used to build a multilayer Cu-CMP model and study hotspots.

3.1 CMP testchip plan

CMP testchip planning and designing is the first and crucial step for CMP model development. CMP testchip should cover various line width supported by the respective technology, ranging from low to high density patterns as shown in Figure 4.

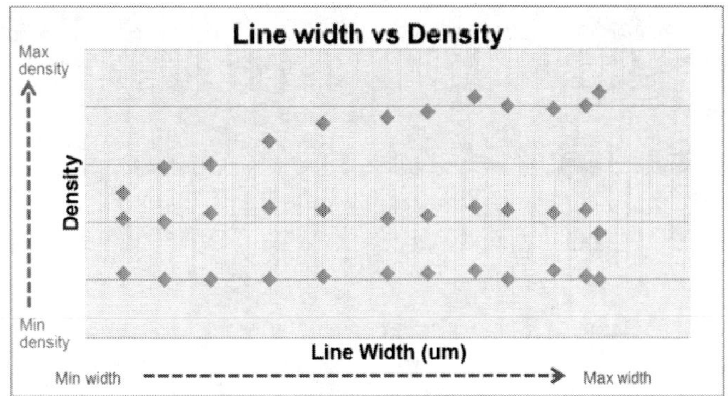

Figure 4: Sample line width and density specification of CMP testchip

CMP process can be influenced by local and global CMP variations. CMP macros should be widely spaced apart to minimize the interference from neighboring structures as shown in Figure 5. Testchip array blocks should enable CMP Optimize tool to generate uniform width, space and density grids so that measured erosion and dishing properly correspond to designed geometry. Band lines without dummy fill through the middle of the field region will allow good line scan signal for erosion measurement as shown in Figure 6.

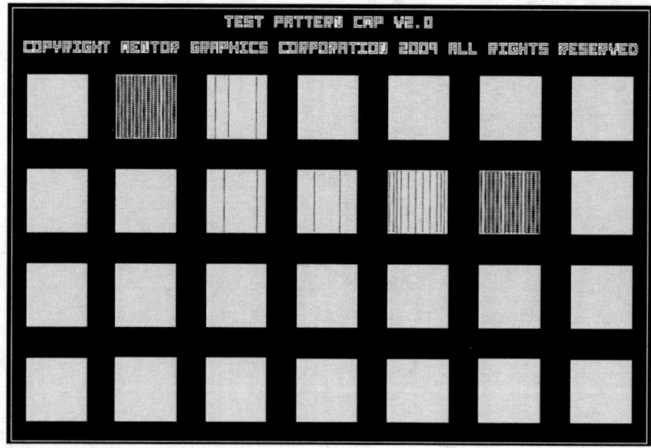

Figure 5: CMP testchip sample floorplan

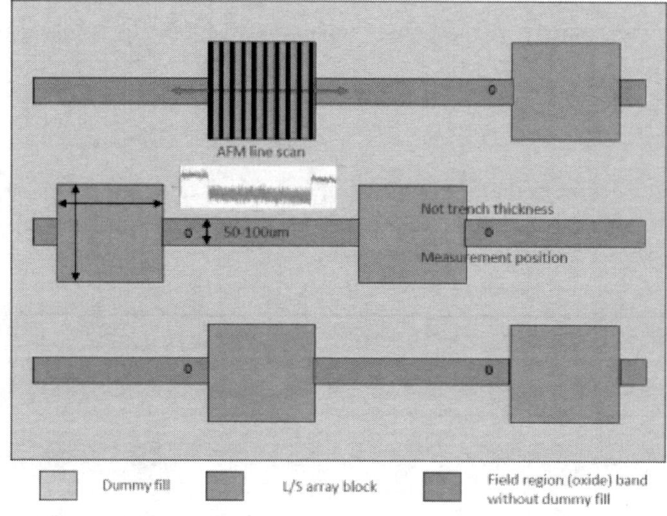

Figure 6: Picture showing band line for profile scan improvements

3.2 CMP testchip metrology collection

Next step for CMP modeling is to collect the metrology data from the CMP testchip during fabrication stage. Metrology data need to be collected for all the modeling steps. In general, one metal layer CMP modeling requires calibration of electrochemical deposition, platen1, platen2, and platen3 CMP models as shown in Figure 7.

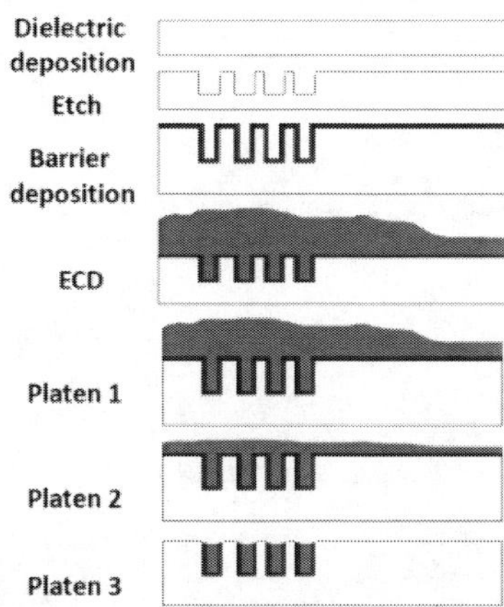

Figure 7: Metal layer fabrication flow schematic view

Metrology data need to be collected for all the modeling steps. CMP modeling needs profile scan data for Erosion and Dishing and x-section (TEM, SEM) data for absolute thickness measurements as shown in Figure 8: TEM measurements are preferred than SEM in advanced technologies for accurate measurements.

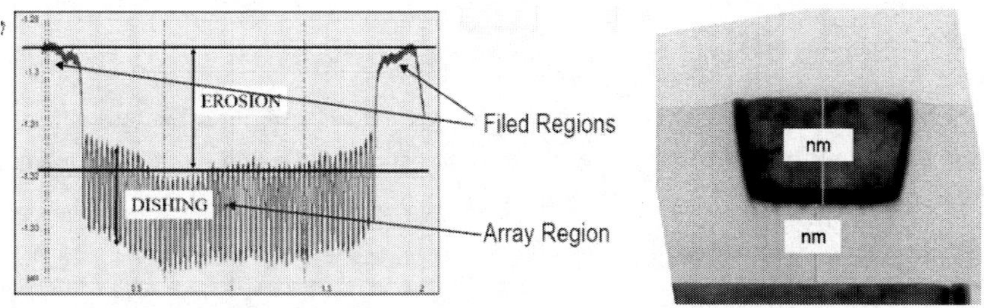

Figure 8: CMP test structures sample profile scan and TEM x-section metrology

Si metrology data need to be very accurate for high quality CMP model development especially for leading edge applications. Smaller feature sizes in advanced technology node poses additional challenges to maintain accurate data. Advanced surface profile topography metrology tools with high resolution and high throughput required for accurate measurement of nano-scale features CMP dishing and erosion measurements.

As aspect ratios become higher, features become smaller and process variations increases in advanced technology nodes. To minimize with-in wafer variation affecting the metrology data, collect multiple dies profile scan instead of only single die data for CMP modeling. Analyze multiple dies metrology data and derive systematic variation of erosion and dishing of all the structures. For effective erosion and dishing measurements, always collect the profiles scans from the middle of the array structures.

3.3 Optimizing the metrology requirements

Conventional CMP model calibration flow requires profile scan and x-section measurements at post-deposition and at all the CMP steps. In general, Metal layer CMP process may have 3 major polishing steps. That leads to 4 wafers for single layer CMP modeling. And we need one more additional wafer to validate the CMP model as shown in Figure 9.

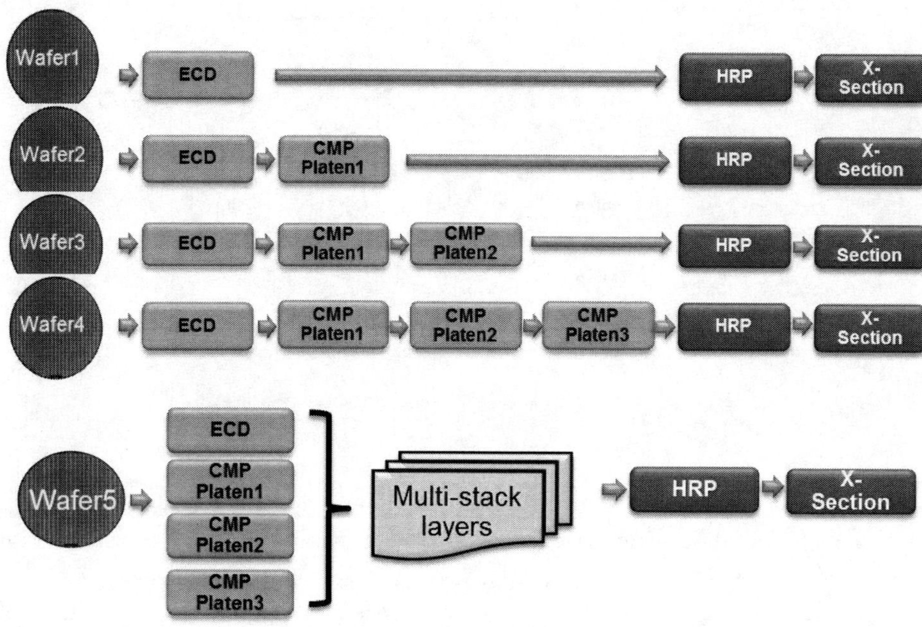

Figure 9: Figure showing the wafer and metrology requirements for CMP modeling

CMP model calibration requires substantial amount of metrology data at pre, intermediate and post CMP polishing steps. The data collection at multiple process steps requires many spare wafers and it is human and machine resource intensive. And the wafer costs are increasing tremendously in advanced process nodes (Figure 10).

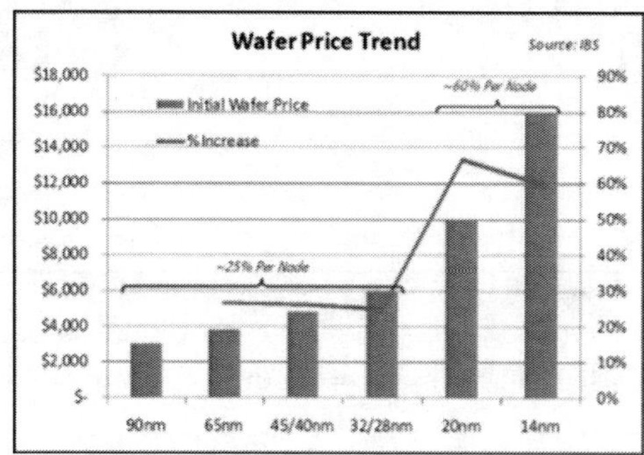

Figure 10: Chart showing Process node vs wafer price trend [4]

To optimize the wafer cost without compromising the quality of the CMP model using Mentor Graphics CMP Optimize tool, we have tried the flow in Figure 11. We used 1 spare wafer for metrology collection instead of 4 wafers. We collected multiple dies metrology from the single wafer to derive systematic CMP process variations. With clear understanding of the CMP process parameters and target conditions, we calibrated the CMP models for all the steps. We modified the modeling parameters and target conditions to correlate with actual process variations. And continuously reviewed and modified all the intermediate and final steps calibration outputs with the CMP process team to achieve accurate CMP simulations. By closely working with the process teams and with Mentor Graphics CMP Optimize support team, we developed the 20nm CMP models with more than 90% accuracy. The final CMP models are able to predict post-CMP topography and Cu thickness variations and follow the process variations trend very closely.

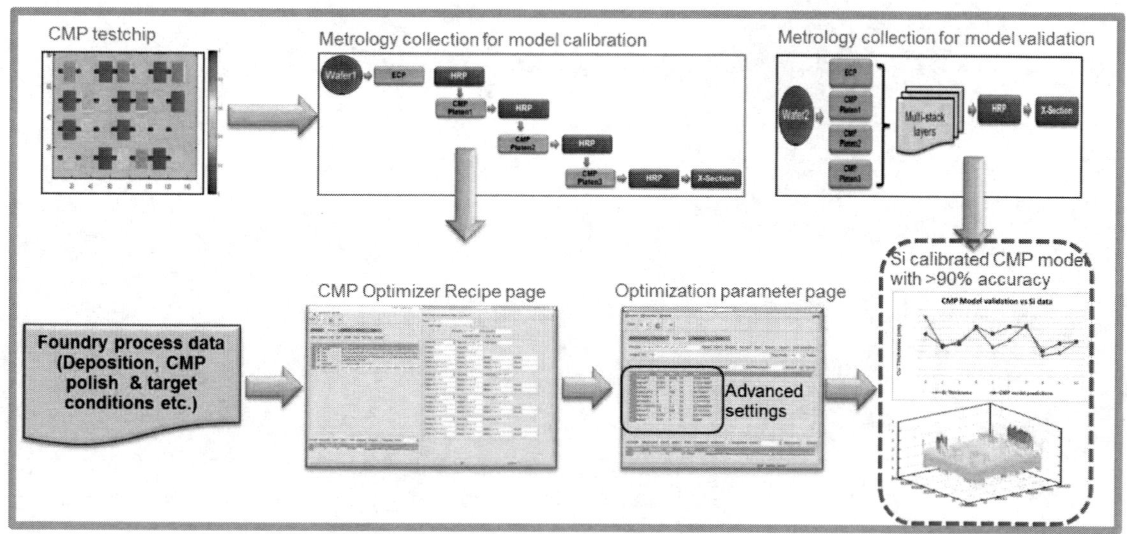

Figure 11: Optimized CMP modeling flow

4. CMP MODEL APPLICATIONS

Si calibrated CMP simulation kit enable designers to visually highlight and examine a variety of CMP manufacturing effects. Detection of design week points or hotspots is one of the key applications of CMP simulation and analysis. IC designers can examine these results layer by layer, and for selected areas in the layout. They can also review specified CMP hotspot checks, such as depth of focus, erosion or dishing. The most commonly studied hotspots are related to erosion, dishing, metal thickness and topography variations. But there are several cases when some non-standard conditions that engineers would like to check in order to identify hotspots. For such kind of hotspots, CMP Analyzer allows user-defined hotspots. Using this feature, users may perform full chip study of post simulation profile and detect all kind of hotspots they are interested in.

Accuracy of the Cu-CMP model has strong impact on the prediction of post CMP hotspots. The created model was successful in fitting the measured data. Validation of the model showed that it follows trends closely for a broad range of metal width and densities on different production designs. Early detection of CMP hotspots with Calibre CMPAnalyzer will improve Design-for-Manufacturability (DFM) robustness for the following design styles:

- High density IP blocks, High to Low density transitions regions (Figure 12),
- Custom design blocks or IP's with low dense dummy fill,
- RF designs with inductors and high density circuits (Figure 13).

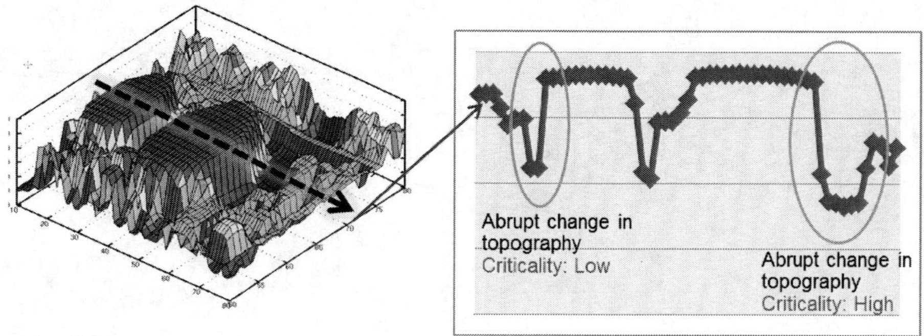

Figure 12: Abrupt topography variations due to high and low density blocks next to each other

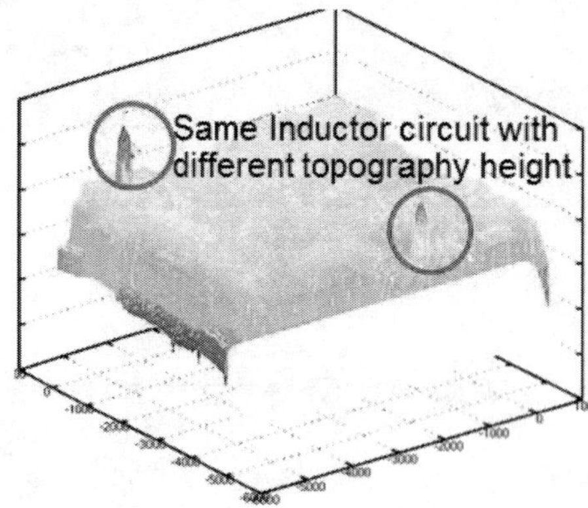

Figure 13: Picture showing RF Inductor circuit topography variation influenced by neighbourhood

Multilayer CMP simulation allows detecting hotspots that arise in DRC clean designs due to multiple layers stacking. Topography non uniformity at lower layers accumulates from layer to layer and eventually creates hotspots at higher layers. In Figure 14, it is shown how over polishing hotspot appear at M4 layer due to topography accumulation at layer M2 and M3 even if at M1 the given area is under polished. Such hotspots can be easily identified by CMP Analyzer hotspot detection methods.

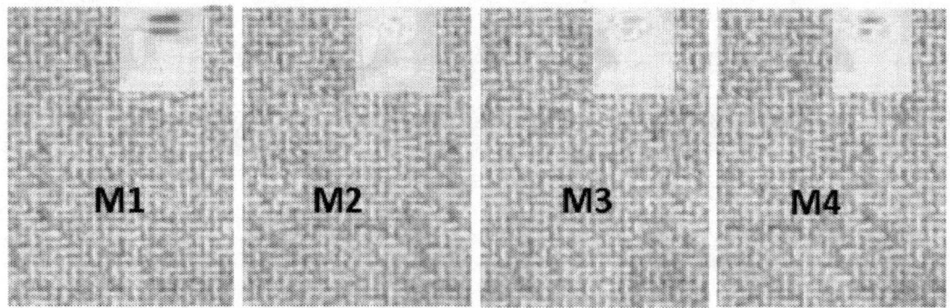

Figure 14: Picture showing topography hotspot generation at higher layers due to multilayer stacking

Another important application for a CMP model is to improve the yield of the design by enhancing the layout. The step in the design process that has a significant impact on the planarity is the insertion of dummy fill. The key is to use a smarter approach when inserting the fill. The CMP model is used to identify hotspots from erosion or dishing but it can also produce general thickness data. The CMP model provides a low cost way to analyze multiple layouts and design styles. The CMP model is used to correlate layout parameters, such as density and perimeter values, with the CMP data. This information is used to establish and improve design rules associated with planarity. The key to a smarter approach to fill is to combine the analysis of these layout parameters when adding fill. This analysis capability needs to be able to examine not only the layout parameters but have the flexibility to optimize these parameters across the design. The effectiveness of this analysis increases when coupled with the ability to span multiple layers so any accumulative effects of the CMP process can be minimized. The CMP model can be used to improve the uniformity of the design when coupled with an analysis driven fill solution like the SmartFill functionality in Calibre YieldEnhancer.

The next phase of this project is to close the loop by using the CMP model to improve the insertion of the dummy fill shapes. With a quality CMP model and a correct by construction fill solution this work can be fairly straight forward. However, the current advanced technologies complicate the complexity of fill because there are a number of manufacturing processes that influence fill methodology to improve the robustness and uniformity of the design.

5. CONCLUSION

In this paper, we presented the work done in GLOBALFOUNDRIES for 20nm CMP model. We discussed following topics in this paper:

1) Accurate CMP model calibration with optimized metrology data using Mentor Graphics CMP Optimize tool.
2) We provided comparison of rule-based and model-based approach and showcase the advantages of model-based analysis.
3) We showed how small profile variations at low metal layers may propagate through multi-layer stacking and generate hotspots at high metal layers. We also give recommendations on how to avoid such hotspots.
4) Finally, we provided DFM recommendations to the designers on hotspots fixing methodologies.

REFERENCES

[1] Parshuram B, Zantye, Kumar A, et al.,

- "Chemical mechanical planarization for microelectronics applications". Mater Sci Eng, 2004, 45(3–6): 89

[2] Ushasree Katakamsetty, Narayana Samy Aravind, et al.,

- "Scanner correction capabilities aware CMP Lithography hotspot analysis", in Proc. SPIE Advanced Lithography 2014

[3] Ruben Ghulghazaryan, Jeff Wilson, Kyoko Izuha, et al.,

- Using a Highly Accurate Self-stop Cu-CMP model in the Design Flow, in Proc. SPIE Vol. 7641, 2010.

[4] https://www.semiwiki.com/forum/content/3420-semiconductor-cost-models-boring-but-crucial.html

TOPOGRAPHY AWARE DFM RULE BASED SCORING FOR SILICON YIELD MODELING

Vikas Tripathi[1], Ushasree Katakamsetty[1], Yeo Sky[1], Hui Colin[1]

[1] GLOBALFOUNDRIES, 60 Woodlands Ind. Park D Street 2, 738406, Singapore

ABSTRACT

DFM rule based scoring is associated with manufacturability rules checking and applying the scoring to predict the yield entitlement for an IC chip design. Achieving high DFM score is one of the key requirements to get high yield. The DFM scoring methodology is currently limited to DFM recommend rules and their associated failure rates. In contrast to failure mechanism, chemical-mechanical polishing (CMP) step topography variations places an important role to it. In this paper, we present an advanced DFM analysis flow to compute DFM score that incorporate topography variation along with recommend rule scoring using complex scoring model to increase silicon yield correlation.

1. INTRODUCTION

DFM checks are one of the critical steps in IC design flow. These checks are important and imposed by foundries to improve yield. Foundries provide a list of critical DFM rules for IC designers to evaluate the design using a DFM analysis flow. The DFM score reflects the overall design quality and help to predict the yield entitlement. These scores vary from 0 to 1 where 1 considers as highest score and 0 is the lowest score. Higher score leads to better yield. Hence it is very critical to have accurate scoring mechanism to reflect the yield entitlement.

Similarly, foundries also provided CMP simulation kits to identify and report CMP manufacturing step hotspots. The variation in pattern density can cause planarity issues and hence can impact the overall failure mechanism of layout features with DFM violation.

It is one of the challenges to have advanced DFM scoring mechanism which can incorporate topography variation along with DFM critical rules. In this paper, we are presenting the flow developed by GLOBALFOUNDRIES that take CMP topography results along with recommended rules to compute optimized DFM score. The paper also shows correlation of DFM score with CMP topography variation. As a result, the same layout feature falling in different topography region will be shown to have different DFM score. The DFM violations under extreme topography regions are more susceptible for failure and given low DFM score. This flow enables designer to address the critical DFM violations to avoid yield issues and achieve DFM score correlation with the silicon yield.

1.1 DFM rules scoring and analysis

At the advanced technology nodes, design process is getting more complex. Issues related to systematic process variations increases and foundries rely on DFM checks to mitigate these systematic related defects[1]. Foundries also provide list of DFM rules to avoid some of these systematic failures. These DFM rules require more margin than DRC. Any layout feature which does not meet the DFM rule margin considered as DFM violation. Each DFM violation has a DFM score associated with it. The DFM score depends upon the magnitude of DFM violation.

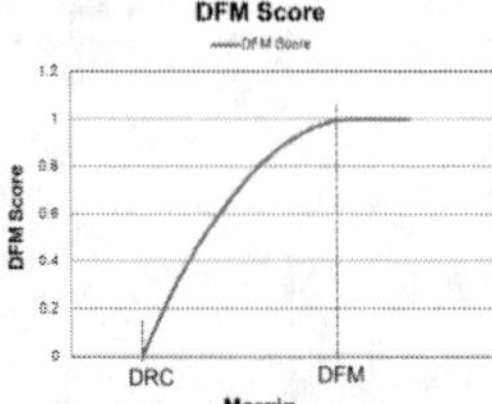

Fig 1: Graph showing DRC to DFM score variation

Design-Process-Technology Co-optimization for Manufacturability IX, edited by John L. Sturtevant, Luigi Capodieci, Proc. of SPIE Vol. 9427, 94270V · © 2015 SPIE · CCC code: 0277-786X/15/$18 · doi: 10.1117/12.2085733

The DFM score value tends to 0.0 when the drawn value is close to DRC threshold value and score value tends to 1.0 when drawn value is close to DFM threshold value. Fig 1 explains the relation of DFM score with respect to DRC and DFM threshold values.
These DFM violations and their respective score are used to derive total DFM score of an Integrated Circuit (IC) design. The total DFM score primarily depends upon the DFM violation density over the complete design area. Higher the violation density lower be the total DFM score.
However, every rule has corresponding failure rate associated with it. This failure rate reflects the sensitivity of the rule for process variation and further used to compute total DFM score. The higher the failure rate, lower is the total DFM score. Fig 2 describes the relationship between total DFM score and error density for two different rules of different failure rates.

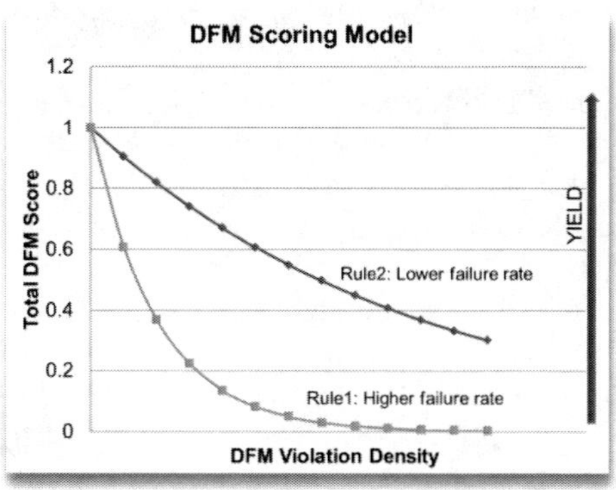

Fig 2: Figure showing total DFM score vs DFM violation density graph for two rules of different priority

As shown in the Fig 2, rule 2 has lower failure rate with respect to rule 1 and for the equal DFM violation density, rule 2 having better total DFM score then rule1. For high total DFM score, DFM error density should be less as well corresponding DFM scores for each DFM violations should be high. Below is the equation (Eq. 1) to represent total DFM score and its depended components.

$$f(\text{Total DFM score}) \cong \sum_{n=1}^{\substack{violation\\count}} (dfm\ score)(failure\ rate)(design\ area)$$

Eq 1: representation of total DFM score equation

2. MOTIVATION

2.1 CMP Topography

Chemical-mechanical polishing (CMP) is a very common manufacturing step for planarization. CMP plays an important role in reducing local step height and achieving global planarization. However, CMP processes have been hampered by layout pattern dependent variations in the Cu and inter-level dielectric (ILD) thickness which can reduce yield [2]. Localized and global topography variations is one of the leading cause to impact the Lithography best focus, resulting in bridging or pinching weak points. As the CMP topography accumulated effects increases at higher layers, best focus variations will also increase [3]. Early analysis of the printability weak points based on topography profile will help to minimize the yield loss.

2.2 Requirement of topography aware DFM scoring flow

CMP topography variation can create printability challenges and hence it can impact the failure rate, primarily for the layout features at extreme topography regions. Currently the DFM Scoring methodology does not incorporate CMP topography information. This can lead to challenges in correlating DFM score with silicon yield. Layout features falling in extreme topography region should have lower DFM score with respect to the layout feature falling in average topography region.

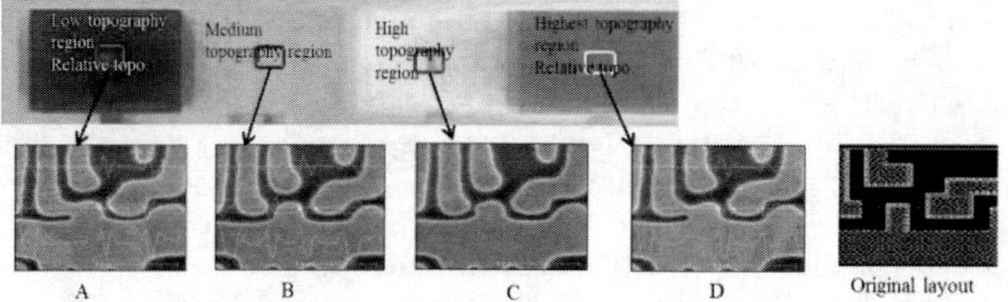

Fig 3: Figure showing the CMP topography variation vs printability weak points

Fig 3 shows same layout feature placed at different topography region. The instance A and D were placed in extreme topography region while instance B and C were placed in medium and high topography region. The silicon images shows that the instance A and D are more sensitive for bridging weak point while instance B and C are clean. This is due to impact of topography variation on printability [4].

On the other hand, since all these layout features are identical, so their DFM scores are also be the same but their silicon yield results are different. This creates mismatch between DFM score and yield. Hence to overcome this issue, there is a need for topography aware DFM scoring mechanism which can read in the topography variation and optimize the DFM score accordingly.

Similarly, it's always an interest to know critical layout weak points which are more susceptible towards failure and address them before actual tapeout to increase the yield and overall cost. Moreover it is also requirement to have appropriate DFM score sign-off criteria for full chip design as to predict the yield entitlement. Further to support these requirements, DFM scoring flow should reflect the yield trend. Finally a solution is required which can incorporate the topography variation into DFM scoring model and provide better yield correlation.

In the next section, we will present the proposed flow to incorporate Topography information in to DFM scoring flow and correlation against silicon results.

3. TOPOGRAPHY AWARE DFM SCORING FLOW

3.1 Concept and Flow Overview

In the Fig 4, we are presenting the topography aware DFM scoring flow. The flow highlights all the input and respective steps. To enable topography aware DFM scoring, the flow takes DFM and DRC rule set as input along with CMP results.

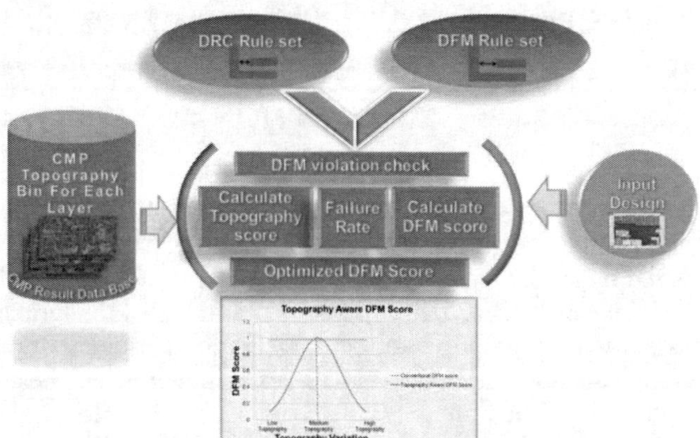

Fig 4: Figure showing topography aware DFM scoring flow overview

The flow first identifies the DFM violation based on given DFM rule set and then calculate preliminary DFM sore as described in section 1.1.

In the second step, the flow calculates the Topography score for each DFM violation using the CMP results. The CMP results contain information related to topography variation across complete design for each layer which is further divided in to various bins. For example, the region with lowest topography region binned as bin0 and so on. Along with this, CMP results also include the average related topography for respective layers. Flow analyzes the DFM violation and corresponding topography bin under which the DFM violation is falling for respective design layer. Using the corresponding topography value for that bin flow further calculates the deviation in topography with respect to average relative topography value for respective design layer. This deviation is used in topography score calculation using topography scoring model. Higher the deviation lesser be the score. Similar to DFM score, the topography score also varies from 0.0 to 1.0. Below is the graph (Fig 5) showing topography score model. We can observe that topography score dependency on relative topography variation. As the topography deviates from its relative average value the score starts to fall down.

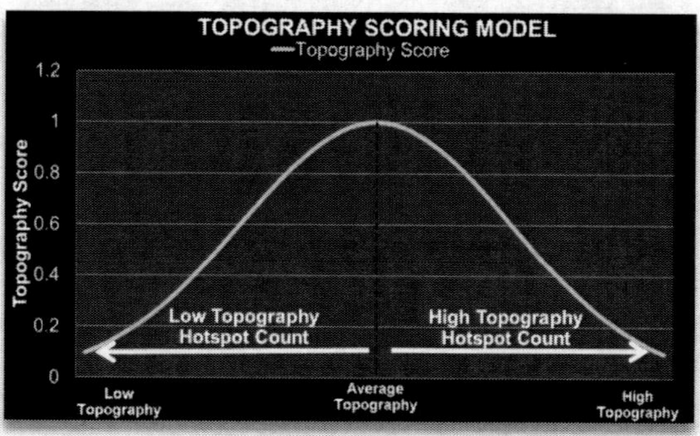

Fig 5: Representation topography scoring model

This topography score is further used to optimize the DFM score for each DFM violation. This optimization enables the topography impact on final score. Eq. 2 is the representation of topography aware DFM score model

$$f(Topography\ Aware\ DFM\ Score) \cong f(orginal\ dfm\ score)\, f(topography\ score)$$

Eq 2: representation topography aware DFM score

This score optimization enables designer to include topography aware DFM scoring. DFM violations with same violation magnitude can have different scores based on what topography region they are falling into. The DFM violations with high violation magnitude if falling under average topography region then their DFM score will improve and if DFM violations falling under extreme topography region the DFM score will degrade. Fig 6 explains the matrix defined for relationship of DFM Score with respect to DFM violation magnitude and topography variation. Green color represents the high score and red color represents low score. To achieve Higher DFM score the magnitude of DFM violation and topography variation should be less.

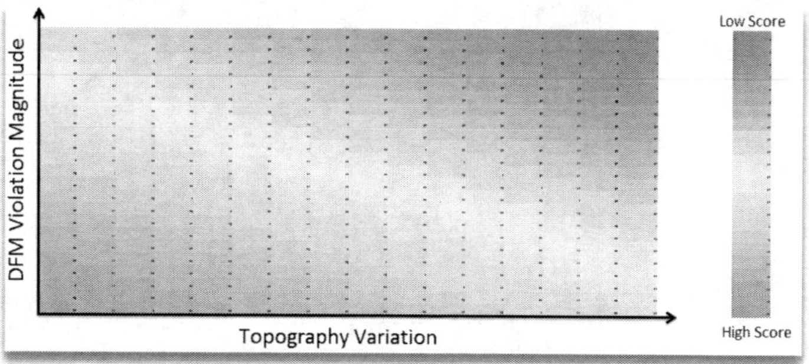

Fig 6: Figure showing DFM scoring matrix

3.2 Results

A full chip design, manufactured at GLOBALFOUNDRIES 28nm technology process has been chosen to verify the topography aware DFM scoring flow. The physical dimension of the design is 3.6mm x 3.10mm and the metal usage is from metal 1 to metal 5. The full chip CMP simulation results available for these 5 metal layers. Below is the figure (Fig 7) showing topography results for M1-M5.

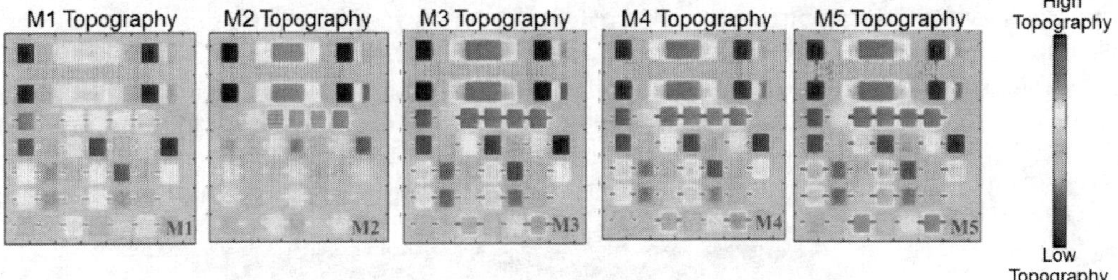

Fig 7: Figure showing topography results for Metal1 – Metal5

In this design, identical layout features were placed at different topography regions. The layout features under extreme topography region shows silicon failures and to verify the flow it is required that DFM score should show sensitivity toward topography variation.

To analyze the results we have classified some of the yield detractor patterns and compared the topography aware DFM score with conventional DFM score. Comparison results are shown below.

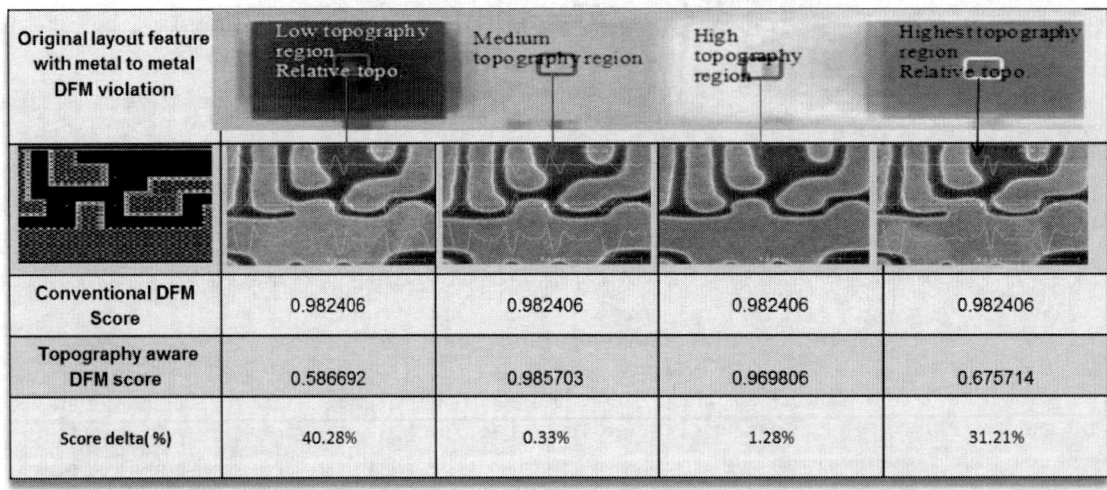

Original layout feature with metal to metal DFM violation	Low topography region Relative topo	Medium topography region	High topography region	Highest topography region Relative topo
Conventional DFM Score	0.982406	0.982406	0.982406	0.982406
Topography aware DFM score	0.586692	0.985703	0.969806	0.675714
Score delta(%)	40.28%	0.33%	1.28%	31.21%

Fig 8(a): Figure showing silicon images for different topography region and DFM scoring results

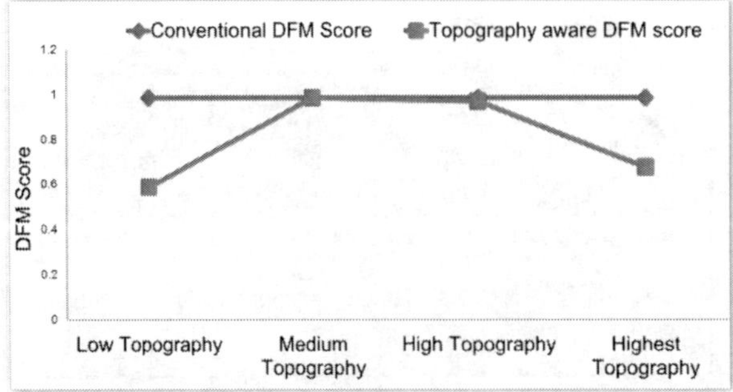

Fig 8(b): Figure showing graph between conventional and topography aware DFM Score

Fig 8(a) shows the silicon image of identical layout features placed at different topography regions and their respective topography aware DFM score and convention DFM scores. We observe that the topography aware score varies for the entire layout feature lying in different topography region while the conventional DFM score is constant for all. The layout feature lying under low and highest topography region is having up to 40% lower score in comparison to feature falling under medium topography region. Fig 8(b) shows the comparison graph to highlight the difference between their respective scores.

We further verified the flow on three different full chip designs manufactured at GLOBALFOUNDRIES 28nm technology process and we named them as designed A, B and C for reference. The yield for design B is highest and for design C is the lowest. We also performed CMP simulation on theses three design to get the CMP results and the total number of topography hotspots. Fig 9(a) to show the respective yield for design A,B and C with their corresponding topography hotspot. The topography hotspot count for design A is the highest where B is the lowest. Where as in the Fig 9(b) the total DFM score for conventional and topography aware DFM has been shown.

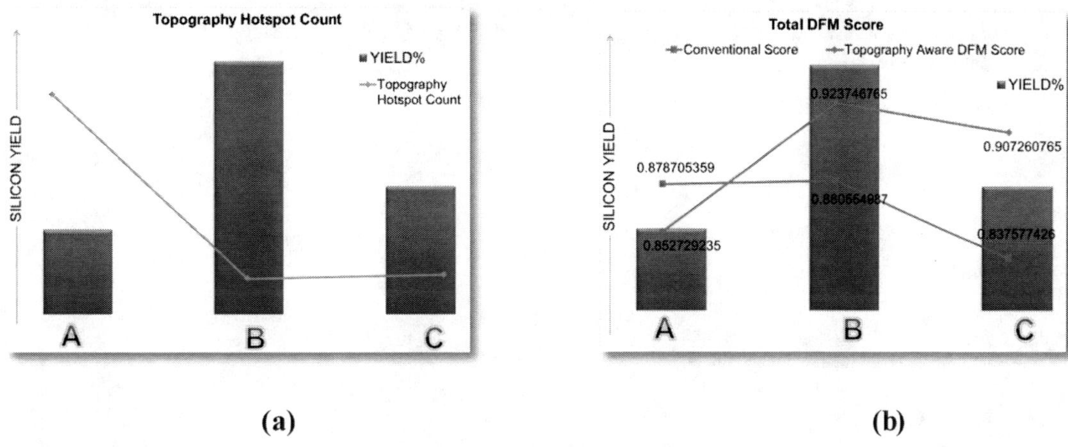

(a) (b)

Fig 9: (a) showing yield vs topography hotspot count, (b) showing plot for conventional and topography aware DFM score

We observe the conventional DFM score does not completely correlating with yield trend where as the topography aware DFM score much closer to it. This is explained due topography hotspot in design A is much higher then B and C which reduced the DFM score for design A and increase for design B and C. This meet our goal to include CMP impact on DFM score.

4. CONCLUSION

This paper presented the new topography aware DFM scoring model which has been verified on full chip design manufactured at GLOBALFOUNDRIES 28nm technology process. The results are closer to silicon yield and reflect the topography variation impact on DFM score. The results also fulfills the requirement to filter out critical layout week points with low DFM score and helps to set up more accurate DFM Sign-off score requirement for full chip designs.

5. REFERENCES

[1] Piyush Pathak, Sriram Madhavan, Shobhit Malik, Lynn T. Wang, Luigi Capodieci., "Framework for identifying recommended rules and DFM scoring model to improve manufacturability of sub-20nm layout design"

[2] Colin Hui, W.X. Bin, H. Huang, et al., "Hotspot detection and design recommendation using silicon calibrated CMP model", in Proc. SPIE Design for Manufacturability through Design-Process Integration, 72751R, 2009

[3] Aravind Narayana Samy, Roberto Schiwon, Rolf Seltmann, et al., "Challenges in process marginality for advanced technology nodes and tackling its contributors", Proc. SPIE EMLC, Vol.8886, 2013

[4] Ushasree Katakamsetty, et al., "Scanner correction capabilities aware CMP Lithography hotspot analysis ", in Proc. SPIE, Design-Process-Technology Co-optimization for Manufacturability VIII, 905312, 2014

A Compact Model to Predict Pillar-Edge-Roughness Effects on 3D Vertical Nanowire MOSFETs Using the Perturbation Method

Pu Wang, Chuyang Hong, Qi Cheng, Yijian Chen*

School of Electronic and Computer Engineering
Shenzhen Graduate School, Peking University
Shenzhen 518055, Guangdong, China

ABSTRACT

In this paper, we present a compact model to predict the pillar-edge-roughness (PER) effects on 3D vertical nanowire MOSFETs using the perturbation method. An analytic solution to 3D Poisson's equation in the cylindrical coordinate with a perturbed boundary is obtained to describe the PER effects on the vertical channel potential. The induced variations of drain current, threshold voltage (V_{th}), and sub-threshold slope (SS) are calculated using the developed model. We also investigate the PER phase and frequency dependent behavior of the nanowire MOSFETs, and find that both phase and (angular) frequency of the PER function will significantly affect the device performance. Our model calculation results are compared with TCAD simulations and a good agreement between them is found. It is suggested that our metrology society needs to develop relevant measurement methodology to characterize the nanowire pillar-edge roughness at deep nanoscale.

Keywords: 3D vertical nanowire MOSFETs, pillar-edge roughness (PER), Poisson's equation, Laplace equation, perturbation method

1. INTRODUCTION

Nanowire MOSFET is widely considered as a promising device structure for future CMOS scaling because of its excellent control of the channel and improved immunity to the short-channel effects (SCE). Among many possible nanowire MOSFET schemes, 3D vertical nanowire MOSFET is especially attractive as it provides a perspective solution to break through the physical limits of horizontal scaling. However, process variations such as pillar-edge roughness (PER) produced in the lithography and etching processes will be a serious concern as the device dimension is scaled down to sub-10nm region [1-3].

Shrinking randomly distributed pillar or contact CD/pitch has become increasingly difficult due to its extremely small process window. Even considering EUV and DSA technologies aided with a tone-reversal process, scaling pillar CD and half pitch to sub-10nm still seems to be prohibitive. PER variations can significantly change the nanowire/channel shape and potential distribution, which consequently will impact the device behavior. The line-edge roughness for planar and FinFET devices have been widely studied using analytical models [4-6] and TCAD numerical simulations [7-8]; however, little progress has been made in developing an efficient compact model capable of guiding IC engineers to understand the PER related device behavior. Statistical TCAD simulations are both expensive and time-consuming, and more importantly, only able to provide limited physical insight. Therefore, it will be beneficial to develop a compact model to describe the PER effects for future integration of 3D vertical nanowire MOSFETs.

In this paper, an analytic model to correlate PER to vertical nanowire MOSFET device variability using the perturbation method is presented. In 3D Poisson's equation, the PER effect is described by a small deviation (e.g., $t(\theta)$ as a function of the angle) of the nanowire "radius" from the ideal geometry (unperturbed cylindrical column with a constant radius), as shown in Fig. 1. Apparently, the usually assumed device symmetry and smoothness are destroyed in this condition and a new modeling methodology is necessary. Here, the perturbation method will be applied to obtain the analytic solution to 3D Poisson's equation, wherein the electric potential is divided into the zeroth-order solution and first-order correction. The zeroth-order solution corresponds to the no-PER nanowire MOSFET device behavior and the first-order correction describes the PER effect. It is found that both the (angular) frequency and phase of the PER function will affect the device behavior, and thinner gate oxide will help to reduce the PER effects.

*Phone: (+086)755-21536236, Email: chenyj@pkusz.edu.cn

Design-Process-Technology Co-optimization for Manufacturability IX, edited by John L. Sturtevant, Luigi Capodieci,
Proc. of SPIE Vol. 9427, 94270W · © 2015 SPIE · CCC code: 0277-786X/15/$18 · doi: 10.1117/12.2085919

2. VERTICAL NANOWIRE MOSFET MODEL

The device structure of a (p-type doped channel) 3D vertical nanowire MOSFET with pillar-edge roughness (PER) is shown in Fig. 1. Here, L is the channel length, t_{ox} is the thickness of the gate oxide, R is the radius of an ideal channel without PER, and $t(\theta)$ (as a function of the angle) is the PER function of the nanowire surface defined to be the deviation of the nanowire "radius" from its ideal value R. The governing 3D Poisson's equation for short-channel vertical nanowire MSOFETs in the sub-threshold region is:

$$\frac{\partial^2 \varphi}{\partial r^2} + \frac{1}{r}\frac{\partial \varphi}{\partial r} + \frac{1}{r^2}\frac{\partial^2 \varphi}{\partial \theta^2} + \frac{\partial^2 \varphi}{\partial z^2} = \frac{qN_a}{\varepsilon_{si}} \tag{1}$$

where N_a is the channel doping concentration, ε_{si} is the permittivity of silicon, φ is the electric potential.

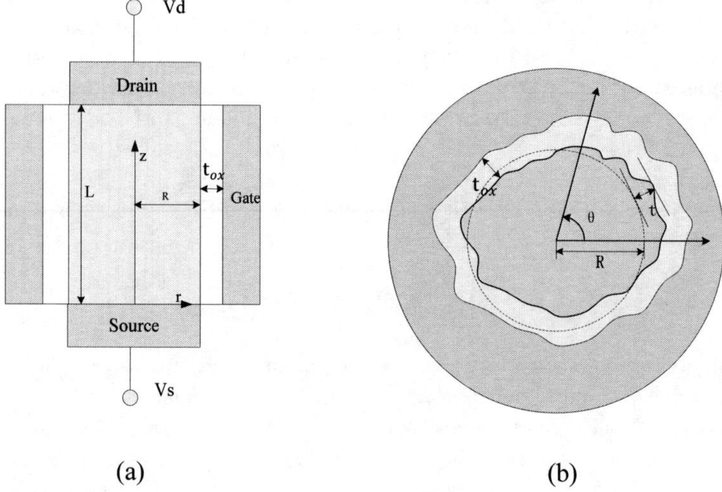

Fig. 1. (a) A schematic illumination of a vertical nanowire MOSFET, (b) the cross-section view (horizontally cutting through the nanowire) of a nanowire MOSFET with pillar-edge roughness (PER).

According to the boundary perturbation method, the electric potential φ is written as the sum of two parts: $\varphi = \varphi_0 + \varphi_1 (\varphi_1 \ll \varphi_0)$ [5-6]. Correspondingly, equation (1) can be divided into two equations (e.g., the zeroth and first orders). The zeroth-order solution (φ_0) describes the no-PER device behavior and the first-order solution (φ_1) describes the PER boundary effects. Poisson's equation (1) (with PER) is separated into two equations as below:

$$\frac{\partial^2 \varphi_0}{\partial r^2} + \frac{1}{r}\frac{\partial \varphi_0}{\partial r} + \frac{1}{r^2}\frac{\partial^2 \varphi_0}{\partial \theta^2} = \frac{qN_a}{\varepsilon_{si}} \tag{2}$$

$$\frac{\partial^2 \varphi_1}{\partial r^2} + \frac{1}{r}\frac{\partial \varphi_1}{\partial r} + \frac{1}{r^2}\frac{\partial^2 \varphi_1}{\partial \theta^2} + \frac{\partial^2 \varphi_1}{\partial z^2} = 0 \tag{3}$$

2.1 PER boundary conditions

The structure of a typical vertical nanowire MOSFET with PER is shown in Fig. 2. The PER function t is introduced to describe the channel surface roughness which is equal to zero in an ideal vertical nanowire MOSFET. A PER boundary condition considering the electric field perpendicular to the PER surface is derived as below:

$$C_{ox}\left(V_g - V_{fb} - \varphi_s\right) = \varepsilon_{si}\frac{\partial\varphi}{\partial P} = \varepsilon_{si}\sqrt{1+\left(\frac{1}{r}\frac{\partial r}{\partial\theta}\right)^2}\frac{\partial\varphi}{\partial r}\bigg|_{r=R+t} \tag{4}$$

C_{ox} is the surface density of gate oxide capacitance which is approximated by replacing the constant radius R in the oxide formula (6) with an angular function $R + t(\theta)$, φ_s is the surface potential, V_g is the gate voltage and the V_{fb} is the flat-band voltage. In deriving the above equation, the perturbed electric field is still perpendicular to the PER surface and can be calculated as:

$$\frac{\partial\varphi}{\partial P} = \sqrt{\left(\frac{\partial\varphi}{\partial r}\right)^2 + \left(\frac{1}{r}\frac{\partial\varphi}{\partial\theta}\right)^2} = \sqrt{1+\left(\frac{1}{r}\frac{\partial r}{\partial\theta}\right)^2}\frac{\partial\varphi}{\partial r} \tag{5}$$

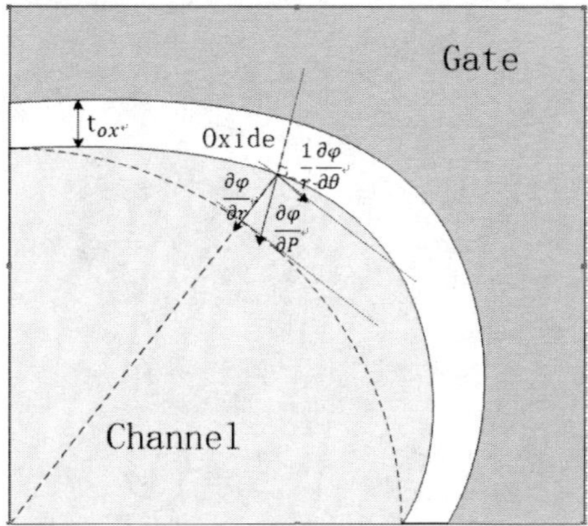

Fig. 2. A schematic illumination of the electric field on the boundary.

The zeroth-order boundary condition (with subscript "0") without PER is given as:

$$C_{ox0}\left(V_g - V_{fb} - \varphi_s\right) = \varepsilon_{si}\frac{\partial\varphi_0}{\partial r}\bigg|_{r=R}, \qquad C_{ox0} = \frac{\varepsilon_{ox}}{R\ln(1 + t_{ox}/R)} \tag{6}$$

Substituting (6) into (4), we have:

$$\frac{\partial\varphi}{\partial r}\bigg|_{r=R+t} = \frac{C_{ox}}{C_{ox0}}\frac{\left.\frac{\partial\varphi_0}{\partial r}\right|_{r=R}}{\sqrt{1+\left(\frac{1}{R+t}\frac{\partial r}{\partial\theta}\right)^2}} \tag{7}$$

Using Taylor expansion at the operating point $r = R$ and keeping the zeroth-order (with subscript "0") and first-order (with subscript "1") terms, $(\partial\varphi/\partial r)|_{r=R+t}$ in equation (4) can be approximated as:

$$\frac{\partial\varphi}{\partial r}\bigg|_{r=R+t} \approx \frac{\partial\varphi_1}{\partial r}\bigg|_{r=R} + \frac{\partial\varphi_0}{\partial r}\bigg|_{r=R} + t\frac{\partial^2\varphi_0}{\partial r^2}\bigg|_{r=R} \tag{8}$$

Form equations (7) and (8), we have:

$$\left.\frac{\partial\varphi_1}{\partial r}\right|_{r=R} \approx \frac{C_{ox}}{C_{ox0}} \frac{\left.\frac{\partial\varphi_0}{\partial r}\right|_{r=R}}{\sqrt{1+\left(\frac{1}{r}\frac{\partial r}{\partial\theta}\right)^2}} - \left.\frac{\partial\varphi_0}{\partial r}\right|_{r=R} - t\left.\frac{\partial^2\varphi_0}{\partial r^2}\right|_{r=R} \tag{9}$$

2.2 The zeroth-order solution

The zeroth-order solution φ_0 of equation (2) corresponds to the potential distribution in an ideal vertical nanowire MOSFETs (without PER), as shown in Fig. 3. A parabolic solution can be assumed in the sub-threshold region [9].

$$\varphi_0(r,z) \approx C_0(z) + C_1(z)r + C_2(z)r^2 \tag{10}$$

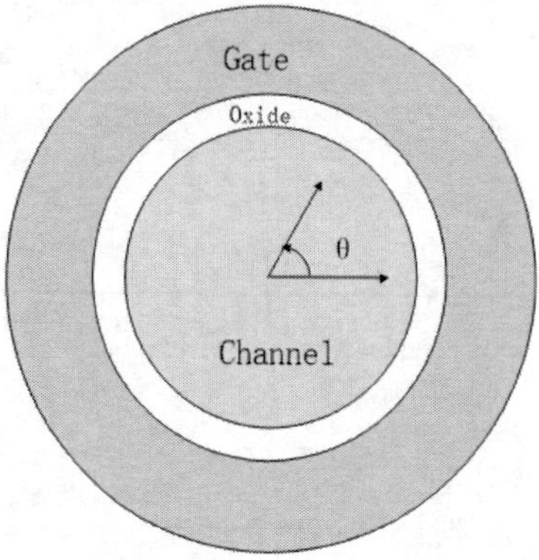

Fig. 3. The cross-section view of an ideal vertical nanowire MOSFET with a smooth channel surface.

The boundary conditions are listed as below:

1) The central potential φ_c is a function of z only:

$$\varphi_0(0,z) = \varphi_c(z) = C_0(z)$$

2) The electric field at the center of nanowire is zero.

$$\left.\frac{\partial\varphi_0}{\partial r}\right|_{r=0} = 0 = C_1(z)$$

3) The oxide-interface boundary condition is:

$$\varepsilon_{si}\left.\frac{\partial\varphi_0}{\partial r}\right|_{r=R} = \frac{\varepsilon_{ox}\left(V_g - V_{fb} - \varphi_s\right)}{Rln\left(1 + t_{ox}/R\right)} = 2\varepsilon_{si}RC_2(z)$$

The resultant solution of $\varphi_0(r,z)$ is:

$$\varphi_0(r,z) = \varphi_c(z) - \left(\frac{\varepsilon_{ox}r^2\left(\varphi_c(z) - V_g + V_{fb}\right)}{2\varepsilon_{si}R^2ln\left(1 + t_{ox}/R\right) + \varepsilon_{ox}R^2}\right) \tag{11}$$

Inserting (11) back into (3), Poisson's equation can be solved at the nanowire center $r = 0$.

$$\frac{\partial^2 \varphi_c(z)}{\partial r^2} + \frac{V_g - V_{fb} - \varphi_c(z)}{\lambda^2} = \frac{qN_a}{\varepsilon_{si}}$$

where $\lambda = \sqrt{[2\varepsilon_{si}R^2 ln\ (1 + t_{ox}/R) + \varepsilon_{ox}R^2]/4\varepsilon_{ox}}$. With the boundary conditions $\varphi_c(0) = V_{bi}$ and $\varphi_c(L) = V_{bi} + V_{ds}$, we can obtain the central potential:

$$\varphi_c(z) = V_{SL} + (V_{bi} - V_{SL})\frac{sinh\left(\frac{L-z}{\lambda}\right)}{sinh\left(\frac{L}{\lambda}\right)} + (V_{bi} + V_{ds} - V_{SL})\frac{sinh\left(\frac{z}{\lambda}\right)}{sinh\left(\frac{L}{\lambda}\right)} \tag{12}$$

where $V_{SL} = V_g - V_{fb} - \frac{qN_a}{\varepsilon_{si}}\lambda^2$ and $V_{bi} = \frac{E_g}{2q} + \frac{KT}{q} ln\left(\frac{N_a}{n_i}\right)$.

2.3 The first-order correction

The 3D Laplace equation (3) in the cylindrical coordinate wherein the doping term is removed can be solved by the method of separation of variables. It assumes that $\varphi_1(r, \theta, z)$ can be written as:

$$\varphi_1(r, \theta, z) = F(r)G(\theta)H(z) \tag{13}$$

Substituting it into (3) we have:

$$r^2\frac{F''(r)}{F(r)} + r\frac{F'(r)}{F(r)} + \frac{G''(\theta)}{G(\theta)} + r^2\frac{H''(z)}{H(z)} = 0$$

The above equation can be satisfied only if:

$$\frac{G''(\theta)}{G(\theta)} = -[r^2\frac{F''(r)}{F(r)} + r\frac{F'(r)}{F(r)} + r^2\frac{H''(z)}{H(z)}] = -m^2$$

$$\frac{H''(z)}{H(z)} = -\beta^2$$

where m and β are constants. The above equations can be reformulated as:

$$G''(\theta) + m^2\, G(\theta) = 0$$
$$H''(z) + \beta^2\, H(z) = 0$$
$$r^2F''(r) + rF'(r) + (-\beta^2r^2 - m^2)F(r) = 0$$

Notice that φ_1 corresponds to different boundary conditions from the zeroth-order solution φ_0.

$$\varphi_1|_{z=0} = 0; \ \varphi_1|_{z=L} = 0 \tag{14}$$

And the other two boundary conditions are:

$$\varphi_1|_{\theta=0} = \varphi_1|_{\theta=2\pi} \tag{15}$$

$$\varphi_1|_{r=0} = \ finite\ value \tag{16}$$

Using the boundary conditions described above, $F(r)$, $G(\theta)$ and $H(z)$ can be solved analytically [10]:

$$G(\theta) = A_m \cos m\theta + B_m \sin m\theta$$
$$H(z) = D_n sin\beta_n, \quad \beta_n = \frac{n\pi}{L}$$
$$F(r) = E_{mn} I_m(\beta_n r)$$

where I_m is the modified Bessel function of the first kind. Therefore, the first-order potential can be expressed as:

$$\varphi_1(r,\theta,z) = \sum_{n=1}^{\infty}\sum_{m=0}^{\infty} I_m\left(\frac{n\pi}{L}r\right) sin\frac{n\pi}{L}z\left(A_{mn}\cos m\theta + B_{mn}\sin m\theta\right)$$

where E_{mn}, D_n, A_m, B_m are constants, and $A_{mn} = E_{mn}A_mD_n$, $B_{mn} = E_{mn}B_mD_n$. Apply the above form of the first-order solution to the oxide-interface boundary condition (9), we obtain:

$$\frac{C_{ox}}{C_{ox0}}\frac{\left.\frac{\partial\varphi_0}{\partial r}\right|_{r=R}}{\sqrt{1+\left(\frac{1}{r}\frac{\partial r}{\partial\theta}\right)^2}} - \left.\frac{\partial\varphi_0}{\partial r}\right|_{r=R} - t\left.\frac{\partial^2\varphi_0}{\partial r^2}\right|_{r=R}$$
$$= \sum_{n=1}^{\infty}\sum_{m=0}^{\infty}\frac{n\pi}{2L}\left[I_{m-1}\left(\frac{n\pi}{L}R\right) + I_{m+1}\left(\frac{n\pi}{L}R\right)\right] sin\left(\frac{n\pi}{L}z\right)\left[A_{mn}\cos(m\theta) + B_{mn}\sin(m\theta)\right]$$

where A_{mn}, B_{mn} can be calculated from the below integrals:

$$A_{mn} = \frac{\frac{2}{L}\int_0^L\left[\frac{1}{\pi}\int_0^{2\pi} f(\theta,z)\cos m\theta\, d\theta\right] sin\frac{n\pi}{L}zdz}{\frac{n\pi}{2L}\left[I_{m-1}\left(\frac{n\pi}{L}R\right) + I_{m+1}\left(\frac{n\pi}{L}R\right)\right]}$$
$$B_{mn} = \frac{\frac{2}{L}\int_0^L\left[\frac{1}{\pi}\int_0^{2\pi} f(\theta,z)\sin m\theta\, d\theta\right] sin\frac{n\pi}{L}zdz}{\frac{n\pi}{2L}\left[I_{m-1}\left(\frac{n\pi}{L}R\right) + I_{m+1}\left(\frac{n\pi}{L}R\right)\right]}$$

2.4 Sub-threshold drain current model

In this sub-section, the drain-current model will be developed by a widely adopted approach [11-12]. It assumes that the electron quasi-Fermi potential $V(z)$ is a function of z (channel direction) only. The drain current can be expressed as:

$$I_{ds} = \mu q Q_{inv}\frac{dV(z)}{dz} \tag{17}$$

where Q_{inv} is the number density (per unit length in the channel direction) of the inversion/mobile charge. It should be reminded that the main path of leakage current in a vertical nanowire MOSFET (in the subthreshold region) is located at the center of the channel [13]. Thus the fluctuation of the inversion charge near the silicon/oxide interface due to the PER effect is ignored even its perturbation on the channel potential is taken into account. Therefore, an approximation to calculate Q_{inv} is made:

$$Q_{inv} = \frac{n_i^2}{N_a}\int_o^{2\pi}\int_o^{R+t(\theta)} e^{\frac{q}{kT}(\varphi_0+\varphi_1-V)}\, r\, dr\, d\theta \approx \frac{n_i^2}{N_a}e^{-\frac{q}{kT}V}\int_o^{2\pi}\int_o^{R} e^{\frac{q}{kT}(\varphi_0+\varphi_1)}\, r\, dr\, d\theta$$

(17) can be simplified using Taylor expansion ($\varphi_1 \ll \varphi_0$) as:

$$I_{ds}dz = \mu q \frac{{n_i}^2}{N_a} e^{-\frac{qV}{kT}} \int_o^{2\pi} \int_o^R e^{\frac{q}{kT}(\varphi_0+\varphi_1)}\, r\, dr\, d\theta\, dV \approx \mu q \frac{{n_i}^2}{N_a} e^{-\frac{qV}{kT}} \int_o^{2\pi} \int_o^R (1+\Delta) e^{\frac{q}{kT}\varphi_0}\, r\, dr\, d\theta\, dV$$

$$\Delta = \frac{q}{kt}\varphi_1 + \frac{1}{2}\left(\frac{q\varphi_1}{kt}\right)^2 + \cdots$$

Integrating the above equation in z direction allows us to write I_{ds} as the sum of I_{dso} (ideal drain current without considering the PER effects) and ΔI_{ds} (deviation of the actual drain current from I_{dso}):

$$I_{dso} = \frac{\mu q \frac{{n_i}^2}{N_a} \int_0^{V_{ds}} e^{-\frac{qV}{kT}} dV}{\int_0^L \frac{1}{\int_o^{2\pi} \int_o^R e^{\frac{q}{kT}\varphi_0}\, r\, dr\, d\theta} dz} \tag{18}$$

and

$$\Delta I_{ds} = \frac{\mu q \frac{{n_i}^2}{N_a} \int_0^{V_{ds}} e^{-\frac{qV}{kT}} dV}{\int_0^L \frac{1}{\int_o^{2\pi} \int_o^R \Delta\, e^{\frac{q}{kT}\varphi_0}\, r\, dr\, d\theta} dz} \tag{19}$$

3. MODEL RESULTS AND DISCUSSION

3.1 Device parameter setting

Table 1. Parameters used in our TCAD Simulations.

Node (nm)	15	10	7
EOT (nm)	0.64	0.59	0.5
Na (cm^-3)	1e18	1e17-1e18	1e18
R (nm)	3	2.5	2
A (nm)	1	0.8	0.6

In this section, model calculations and TCAD simulations for three nodes (15nm, 10nm, 7nm) are compared and the device parameters are listed in Table 1. The PER function t is described by a sine function with various amplitudes:

$$t(\theta) = A\, sin(h \times \theta)$$

where A is the amplitude of a PER function, and $2\pi/h$ is the (angular) period, i.e., $f = h/2\pi$ is the (angular) frequency. Therefore, the physical meaning of $h = 2\pi f$ is the number of (angular) periods contained within the whole circular surface of a nanowire.

3.2 Analysis of model accuracy

Figs. 4-5 show the calculated I_{ds} and ΔI_{ds} verified with TCAD simulations for different t_{ox} at several nodes (15-7 nm). A good agreement between model results and TCAD simulations is evident. However, the model accuracy is lower at smaller nodes (e.g., 7 nm). It is shown in Fig. 4 that a negative PER function leads to a steeper sub-threshold slope (SS) as a thinner channel can reduce the leakage current and simultaneously improve the SS value. It is also found that thinner gate oxide will help to reduce the PER effect.

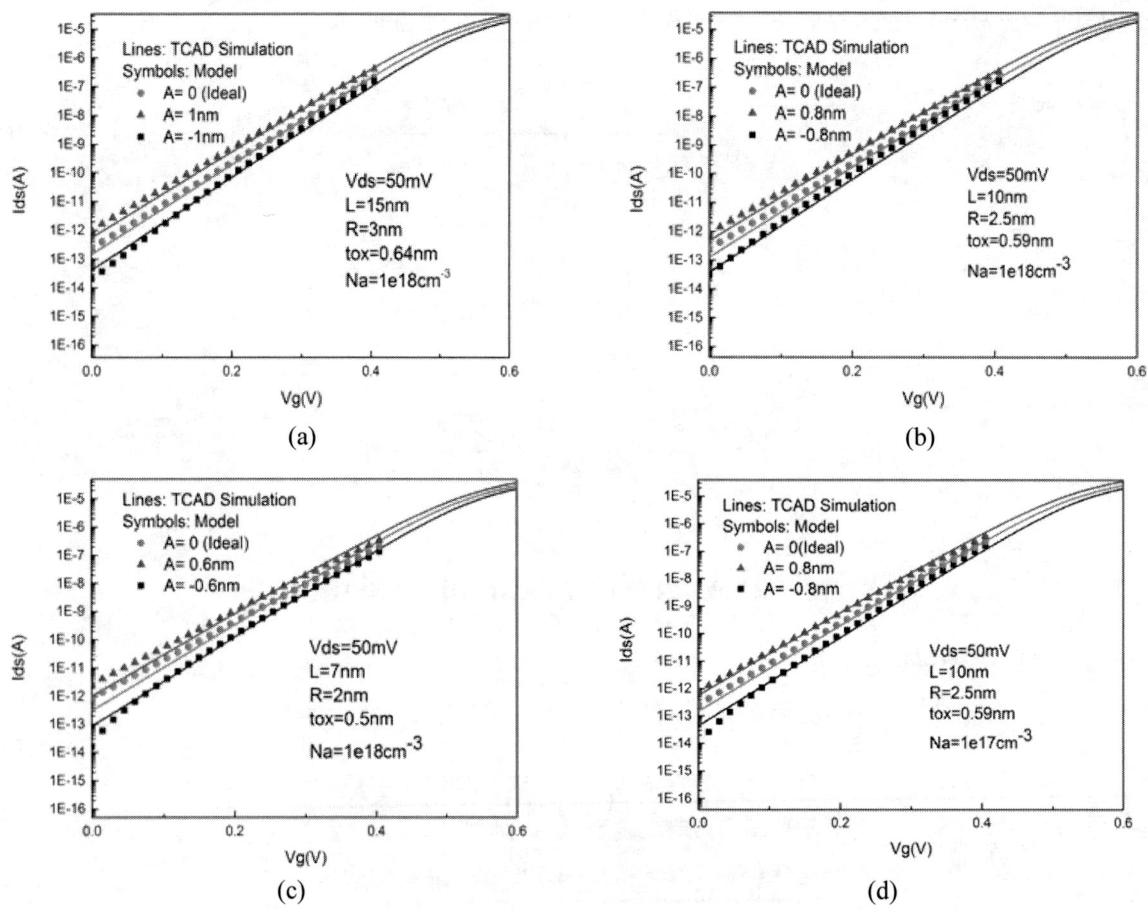

Fig. 4. PER effect incorporated Ids for different amplitudes and doping concentrations (h=0.5). (a) L=15nm, Na=1e18cm^{-3}, (b) L=10nm, Na=1e18cm^{-3}, (c) L=7nm , Na=1e18cm^{-3}, (d) L=10nm, Na=1e17cm^{-3}.

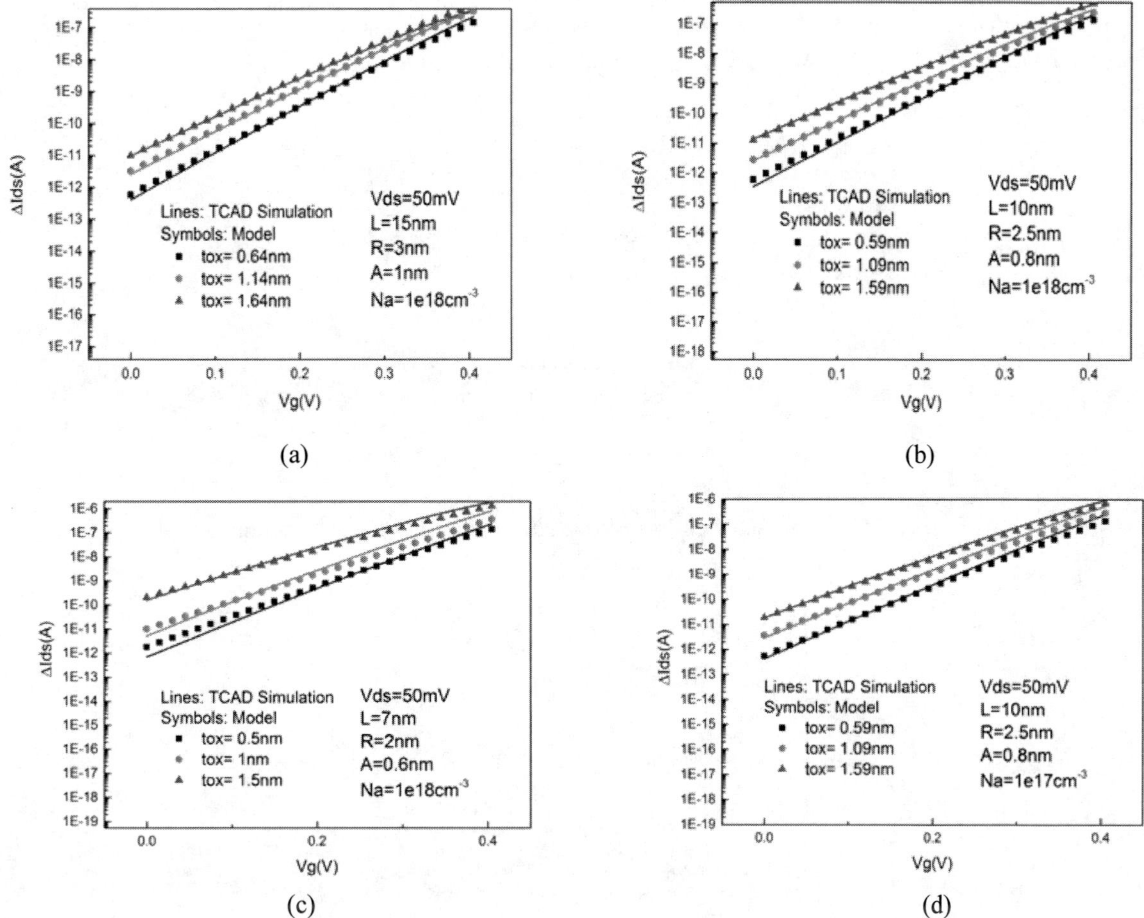

Fig. 5. PER effects induced ΔIds for different amplitudes and doping concentrations (h=0.5). (a) L=15nm, Na=1e18cm^{-3}, (b) L=10nm, Na=1e18cm^{-3}, (c) L=7nm ,Na=1e18cm^{-3}, (d) L=10nm, Na=1e17cm^{-3}.

Fig. 6 shows the calculation results and TCAD simulations of ΔI_{ds} for varying angular frequency h (h=0.1-5). In this paper, we only examined the relatively low-frequency spectrum as the high-frequency components are often considered to be non-critical in MOSFET devices. It is found that PER induced sub-threshold current peaks at certain (angular) frequency and gradually decays when the spectrum moves to the lower- and higher-frequency regions. Fig. 7 shows the threshold voltages V_{th} ($V_{ds} = 50\ mV$) extracted from our model and by TCAD simulation at various nodes. V_{th} is defined to be the gate voltage at which the sub-threshold current reaches $300nA \times 2\pi R/L$ [6]. A "relative error" index of the threshold voltage is also defined as below to quantify the model accuracy; and apparently, the relative error of threshold voltage increases when the node CD is scaled down.

$$Relative\ error = \frac{\sum|V_{th}(TCAD) - V_{th}(model)|}{\sum V_{th}(TCAD)}$$

In Fig. 8, the SS values calculated using our model are compared with those obtained from the TCAD simulations for three different A values (1nm, 0, -1nm) and at three nodes (15nm, 10nm, 7nm). We can see that SS values predicted by this model is less accurate when the channel length becomes shorter. More importantly, it confirms again that both phase and frequency of the PER function will affect the device behavior, and it is important for the metrology society to develop relevant measurement technologies to characterize PER function and extract both information.

In general, when the PER effect is taken into account, the conventional model (without PER effect) is not accurate enough, especially at smaller nodes (e.g., 10nm and 7nm). Therefore a more accurate compact model incorporating the PER effect is critical for future 3D nanowire MOSFET device simulation and IC design.

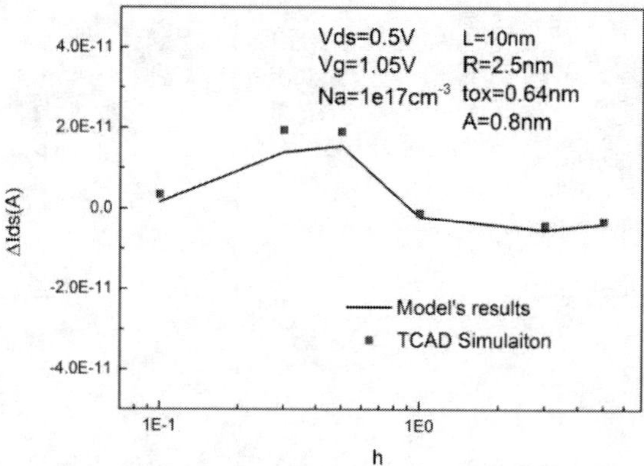

Fig. 6. ΔI_{ds} vs. h calculated from our model and by TCAD simulation.

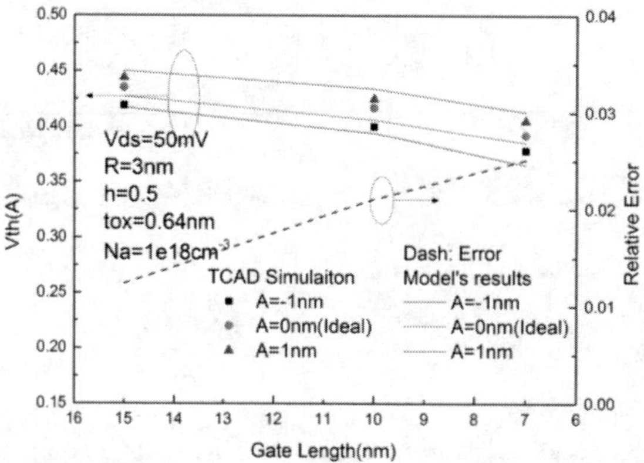

Fig. 7. V_{th} extracted from our model and TCAD simulation for three A values (1nm, 0, -1nm) at three nodes (15nm, 10nm, 7nm). The dashed line is the relative error.

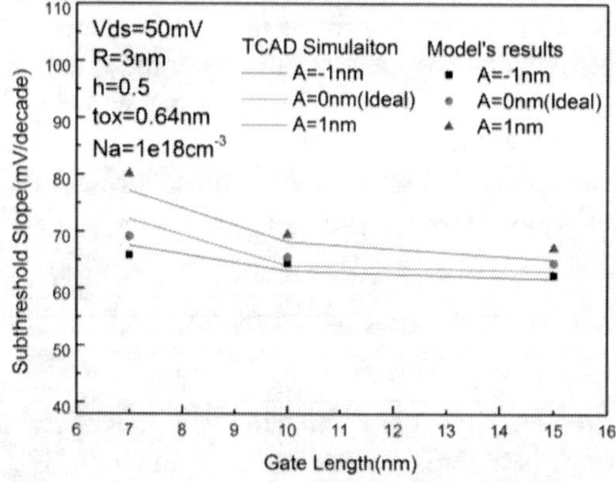

Fig. 8. The subthreshold slopes for three different amplitudes (1nm, 0, -1nm) at three nodes (15nm, 10nm, 7nm) compared with TCAD simulation results.

4. CONCLUSION

In this paper, a compact model to predict the PER effects on 3D vertical nanowire MOSFETs has been developed using the perturbation method. The analytic solution (of the electric potential) to 3D Laplace's equation in the cylindrical coordinate is obtained to calculate the PER induced variations of drain current. It is found that both phase and frequency of the PER function will significantly affect the device behavior, e.g., a negative PER function leads to a thinner channel and steeper sub-threshold slope (SS). Our calculation and TCAD simulation also show that thinner gate oxide will help to reduce the PER effect. The conventional model (e.g., without PER effect) in general is not accurate enough to describe the 3D device variability due to process variations such as PER, especially at smaller nodes. Therefore, a more accurate compact model incorporating the PER effect is required for 3D nanowire MOSFET device simulation and IC design.

ACKNOWLEDGEMENT

This research is supported by the Shenzhen City Fund for Fundamental Research and the start-up research fund from Shenzhen Graduate School of Peking University.

REFERENCES

[1] C. H. Diaz, H. J. Tao, Y. C. Ku, A. Yen, and K. Young, "An experimentally validated analytical model for gate-line-edge roughness (LER) effects on technology scaling," *IEEE Electron Device Lett.*, vol. 22, no. 6, pp. 286-289, June, 2001.

[2] ITRS report-Lithography, 2013 Edition.

[3] B. Yang, K. D. Buddharaju, S. H. G. Teo, N. Singh, G. Q. Lo, D. L. Kwong, "Vertical silicon-nanowire formation and gate-all-around MOSFET," *IEEE Electron Device Lett.,* vol. 29, no. 7, pp. 791-794, July. 2008.

[4] N. Kenji, "Compact modeling of ballistic nanowire MOSFETs," *IEEE Trans. Electron Devices,* vol. 55, no. 11, pp. 2877-2885, Nov.2008.

[5] Q. Cheng, W. Kang, Y. Chen, "Compact modeling of Fin-width roughness induced FinFET device varibility using the perturbation method," *Proc. SPIE*, vol. 8684, pp. 86840I-86840I, March, 2013.

[6] Q. Cheng, J. You, Y. Chen, "A Generalized Mode to Predict Fin-Width Roughness Induced FinFET Device Variability Using the Boundary Perturbation method," Proc. SPIE, vol. 9053, pp. 9030U-9030U, March, 2014.

[7] E Baravelli, M Jurczak, N Speciale, K D Meyer, and A Dixit, "Impact of LER and random dopant fluctuations on FinFET matching performance," *IEEE Transactions on Nanotechnology*, vol. 7, no. 3, pp. 291-298, May, 2008.

[8] K Patel, T-J K Liu and C J. Spanos, "Gate line edge roughness model for estimation of FinFET performance variability," *IEEE Trans. Electron Devices*, vol. 56, no. 12, pp. 3055-3063, Dec, 2009.

[9] C. P. Auth, J. D. Plummer, "Scaling Theory for Cylindrical Fully-Depleted, Surrounding-Gate MOSFETs," *IEEE Electron Device Lett.*, vol. 18, no. 2, pp. 0741-3106, Feb, 1997.

[10] A. M. Wazwaz, *Partial Differential Equations and Solitary Waves Theory*, 1st ed. Higher Education Press, 2009, pp. 259-265.

[11] J. P. Colinge, *FinFETs and Other Multi-Gate Transistors*, 1st ed., Berlin, Springer-Verlag, pp. 138-142, 2007.

[12] S. Yu, Y. Zhao, Y. Song, G. Du, J. Kang, R. Han, X. Liu, "3-D simulation of geometrical variations impact on nanoscale FinFETs," *IEEE Solid-State and Integrated-Circuit Technology*, vol. 29, no. 7, pp. 791-794, July 2008.

[13] X. Liang, Y. Taur, "A 2-D analytical solution for SCEs in DG MOSFETs," *IEEE Trans. Electron Devices*, vol. 51, no. 9, pp. 138-139, Sep. 2004.

Efficient Etch Bias Compensation Techniques for Accurate On-wafer Patterning

Mohamed Salama[a] and Ayman Hamouda[b]

[a]GLOBALFOUNDRIES Inc., Malta, NY, 12020

[b]GLOBALFOUNDRIES Inc., Hopewell Junction, NY, 12533

ABSTRACT

As technology development advances into deep submicron nodes, it is very important not to ignore any systematic effect that can impact CD uniformity and the final parametric yield. One important challenge for OPC is in choosing the proper etch process correction flow to compensate for design-to-design etch shrink variations. Although model-based etch compensation tools have been commercially available for a few years now, rules-based etch compensation tables have been the standard practice for several nodes. In our work, we study the limitations of the rules-based etch compensation versus model-based etch compensation. We study a 10nm process and provide the details of why using Model-Based Etch Process Correction can achieve up to 15% improvement in final CD uniformity. We also provide a systematic methodology for identifying the proper etch correction technique for a given etch process and assessing the potential accuracy gain when switching to the model-based etch correction.

Keywords: OPC, Model based Etch Correction, Etch Modeling, Rule Based Etch Correction

INTRODUCTION

Design scaling is one of the most critical goals in the Semiconductor technology. This is becoming very challenging in the deep submicron nodes, where the design-to-wafer fidelity cannot be achieved without applying very aggressive Resolution Enhancement Techniques (RET) and Optical Proximity Correction (OPC) flow [1]. A very challenging step in the OPC and design to mask flows would be etch process correction where final patterns CD (Critical Dimension) on wafer variations is compensated for. As technology advances, the challenge for OPC engineers is not to hit their post litho targets anymore but rather to converge on their final post etch CD due to systematic etch induced variations. That is why OPC, litho and etch engineers should work closely together to make sure the proper etch compensation technique is applied.

Etch process compensation can be done using a constant etch bias or variable etch bias, and it can be model based or rule based. In older technologies, the standard was to apply either a constant bias or a table based rule set bias. As technology advances with dimensions shrinking and densities increasing, we are seeing the necessity of using etch models for not only verification but also compensation due to the benefits seen in the CD uniformity even with the complexity of its implementation and the turnaround time penalty seen in the model based solution. In our study, we are showing the challenges facing the rules-based etch compensation and why it is becoming essential to move to a model based approach.

Design-Process-Technology Co-optimization for Manufacturability IX, edited by John L. Sturtevant, Luigi Capodieci,
Proc. of SPIE Vol. 9427, 94270X · © 2015 SPIE · CCC code: 0277-786X/15/$18 · doi: 10.1117/12.2085956

Figure 1: Etch Process Correction Flow

Shown in Figure 1 is a common etch process correction flow where an etch process is characterized using a comprehensive set of test patterns that reflect what the input design would be like. Wafer data is then analyzed and a decision is made on whether to use a rule based solution versus a model based one based on the layer and etch process characterized and the data analyzed. The accuracy and success of model based etch correction approach isn't in question here as shown in previous publications [2] [3]. Our study here is under the assumption of having an accurate wafer verified etch model.

Etch Process Correction: Advanced Nodes Requirements

In advanced nodes, etch bias error tolerance and final CDU (Critical Dimension Uniformity) variation allowance is very limited as the final target CD for patterns is lower in size, so the more the variations are the higher the variations % normalized to target CD is. Also, the etch bias variations in nm's are higher as we show in our study compared to older technologies as seen in [5]. This adds to our goal of ensuring the control of any induced etch variations on wafer, and that drives to study the accuracy of rule based versus model based etch compensation. Etch bias mainly depends on two factors: 1-The aperture effect and 2-The microloading effect.

The aperture effect mainly depends on the width and space seen by a given pattern, as for the microloading effect it mainly depends on the pattern density seen. It is widely accepted and well understood that etch processing impacts CD variability through microloading [4]. Rule Based Etch compensation for a given pattern is done per edge based on the polygon's width and space seen by the edge which covers the aperture effect only and doesn't take the density effect into consideration. As feature size and dimensions decrease, the density effect increases compared to the Aperture effect; hence, the width and space of the polygon/edge are not the dominant factor that defines the etch bias any more. This creates a limitation on efficient usage of rule based compensation and results in higher final CDU variations. In our experiments in 10nm layers, we are studying the pattern density effect on different etch processes for one-dimensional and two-dimensional structures and how with the application of model based etch compensation we can achieve higher CDU versus rule based etch compensation.

Experimental Results

The results of the study for the different 10nm etch processes will be divided into

- One-dimensional pattern challenges.
- Two-dimensional pattern challenges.
  - Metal two-dimensional Features.
  - Vias and Contacts.

Figure 2: 1D test patterns for etch study

In our 10nm experiment, an etch model that was calibrated and verified to match the etch process under study was used. Different test patterns were designed as shown in Figure 2 and passed to the etch model as an input to estimate the etch bias of these designs. The width and space seen by a given polygon edge were kept constant; this was done to make sure the aperture effect is the same for the different patterns as we are studying only the etch variations induced by density effects here. The etch bias and final CD variations were observed for this given structure of same width and space in the different densities and the normalized CD variation % was plotted as shown in Figure 3.

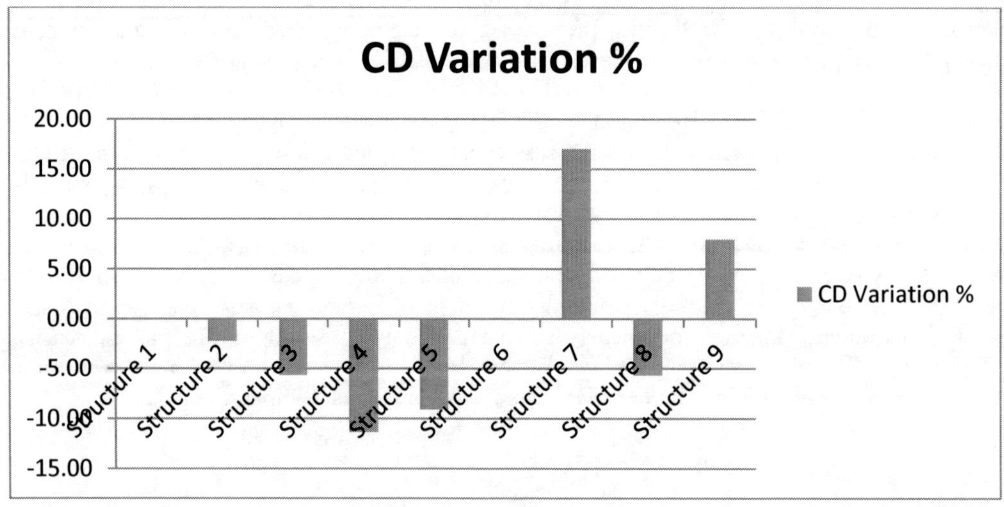

Figure 3: Final post etch CD Variation % for 1D structures

Since the edge of the relevant structure sees the same width and space, it would have gotten the same etch process compensation bias value if we use a rule based etch table. The results plotted using model simulations show that this isn't optimal since the range of final CD % variations can be up to 28% post Etch, which is beyond error tolerance and isn't acceptable for 10 nm technology. That proves the concept and shows that it is necessary to use model based etch compensation to avoid this high % variation, and it also shows that rule based Etch compensation can't be used in this given etch process as it would apply the same compensation bias for this feature regardless of the neighboring density.

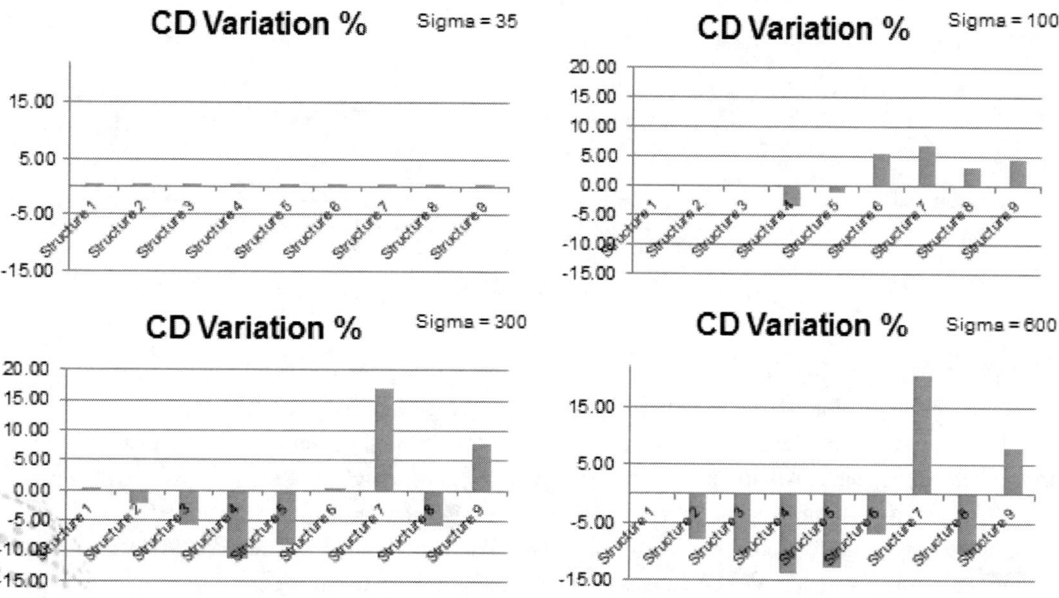

Figure 4: Final post etch CD Variation % for 1D structures for different etch processes

Figure 4 shows the normalized CD variations % plot of the same exact experiment of maintaining the same width and space seen by the given structure and observing the etch bias for it in different densities. This was done on 4 different etch processes each of which has its own etch microloading effect range (Sigma). Sigma is a parameter that is used to express how much an etch process is dominated by density and microloading effects; the higher the sigma is the more the effect of density is on your etch bias.

Based on the final CDU results for the different etch processes, it was noted that as sigma increases CD variations % error range and standard deviation increases. Density effects that lead to higher CDU are directly proportional to Sigma as shown in Figure 5. Hence, a way of predetermining whether it is essential to use an Etch model versus a rule based table would be noticing sigma values in the etch model that is calibrated and matches the etch process. The higher the sigma the more essential the usage of etch model for compensation is.

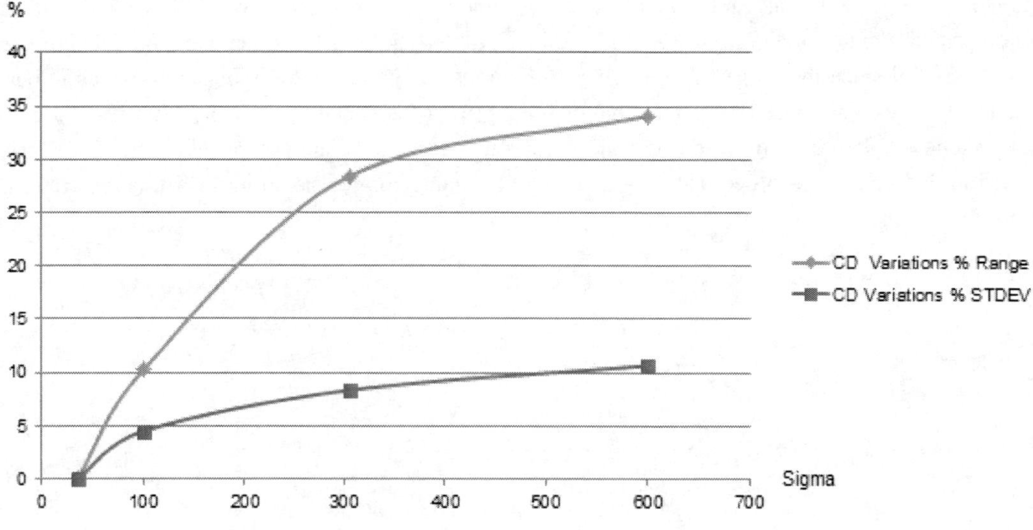

Figure 5: Sigma effect on CD variations %

One-Dimensional RB Etch compensation is done per edge based on width and space seen by a given edge. That creates a limitation for RB two-Dimensional etch compensation and the ability of taking into consideration different design patterns and density effects based on tip to tip, tip to side, length, width and space of the structure. This is another factor that shows the importance of using an Etch model for compensation; namely, the benefits it has when it comes to two-dimensional compensation. 10nm Metal Line-end etch behavior was simulated using a calibrated and wafer verified etch model for different line end widths. The data shows that the etch bias significantly varies as a function of the line end width as seen in Figure 6 and Figure 7.

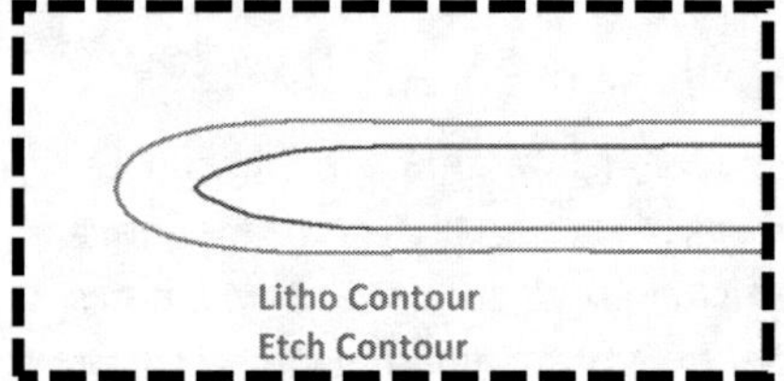

Figure 6: 10nm Metal line end etch bias simulation

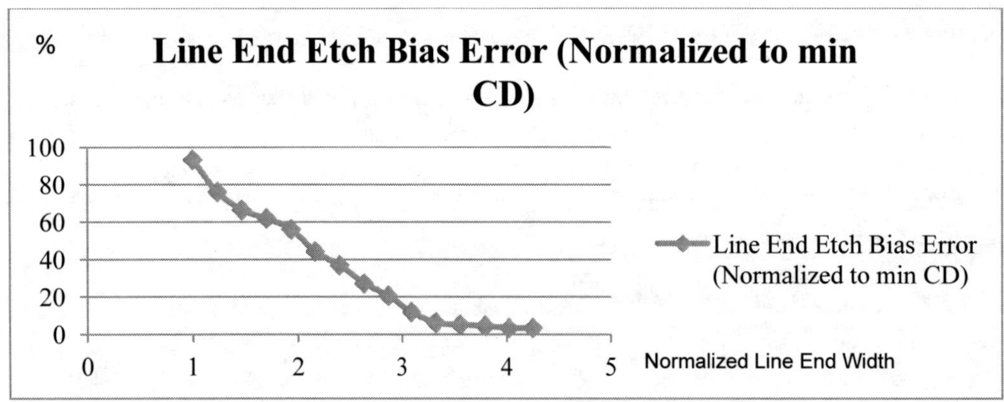

Figure 7: 10nm Metals line end etch behavior

The 2nd two-dimensional experiment was an etch bias behavior study done on a 10nm via layer to assess the need of using an etch model for compensation versus a constant bias as it isn't easy to map via structures and patterns into rule based etch compensation table or rule sets due to the complexity of the design variations. Via arrays can vary from orthogonal, staggered among other. Examples of arrays of patterns used are shown below in Figure 8.

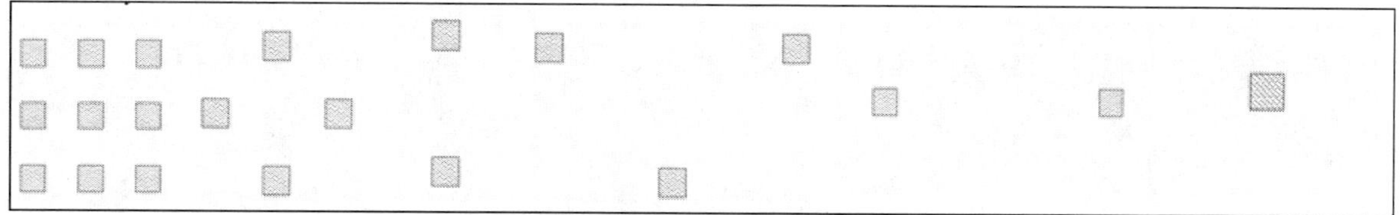

Figure 8: 2D via test patterns for etch study

Etch bias data collected from these structures and final via are post etch was calculated and plotted as shown in Figure 9. Final post etch area variation range is 28.7% which isn't acceptable. This shows the essential need of using Etch model compensation in 10nm technology for via like layers.

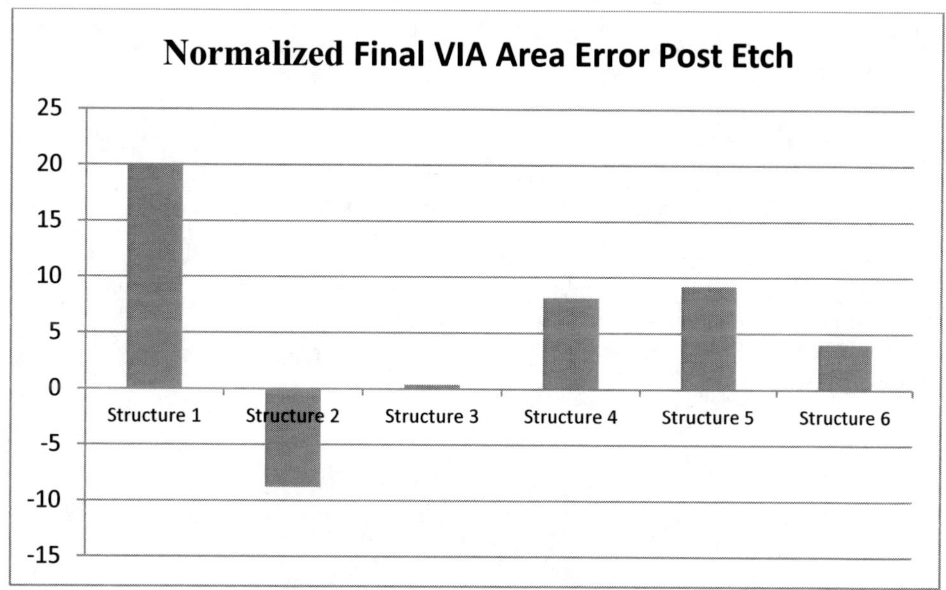

Figure 9: 10nm vias final post etch area error

CONCLUSION

A systematic study was done on etch process correction limitations and challenges for advanced nodes. In 10nm experiments, it was noted that it is essential to consider etch model compensation to achieve required final post etch CDU accuracy and meet CDU variation % tolerance. In one-dimensional patterns, the experiments show final CD % variation up to 28.4% in the absence of Etch model compensation. Final post etch results also show high dependence on the micro-loading effect range (Sigma). Sigma value can be an indicator whether model based etch compensation is necessary or not. In two-dimensional structures, experiments show for via like layers final post etch area error up to 28.7% and it also shows for a significant line end etch bias variation for metal like layers.

Due to the complexity and inability of coming up with a robust rule based Etch compensation solution that covers both density and aperture effects in one dimensional and two dimensional structures, model based Etch compensation usage is necessary in advanced nodes. Using rule based etch compensation only can have high risks.

REFERENCES

[1] Ayman Hamouda; Mohamed Salama; "Efficient model-based dummy-fill OPC correction flow for deep sub-micron technology nodes" Proc. SPIE 9235, Photomask Technology 2014, 92351W (September 16, 2014).

[2] Liu,Q., Cheng,R., Zhang, L., "Study of model based etch bias retarget for OPC", *Proc. SPIE* 7640, 76402T (2010).

[3] Dunn,D.D., Mansfield,S., Stobert,I., et al., "Etch aware optical proximity correction: a first step toward integrated pattern engineering", *Proc. SPIE* 7274, 727412 (2009).

[4] Yuri Granik; "Dry etch proximity modeling in mask fabrication", Proc. SPIE 5130, Photomask and Next-Generation Lithography Mask Technology X, 86 (August 26, 2003).

[5] Ian Stobert, Derren Dunn;"Etch correction and OPC, a look at the current state and future of etch correction" Proc. SPIE 8685, Advanced Etch Technology for Nanopatterning II, 868504 (March 29, 2013).

An Efficient lithographic hotspot severity analysis methodology using Calibre PATTERN MATCHING and DRC application

ZeXi Deng[a], ChunShan Du[b], Lin Hong[a], LiGuo Zhang[b] , JinYan Wang[a]

[a]Semiconductor Manufacturing International (Shanghai) Corp. 18 Zhangjiang Road, Pudong New Area, Shanghai 201203, P.R.C.

[b]Mentor Graphics(Shanghai) Electronic Technology Co., Ltd. Rm.2901, Jin Mao Tower, Shi Ji Da Dao(Century Boulevard) Pudong New Area, Shanghai 200120, P.R.C.

Abstract

As the IC industry moves forward to advanced nodes, especially under 28nm technology, the printability issue is becoming more and more challenging because layouts are more congested with a smaller critical feature size and the manufacturing process window is tighter. Consequently, design-process co-optimization plays an important role in achieving a higher yield in a shorter tape-out time.

A great effort has to be made to analyze the process defects and build checking kits to deliver the manufacturing information, by utilizing EDA software, to designers to dig out the potential manufacturing issues and quickly identify hotspots and prioritize how to fix them according to the severity levels.

This paper presents a unique hotspot pattern analysis flow that SMIC has built for advanced technology to analyze the potential yield detractor patterns in millions of patterns from real designs and rank them with severity levels related to real fab process. The flow uses Mentor Graphics Calibre® PM (Calibre® Pattern Matching) technology for pattern library creation and pattern clustering; meanwhile it incorporates Calibre® LFD (Calibre® Litho-Friendly Design) technology for accurate simulation-based lithographic hotspot checking. Pattern building, clustering, scoring, ranking and fixing are introduced in detail in this paper.

Keywords: Pattern Matching, pattern library, Litho DFM, Scoring, hotspot fixing

Design-Process-Technology Co-optimization for Manufacturability IX, edited by John L. Sturtevant, Luigi Capodieci, Proc. of SPIE Vol. 9427, 94270Y · © 2015 SPIE · CCC code: 0277-786X/15/$18 · doi: 10.1117/12.2086041

1. INTRODUCTION

In advanced technology nodes, the lithographic hotspot issue becomes a critical yield killer, especially under 28nm technology which has a tighter process window. Foundries provide litho DFM kits to make it possible for designers to check and fix potential issues at the design stage, on the one hand this eases foundries' manufacturing burden to achieve higher yield, and on the other hand, it can shorten the tape-out time and improves chance of one-time success [1].

DFM team has to take part of the challenging work. The 28nm and beyond technology nodes have a huge number of hotspot patterns to manage and the importance of efficient lithographic hotspot pattern severity analysis is increasing dramatically. The hotspots patterns sources partly come from wafer measurement; however, more are from LFD simulation results. With these millions of patterns availability, the final target is to accurately describe them, classify them and clearly rank them to obtain a minimum pattern set which can catch all the patterns in concern. Runtime, automation, easy maintenance etc. have to be taken into consideration.

This paper presents key steps in SMIC advanced technology pattern analysis flow; it includes pattern capture, clustering, scoring. The flow is based on Mentor Graphics Calibre platform and all the steps are automated and connected with scripts. The outcome to present is Calibre PM/FLFD (Fast LFD) kits for designer to use.

This paper also introduces a Calibre RVE™ utility to convert Calibre® ASCII results database (RDB) format to Synopsys IC Compiler DRC error markers and route guides to implement the fixing in P&R tool.

2. CHARACTERIZE A HOTSPOT PATTERN WITH A NOVEL PATTERN CAPTURE METHODOLOGY

Typically the front end layers such as Active and Poly in a design do not have very sophisticated patterns; most of the polygons on the front layers are regular shapes (Figure 1), so it is not a big concern in terms of layout friendly, however, the complexity is increasing dramatically in the back end metal layers (Figure 2). Challenge for an OPC recipe engineer is to struggle with congested and dense areas that are prone to at least two of the common hotspot types: pinch, bridge and enclosure (Figure 3). From the perspective of DFM, it would be necessary to build a kit to capture these patterns and provide the kit to the layout designers so that these hotspot patterns can be avoided in the design stage. Simply capturing an exact pattern is no big deal; the difficulty in practice is that several important factors have to be taken into consideration at the same time in order to create a more meaningful pattern, such as to capture key layout information that causes the hotspot and bypass irrelevant information, fuzz the pattern to improve the coverage and avoid over-capture, simplify pattern capture procedure for easy maintenance, reasonable runtime and performance.

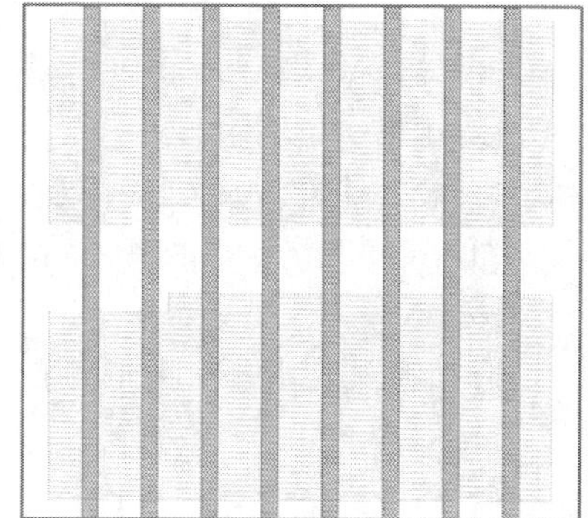

Figure 1 Active/Poly layout example

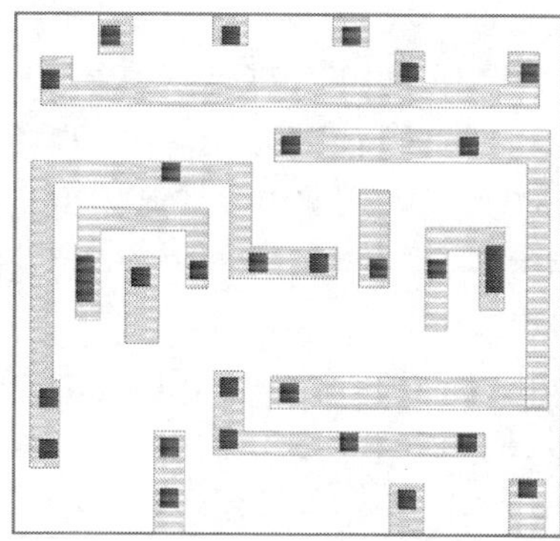

Figure 2 Metal1/Contact layout example

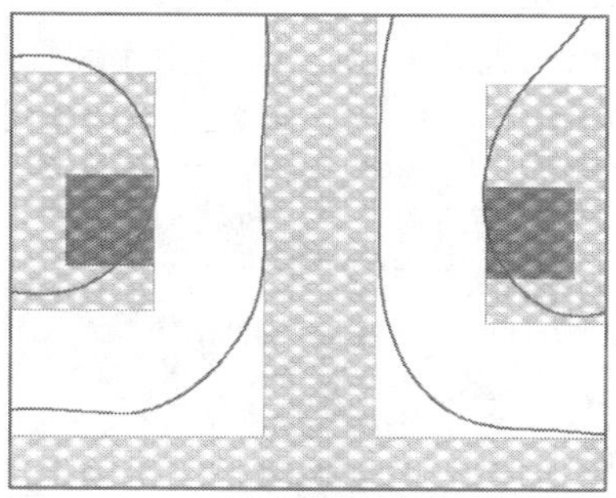

Figure 3 (a)

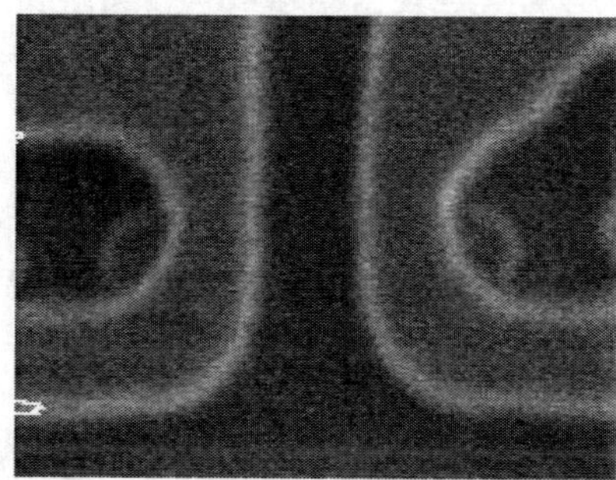

Figure 3 (b)

Metal1 pinch/bridge/enclosure risk review with layout vs. contour (Figure 3(a)) and silicon result (Figure 3 (b))

2.1 Pre-process hotspot error markers before capture patterns

Calibre Pattern Matching is used for capturing patterns through an SVRF rule file executed in batch mode. Either a pattern mask layer that defines the pattern boundary or a hotspot marker layer plus pattern halo is required for pattern capture. A hotspot marker layer is preferred for SMIC advanced pattern analysis flow as it can easily define the same pattern sizes and pattern center location. Extra work to pre-process hotspot markers should be done before capturing patterns as hotspot markers from simulation results are usually irregular shapes and it brings uncertainty in locating the center. If not accounted for, this could result in unnecessary pattern duplicates due to the uncertainty of the pattern center.

DRC operations are performed to relate the irregular shapes to the drawn shapes and generate new marker substitutions. It simply replaces the enclosure hotspot markers with the drawn via or contact layer interacted with the hotspot marker. Pinch or bridge hotspot marker processing follows the similar steps:

1) Get the original marker extent and size up the extent as the search range;
2) In the search range, search for the nearest corners or line ends;
3) If a line end is found, then measure the external opposite distance between this line end and the draw layer to get a polygon region which is the new marker substitution;
4) If only a corner is found then select an edge segment parallel with the marker and repeat step 3;
5) If no corner or no line end, size up the extent a bit more until find a corner and repeat step 3 or step 4

Below (Figure 4) are some examples for the new marker substitutions generations. With this pre-processing, duplicated patterns can be avoided as much as possible.

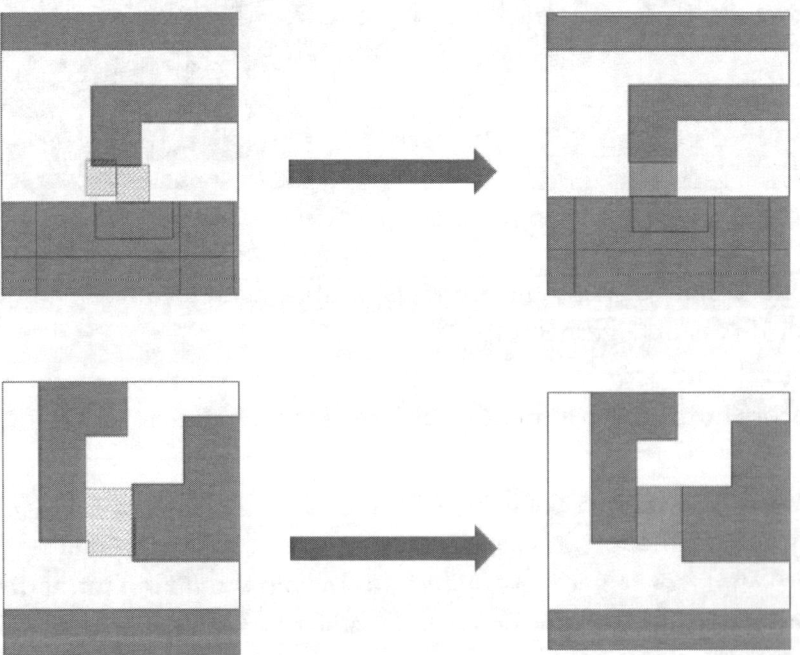

Figure 4 process of hotspot marker reshaping

2.2 Split a complex multiple-layer pattern into two sub-patterns

For the back end layers we focus on, hotspots patterns usually involve multiple layers: metal layer and hole layers above and below the metal layer. What is the reason behind then? Let's recall how OPC corrections are carried out. The OPC corrections never change the hole locations or perform edge target movements obviously for the circuit connectivity considerations, basically OPC processing in hole layers is independent of other layers. No matter what techniques are taken, the target is to print out the hole layers to the most extent as they are drawn. However, the metal layers OPC processing has to refer to the hole locations for enclosure consideration in addition to satisfying pinch and bridge specs in advanced technologies. That means metal layers OPC processing has less freedom when there are constraints both from hole layers and shapes themselves. From patterning view of point, hole layers can be thought to be a blockage that limits the metal layer edge movement so that in a congested area, a balance among pinch, bridge and enclosure becomes harder to achieve.

As stated above, interest of area for different layers should be considered differently in terms of patterning. The metal layer defines the basic pattern structures, which should have a relatively larger extent size. The hole layers only limit the metal edge movements, so a smaller size that could cover the holes segments that blocks relevant metal edges movements should be fine enough to characterize the hotspot pattern. With this in mind, a complex pattern can be safely split into two patterns: one larger single-layer pattern and one smaller multiple-layer pattern.

The single-layer pattern is centered on the new hotspot error marker substitution center and has a halo around 3-4x minimum DRC space. To account for lingering center shift or rotation independent pattern similarities, the Calibre Pattern Matching cluster utility will analyze the patterns that can be represented by one slightly fuzzy pattern.

The single-layer pattern has an output marker layer which defines the pattern search range and pattern capture area for smaller multiple-layer patterns, this marker layer is called assistant marker layer. The assistant marker layer is a key point in this pattern capture methodology. The assistant marker layer is defined based on the hotspot marker but is not necessarily centered on the hotspot error marker center and does not necessarily grow equally to all the directions, as what is said earlier, a type of hotspot usually goes along with other types of hotspots. You have to rely on this assistant marker layer to have the context in concern included. For example, in pinch type, the hotspot marker (in blue as in Figure 5b) is grown more toward the neighbor polygon while for the enclosure type as Figure 5c, and the assistant marker layer extends more toward the edges that are opposite to the edges more closely enclosing this hole.

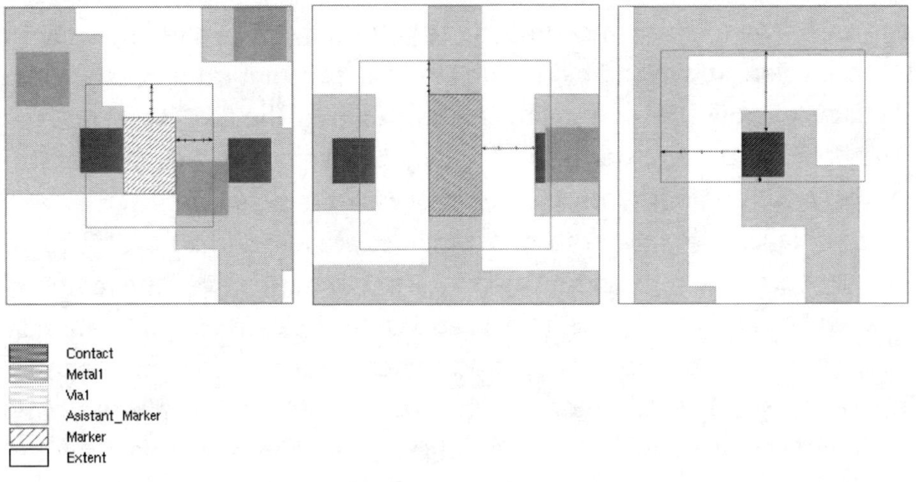

Figure 5 a) Bridge Figure 5 b) Pinch Figure 5 c) Enclosure

2.3 Pattern Fuzziness considerations

As what is said previously, only holes touching the assistant layers are required to take into considered to build the pattern and others are not the key factor that causes the hotspots and can be ignored. The hole, as part of the hotspot pattern, does not necessarily have a fixed position relative to the metal; a reasonably directional edge movement should be allowed in order to capture more similar pattern variants.

The drawn hole layer will not be a good candidate that is used as the pattern capture input, instead, some DRC operations are performed to directionally size down the hole layer to get a smaller hole layer representative and surround it with a DO-NOT-CARE regions (Figure 6), it is equivalent to move the hole into the desired direction, this is obviously can catch more situations (Figure 7).

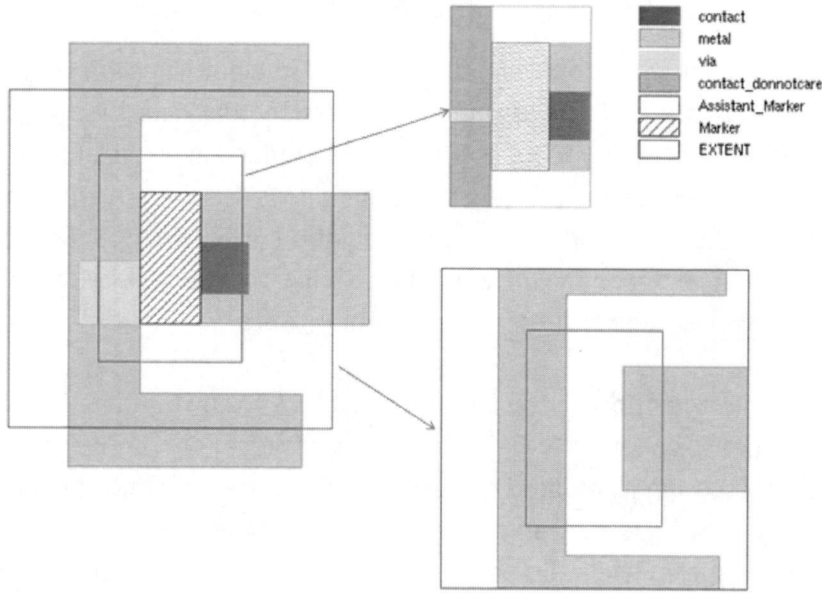

Figure 6 illustration of pattern splitting

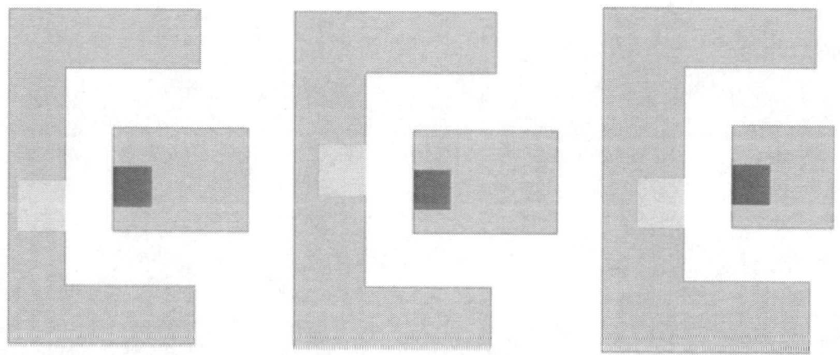

Figure 7 More patterns can be caught through fuzziness introduction

With these two pattern libraries generated, you can customize your run deck. In the run deck, you call single-pattern library first to get potentially hotspot regions and output the assistant maker to define the search region of multiple-layer search. The advantage is that this will make the pattern matching rather fast as the first one-layer pattern (exact pattern or patterns with limited constraints) will serve as the filtering. The second matching will only be run on the output of the first search results.

3. HOTSPOT PATTERN SEVERITY SCORING

To give designers a clear severity level to decide the hotspot fixing while balancing the circuit performance and process margin, pattern scoring is necessary to prioritize patterns in the pattern library to be delivered to designers. In terms of how to score a pattern, two factors are taken into consideration, one is the pattern's intrinsic geometry information, and the other is the simulation or silicon data. The former evaluates the complexity for an OPC engineer to solve the hotspot issue potentially; the latter reflects the wafer results in the current process.

In the geometry information, we consider the distance to surrounding polygons, number of the dense space surrounding polygons, via locations, corner number, etc. Each has its own scoring formula so that we can differ one from another, take the following one (Figure 8) as an example, the layout on the left will get 2 effective number of dense space, while the one on the right will get 3 effective number of dense space per DRC-like calculation, then the latter will be assigned a relative higher score to reflect that it is a more severe pattern in terms of dense space factor. This is reasonable as the higher the score is, the harder the hotspot can be fixed. Following this concept, we can score other factors.

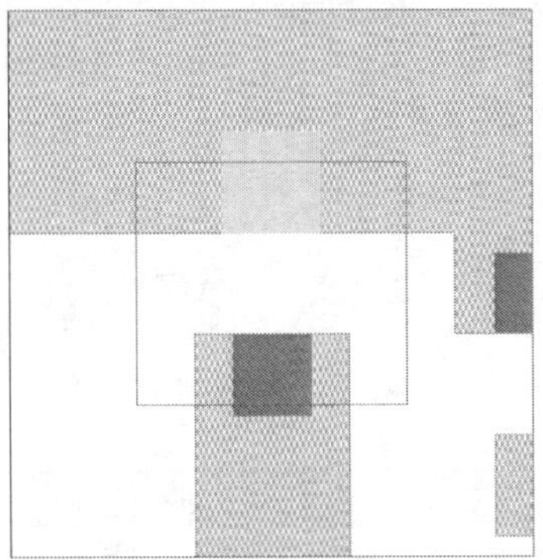

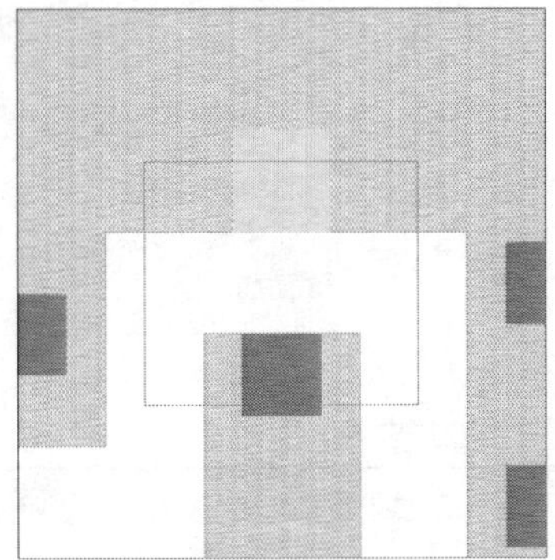

Figure 8 a) effective number of dense space:2 b) effective number of dense space:3

In the simulation or measured contour data analysis, there might be a lot of matched points in certain layout, and the wafer or simulation result has a little variation due to tiny difference in the area of optical diameter (>1um) or mask making, we first plot each pattern's hotspot properties distribution, and then select the mean value as that pattern's contour data, as in the following in Figure 9, most of property values are around 0.98 (normalized) in this distribution, so 0.98 will be a good value to be used as this pattern's contour data, and use this value to compare with our spec to assign the score for the contour factor. Of course, if the distribution is not what we expect, it is a good hint for us to review the specific pattern and consider whether we need to redefine them or split them into two or more.

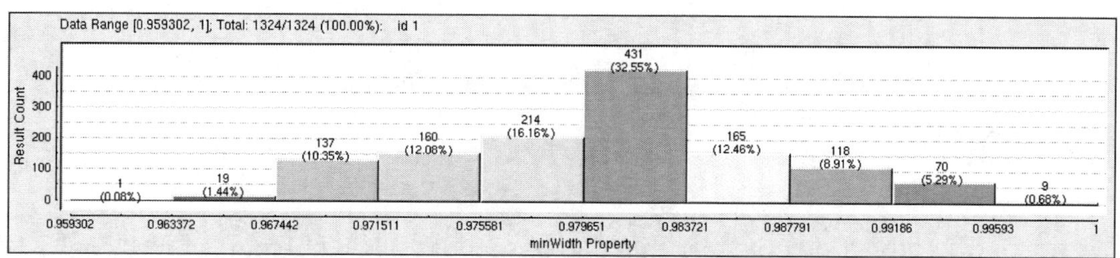

Figure 9 A pinch hotspot normalized CD distribution plot

After the scoring of each factor is assigned, now we apply the scoring formula to get a total score:

$$Score(Total) = \sum_{i=1}^{n} Weight(i) * Score(i)$$

Each pattern now has a score; foundry now can easily select patterns and decide how to select patterns and incorporate them in the PM and FLFD kits. Moreover, group them into different levels according to the scores so that designers can also refer to the level to prioritize how to fixing the hotspots.

4. HOTSPOT FIXING DISCUSSION

Finding hotspot patterns is one thing; how to fix them efficiently is the other thing. Layout engineers always like to fix hotspots in their preferred design environment. It becomes a necessity that foundry set up the integration solution and reference flow for the major P&R platform by linking data between different EDA tools. Calibre PM and LFD output is an ASCII format result database which has the check name and hotspot coordinates. In order to read the information from this RDB format, it's better to convert it to the format that can be recognized by P&R tool.

Calibre RVE™ provides a utility to convert Calibre ASCII results database (RDB) format to Synopsys IC Compiler DRC error markers [2], the conversion process automatically imports the error markers to the Synopsys IC Compiler session, where you can view them in the Error Browser:

```
icc_shell> Calibre::convert_to_drc_errors <calibre_drc_results.rdb>
```

or to convert Calibre ASCII results database (RDB) format to Synopsys IC Compiler route guides, the conversion automatically imports the route guides to the Synopsys IC Compiler session, where you use Zroute with the new route guides to repair violations.

```
icc_shell> Calibre::convert_to_route_guides <calibre_drc_results.rdb> -check_map_file < check_map_file>
```

We applied this function on a test chip database; both pinch hotspots and bridge hotspots were removed after the chip underwent this fixing function in ICC platform (Figure 10, Figure 11), while the local changes are very tiny and not negatively impact the chip performance [3].

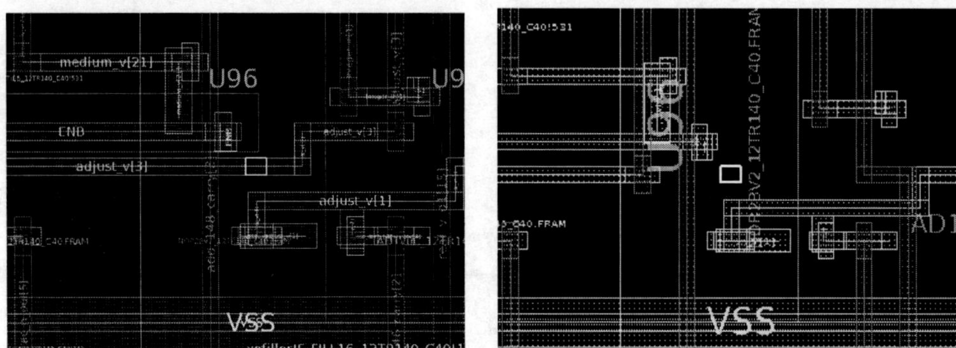

Figure 10 pinch hotspot was fixed by the fixing function

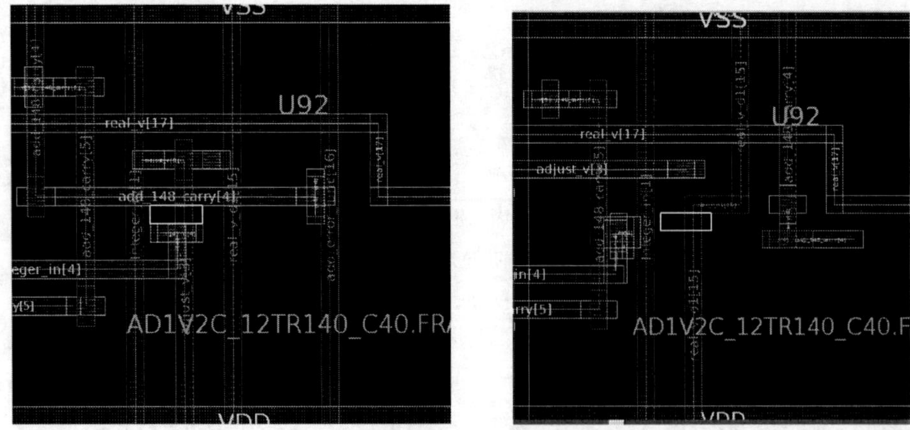

Figure 11 bridge hotspot was fixed by the fixing function

5. CONCLUSION

Thiss paper introduced a pattern analysis flow for DFM kit development. It covered all the key technique details that need to be carefully handled, from pattern capture to pattern scoring and also pattern fixing. This is a very efficient approach for pattern analysis and kit development. A system was built to connect all the steps for the automation. The pattern split is a new approach for accurate pattern definition and a good runtime performance. This is the system SMIC uses for DFM kit development. Experiments show it is very effective way to develop and maintain the kits.

6. ACKNOWLEDGMENT

The authors would like to thank Mentor Graphics foundry team and SMIC DFM team for their outstanding support to make this flow implemented and applied in the project.

7. REFERENCES

1 Seung Weon Paek, Jae Hyun Kang, Naya Ha, Byung-Moo Kim, Dae Hyun Jang, Junsu Jeon, DaeWook Kim, Kun Young Chung, Sung-eun Yua, Joo Hyun Park, SangMin Bae, DongSup Song, WooYoung Noh, YoungDuck Kim, HyunSeok Song, HungBok Choi, Kee Sup Kim, Kyu-Myung Choi,Woonhyuk Choi, JoongWon Jeon, JinWoo Lee, Ki-Su Kim, SeongHo Park, No-Young Chung, KangDuck Lee, YoungKi Hong, BongSeok Kim, "Yield enhancement with DFM", Proc. of SPIE Vol. 8327 832704 (2011)

2 Mentor Graphics, "Calibre Interactive and Calibre RVE User's Manual", 2014.4

3 Charlotte Beylier, Clement Moyroud, Fabrice Bernard Granger, Frederic Robert, Emek Yesilada, Yorick Trouiller, Jean-Claude Marin, "Fully integrated litho aware PnR design solution", Proc. of SPIE Vol. 8327 83270A (2012)

A holistic methodology that drives to process window entitlement and its application to 20 nm logic

Lalit Shokeen[a], Ayman Hamouda[a], Mark Terry[a], Dan J. Dechene[b], Stephen Hsu[b], Michael Crouse[b], Pengcheng Li[b], Keith Gronlund[b], Gary Zhang[b], [a]GLOBALFOUNDRIES, 400 Stone Break Extension, Malta, NY 12020, [b]ASML Brion, 4211 Burton Drive, Santa Clara, CA 95054, USA

ABSTRACT

Early in a semiconductor node's process development cycle, the technology definition is locked down using somewhat risky assumptions on what the process can deliver once it matures. In this early phase of the development cycle, detailed design rules start to be codified while the wafer patterning process is still being fine-tuned. As the process moves along the development cycle, and wafer processes are dialed-in, key yield improvement efforts focus on variability reduction. Design retargeting definitions are tweaked and finalized, and the use of finely tuned etch models to compensate for process bias are applied to accurately capture the more mature wafer process. The resulting mature patterning process is quite different from the one developed during the early stages of the technology definition.

In this paper we describe an approach and flow to drive continuous improvement in the mask solution (OPC and MB-SRAF) later in the process development and production readiness cycle stage. First, we establish the process window entitlement within the design-space by utilizing advanced mask optimization (MO) combined with the baseline process (i.e., model, etch compensation, and design retargeting). Second, gaps to the entitlement are used to identify and target issues with the existing OPC recipe and to drive continuous improvements to close these performance gaps across the critical design rules. We demonstrate this flow on a 20 nm contact layer.

Keywords: RET, Mask Optimization, Process window entitlement

1. INTRODUCTION

Technology development cycles for advanced technology nodes (i.e., 20 nm and below) can be broadly classified into three distinct phases: process definition, process development, and production readiness, each with varying degrees of flexibility in allowable changes to the process. Figure 1 shows a summary of the three different stages of technology development with different degrees of flexibility. While we briefly describe each of these hereafter, the major focus of this paper falls within the last two stages.

1.1 Process Definition

The process definition phase has the most flexibility to meet process requirements. During this phase of process development, design rules, patterning schemes, as well as key process specifics are being evaluated. In the context of resolution enhancement techniques/optical proximity correction (RET/OPC), it is in this stage where approaches such as source-mask optimization (SMO) [1] and design rule optimization (DRO) [2] are performed to obtain an optimal source pupil and retargeting shapes to meet process specifications. Design rules and standard cell designs are highly co-optimized in this stage through collaborative efforts often referred to as Design Technology Co-Optimization (DTCO).

1.2 Process Development

During the process development phase key technology identifiers such as the patterning scheme and key design rules have been determined in the context of RET/OPC, at this stage the majority of the focus is on accurately modeling the process (both post lithographic exposure, and post etch), as well as tuning the OPC solution to accurately print wafer targets while delivering maximum process window. The latter is accomplished through use of model based sub-resolution assist features (MB-SRAF) and the application of process window aware OPC solvers (such as PW-OPC).

Design-Process-Technology Co-optimization for Manufacturability IX, edited by John L. Sturtevant, Luigi Capodieci, Proc. of SPIE Vol. 9427, 94270Z · © 2015 SPIE · CCC code: 0277-786X/15/$18 · doi: 10.1117/12.2086101

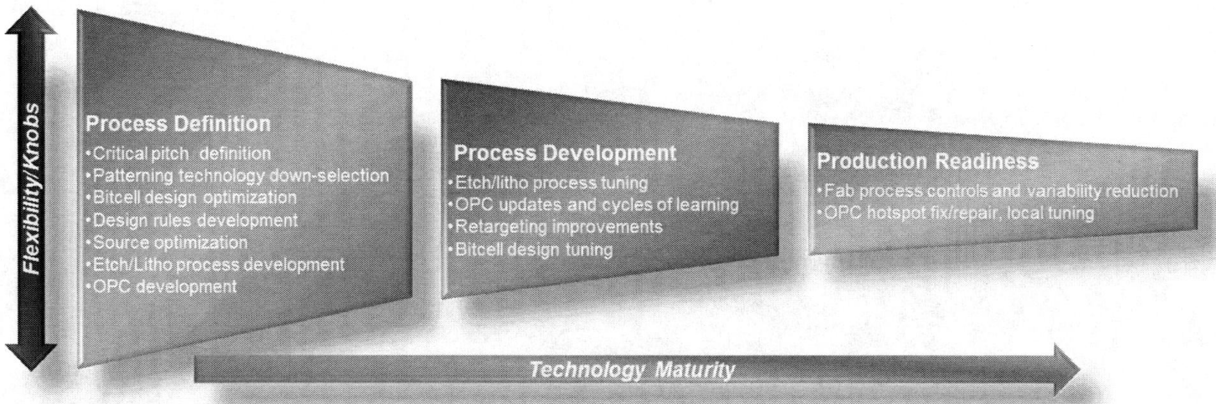

Figure 1. Process development flow for advanced technology nodes. From left to right, as technology development drives towards maturity, the flexibility to change the process (or knobs) is reduced.

1.3 Production Readiness

As the technology approaches maturity, process development focus shifts to yield ramp. Typical methods include identifying and fixing systematic yield detractors, as well as tightening process controls. On the lithographic scanner side, these process controls can include a number of techniques such as lens heating correction and scanner matching depending on the sensitivity and specs required by the process.

2. METHODOLOGY AND APPROACH

The major focus of this paper is later in the second and third stages of development where it is assumed that at this time, design rules, retargeting definitions and patterning technology has been locked in and an accurate process model is available. For the remainder of this paper, the term "Process Window Entitlement" is used to defined the maximum performance available (i.e., in terms of depth of focus) by changes induced via RET/OPC techniques.

2.1 Process Window Entitlement

Process window entitlement is defined as the targeted performance on a given process with a given set of constraints. This can be quantified in a number of ways. For the sake of simplicity, we define the process window entitlement (PW entitlement) for a given process as the depth-of-focus (DOF) in nanometers at 5% exposure latitude. With this definition and a stable wafer process, PW entitlement of an arbitrary geometry "z" is determined by computing the mask solution that maximized depth of focus. With this definition, the use of inverse lithographic solution (on a full-mask scale) can be used to approximate the process window entitlement for a given feature. It is assumed that such solutions cannot be used for this node in production environments on full designs due to practical OPC runtime considerations [1]. However, such approaches can be used to evaluate PW entitlement on systematic configurations and hotspots. Once the PW entitlement has been quantified for a given geometry, knowledge of the mask solution can be used to feedback into OPC recipe tuning. This approach allows us to focus on closing systematically large gaps where there is a high probability of reducing the gap with traditional OPC techniques. The aforementioned flow is described below.

2.2 Methodology Flow

A high-level flow diagram is shown in Figure 2. Here, design constructs such as logic standard cells, SRAM bit cells, systematic layouts and hotspots are fed into the system. These layouts are processed through two branches. The first upper branch in Fig. 2 quantifies the process window entitlement for a given geometry, while the lower branch quantifies the performance. A comparison is made on the performance of the two solutions. For geometries where the gap in performance between the two solutions is large, the existing recipe is tuned to improve the solution. Once the gap has been closed to within an acceptable tolerance, the performance is

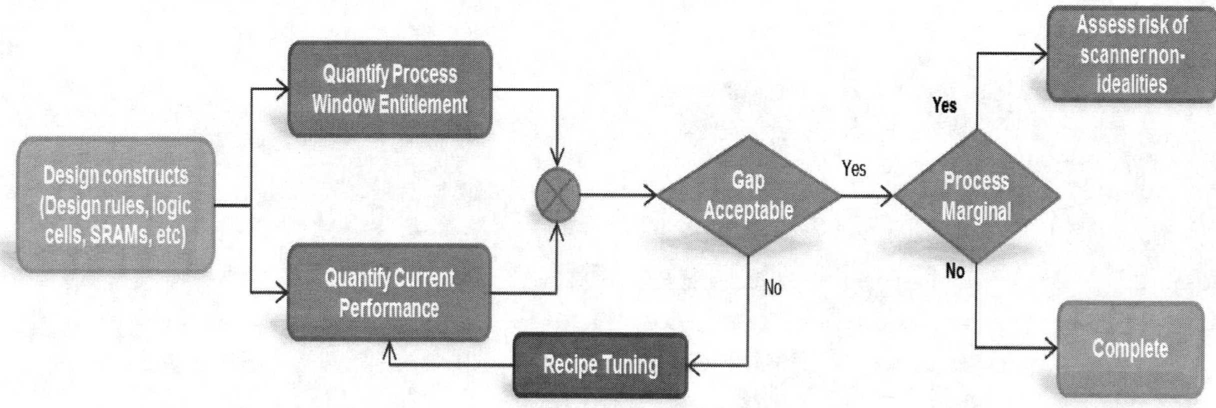

Figure 2. High level flowchart on methodology to close gap on process window entitlement

compared to the specifications required by the process. This portion of the flow is described in Section 3. For structures that have marginal process window, further analysis can be done on the geometry to assess the sensitivity to non-idealities for individual lithographic scanners within the fab. This portion of the flow is detailed in Section 4.

2.3 Tools Suite

The ASML Brion Tachyon software toolset is used to apply the methodology described above and is shown in Figure 3. Here, a calibrated FEM+ resist model is available and used to accurately simulate wafer level critical dimensions (CD). Geometries are taken from two sources: Systematic layout designed to assess ground rules and systematic deficiencies, and a pattern library containing known hotspots. The input geometries have been treated with all retargeting operations to provide the final lithographic target as input into the flow.

The targeted performance (or process window entitlement) is found by using Tachyon SMO's mask-only optimization, which provides a mask solution that also satisfies mask manufacturing rules. The "current performance" is obtained by running OPC using an existing recipe with Tachyon OPC+.

The output of both mask solutions is fed into an identical Tachyon LMC (process verification) recipe. The LMC recipe computes performance of both mask solutions based on a set of process assumptions. The set of process assumptions includes specs on the maximum tolerance on CD variation, and a set of process conditions describing the expected variation to be encountered by the process. The output of the two LMC runs summarizes a set of performance criteria which includes process variation bands (PV bands) and depth-of-focus (process window). A comparison is performed between both the mask optimization and existing mask solutions to identify the delta in performance. Geometries (or sets of geometries) with a large gap in performance (i.e., large gap in process window entitlement) are identified. Assist feature placement and OPC solution results from the mask optimization solution is fed back and used to tune the existing recipe (through both rule placement, MB-SRAF parameter tuning and OPC tuning). This process is continued until an acceptable gap has been achieved.

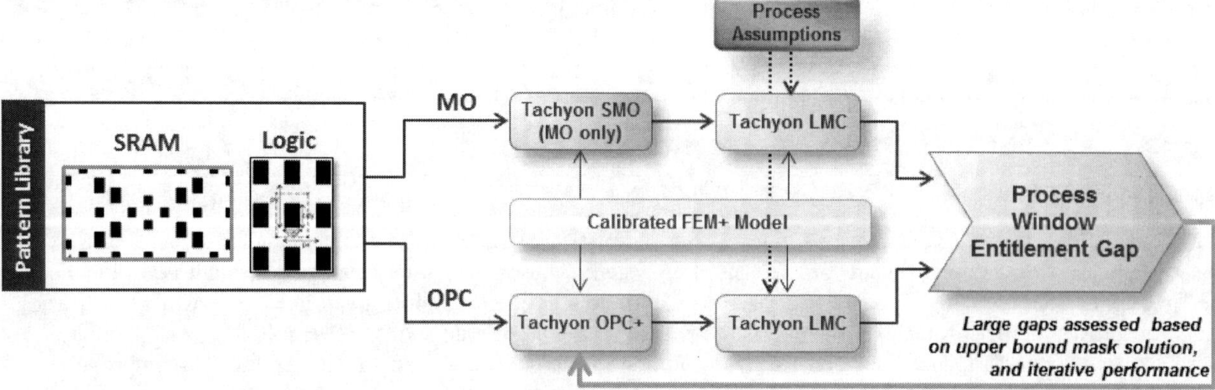

Figure 3. Detailed flow diagram of process window entitlement iterative loop

3. APPLICATION AND RESULTS

We apply the approach described in Section 2 on a 20nm via layer. An existing FEM+ model and OPC+ recipe are available. Process assumptions including necessary CD tolerance at litho, dose and focus control deliverable by the tool, and mask specifications are used to create an LMC recipe for evaluation. Retargeting is performed to obtain an overall lithographic target for evaluation. A case for systematic layouts and hotspots is presented as follows.

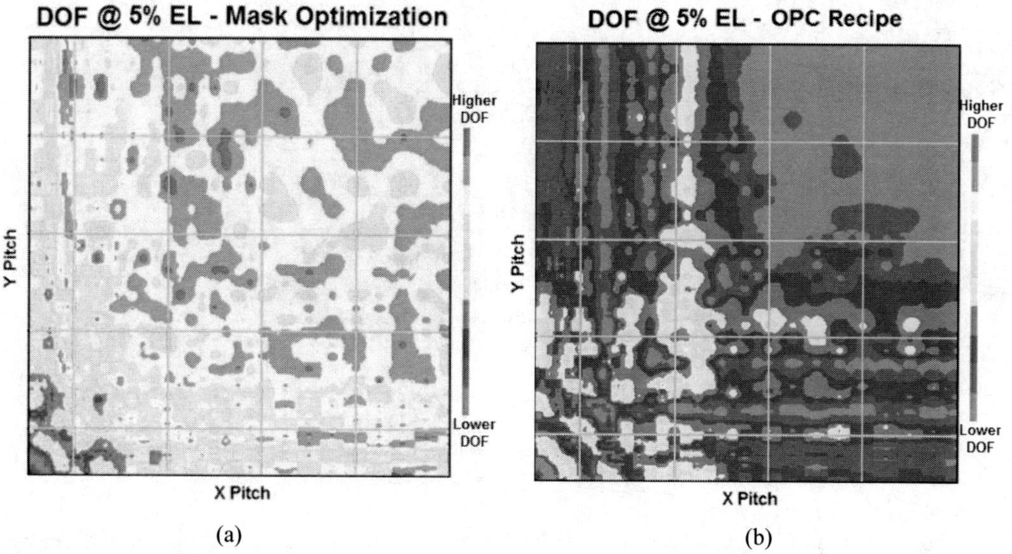

Figure 4. Depth-of-focus performance of the systematic array. a) with full mask optimization and b) with existing recipe.

3.1 Systematic Layouts

Figure 4a) and b) show the DOF for one of a given set of layouts with a variation in x and y-pitch with full mask optimization and existing solution respectively while Figure 5 shows the delta between the two graphs shown in Figure 4. From the latter; we observe the regime for which there is opportunity for improvement (i.e., the gap is large). By comparing the mask results between both solutions, we observe additional assist feature placements with common run length on the edges which is not present with the existing solution. Using this information, we perform the following:

a) Tag polygon edges satisfying this systematic geometry
b) Tune MB-SRAF extraction and cleanup parameters limiting placement along these edges

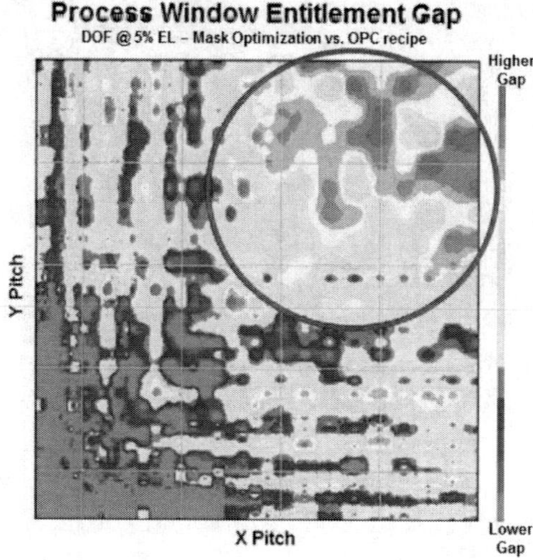

Figure 5. PW entitlement gap. Comparing Figs. 4a) and 4b) highlights systematic gaps to target tuning efforts. The scale from green to red shows regions with a small to large opportunity for improvement with tuning

Figure 6 shows the process window entitlement gap before and after recipe improvements. Figure 6a shows the large systematic gap has been improved via recipe tuning; while Figures 6b and 6c show the before and after solution.

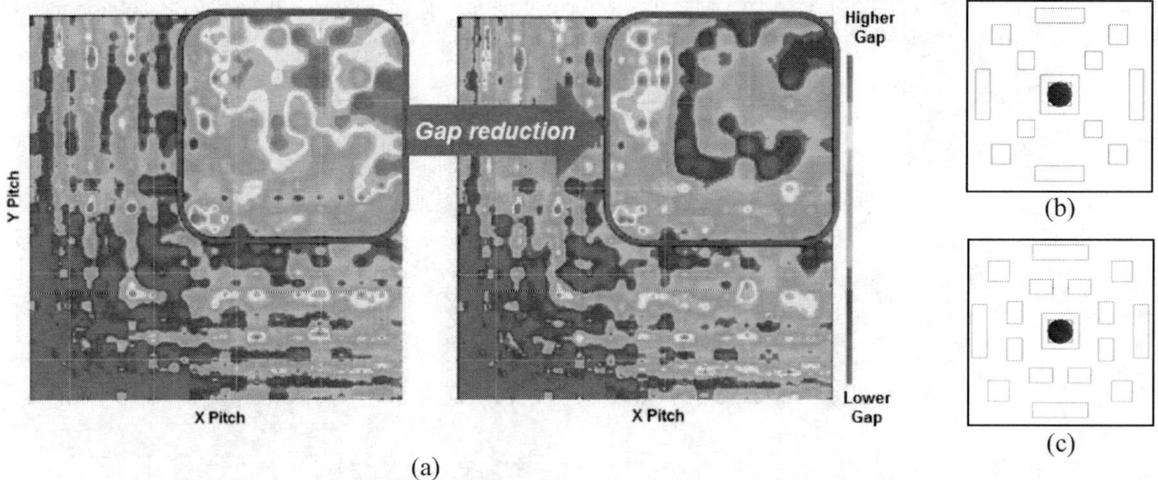

Figure 6. Recipe tuning on systematic array. a) overall process window entitlement gap before and after recipe optimization b) isolated SRAF solution before optimization, and c) isolated SRAF solution after optimization

Another case is shown in Figure 7. Here, we observe for bar structures, there is a large systematic region where process window is reduced. Further analysis of the full mask optimization solution shows that these structures benefit from support of the minor CD (shorter edge) with assist features. Here, we can isolate this systematic region with tagging and by filtering based on space. The model based assist parameters of the tagged structures are then tuned to provide additional support to the minor CD. Figure 8 shows the improvement obtained based on tuning. Figures 8a. and 8b. show the mask solution before and after tuning respectively. In Figure 8b, we can clearly observe the additional assists parallel to the minor bar CD. Figures 8c. shows the process window improvement from tuning the mask solution. From the EL/DOF plot, we can clearly observe a dramatic improvement in the common process window with the improved mask solution.

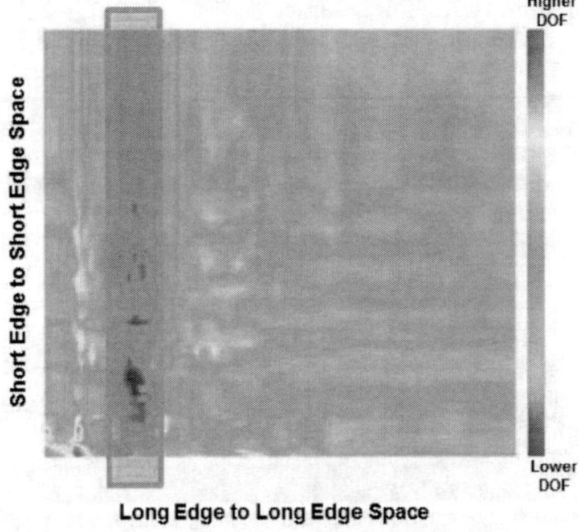

Figure 7. Another case of limited process window in a systematic configuration for bar structures

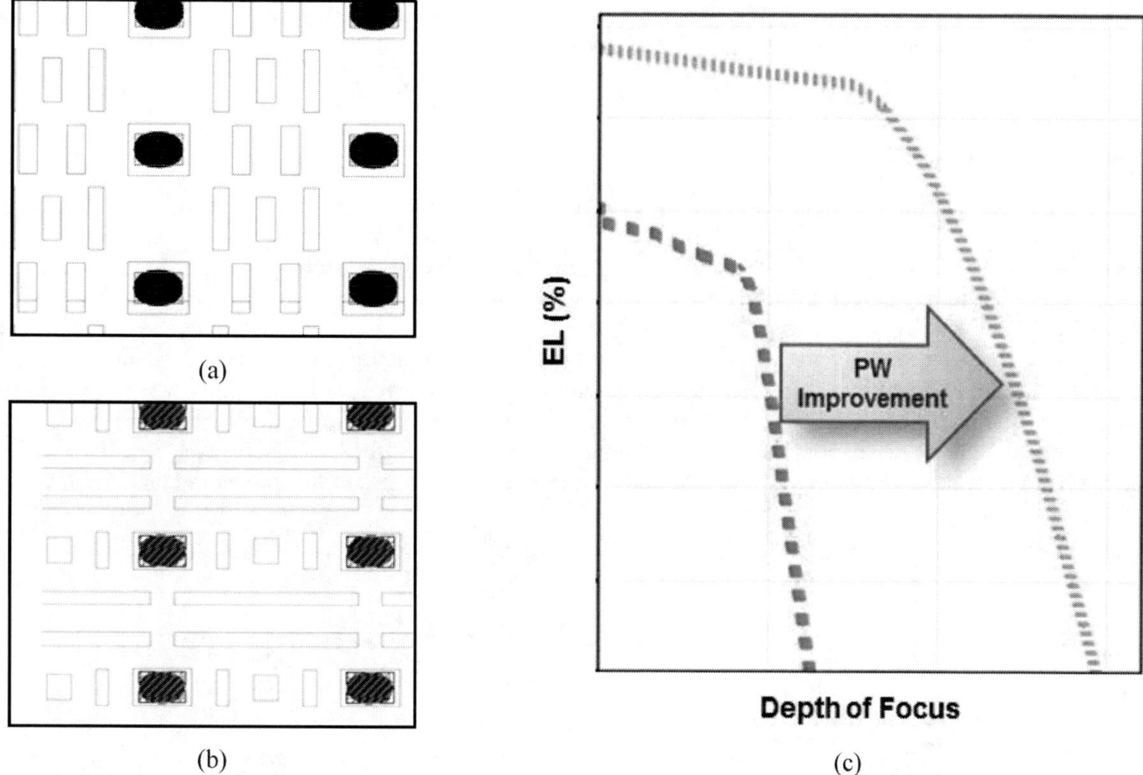

Figure 8. Systematically process window limited structures. Here a) shows the solution before recipe optimization, b) shows the solution after recipe tuning using full mask optimization results and c) shows the process window (PW) improvement before and after tuning.

The described approach above is applied to a number of systematic layouts within the design space.

3.2 Application to Random Logic Configurations

Once systematic performance gaps have been closed, the focus is on random logic hot spots including critical constructs, and SRAMs cells. Application to these specific constructs requires more specific approaches to identify and tag critical structures. In addition to traditional DRC, the use of pattern matching or upstream marker layers can also be used. Figure 9, we show an example of one such configuration and its current mask solution.

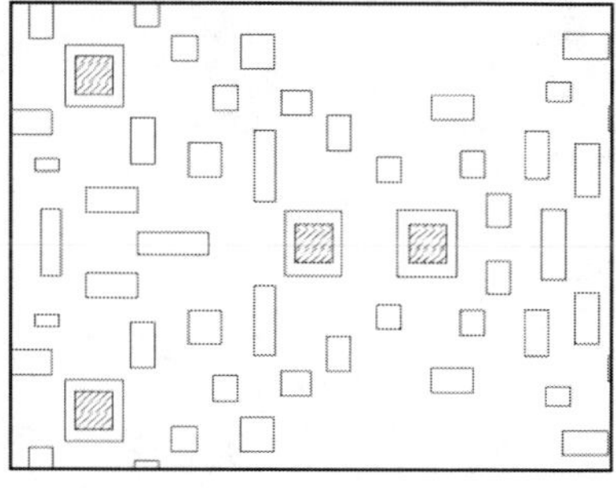

Figure 9. A non-systematic (logic) hotspot clip to be optimized.

To apply the described methodology, the input litho targets are run through full mask optimization to evaluate the process window entitlement. By comparing both solutions, there are several key observations

- Assist density differences between the two solutions
- Assist placement differences between both solutions
- Target differences between both solutions (limiting common process window)

The process window entitlement gap is shown in Figure 10a. In round 1, we address the first and second observations by tuning the limiting MB-SRAF placement parameters and increasing the extraction search range for this hotspot. The results of the first round of optimization are shown in Figure 10b. We observe an improvement in the exposure latitude. Comparison of the round 1 solution still shows denser SRAF solution. To mimic this, we further reduce the limitations on SRAF to SRAF spacing which provides more flexible placement. The result of this tuning stage is shown in Figure 10c. Here, we see a clear improvement in the focus latitude. Comparing the results of round 2, we observe the common process window of the vertical and horizontal CD as the process window limiting features when compared to mask optimization solution. For round 3, we adjust the control points on the H and V edges separately to improve the common process window. The result of this third stage of tuning is shown in Figure 10d. Here, the overall focus latitude is improved with a small impact to the exposure latitude. At this stage, we consider the gap closed and progress to the next hotspot.

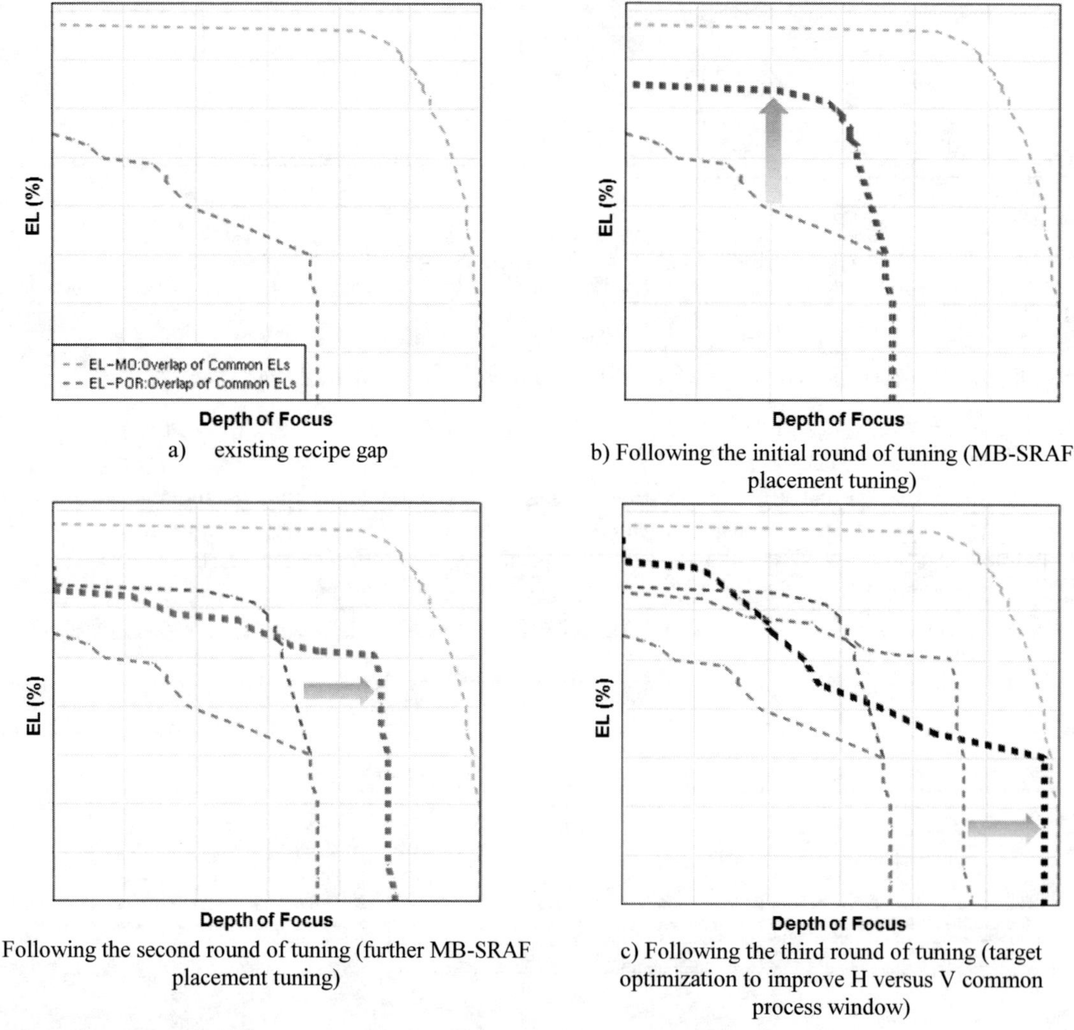

a) existing recipe gap

b) Following the initial round of tuning (MB-SRAF placement tuning)

c) Following the second round of tuning (further MB-SRAF placement tuning)

c) Following the third round of tuning (target optimization to improve H versus V common process window)

Figure 10. Iterative improvements with recipe optimization to drive to process window entitlement by applying learning from mask optimization

The approach is applied continuously on all hotspots within a level. Hotspots with a small process window entitlement gap are further analyzed on specific lithographic scanners within the fleet.

4. SCANNER SPECIFIC EVALUATION

Typically, resist models for OPC are calibrated using an average model of the process (i.e., an average of the lithographic scanners within the fleet). However, features with small process window may be more susceptible to non-idealities introduced by one or more of scanners within the fleet. Following the improvements described in Section 3, there may still exist some configurations with a small process window entitlement gap that are marginal in terms of process window performance. For example, examining Figures 4 and 5, we observe some semi-dense configurations where a) process window is lower and b) the process window entitlement gap is small. For these configurations, the impact of individual scanner non-idealities of specific scanners within the fleet can be examined to determine if there is any impact to these features.

The assessment of scanner specific non-idealities is performed using the flow shown in Figure 11. Here a scanner specific model is a snapshot at a specific time. Non-idealities including abberations, measured source pupil, mechanical stage dynamics and measured laser bandwidth can be used to create a scanner specific FEM+ model. We note however that the scanner specific model is not used to perform OPC, rather this is still done with the existing OPC recipe (i.e., the average scanner model).

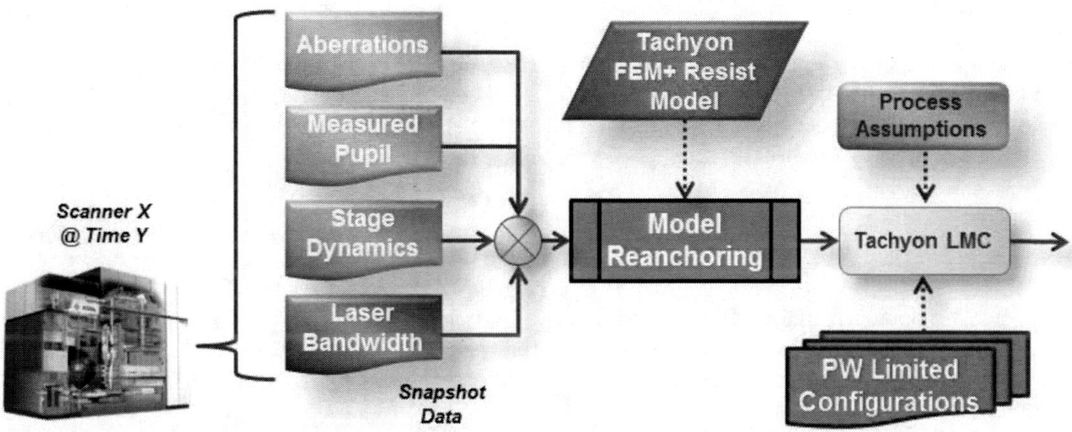

Figure 11. Flow chart on scanner specific FEM+ model generation and evaluation

The DOF for the average scanner, Scanner A and Scanner B is shown in Figure 12 for select orthogonal configurations from the test case shown in Figure 4. For these structures, we observe several key findings. First, the performance of the two scanners on these structures has very little negative impact on the limiting structure (i.e., the semi-dense pitch configuration). Second, we actually observe some small improvement in process window for some configurations versus the average scanner. The rationale here is as follows. OPC is performed with the average scanner model, while simulation is performed with individual models. While models utilize the same dose anchor, structures outside the anchor could experience a small dose offset (i.e., print slightly off nominal) due to variations in aberration and stage dynamics. For focus sensitive structures, the slight increase in nominal CD can actually marginally increase the process window of this structure (i.e., similar to how process window OPC for contacts prints slightly larger than nominal).

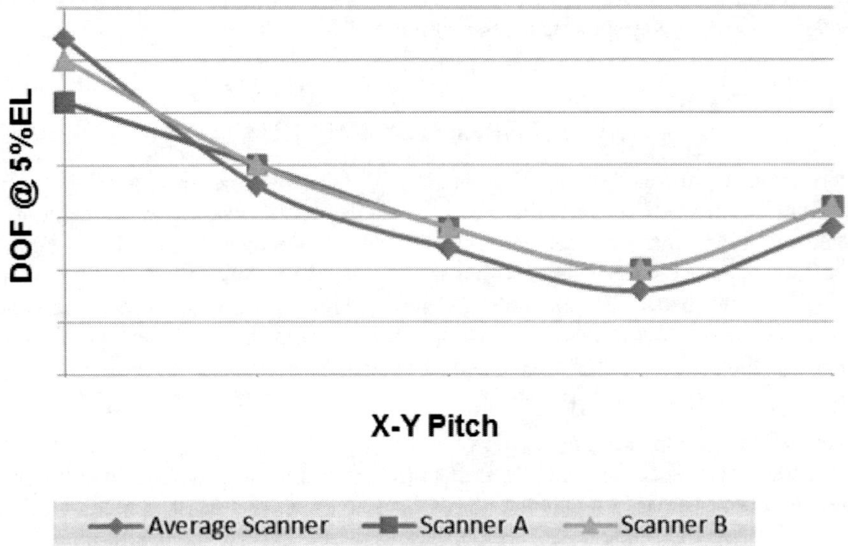

Figure 12. Performance of semi-dense orthogonal configurations on the average scanner (i.e., existing FEM+ model) and two specific scanners a given time.

For configurations that are shown to be sensitive to one or more specific scanners, there are several approaches that can be taken. For example, Pattern Matcher Full Chip (PMFC) could be employed. PWO can be used combined with focal maps to improve common process window by tuning focus. If a specific scanner shows high sensitivity, PMFC can be used to reduce scanner to scanner variation.

5. CONCLUSION

We presented a holistic approach that can be used to drive towards meeting process window entitlement in later stages of technology development. The use of full mask optimization was employed to identify systematic and random logic configurations with a large process window entitlement gap for targeted recipe improvements. Learning was iteratively fed back to improve the existing OPC recipe. Configurations with little to no entitlement gap were also found with this method and we showed that further analysis can be performed to better understand if any sensitivity on specific scanners when considering scanner non-idealities. A subset of results was shown demonstrating this approach on a 20nm logic layer.

REFERENCES

[1] DongQing Zhang, et al., "Source mask optimization methodology (SMO) and application to real full chip optical proximity correction," Proc. SPIE 8326, (2012).

[2] Chung, No-Young, et al. "Smart source, mask, and target co-optimization to improve design related lithographically weak spots." Proc. SPIE 9053, (2014).

Practical DTCO Through Design/Patterning Exploration

Neal Lafferty[a], Jason Meiring[a], Mohamed Bahnas[b], Joseph ONeill[c], Toshikazu Endo[d], Dan Schumacher[c], James Culp[e], Glenn Wawrzynski[f], Gurpreet Singh Lamba[g], Kostas Adam[d], John Sturtevant[a], Chris McGinty[c]

[a]Mentor Graphics Corporation, 8005 SW Boeckman Road, Wilsonville, OR, USA;
[b]Mentor Graphics Corporation, 15 Independence Boulevard, Warren, NJ, USA;
[c]Mentor Graphics Corporation, 1 Lewis Wharf, Boston, MA, USA;
[d]Mentor Graphics Silicon Valley HQ, 46871 Bayside Parkway, Fremont, CA, USA;
[e]IBM Microelectronics, 2070 Rt. 52, Hopewell Junction, NY, USA;
[f]Mentor Graphics Corporation, 1 Galleria Tower, Dallas, TX, USA;
[g]Mentor Graphics Corporation, 15/1A Sarjapur Outer Ring Road, Bangalore, India;

ABSTRACT

Design Technology Co-Optimization (DTCO) becomes more important with every new technology node. Complex patterning issues can no longer wait to be detected experimentally using test sites because of compressed technology development schedules. Simulation must be used to discover complex interactions between an iteration of the design rules, and a simultaneous iteration of an intended patterning technology. The problem is often further complicated by an incomplete definition of the patterning space. The DTCO process must be efficient and thoroughly interrogate the legal design space for a technology to be successful. In this paper we present our view of DTCO, called *Design and Patterning Exploration.* Three emphasis areas are identified and explained with examples: Technology Definition, Technology Learning, and Technology Refinement. The Design and Patterning Exploration flows are applied to a logic 1.3x metal routing layer. Using these flows, yield limiting patterns are identified faster using random layout generation, and can be ruled out or tracked using a database of problem patterns. At the same time, a pattern no longer in the set of rules should not be considered during OPC tuning. The OPC recipe may then be adjusted for better performance on the legal set of pattern constructs. The entire system is dynamic, and users must be able to access related teams output for faster more accurate understanding of design and patterning interactions. In the discussed example, the design rules and OPC recipe are tuned at the same time, leading to faster design rule revisions, as well as improved patterning through more customized OPC and RET.

Keywords: Design Technology Co Optimization

1. INTRODUCTION

Design Technology Co-Optimization refers to the joint development of a technology between designers and technologists.[1] The term has different implications depending on the application. In use by digital designers, for example, the DTCO process can involve standard cell design, routing using exploratory rules, and chip level electrical simulation.[2–4] The output of the DTCO exercise can be simulated for speed, power consumption, routability, and layout area to determine if the technology meets key scaling criteria. Alternatively, there are also process-based approaches to DTCO, such as Ref. 5, where layouts are mined for simulated hotspots, and then they are characterized and stored. Regardless, each DTCO approach takes a core viewpoint and builds around it. For example, Ref. 3 uses a fast first-order patterning simulation for speed reasons, while also considering the full standard cell architecture, such as track layout and other core technology decisions. As such, the approach requires a high level of designer input, with less certainty on patterning variations. The output is used as guidance on very important fundamental technology setup. In Ref. 5, the process discussed builds a database of poorly performing patterns which can then be tracked and utilized for improving yield ramp, among others.

Send correspondence to N.V.L.:
E-mail: neal_lafferty@mentor.com; Phone: 1 503 685-1018; http://www.mentor.com

Design-Process-Technology Co-optimization for Manufacturability IX, edited by John L. Sturtevant, Luigi Capodieci, Proc. of SPIE Vol. 9427, 942710 · © 2015 SPIE · CCC code: 0277-786X/15/$18 · doi: 10.1117/12.2086355

Although each approach has a key use group, various degrees of co-optimization occur when proposed changes are vetted by technologists. On the process side, initial design rules must be proposed, often derived from previous generations. These rules must then be tested by design teams to determine if the rules are sufficient for economical scaling, usually on a minimum set of important constructs, as in Ref. 1. If changes are necessary, ideally they should be detected early, before process development is too far along. Also, proposals must be thoroughly considered by technologists before being accepted, counter-proposed, or denied.

Practically speaking, early in development these tasks are complicated because designs are also not typically available. Given the large variety of possible design styles it is impractical to manually do a thorough exploration. Once proposed, technologists often must make design rule decisions with certainty before exhaustive layout is available; complex full featured layouts cannot be created for each design perturbation. Lithography or OPC engineers are often involved in DRC revision due to knowledge of the process and Resolution Enhancement Techniques (RETs). Lack of detailed exploration of design changes can contribute to sub-optimal design manual decisions if a critical or context sensitive failure occurs as a result. Even worse, late detection of failures results in rework of layout and designer inefficiency. In the worst case, it may not be possible to adjust design rules, and under-performing patterns may become part of a technology.

To bypass this limitation, this work will describe a DTCO approach using guided random enumeration as suggested by Ref. 6. These large scale synthetic layouts can then be analyzed and the full impact of design perturbations considered on a layer by layer basis. The proposed approach focuses on a detailed layer-by-layer exploration of layout, attempting to characterize all legal patterns. The guided random layout can be iterated rapidly with design rule updates, and problem patterns stored in a database for tracking over time. Since the layout iterations can be performed quickly and early in the technology setup, problems can be detected and solved early. The focus of this paper is on enabling efficient development of technology as early as possible using guided random layout to mimic design input. One measure of the success of this approach is the amount or type of yield limiting patterns detected on successive tape outs of a technology. If pattern exploration is successful, systematic faults should be fully characterized and corrected before the first tape outs happen with real design content.

2. TOOL/FLOW COMPONENTS

The DTCO flow uses a series of tools and processes that are depicted in Fig. 1. To start a large, design rule clean, random layout is generated using a tool called Layout Schema Generator. Next, interesting design topologies (patterns) are extracted from this layout and placed in a pattern library; simultaneously, the layout is run through OPC and litho process verification. The verification data is then used to annotate the patterns with data about printability. The annotated patterns are then analyzed, the results of which can drive various actions including OPC recipe refinements and DRC refinements. This entire process is iterative as refinements in OPC and design rules drive changes to the layout generation and verification data. The individual components of this flow will be described in detail.

2.1 Layout Schema Generator

As technology development progresses from the conception phase to hardware learning to manufacturing, there is an ever increasing need for high quality patterns that can be used to tune OPC models and recipes and refine design rules. In the early stages a few simple, highly synthetic patterns may be sufficient to elucidate basic design limits and the optimal source and mask conditions (the latter being handled by source mask optimization, or 'SMO', tools). However, these patterns are poor proxies for the real layouts upon which a manufacturable process will be judged. Unfortunately, large-scale design layouts typically only become available near the end of the technology development, making it difficult to react to surprises that may impact yield. To compensate for the lack of real layout, we turn to a pattern enumeration engine called Layout Schema Generator (LSG) to create interesting, design-like content.[7]

Pattern enumeration describes the process by which design rules are translated algorithmically to a layout. We can consider enumeration algorithms to exist on a spectrum with highly regular, highly constrained patterns on one side and highly random, unconstrained patterns on the other (Fig. 2). In the former case, patterns are generated via a table of parameters that vary systematically. Typically each class of pattern requires a separate

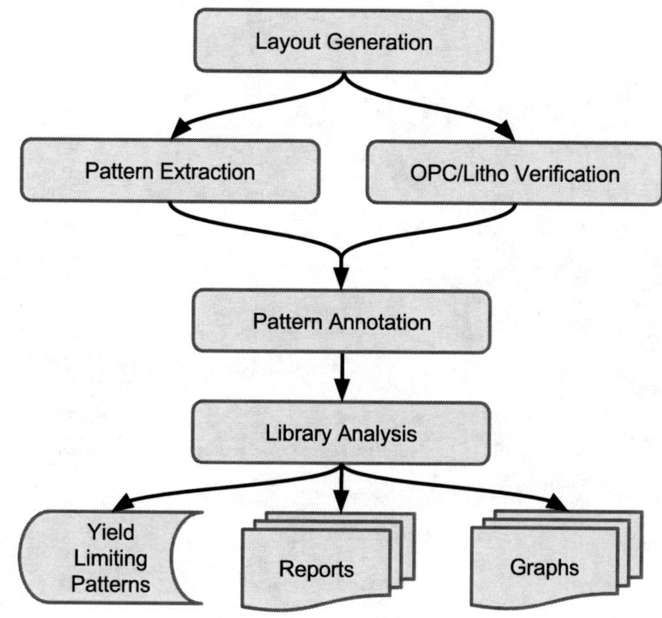

Figure 1: Design and Patterning Explorer component flow

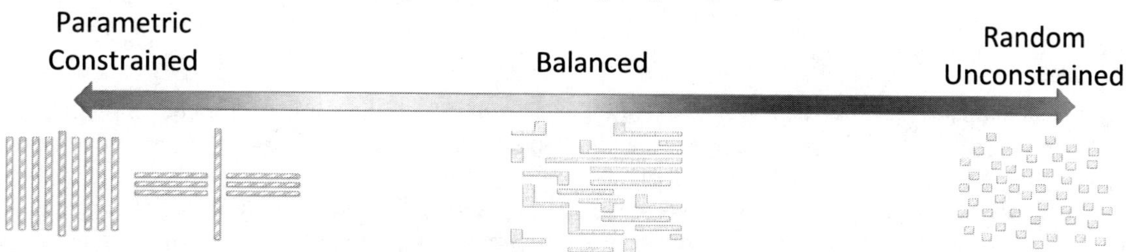

Figure 2: Pattern enumeration spectrum

generator function. For example, a tip-to-side pattern generator may include inputs for attacker width, victim width, and tip-to-side space. While these types of patterns are relatively straightforward to generate and define (once the generators are written!), they are generally very unrealistic from a design perspective. In addition, they require creativity on the part of the user to try and capture all the pattern variations that a set of design rules may allow. On the other side of the spectrum lies fully random patterns. (Consider, for example placing squares at random coordinates to simulate a contact array.) In this case, we can expect highly varied layout, but at the cost of a high degree of low quality patterns that might violate design rules or be unrealistic. Clearly a balance between the two extremes is needed.

LSG seeks to strike this balance by producing random layout, but doing so in a constrained system that attempts to mimic the same kind of constraints designers encounter when doing actual layout. Simple probabilistic rules are written to prohibit or encourage simple layout constructs that are placed via a Monte Carlo algorithm. By changing the probabilities, different styles of layout can be created, from metal to contact. In addition, predefined layout constructs can be placed among the randomly generated content, giving them realistic context. Random patterns can also be generated inside template layouts to simulate fixed power rails, for example. This approach can be extended to multiple patterning levels by writing rules and probabilities that include layer-to-layer constraints. Checks are done during and after the generation process to ensure that the layout is fully design rule compliant. Fig. 3 shows a realistic metal-style random layout that was generated with LSG.

In developing the ideal set of LSG constraints that mimic the target design style, there are several important

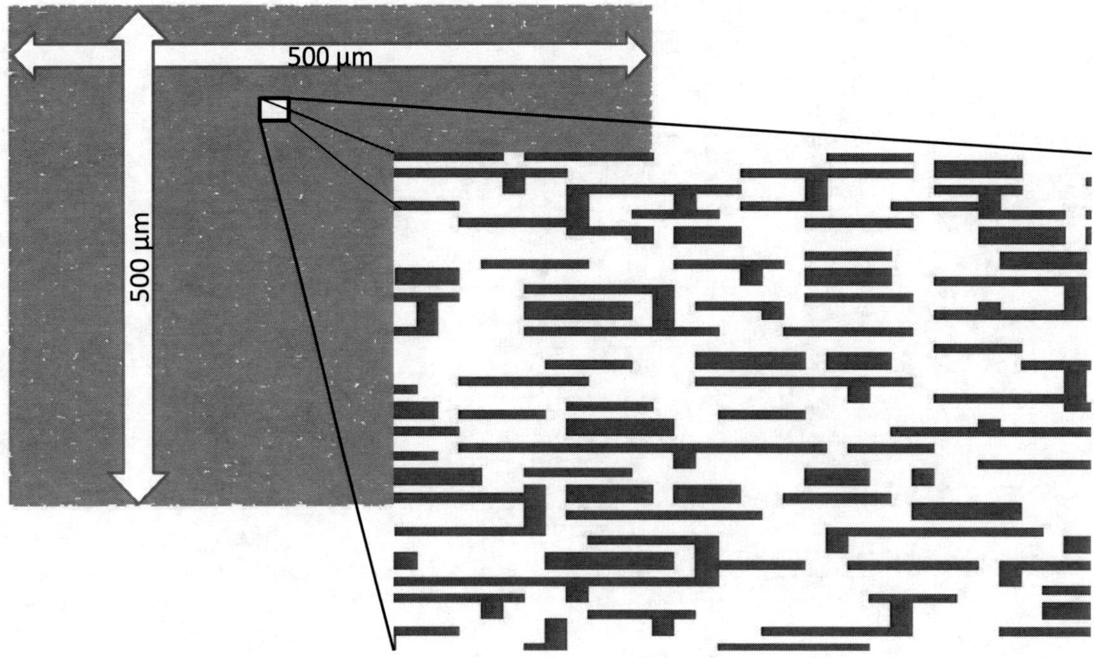

Figure 3: Example large scale random M1-style layout generated with LSG

quality metrics to consider. First, the resulting layout should contain a large number of unique pattern topologies. This can be quantified by clipping out windows around certain interesting pattern geometries (like corners) and summing up the number of unique clips found by geometric pattern matching. Second, the patterns extracted from this pattern matching step should completely match the patterns in the original layout. Using a design rule checker (DRC) with pattern matching against the LSG layout we can compute the fractional area of matched patterns vs. the entire layout area. We call this extent of matching coverage and ideally it should approach 100%. Third, the extracted patterns should show a high degree of coverage towards real design. This is typically lower than 100% since real design often contains lithographically uninteresting or irrelevant features. (This last metric may not be obtainable early in a technology flow when real design does not exist.)

2.2 Pattern Classification

To detect the yield limiting or critical patterns, the OPC and simulation verification recipes are executed on the generated layout. Here we define yield limiting patterns as those that exceed process tolerance for Process Variation Bands (PV Bands). However the approach can be extended to any measurable image quantity, such as resist toploss, Normalized Image Log-Slope (NILS), or Depth of Focus (DOF). The OPC recipe can be either pre-existing model-based OPC code or Inverse Lithography Technology (ILT) code. While traditional model-based OPC is a manufacturing solution that may require tuning, early cycles during technology setup will not have sufficiently mature model-based OPC setups. Robust ILT-based techniques are essential for evaluating early design rule proposals when no other recipe is available. Although ILT may be optimistic, it is a good prediction when realistic constraints are in place. A simulation verification recipe will be mainly focused on PV Band based printing issues (i.e. bridging/pinching), and can also include metrics such as MEEF and NILS. According to the user selected thresholds in the simulation checks, the critical patterns will be marked for further analysis. The marked patterns will be annotated with all measured properties that will be used in the analysis and classification afterwards.

2.3 Pattern Annotation

To ensure the generation of diversified spectrum of random patterns, the LSG layout needs to be of large area to include critical patterns with many different optical contexts. However, handling all of these marked patterns run through simulation verification will not be easy for the users visual review and analysis. Accordingly, we need to reduce the overall number of reviewed clips by applying different identification aspects on the patterns. This will lead to creating a patterns library with a manageable size. The first step in filtering the patterns is applying the pattern matching and selecting one representative per matching geometry in the identified halo size. This is implemented using geometric pattern matching software tool. For further downsizing of the representatives list, we need to apply some classifications with respect to the topology. First, we can group similar patterns that match each other within a range of edge displacements. Second, we can group patterns based on their design topologies even if the edge geometries do not match. Ultimately there is a trade-off between the number of groups and the similarity of the patterns within a group. While having fewer groups facilitates visual analysis, it leads to more dissimilar patterns and (potentially) overlooked topologies. These two steps are controlled by several parameters giving the user the flexibility to find the right trade-off for their application.

2.4 Pattern Analysis

Focusing on these representative patterns in the proceeding steps of the flow can be useful in saving both the processing runtime and the user review. These patterns can be used in both the OPC recipe and DRC rules refinement processes. The annotated properties of these patterns are to be used for analyzing the criticality level of the patterns and scoring them accordingly. These properties are statistically reported and analyzed. Visualizing these properties is done in a contemporary framework with ease-of-use exploration capabilities.

3. YIELD LIMITING PATTERNS FLOW

Design and Patterning Exploration is built around the so-called Yield Limiting Pattern Flow composed from components in Section 2. It is the goal of this flow to detect problem patterns on a synthetic layout, forecasting those that are likely to be detected later in the technology cycle. Once these poorly performing problem patterns are detected, various corrective action can be taken.

As an example of this flow, a 1.3x routed-metal layer from a recent technology node was studied. Using the levels DRC deck, and a set of rules to guide LSG, we have created a layout similar in topology to what is expected for a 1.3x metal. The guided random fashion of this layout has the advantage of including patterning variations which are legal, but may be unanticipated by a designer or RET engineer. The guided random enumeration can also detect patterns that are sensitive to context, since they are selected from a much larger layout. As an example, consider Figure 4. These patterns are extracted from a much larger layout and have different neighborhoods. The patterns context determines whether the 10 nm pinch region occurs or not, and is vital to capture to avoid yield surprises.

Because the design rules are envisioned to be rapidly changing during early technology setup, this tool can be executed with each iteration, checking for closure of previously detected yield limiters and any new hotspots that may occur due to those design manual changes.

4. OPC RECIPE REFINEMENT

The results of Section 3 are then used to refine an OPC recipe. DRC adjustment is an effective method for excluding yield limiting patterns, however it can only be applied to constructs of low design value. For high value design topologies, OPC recipe refinement must be used to improve problem constructs. The refinement can be performed manually or automatically using an optimizer and feedback loop. For proof of concept, this paper illustrates a case in which the OPC recipe was manually adjusted.

The illustrated case is dealing mainly with soft pinching and via coverage issues. The manual adjustment for the OPC recipe is related to the trade-off optimization between soft pinching and bridging issues. The optimization is to gradually increase the weight of the pinching process window condition over the bridging one. The following results are comparing the original recipe with two optimized recipes, where Recipe 1 is giving higher weight to the pinching condition than Recipe 2.

According to the count of the verification markers shown in Table 1, the optimization experiments were able to effectively tackle the pinching, the via coverage issues, and the negative EPE, where all of these issues have the same root cause of less weight of the pinching PW condition.

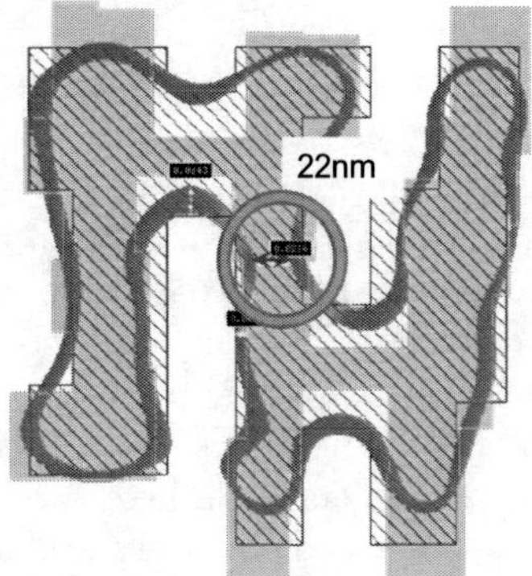

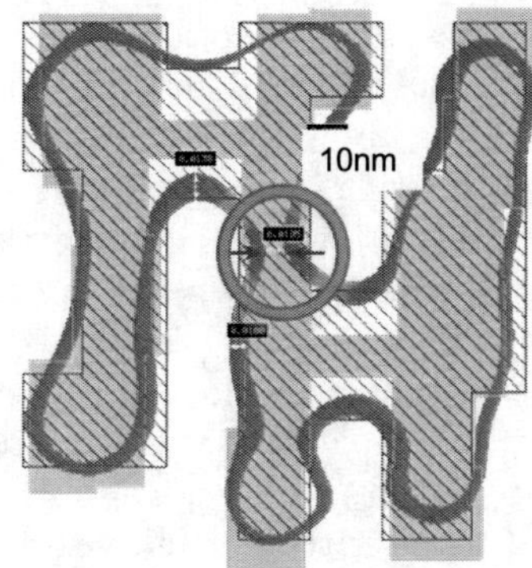

Figure 4: Example of two identical clips exhibiting context-dependent failure: this pattern prints acceptably (left) or fails (right) depending on its surroundings.

Table 1: Simulation check errors/markers count per each OPC recipe experiment

Simulation Checks	**OPC Recipe Experiments**		
	Original	**Recipe 1**	**Recipe 2**
Soft Pinch	750	215	2
Via Coverage	97	11	6
Large Negative EPE	514	225	182
Soft Bridge	318	319	320

As further analysis beside the verification markers count, the histogram distribution of the minimum width property of the inner PV Band contour is plotted for the residual soft pinch markers, for both original and optimized recipes 1 and 2 in Fig. 5. The mean value of the distribution is shifted towards the bigger width, which concludes a better resolution for the minimum printing dimensions.

The following figures (6–8) are illustrating the enhancement in the different verification aspects, related to the markers count in Table 1. The figures are showing the target (hashed) and post-OPC PV Band (dotted fill) layers. The figures are for the original and optimized OPC recipes.

The aim of the OPC recipe refinement approach is to try as much as possible to enhance the resolution of all reported yield limiting patterns, so as not to put any extra limitations on the design rules. This approach is assuring better stability for manufacturing newly introduced design patterns, and simultaneously keep adequate flexibility and wider scope for the layout design styles. Consequently, if some yield limiting patterns are not resolved by refining the OPC production recipe, then a specific local enhancement treatment is applied using the ILT OPC engine. The ILT OPC solution is an optimum approach for such critical spots without modifying the original OPC solution in the vicinity. However, if the OPC verification issues are still persisting after applying the pre-mentioned approaches, then the following technique is to revisit the design rules related to this pattern or topology.

5. DRC REFINEMENT

The results of Section 3 can also be used to adjust design manuals. The advantage in this case is that through early problem detection, the manuals may be adjusted earlier, resulting in reduced wasted design effort, test mask tape-outs, etc. Design rule refinements may occur throughout the technology development timeline but generally the farther along a process is in development, the less latitude there is for major changes to the rules. Thus, in the latest stages any yield problems may be pushed fully into process and/or OPC refinement efforts. Nonetheless, design rule decisions in the early development stage may be just as problematic, because an apparent yield problem may vanish as the technology improves and process windows tighten. Ruling out a design construct too early may end up making a technology less competitive from a design customer standpoint.

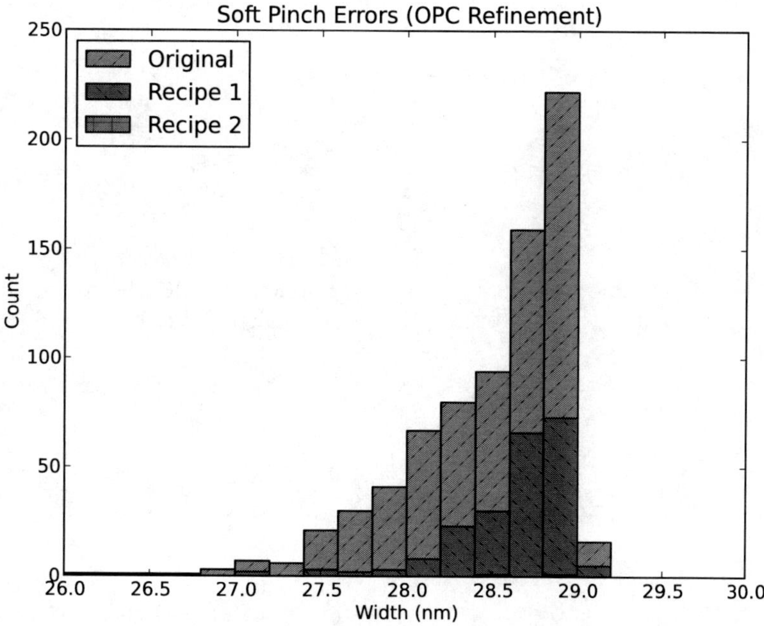

Figure 5: Histogram of soft pinching minimum width value, for original and optimized OPC recipes.

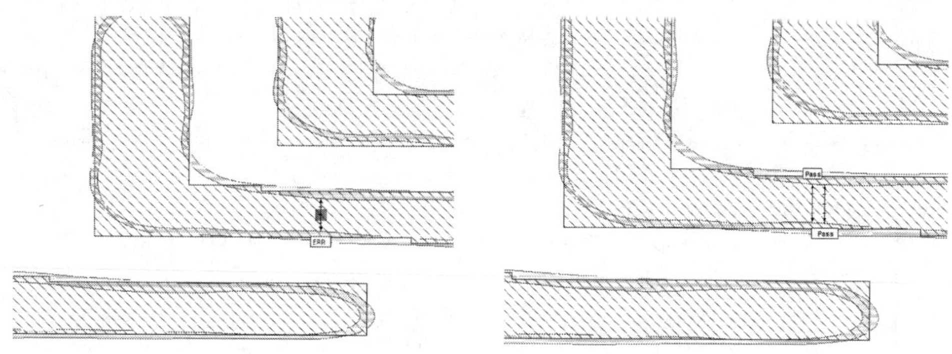

(a) original OPC Recipe (pinching) (b) optimized OPC Recipe (no pinching)

Figure 6: PV Band simulation (soft pinch check) for original and optimized OPC recipes.

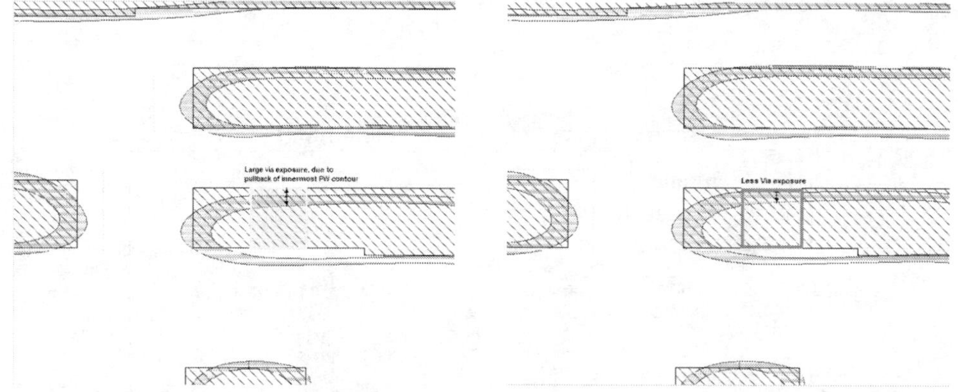

(a) original OPC Recipe (via exposed) (b) optimized OPC Recipe (via covered)

Figure 7: PV Band simulation (via coverage check) for original and optimized OPC recipes.

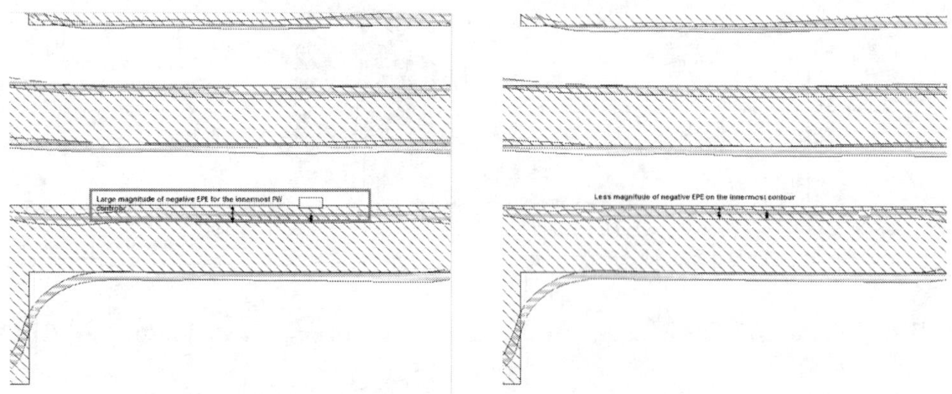

(a) original OPC Recipe (large EPE) (b) optimized OPC Recipe (small EPE)

Figure 8: PV Band simulation (large magnitude negative EPE check) for original and optimized OPC recipes.

One way to reduce the reliance on early process data is to make decisions using best case simulation data, i.e. with reduced sigma process variation conditions and constant threshold models with ILT in place of OPC. Engineers must strike a balance between too conservative and too aggressive simulation conditions. Random pattern generation can help this process by adding layout context to the simulation axes. A pattern that prints poorly in a variety of different layout contexts makes a strong case to be ruled out, whereas a pattern that fails in only certain contexts may be enabled with OPC recipe refinements.

Although this stage is highly interdependent on negotiations between the design and process teams, in this example we have identified two topologies that could be further stipulated in the DM in order to reduce the probability of creating yield limiting constructs on real layout. The advantage is that the known bad, or legal yet unrealistic patterns are identified, and early corrective action can be taken before problems become surprises.

Figure 9 is showing two examples of yield-limiting layout patterns with low design value that should be ruled out of the design rules. The topology of pseudo u-shapes with nearby corners or line ends will need to be highlighted in the design manual for avoidance. The verification simulation is highlighting the spots of yield limiting patterns after applying the different OPC recipes.

The verification simulation is executed again after excluding the pre-mentioned pattern topology, where Figure 10 is illustrating the enhancement in the histogram distribution using the exact same OPC recipe. For the soft pinch markers, the total count was reduced by 5%.

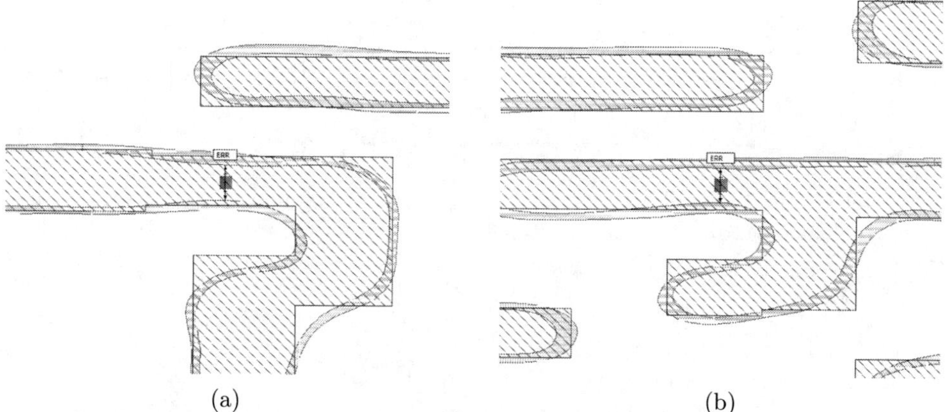

(a) (b)

Figure 9: Yield limiting patterns that need to be ruled out of the recommended design styles.

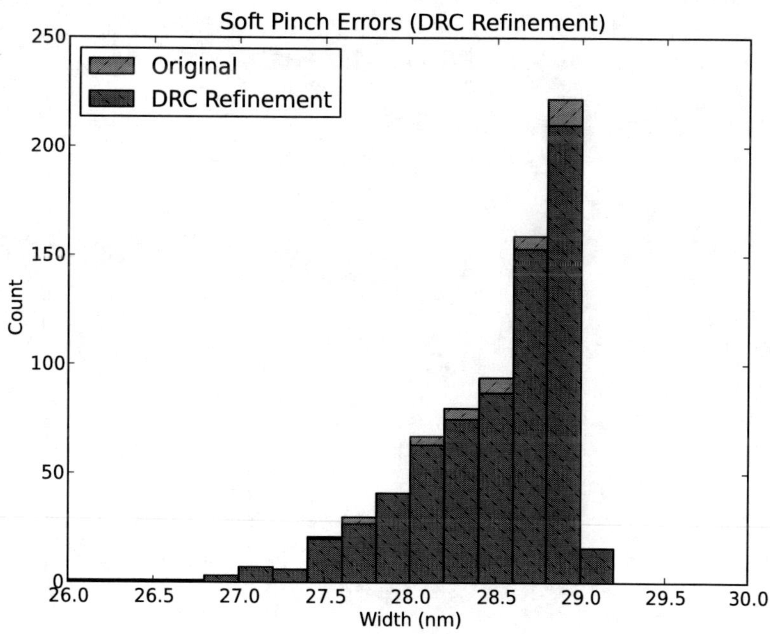

Figure 10: Histogram of soft pinching minimum width value, for remaining and excluded markers respectively.

6. CONCLUSIONS

In summary, we have described our core flow to Design and Patterning Explorer. We successfully generated a realistic 22nm Mx style layout through random generation, extracted patterns, and annotated them with PV band data from OPC and optical verification. Through the application of the results of this core flow to OPC Recipe Refinement and DRC Refinement, we have shown an overall reduction of hotspots by 83% over two OPC recipe refinement iterations, with room for continued improvement. Simultaneously, DRC refinement led to a 5% reductions in hotspots by removing two problem patterns from the allowed set of design rules.

ACKNOWLEDGMENTS

We acknowledge Ioana Graur, Ian Stobert, and Dmitry Vengertsev for assistance with OPC recipes, providing design content, and for many helpful technical discussions.

REFERENCES

[1] Northrop, G., "Design technology co-optimization in technology definition for 22nm and beyond," in [*VLSI Technology (VLSIT), 2011 Symposium on*], 112–113, IEEE (2011).

[2] Yeric, G., Cline, B., Sinha, S., Pietromonaco, D., Chandra, V., and Aitken, R., "The past present and future of design-technology co-optimization," in [*Custom Integrated Circuits Conference (CICC), 2013 IEEE*], 1–8, IEEE (2013).

[3] Ghaida, R. S., Badr, Y., and Gupta, P., "Pattern-restricted design at 10nm and beyond," in [*Computer Design (ICCD), 2014 32nd IEEE International Conference on*], 308–310, IEEE (2014).

[4] Badr, Y., Ma, K.-w., and Gupta, P., "Layout pattern-driven design rule evaluation," *Journal of Micro/Nanolithography, MEMS, and MOEMS* **13**(4), 043018–043018 (2014).

[5] Teoh, E., Dai, V., Capodieci, L., Lai, Y.-C., and Gennari, F., "Systematic data mining using a pattern database to accelerate yield ramp," in [*SPIE Advanced Lithography*], 905306–905306, International Society for Optics and Photonics (2014).

[6] Maeda, S., Ogawa, R., Shibazaki, S., and Nakajima, T., "Novel method for quality assurance of two-dimensional pattern fidelity," in [*Advanced Lithography*], 65211B–65211B, International Society for Optics and Photonics (2007).

[7] See for instance: Mentor Graphics Corporation, "Calibre LFD." `http://www.mentor.com/products/ic_nanometer_design/design-for-manufacturing/calibre-lfd/`.

Comparison of OPC job prioritization schemes to generate data for mask manufacturing

Travis Lewis[1a], Vijay Veeraraghavan[a]
Kenneth Jantzen[b], Stephen Kim[c], Minyoung Park[c], Gordon Russell[c], Mark Simmons[c]
[a]GLOBALFOUNDRIES, 5113 Southwest Parkway, Austin, TX 78735, [b]Mentor Graphics, 5000 Plaza on the Lake, Austin TX, 78746, [c]Mentor Graphics, 46871 Bayside Parkway, Fremont,CA 94538

ABSTRACT

Delivering mask ready OPC corrected data to the mask shop on-time is critical for a foundry to meet the cycle time commitment for a new product. With current OPC compute resource sharing technology, different job scheduling algorithms are possible, such as, priority based resource allocation and fair share resource allocation. In order to maximize computer cluster efficiency, minimize the cost of the data processing and deliver data on schedule, the trade-offs of each scheduling algorithm need to be understood. Using actual production jobs, each of the scheduling algorithms will be tested in a production tape-out environment. Each scheduling algorithm will be judged on its ability to deliver data on schedule and the trade-offs associated with each method will be analyzed. It is now possible to introduce advance scheduling algorithms to the OPC data processing environment to meet the goals of on-time delivery of mask ready OPC data while maximizing efficiency and reducing cost.

Keywords: foundry, OPC, cycle time, resource allocation, job scheduling

1. INTRODUCTION

The design and manufacturing of an integrated circuit chip at the leading edge nodes is a long and complex process with many individual steps that have to be completed for the chip to function. One step in the manufacturing process that is often overlooked is the transformation of the design GDS into mask ready data that can be fractured and written. This process is a very compute intensive process with a single layer requiring thousands of CPU-hours to complete [1]. A challenge for the semiconductor industry is how to manage the need for these very complex full-chip lithographic simulations with the runtime it takes to generate the results and to do this while maintaining the schedules and cycle times expected by the final customer.

In addition to the rising complexity of each simulation run the number of concurrent products in process in a foundry's tape-out environment is also on the rise. Each product's reticles have a required delivery date to the fab that often results in competition for compute resources in the tape-out environment. Manual management of these conflicts in priorities or schedules is no longer a viable solution.

The generation of OPC data for the mask is essentially the first step in the manufacturing process for an advanced semiconductor device. Just as the fab has turn-around time (TAT) requirements that they must meet to deliver the chip on schedule, the generation of mask ready OPC data must also meet their cycle time targets. An integrated circuit requires multiple masks each requiring their own OPC computation. The process of delivering a complete mask set to the fab on schedule requires managing multiple priorities between different masks and priorities between different devices. This paper will look at the application of prioritization schemes using Calibre Cluster Manager (CalCM) and the impact on the compute cluster utilization and cycle time.

1 travis.lewis@globalfoundries.com

Design-Process-Technology Co-optimization for Manufacturability IX, edited by John L. Sturtevant, Luigi Capodieci, Proc. of SPIE Vol. 9427, 942711 · © 2015 SPIE · CCC code: 0277-786X/15/$18 · doi: 10.1117/12.2086927

2. DISCUSSION OF RESOURCE MANAGEMENT SYSTEM

2.1. Challenge of scalability

The challenge of achieving OPC runtimes that meet manufacturing cycle times can only be achieved with a distributed computational architecture. The OPC tool must parallelize the computation so that it can be run on multiple processors. Scalability is the measure of how well an operation utilized all of the valuable compute resources. With a highly scalable operation, the runtime can be reduced by adding more processors to the operation.

In a typical job to generate the OPC mask data, there are a number of different operations used to generate the final data. There is the OPC operation which simulates the lithographic and etch process and moves edges of the design so the final image on the wafer matches the design. There are also Boolean operations, different sizing operations and operations to add sub-resolution assist features to design which are all part of the job to generate the OPC mask data. A typical OPC job may have over a 1000 operations as part of the total run.

From a cycle time perspective it is important to have all of those operations scale as high as possible so that hardware can be used to reduce runtime. However, it is not possible to have all operations scale identically. The result is that in a single OPC job each operation will have a different scalability. Figure 1 shows the scalability for each operation in a typical OPC job. The scalability ranges from around 400 to 3500 cpu cores. Some operations can scale even higher but are limited by the number of cpu cores assigned to the job.

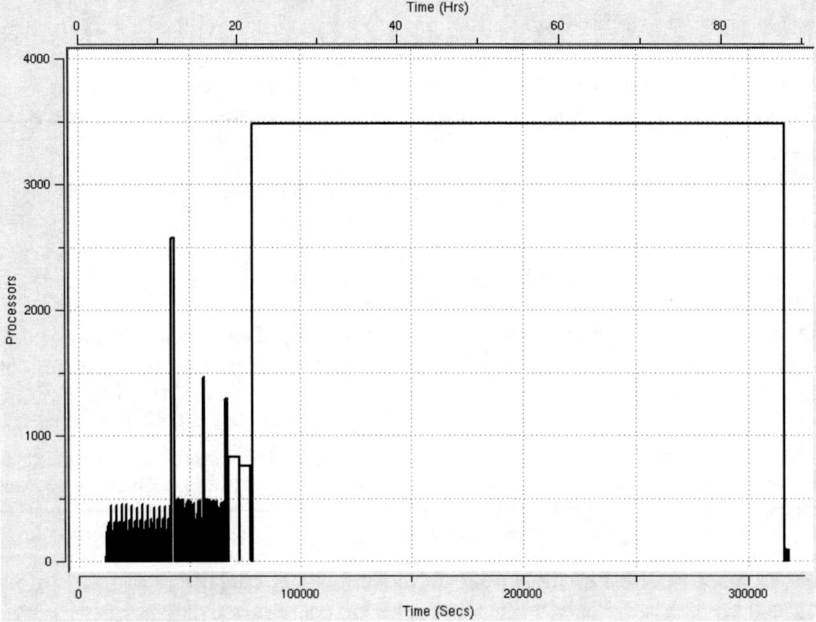

Figure 1 – Scaling of a typical OPC run. Each bar represents the scaling of the specific operation in the OPC run.

2.2. Concept of Demand

CalCM is a tool in the Calibre platform that manages multiple running Calibre jobs in a compute cluster. CalCM operates with a concept of demand which is the difference between the scalability of an operation and the actual number of processors used by the job. A positive demand represents the case where the operation has a greater scalability than the number of processors used by the job and so the job could use additional compute resources. A negative demand occurs when the operation has more compute resources than it requires and therefore resources can be removed from the job without impacting the runtime of the job.

CalCM has the ability to dynamically add and remove compute resources from a running job. Combining this function with CalCM's knowledge of the jobs current demand, the utilization of the entire compute cluster can be maximized. CalCM will remove compute resources from jobs without demand and assign resources to jobs with positive demand. For the job in figure 1, CalCM would only assign compute resources up to the scalability of the operation. So the first 20 hours of the run, the number of resources assigned to the job would be around 500 and when the job reached the highly scalable operation CalCM would assign more resources to the job. The experiment will look at CalCM's effectiveness in optimizing the cluster utilization through a variety of operation scenarios.

2.3. Job Priority and CalCM

Since CalCM is interacting with all running jobs in the compute cluster and adjusting resources based on the specific jobs demand, there is the opportunity to control how those resources are assigned based on a priority mechanism. One operation mode is a fair share mode, where the compute resources are shared between all jobs equally. In priority mode, the job with the highest priority will receive all the resources when it has positive demand. In this mode, a single job will run as fast as possible while the rest of the jobs in the cluster continue to run with the minimum number of resources waiting for the priority job to finish.

3. EXPERIMENTAL DESIGN

The experiment consisted running a total of 16 jobs on a cluster of 1440 cores. The jobs consisted of 8 OPC jobs representing a subset of the critical layers for a design. There are both front-end and back-end layers. In addition to the 8 OPC jobs, each OPC job was followed by the OPC verification job for the same layer. The computer hardware has a great influence on the runtime of the individual job, so for this set of experiments the cores used were limited to a single generation of processor to make the runtime comparisons as accurate as possible.

Four different priority schemes were studied using CalCM. These four schemes were compared to a baseline which represents the current operation mode without CalCM. The baseline represents the current mode of operation in a production mask data prep environment. Jobs are run serially and the cores are assigned at the beginning of the job and remain constant throughout the job. For this set of tests, the baseline jobs were run with 720 cores and there were two parallel threads of jobs. So the baseline was using a total of 1440 cores with a total of two jobs running at any one time. This paper will measure the cluster utilization with each priority method and compare the runtimes to the baseline.

4. RESULTS

4.1. Three thread with ordered priority

In this prioritization method there are three threads of running jobs. The highest priority is on job A. Job B through E have the next highest priority and the jobs F through H are the lowest priority.

The benefit of CalCM is immediately clear from comparing the cumulative runtime of all 16 jobs with CalCM vs. the baseline. With CalCM, the 16 jobs are completed 13.5 hours faster. This is because the final jobs are able to take advantage all 1440 cores, whereas the baseline is limited the number of cores the job began with. Cumulative runtime was also reduced because of the improved cluster utilization. Since the each job enters the highly scalable operations at different times, CalCM can make sure that the compute resources are available for the highly scalable operations which results in faster runtimes. Figure 2 shows the runtime compared to the baseline for this prioritization.

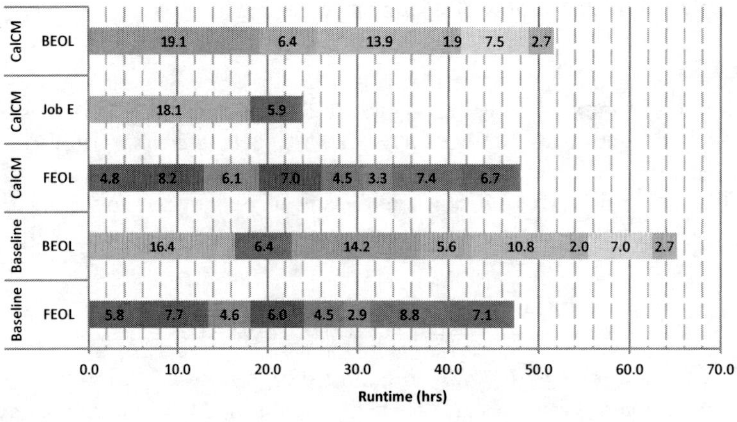

Figure 2 – Runtime vs. baseline jobs

Figure 3 shows the cluster utilization for this priority scheme. The cluster utilization is defined as the assigned CPU hours divided by the total available CPU hours, which is 75% for this case. The graph shows the relationship between the demand and utilization. When the demand is below 0, this indicates that none of the running jobs can effectively use additional cores. Though resources are available there are no jobs that can use the resources so the utilization is low. There are many places where the cluster utilization is low and the cumulative demand is below zero. This situation occurs when jobs end and a new job begins. When a job ends, the compute resources become available, but they are not needed by the next job immediately since typically the initial operations of a job have lower scalability. By increasing the number of jobs in parallel the cluster utilization will be improved because there is a higher chance of at least one job being in a high scalable operation and being able to make use of the resources immediately. This increased utilization leads to a reduction in the cumulative runtime.

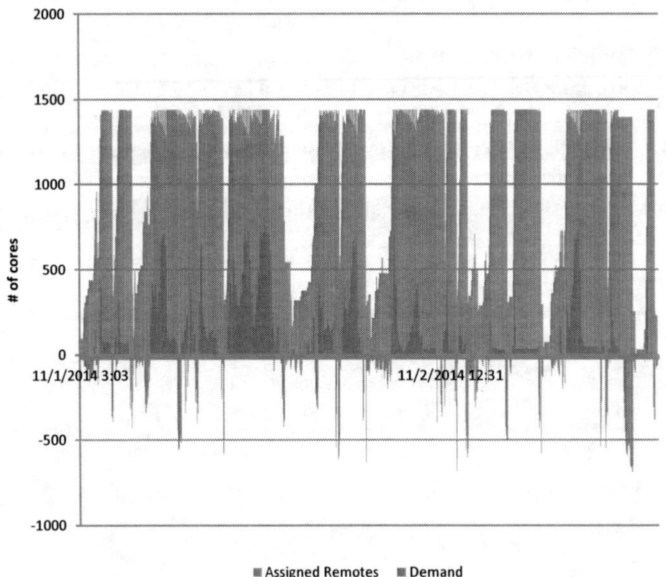

Figure 3 – Cluster Utilization and Demand

4.2. Eight thread with fair share priority

With this prioritization scheme, all eight jobs are started in parallel. No priority is assigned so CalCM will divide the available resources among the jobs equally provided the job has a positive demand.

Figure 4 shows the runtime using the fair share method versus the baseline. The cumulative savings was 32.4 hours compared to the baseline. The trade-off is that individually the jobs ran longer because on average the jobs had less than 720 cores, but the increased cluster utilization lead to an overall runtime savings.

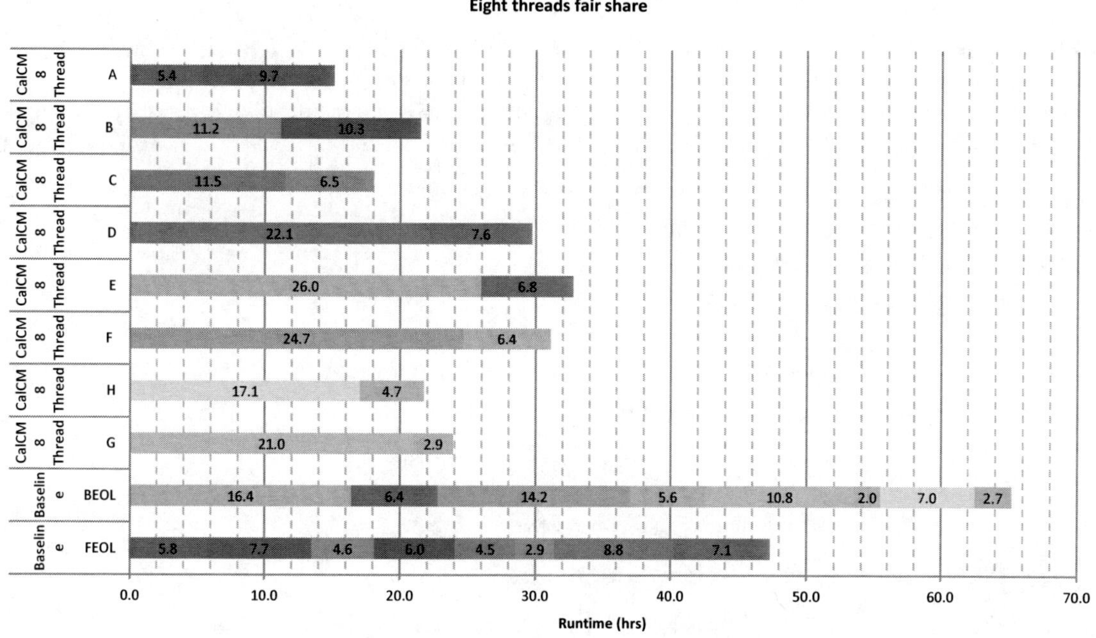

Figure 4 – Runtime with eight thread fair share vs. baseline

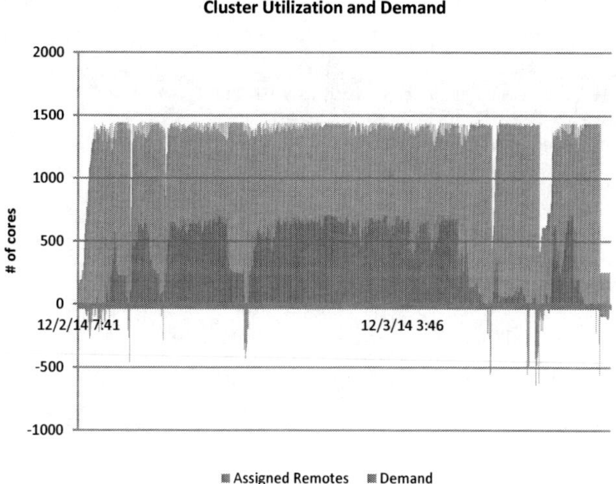

Figure 5 – Cluster utilization with 8 threads and fair share priority

The cluster utilization with eight threads and fair share priority is 93%. This is substantially higher than the previous priority scheme. Increasing the number of jobs in the cluster at one time means that there is a greater overall demand. So if one job doesn't have demand, there will be another job that does. This is shown by the demand curve in figure 5.

The amount of time with negative demand is very short. The demand curve also indicates that increasing the cluster size would benefit the runtimes of the jobs because the jobs could scale to more remotes.

4.3. Eight thread with ordered priority

It is clear that having more jobs in parallel is good for cluster utilization. However the increased runtime the individual jobs are a concern in a production environment where the runtime of the individual jobs matter. So in this case, there were still eight jobs run in parallel, but with the added constraint of the priority from A to G. The result was that the jobs finished roughly in the order of priority while still finishing 29 hours ahead of the baseline.

Figure 7 is the cluster utilization. In this case the utilization is 86%, which is above the utilization of when there are only 3 threads, but below the utilization of the fair share priority scheme. This is expected because using an ordered priority causes more resource shifting between jobs. This shifting has a minor price in the utilization because even though it happens relatively quickly, it is not instantaneous. So shifting resources frequently will lower the cluster utilization.

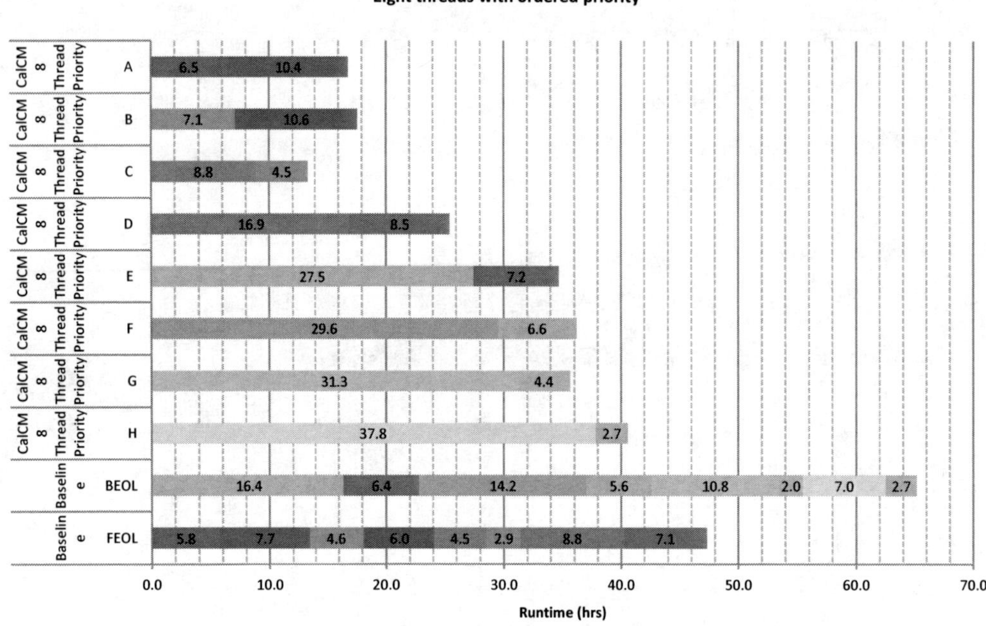

Figure 6 – Runtime with eight threads and ordered priority

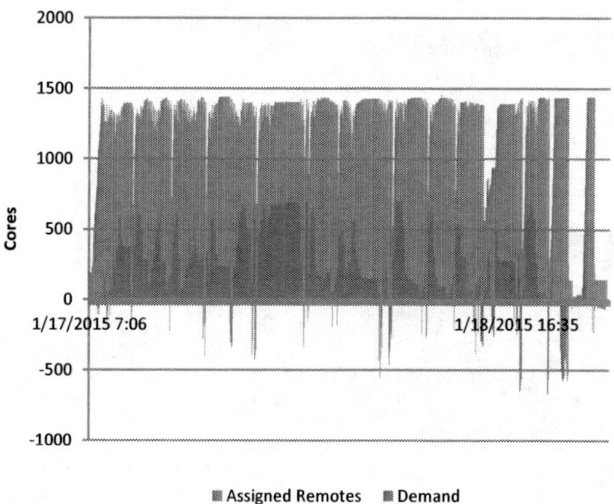

Figure 7 – Cluster utilization eight threads with ordered priority

4.4. Seven thread with single high priority

A very common scenario in the production tape-out operations environment is the introduction of the critical job into the production pipeline. In the current environment without CalCM, the introduction of a critical job is very disruptive to the production flow. Compute resources need to be made available for the critical job which means having to kill running jobs to free up compute resources for the critical job. However, with CalCM the resources can by dynamically shifted between jobs. In this test, the same seven jobs are started in parallel with the same priority so that CalCM is dividing resources among the seven jobs based on a fair share allocation. Three hours after the seven jobs are started a critical job is started in the cluster. This job is started with a higher priority setting which means that while it is running, CalCM prioritizes its need for resources over that of the other 7 jobs. Figure 8 shows the allocation of resources for each job during the experiment. Job E is the priority job and during the OPC and ORC operations, which are highly scalable with a large demand for cores, CalCM ramps the resources assigned to the job until all of the other jobs are running at the minimum number of cores allowed for an individual job.

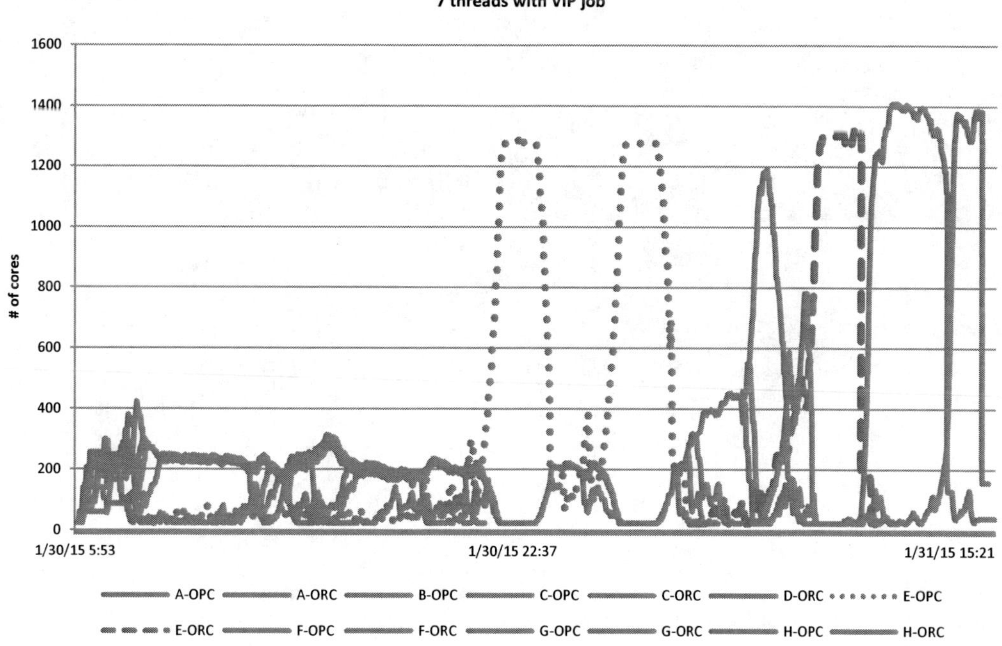

Figure 8 – Resource assignment for a high priority job.

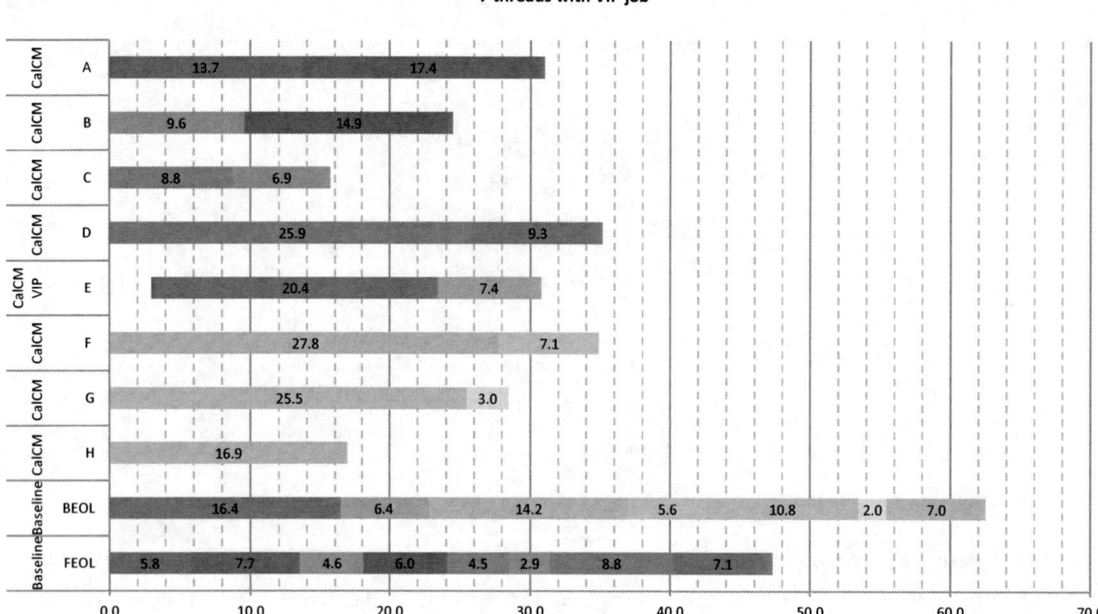

Figure 9 – Seven threads with single VIP priority job

The VIP priority job was 6.9 hours faster than the same job run in the previous fair share mode so the priority mechanisms worked as expected and the runtime decreased because compute resources were assigned to the job because of its priority. The VIP job however was 5 hours slower than the baseline job. We suspect the reason for this is introducing a priority job into a full cluster, the resource are not shifted to the priority job at the start of the job. We suspect that starting the job with a high number of cores would reduce the runtime even further.

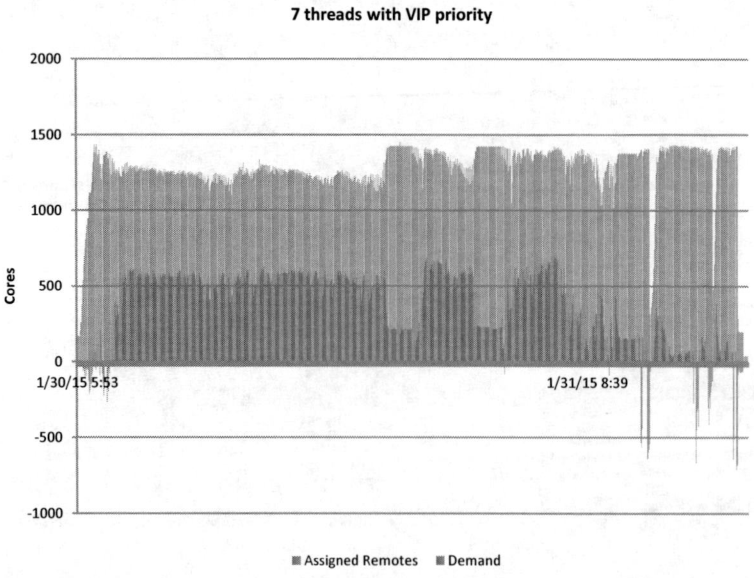

Figure 10 – Cluster Utilization with 7 threads and single critical job

4.5. Summary of results

This set of priority experiments show the relationship between priority, cluster utilization and cycle time savings. The higher the cluster utilization the larger the runtime savings is. This is important because the runtime savings represents time that the compute resources can be spent running other jobs. Running more jobs in parallel is beneficial from a utilization perspective. However, adding to many jobs to a cluster of fixed size will eventually lead to overall slower runtimes because the compute resources will be spread among too many jobs. Finally, priority should be used judiciously because there is a penalty associated with constantly shifting resources from one job to another.

Table 1 – Summary of runtime savings and cluster utilization

	3 Threads & ordered priority	8 Threads & fair share	8 Threads & ordered priority	7 threads and single high priority
Runtime Savings vs. baseline	13.5 hrs	32 hrs	29 hrs	30 hrs
Cluster Utilization	75%	93%	86%	88%

5. CONCLUSION

Controlling cycle time in the post tape-out production flow is critical for meeting IC manufacturing cycle times. One method to help control cycle times is with a resource management tool that can improve the utilization of the available compute resources and control the priority of running jobs.

REFERENCES

[1] Spence, C., Goad, S., "Computational requirements for OPC," Proc. SPIE. 7275, Design for Manufacturability through Design-Process Integration III, 50-59 (2009)

VLSI Physical Design Analyzer: A profiling and data mining tool

Shikha Somani[1]*, Piyush Verma[1], Sriram Madhavan[1] Fadi Batarseh[1], Robert C. Pack[1] and Luigi Capodieci[1]

[1]*GLOBALFOUNDRIES Inc, 2600 Great America Way, Santa Clara, CA, USA 95054*

*shikha.somani@globalfoundries.com

Abstract

Traditional physical design verification tools employ a deck of known design rules, each of which has a pre-defined pass/fail criteria associated with it. While passing a design rule deck is a necessary condition for a VLSI design to be manufacturable, it is not sufficient. Other physical design profiling decks that attempt to obtain statistical information about the various critical dimensions in the VLSI design lack a systematic methodology for rule enumeration. These decks are often inadequate, unable to extract all the interlayer and intralayer dimensions in a design that have a correlation with process yield. The Physical Design Analyzer is a comprehensive design analysis tool built with the objective of exhaustively exploring design-process correlations to increase the wafer yield.

1. Introduction

As the complexity of VLSI designs continues to grow, undesirable and unexpected interactions between the geometrical constructs in the physical layout and manufacturing process flows are emerging. These interactions are particularly hard to predict as the characteristic length scale of the defect inducing design geometry is highly variable [1]. Ideally, a physical design verification step should be able to catch all the geometrical constructs in the physical design that can be potential yield detractors. However, the traditional DRC decks lack the sensitivity required to catch all that is undesirable in the physical design. The design rule decks today are based on a large number of pre-decided dimensional configurations that are checked for in a design and assigned either a pass or fail flag. While physical design is very rich in intralayer and interlayer dimensional data, only a very small fraction of this data is really mined and analyzed when performing DRC verification. The key challenges in extracting all relevant dimensional information from a VLSI design include a systematic methodology to enumerate all relevant geometrical constructs, storing the enormous amounts of data in an efficient database system and a measurement tool with low runtimes.

A true physical layout profiling/verification tool would be one that does not make any apriori assumptions about the configurations or their dimensions that it expects to find in the design. It would not only classify known geometries into pass/fail buckets, but also discover new geometries and their dimensions for which no process data is available. Thus, in addition to generating the supervised data sets that the DRC decks attempt to do today, it would also generate input data for unsupervised learning such that new design-process correlations can be built over time. The design analyzer tool presented in this work is built with this aim.

Design-Process-Technology Co-optimization for Manufacturability IX, edited by John L. Sturtevant, Luigi Capodieci, Proc. of SPIE Vol. 9427, 942712 · © 2015 SPIE · CCC code: 0277-786X/15/$18 · doi: 10.1117/12.2087078

2. The Physical Design Analyzer tool

The Physical Design Analyzer consists of a comprehensive rule deck generator, a module to measure physical dimensions (different decks have different corresponding measurement tools), a data transformation and compression module to filter relevant data, a computational module to perform statistical analysis on the data, a database to store data from different designs and a reporting engine to generate custom reports. These various components are shown in figure 1. In addition to reporting all critical intralayer and interlayer dimensions, this tool has the ability to make comparisons across designs and report the key dimensional differences observed between them. Augmented with process data, the tool can be extended to build design scoring models to predict wafer yield. The following sections describe each of the modules of the tool in detail.

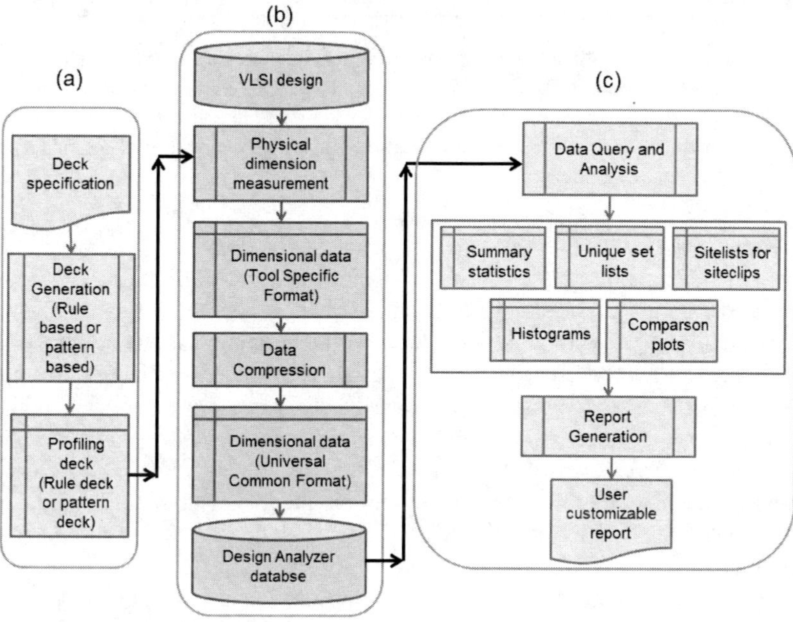

Fig. 1. The flowchart showing various modules within the Physical Design Analyzer tool.

3. Deck generation and execution

The deck is synonymous to the code which when executed with the corresponding software program, outputs the dimensional measurements on a VLSI design. The deck generation module is the first module in the flow shown in figure 1a. Here, we demonstrate a framework to build a deck that comprehensively measures all relevant dimensions in a design. The framework was built based on the process assumptions used to define the technology. It can be understood by making a few simple observations about polygons in a plane.

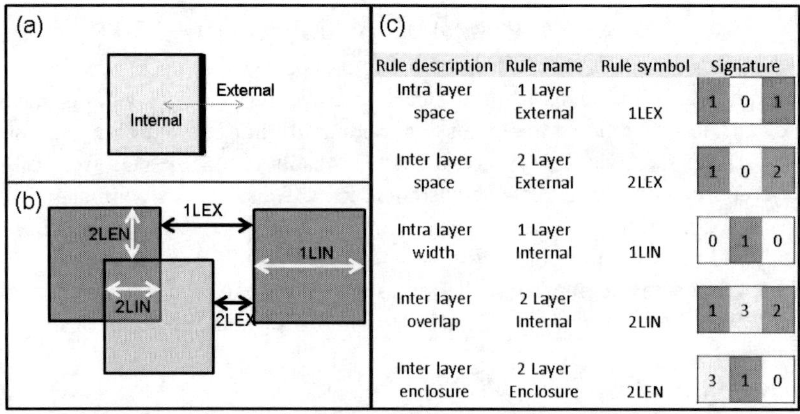

Fig. 2. (a) Every edge of a polygon has two sides, one that faces the interior of the polygon and the other that faces the exterior of the polygon. (b) Various combinations of edges result in internal, external, enclosure and overlap dimensions within and between polygons. (c) All combinations of intralayer and interlayer dimensions, and the corresponding rule names with pattern representation.

Every edge of the polygon has two sides, one side that faces inwards towards the interior of the polygon and the other that faces outwards towards the exterior of the polygon as shown in figure 2a.Various combination of edge orientations with respect to each other give rise to all dimensions in a design. Figure 2b shows all five different combinations of edge orientations within and between two different layers in a design. The dimension names are made up of two parts. The first part is the number of layers involved in the measurement. This is equal to 1L when the measurement is an intralayer measurement and 2L when it is an interlayer measurement. The second part of the name is either 'IN' for internal, 'EX' for external and 'EN' for enclosure. By using this naming convention, the five combinations can be expressed as 1LEX, 2LEX, 1LIN, 2LIN and 2LEN as shown in figure 2c. This figure also shows an equivalent pattern representation of each of those dimensions using a squish signature where Layer 1 is represented as '1', Layer 2 as '2', Both layers as '3' and lack of any layer as '0'.

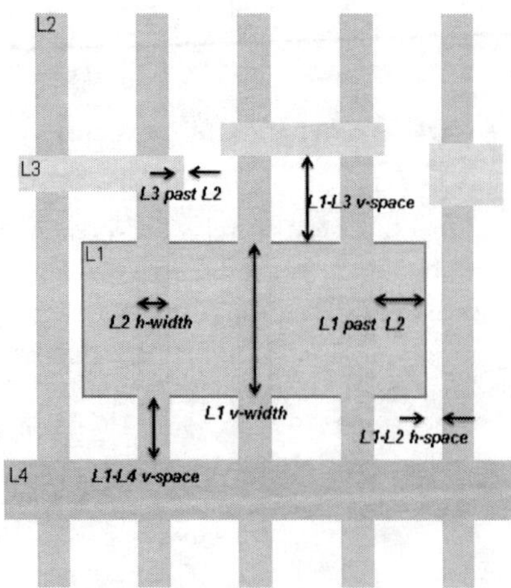

Fig. 3. Relevant dimensions in a design snippet

As an example of relevant dimensions in a physical layout, a snippet of the VLSI design is shown in figure 3. This figure shows a subset of front end layers with relevant dimensions marked. Each of these dimensions can be classified in one of the five categories described in figure 2c. Furthermore, since all edges in the design are orthogonal, for each of the five rules categories, the dimension can either be measured in the horizontal direction or in the vertical direction.

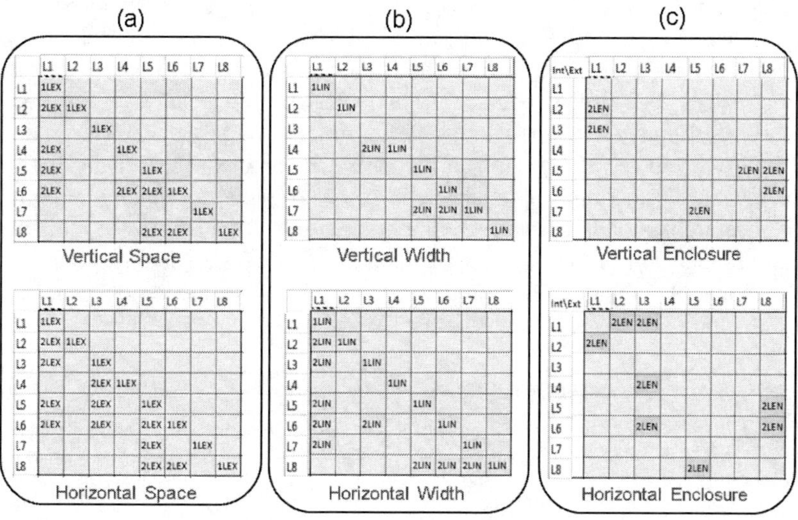

Fig. 4. Critical dimension enumeration matrix for a subset of layers in the physical design. The dimensions are of three kinds - internal (width), external (space) and enclosure; measured along horizontal and vertical directions.

To systematically enumerate all relevant dimensions in a design, a matrix based methodology is used as shown in figure 4. The matrix is built by listing the design layers in process sequence along the rows and columns. One such matrix is built for each of the internal, external and enclosure dimensions, split by horizontal and vertical directions. Each of the cells in the matrix is then filled in by the corresponding rule name if the process assumptions of the technology deem it as a critical dimension. For example, in the matrix on the top of figure 4a which enumerates a vertical space measurement (external), the top left cell is filled in with 1LEX denoting a critical vertical space measurement. The cell in the second row, first column is filled with 2LEX denoting a critical space dimension between the layers L1 and L2 in the vertical direction. The cell in the third row, first column is left blank as the vertical space dimension between layers L1 and L3 is not critical from a process standpoint.

The critical dimension enumeration matrix has the following salient features. The matrix is always square as the same layers are along the row and column directions. The cells along the matrix diagonal are always filled in for the space and width dimensions as single layer widths and spaces are always critical from a process standpoint. The space and width matrices are also symmetrical about the diagonal as these measurements are independent of the measurement sequence - L1 to L2 is same as L2 to L1. The enclosure matrix however is asymmetric as enclosure involves looking internal from one edge towards the external of another. Then external L1 to internal L2 is not the same as internal L1 to external L2.

The deck generation module takes an array of matrices as an input and generates the corresponding rule or 1-dimensional pattern deck. This module supports two leading industry tools for dimension measurements and is capable of generating comprehensive decks for both. The exhaustive deck that is generated using the framework above is not the only deck that can be plugged into this flow. In fact, any complex pattern-based or complex rule-based deck can also be used equally conveniently in the flow without changing any other components.

4. Data extraction and storage

The output of running the design profiling deck on a given design consists of a polygonal marker at the design coordinate where a specific dimension is measured and the value of the dimensional measurements at the said marker.

The objective of the data extraction module is to extract only the relevant information from the profiling output such that (a) data from large designs across all layers can be stored in a reasonably sized database; (b) no statistical information is lost and yield predictions can be made and (c) a representative clip can be generated for subsequent analysis (simulations and model testing) Furthermore, since the deck generator module supports multiple tools for dimension measurement, the data extraction module needs to transform the tool-specific data to a standard format called as UCF (universal common format) shown in the figure 5. The data extraction module filters the profiling output such that the above mentioned goals are met and stores the normalized data (UCF) in a database.

Unique Dimensions of interest (DOI)					Representative Location (BBOX)				Count
Dim1	dim2	dim3	...	dimN	llx	lly	urx	ury	Count
...	...	...	...	...	...	...	...	...	...

Fig. 5. Universal Common Format

5. Data Analysis

The normalized data is used for subsequent data analysis including exploratory, confirmatory and predictive data analysis. For the example that we have used here to illustrate various aspects of design analyzer, the numerical value of the dimension is normalized and reported as 'Hotspot DFM Index' such that the value of the index is 0 at the DRC value and 1 at the DFM value.

One main component of statistical analysis is the graphical visualization of the data. Various different types of graphs are used to highlight different statistical aspects of the data as shown in figure 6. For example, box plots are used to show the variance of the data, bar graphs are used to show the frequency of each dimensional value, density plots are used to show normalized distributions etc. The graphical visualizations help clearly see the dimensional trends in the design.

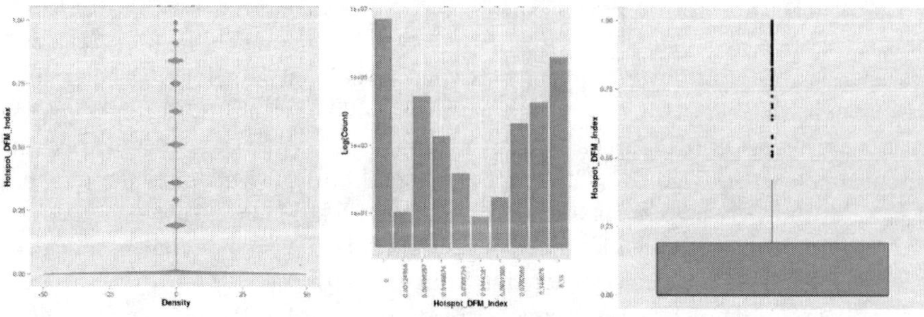

Fig. 6. Graphical visualization of dimensional data

The other important component of the analysis is the computation of statistical measures like mean, median, mode, minimum value, maximum value, quantiles, standard deviation, variance, number of unique counts etc. as shown in figure 7 These numerical measures help quantify the dimensional data variation. For example, the 'minimum value' of a dimension signifies how close the design is to DRC values with respect to that dimension, the 'mode value' depicts the most commonly drawn dimension, the 'number of unique values' signify how uniformly is a dimension used in the design, 'total count' signifies the usage frequency for a given dimension etc.

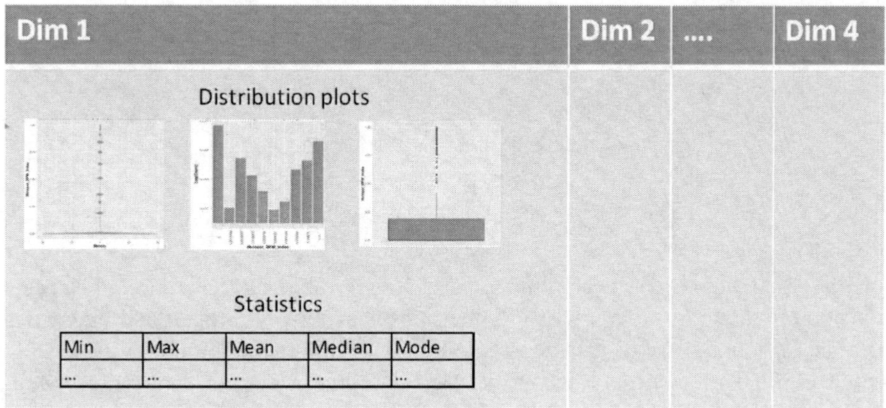

Fig. 7. Graphical visualization and statistical measures for different dimensions in a rule in a design

Another module in the design analyzer explores the relationship between different dimensions. Note that these dimensions can also be derived from the original dimensions specified in the rule. The module can be used to find correlations between different dimensions (derived or default) within a rule and across rules. One such example is combination of gate lengths and widths (say unique sets of (length, width)) in a design. Such combinations or correlations help analyze patterns with a much higher complexity than regular DRCs.

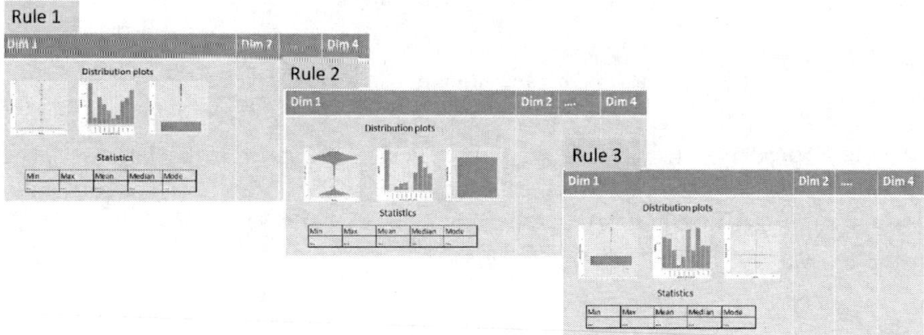

Fig. 8. Comparison between different rules in a design

The sitelist module of design analyzer generates a sitelist based on a certain criteria that depends on the value of the set of dimensions. The default criteria is the unique combination of all dimensions, but the custom criteria for different applications can also be used. The sitelists are then used to generate clips of the design for subsequent analysis like simulations or library generation [2].

Another module in the design analyzer compares the dimensional distributions across different designs. This helps not only in detecting key design style differences, but also helps in detecting which design constructs might be driving the yield in a certain direction when the yields of the 2 designs are different. It also enables the use of design analyzer as a QA tool for any process step that touches or modifies the physical design.

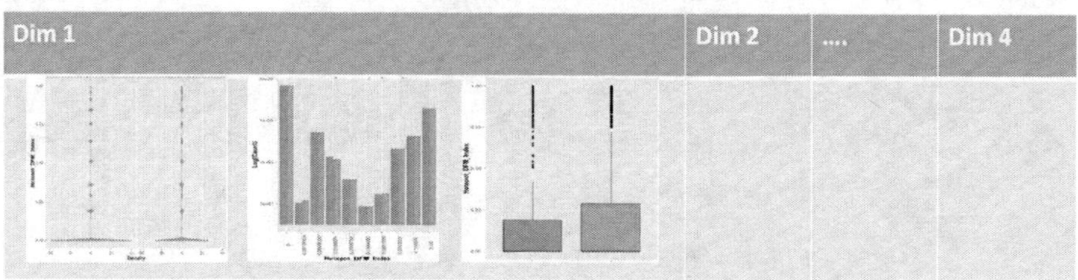

Fig. 9. Differentiating a rule between different designs

Note that each dimension in each rule in each design results in a distribution. Hence, manual detection of the most differentiating dimensions is extremely difficult and impractical. To algorithmically rank all the distributions based on their differences or similarities of interest, standard shifts in the statistical measures like Mean, Median, Mode etc. can be used. Moreover, comprehensive histogram comparison metrics like KL Divergence, Earth Mover's Distance etc come in handy to highlight certain starking differences between any 2 sets of distributions (comparison across rules, or across designs or a combination of the two).

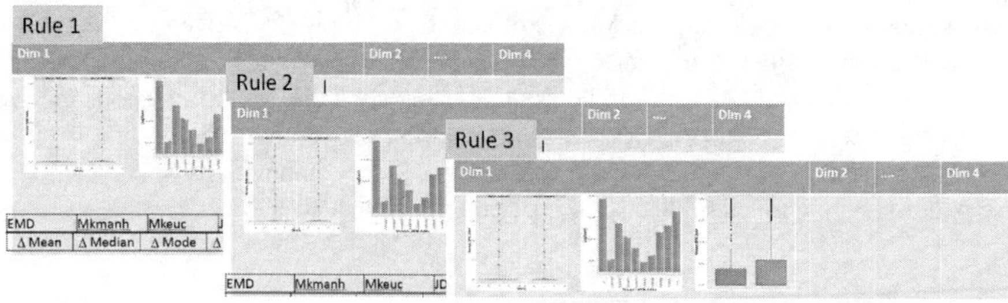

Fig. 10. Comparison between different rules in multiple designs

The basic set of analysis results for a set of (rule, design) are stored in the database. These are used to generate the comprehensive reports for any design. The custom analysis for a particular rule is done either real-time during report generation or triggered separately by user feedback.

6. Reporting

Once the data is analyzed, it may be reported in many formats to the users to address their specific requirements. The reports are generated from the analysis results stored in the database and hence can be generated fairly quickly with some real-time filtering or sectioning implemented in the reporting engine.

Some of the typical reports generated are meant to answer questions like "Which are the rules that are pushed in the given design?, How hard is a given design rule pushed in the given design?, Are there any design rule violations for the given rule?, What is the overall distribution of the dimensions for the given rule?"

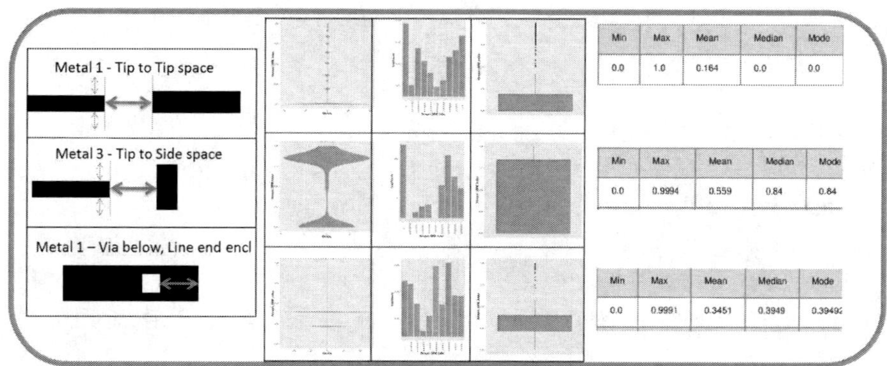

Fig. 11. Default profiling report for a design

Some questions require comparison analysis across different designs - "What is the difference in the design styles for the given rule in different designs?, What are the right thresholds for design constructs using that rule?, Which designs are most similar with respect to a critical rule?, Which design has less margin across all design rules?, What are the overall differences in the design styles?"

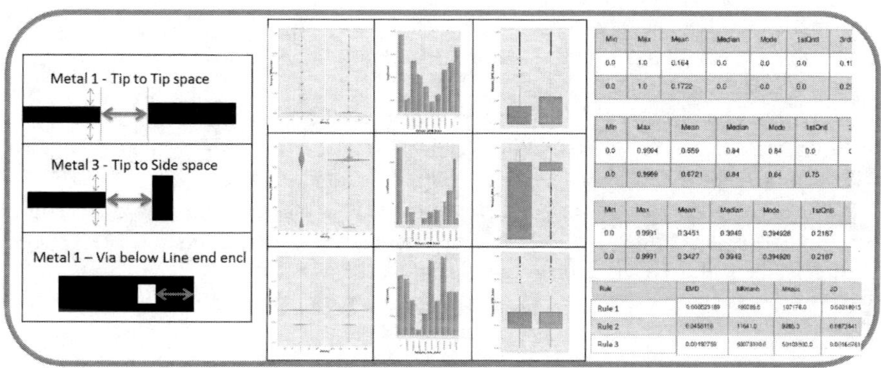

Fig. 12. Default report for comparison between different designs

7. Conclusions

The Design Analyzer is an end-to-end tool with comprehensive rule/pattern deck enumeration and generation capability. On the back-end, data is accumulated over time in the database that enables faster real-time report generation. The statistical engine of design analyzer has modules to analyze and compare histograms across rules, patterns and designs. The user friendly reporting engine generates rich and concise analysis reports. The tool can be extended to develop a comprehensive scoring model for VLSI designs based on all dimensional data. The tool helps capturing typical design styles across different designs for electrical/process model building and characterization. In future, a machine learning module will be plugged in to discover statistical signatures of dimensions of interest that show strong correlations with process yield.

References and links

1. V Dai, J. Yang, N. Rodriguez, L. Capodieci, Design for Manufacturability through Design-Process Integration, Proc. SPIE 6521, 65210A (March 28, 2007)
2. P. Verma, F. Batarseh, S. Somani, J. Wang, S. McGowan, S. Madhavan, Pattern-based pre-OPC operation to improve model-based OPC runtime, Proc. SPIE 9235, Photomask Technology 2014, 923506 (October 8, 2014).

The cell pattern correction through design based metrology

Yonghyeon Kim, Kweonjae Lee, Jinman Chang, Taeheon Kim, Daehan Han, Kyusun Lee, Aeran Hong, Jinyoung Kang, Bumjin Choi, Joosung Lee, Kyehee Yeom, Jooyoung Lee, Hyeongsun Hong, Kyupil Lee, Gyoyoung Jin

DRAM PA Team, Memory Division, Samsung Electronics Co., Ltd, Samsungjeonja-ro 1, Hwaseong-si, Gyeonggi-Do, Korea, Republic of

ABSTRACT

Starting with the sub 2Xnm node, the process window becomes smaller and tighter than before.
Pattern related error budget is required for accurate critical-dimension control of Cell layers. Therefore, lithography has been faced with its various difficulties, such as weird distribution, overlay error, patterning difficulty etc.
The distribution of cell pattern and overlay management are the most important factors in DRAM field. We had been experiencing that the fatal risk is caused by the patterns located in the tail of the distribution. The overlay also induces the various defect sources and misalignment issues.
Even though we knew that these elements are important, we could not classify the defect type of Cell patterns. Because there is no way to gather massive small pattern CD samples in cell unit block and to compare layout with cell patterns by the CD-SEM.
The CD- SEM is used in order to gather these data through high resolution, but CD-SEM takes long time to inspect and extract data because it measures the small FOV. (Field Of View)
However, the NGR(E-beam tool) provides high speed with large FOV and high resolution. Also, it's possible to measure an accurate overlay between the target layout and cell patterns because they provide DBM. (Design Based Metrology)
By using massive measured data, we extract the result that it is persuasive by applying the various analysis techniques, as cell distribution and defects, the pattern overlay error correction etc.
We introduce how to correct cell pattern, by using the DBM measurement, and new analysis methods.

Keyword : Cell, pattern overlay, Design based Metrology, Overlay, CD uniformity, 3sigma, Composite

INTRODUCTION

As semiconductor device shrinks down to sub 2X node, patterning controllability becomes the most important factor for lithography process. Various Kinds of patterning technology has been introduced in recent years not only lithography also double patterning, and etc.

Although there are diverse techniques to compensate the pattern errors, we are faced with the problem that still we cannot overcome specific issue such as the distribution of real cell pattern and overlay mismatch amount.

By using the DBM measurement method, we got the distribution of the Cell blocks, type of pattern defects and wafer uniformity. Moreover, we could analyze data which consider the surrounding block edges environment.
Cell block edge layout drawing issue is one of pattern issues causing factors.

The OPC(Optical Proximity Correction) has to consider surrounding environment, so Cell block edge layout has to be drawn differently compare with cell block center layout. Because of these Layout and environment difference, various errors are occurred in the real cell pattern.
The Cell distribution and the kind of defects were changed according to the each cell block surrounding environment in a chip. So this study is focused on the analysis of cell block edge.

Design-Process-Technology Co-optimization for Manufacturability IX, edited by John L. Sturtevant, Luigi Capodieci, Proc. of SPIE Vol. 9427, 942713 · © 2015 SPIE · CCC code: 0277-786X/15/$18 · doi: 10.1117/12.2085004

If we fail to identify these cell distributions, the process window would appear severely narrower than the real window and we may miss the chance to reinforce cell process window. We should be considered that cell size difference to target layout may have effects on yield of extreme edge chip.

We proceed with an analysis of the CD uniformity with the figure.1

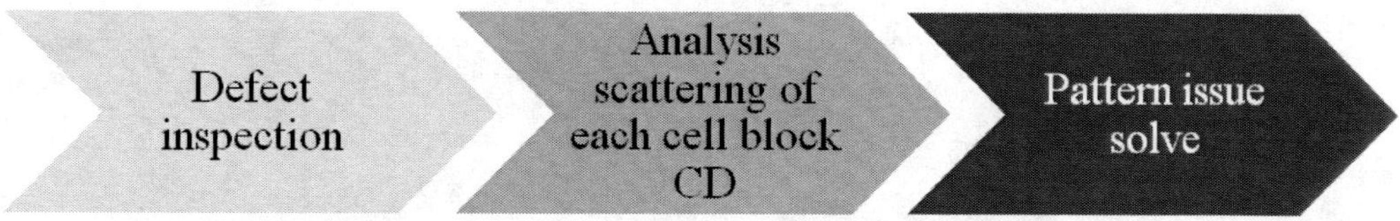

Figure.1 Normal pattern issue solution procedure

In spite of being the cell pattern having the same pitch, Bridge defects were occurred in the specific order of block edge. We tried to solve this pattern bridge by block CD uniformity analysis. But, It was difficult to discover singularities with the current normal solution procedure. Because, the scattering of cell block just showed the average CD and the peak location of the scattering.

We tried several fresh analysis methods such as composite technology, 2D color map, 3 sigma, X/Y direction string CD distribution analysis. After applying these methodologies, we could see a lot of information that we could not see so far. Using new analysis methodology, we found the reason of the pattern bridge from the inspected distribution. This problem was occurred by the pitch walking phenomenon due to DPT.

In order to solve this problem, we tried to optimize architecture process like the photo pattern size modification. Because of this research, the bridge problem appeared to be solved.

After we confirmed the failure analysis results, the pattern bridge problem that between upper layer and lower layer is still remained. This pattern bridge type is caused from this space pattern overlay error. In order to overcome this upper and lower layer bride, we have to develop the pattern overlay error correction technology. (a temporary theme)

While we proceed with this research, we got to know location of the real pattern on the silicon and the drawn layout is not same.

In this paper, we introduce the new analysis methodology and alternative overlay issues by using design based metrology tool. And how to calculate pattern overlay error amount in the cell area can precisely distinguish the mismatch by using in house tool. (Figure.2)

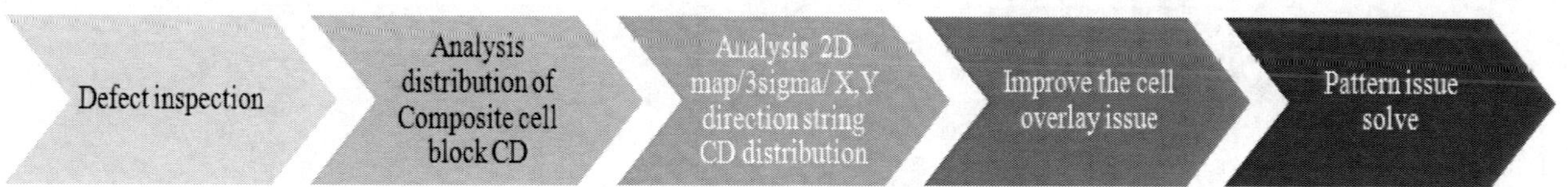

Figure.2 The new solution procedure

1. DBM – (Design based Metrology)

1.1 NGR basic concept

NGR System basic concept and key Technologies (DBM) will be helpful the understanding of our experiments.
Die to Database comparative method is the major concept of NGR system. Because of this concept, we can get the data of DW(Delta width) that comparing layout and exact location of defects.
Moreover, Die to data base overlay calculation is possible because each polygon has its own coordinate.
These concepts of NGR applications were very helpful to know real cell analysis and overlay analysis.
Die to Data base concept is shown in Figure.3. By using this application, we tried to analyze the cell distribution.

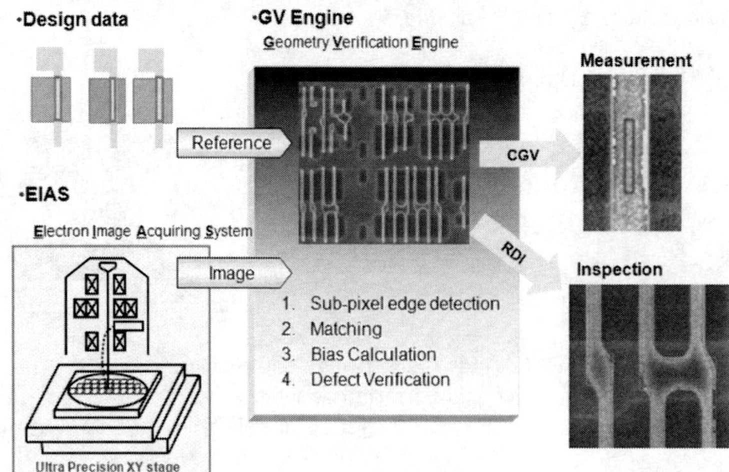

Figure.3 Die to Database comparative Method

1.2 NGR Specification

The main features of the NGR, Die to DB matching and High speed, massive data were used for evaluation of cell pattern correction.

The major concepts for High speed and massive data are shown below
- Wide FOV and High Resolution (High voltage beam based)
- High speed scanning system (45 degree scanning can measure horizontal and vertical pattern same time)
- Parallel image processing (Fast & Massive)
- Die to Database technology (Alignment and matching of design data and SEM images)

Based on these technologies, The NGR can acquire massive of data with high resolution in large FOV. So, we could analyze persuasive results.

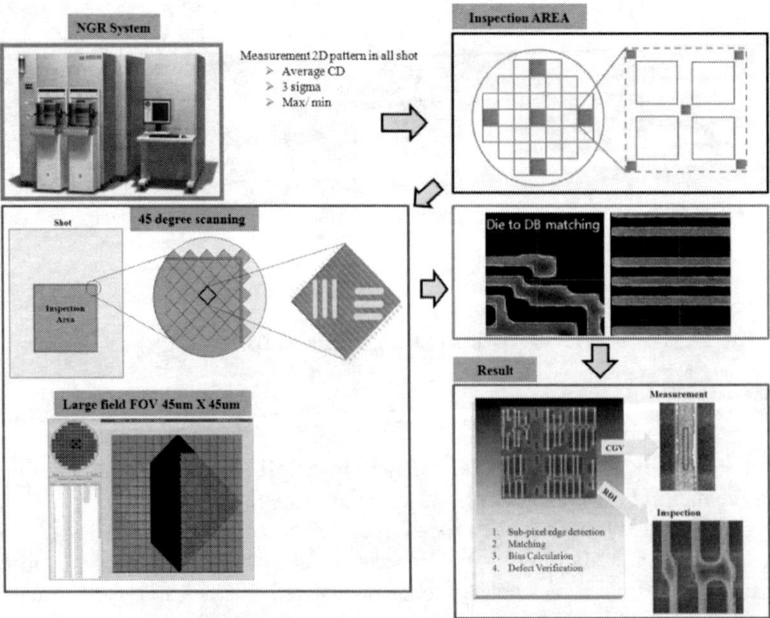

Figure.4 NGR measurement Procedure

1.3 Contour Output

Figure.5 is the major concepts of contour output, provided through NGR cooperation.

By using this contour output, we could calculate the overlay error amount. This feature was used for the cell overlay correction. It is explained this calculation method in the 2-2 experiment part of this paper.

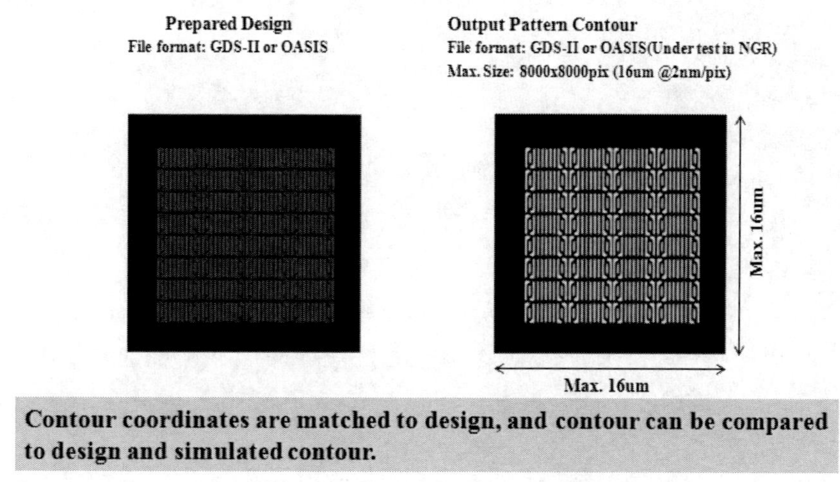

Figure.5 Contour output concept.

2. EXPERIMENT

2.1 RDI inspection and Cell distribution analysis

This experiments progressed in 2X node scale bar and space type layer using DBM tool.

The first experimental procedure of this study was cell defect inspection. It was carried out using RDI (Repeated Defect Inspection) mode. This is a simple method to identify defect analysis

Secondly, cell distribution was reviewed by CGV (Critical Geometry Verification) mode.

Before measuring, NGR contour output must be considered because general output condition cannot extract result contour as GDS files. In this experiment, line and space type layer has been evaluated, and we used massive data as 300 areas data that inspected 5areas in each 12 blocks in each 5chips for the data reliability improvement. And we used composite technology which is data replacement technology. (by using In house Tool) Because, the composited average distribution of measured many cell blocks showed issue point well than each block distribution.

Data composite technique diagram is shown in Figure.6.

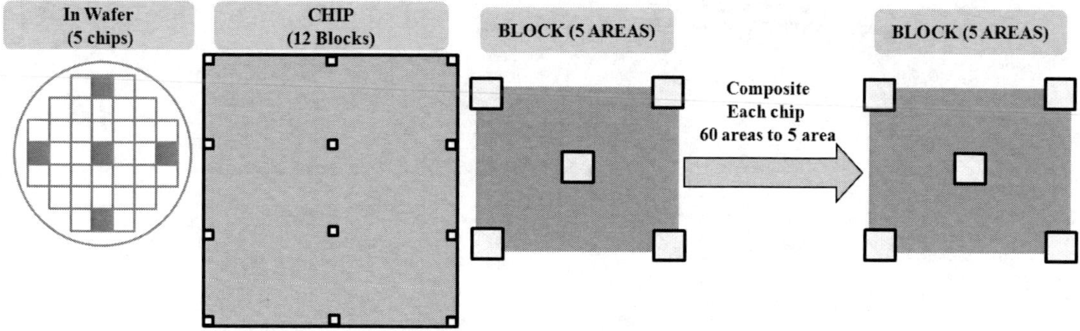

Figure.6 Diagram of cell block composite technique

Result of RDI is illustrated in Figure.7-1 such as the bridge weakpoint between space and space.

And Figure.7-2 showed the CD uniformity of each cell block, there are 2 lines because of DPT patterning. We can confirm 2 kinds of pattern distributions accurately. And Figure.7-3 is the 2D color map of cell blocks. We got to know weakpoint locations such as CD Low patterns or CD High patterns. Figure.7-4 is the 3sigma of each 5chips in the wafer.

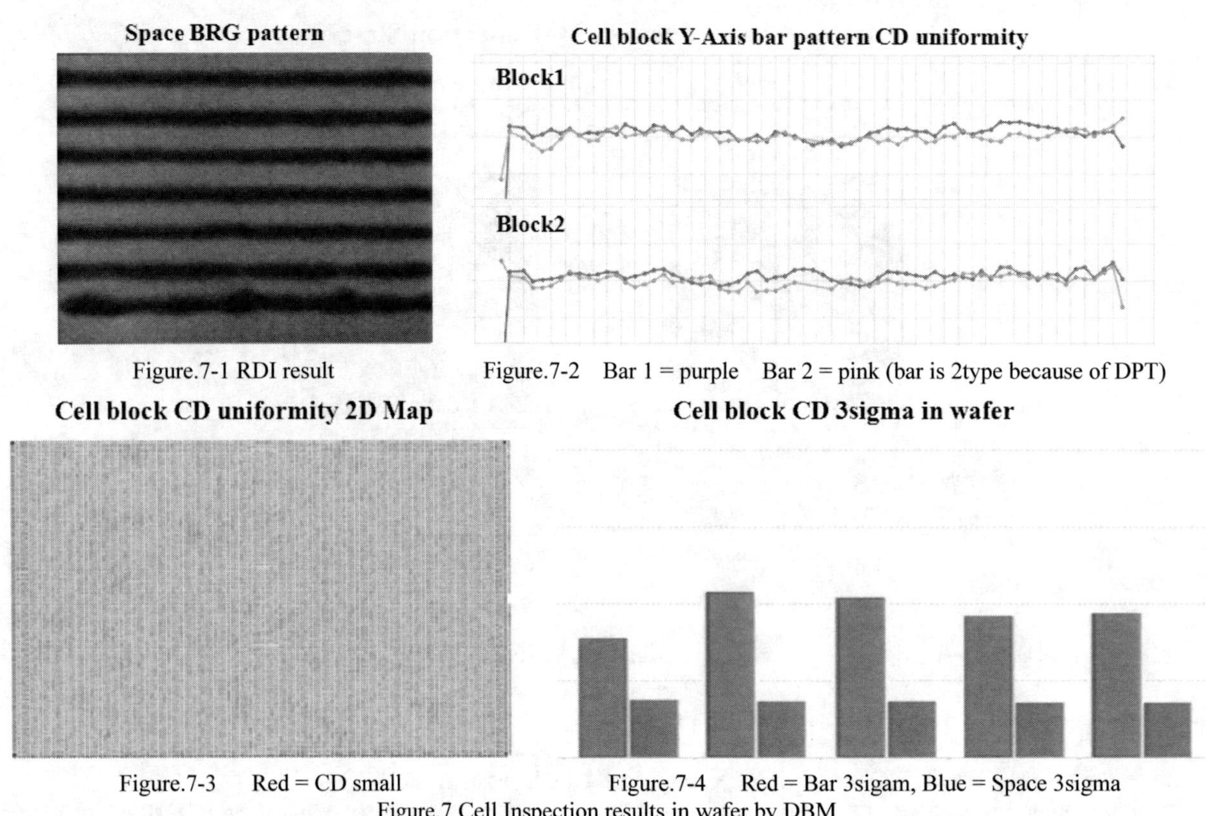

Figure.7-1 RDI result Figure.7-2 Bar 1 = purple Bar 2 = pink (bar is 2type because of DPT)

Figure.7-3 Red = CD small Figure.7-4 Red = Bar 3sigam, Blue = Space 3sigma

Figure.7 Cell Inspection results in wafer by DBM

2.2 DPT pitch walking phenomenon

The cause inducing this bridge defect is various, but I will raise an issue at this paper about CD distribution caused by CAD Layout.

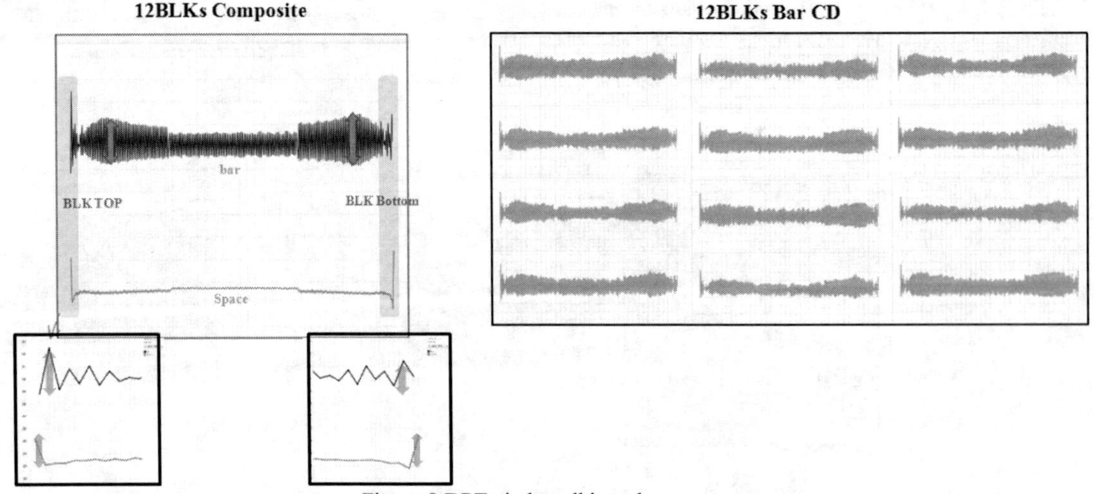

Figure.8 DPT pitch walking phenomenon

For find the reason of this space pattern bridge, we tried to check a bar CD uniformity in the cell block. And we transferred 1 block average data through the composite technology. (Figure.6)
We could draw the CD string distribution as Figure.8. It shows up odd line and even line had different CD value tendency. Odd line CD value is smaller than even line CD value.
This is the pitch walking pair phenomenon due to DPT. The space pattern which both sides of small bar pattern, is vulnerable to the bridge defect.
Figure.9 is illustrated the real space pattern bridge relation according to the photo critical dimension variation.
For improving this phenomenon, Photo specification have to control tighter than general condition. In this case, we judged as the bridge phenomenon due to the Photo CD Low.
After optimized the Photo CD specification and process conditions, the visual bridge defects were improved.

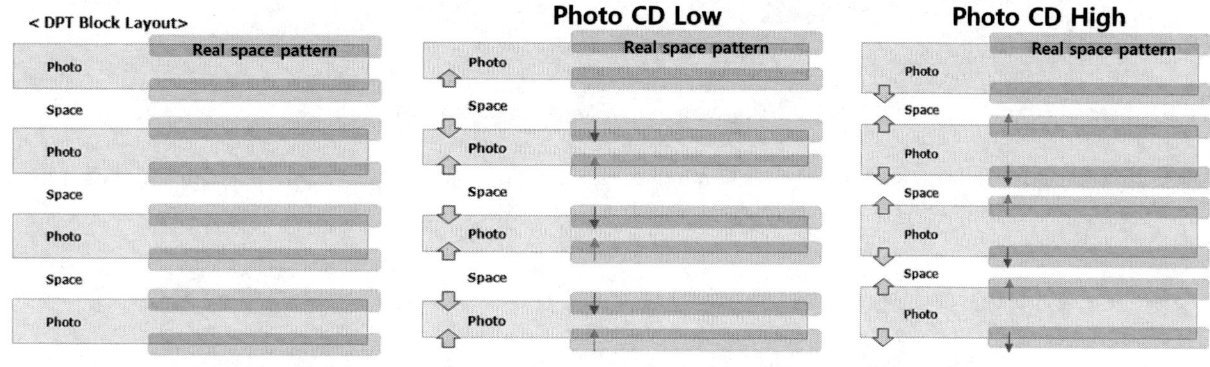

Figure.9 Causal analysis diagram of DPT pitch walking phenomenon due to Photo CD

2.3.1 Pattern overlay Error

Even though the visual defects were improved, bridge related failure that between upper layer and lower layer still remained.
This bridge related failure is caused from this space pattern overlay error. (Figure.10)
We got to know that real patterns are not same location with drawn CAD layout patterns by DBM.
We expected this pattern overlay error induces the pattern bridge phenomenon, in order to overcome this upper and lower layer bridge. So, we challenged in diverse ways in order to find the pattern overlay amount.

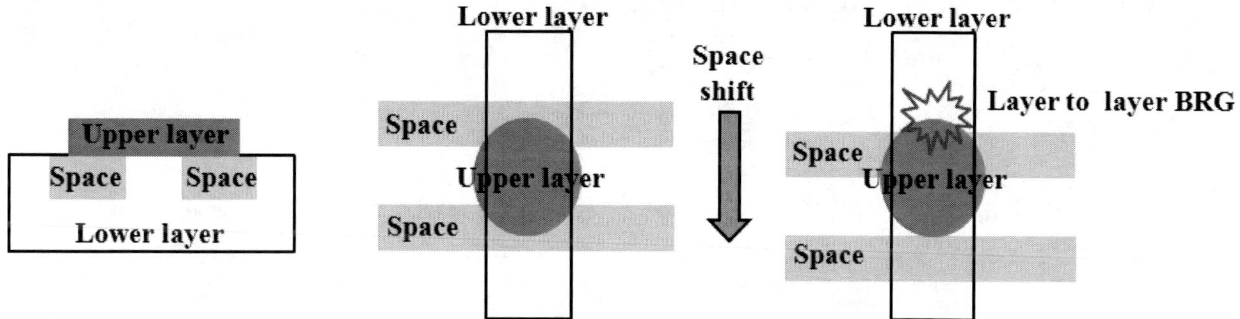

Figure.10 pattern overlay error related bridge failure

For develop the pattern overlay error correction technology, we have defined the difference of the center coordinates between the target layout and the contour outline is amount of pattern overlay error. Calculates the each skew between the central coordinates of the target Layout and the central coordinates of the contour GDS. Then, we extract average skew in the cell block axial direction.
Before measuring, there was pre-process for this experiment. Final pattern on a wafer is different from the drawn photo layout, because of DPT (Double patterning Technology) layer.

So, we had to develop final target layout extractor that consider to real process condition and real target CD. Another concern was the remove wrong data by corner round effect. If we used normal long bar layout, this center coordinates of line end seem like the coordinates representing the whole line. So, we separated target GDS as segments by using the assist GDS. Also, we inspected 12 blocks, and composited all data as 1 block data by using an in house tool.

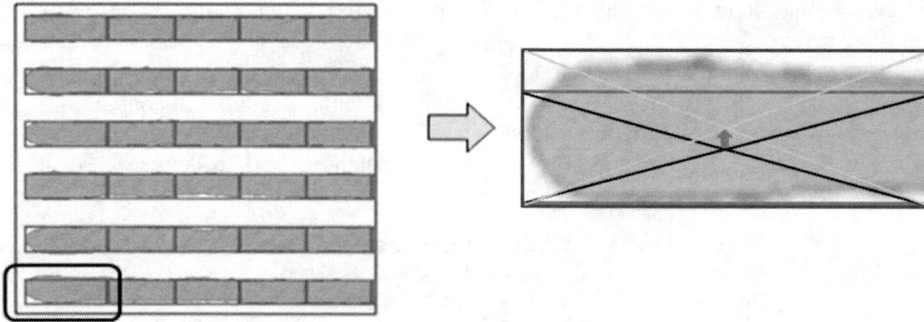

Figure.11 Contour overlay error amount calculate.

After coordinates extract, more than the several hundred of thousand data were collected in the 1chip. Based on these results, pattern overlay error amount was calculated each polygons.

2.3.2 RESULT & IMPROVEMENT

Finally, we got to know real overlay error quantity in cell blocks, and the cell pattern that had the max error quantity induced the pattern bridge through above experiment. (Figure. 12)
Also, we got to know generally edge space patterns of cell block were shifted into an inside in cell blocks
3rd space pattern shift phenomenon at block top area was matched with the result of failure analysis.
Based on this result, we made a new mask which improved layout is applied.

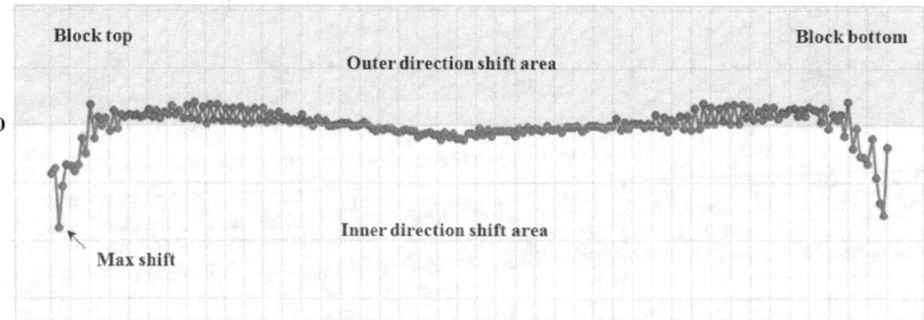

Figure.12 pattern error result of composited cell blocks.

The blue line of Figure.13 is the result in which modified layout was applied.
As we expected, improved result was obtained, and the 3rd space shift related layer to layer bridged patterns also disappeared.

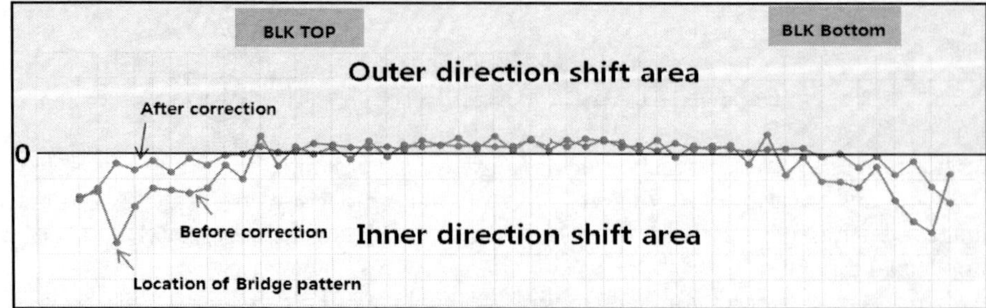

Figure.13 Result of improved pattern overlay error

Through this experiment, we have demonstrated that as an overlay calculating process is reliable.
And we are going to apply this analysis technique to all type of DRAM cell such as line and space type, also contact type.

4. CONCLUSION

In this paper, the DRAM cell analysis method has been introduced, as defect inspection, DPT pitch walking phenomenon, cell overlay improvement.
By using this method, we could understand the chronic cell failure reason well. This study will be helpful, for cell layout modification, and process condition optimization.
Especially, pattern overlay error correction technology will be helpful the layer to layer overlay failure improvement.
We apply this method to the pattern of the hole type not only the bar and space type, it was also improved the pattern bridge due to overlay error.
Real overlay error correction through methodology would be one of an effective way to keep up the shift related failure specification of the next generation device in DRAM manufacturing field.
By using massive measured data, we extracted the result that it is persuasive by applying the various analysis techniques, as cell distribution and defects, the pattern correction etc.
It has become to use as a criterion for yield enhancement through enlarged process window margin. These analysis methods give a high degree of completion of the Cell Pattering and the Distribution. We are certain that it help in the development of the advanced process and the Cell Layout.

5. ACKNOWLEDGMENT

The author thanks GH.Kim from NGR Inc. for helpful data and measurement.
And thanks all co-authors from SAMSUNG Electronics for discussion and advice.

6. REFERENCES

[1] Taeheon Kim et al., "Applications of DBV (Design Based Verification) for steep ramp-up Manufacture" Proc SPIE 7974 (2011)
[2] Taeheon Kim et al., "A study of pattern variability for device performance" Proc SPIE 8327 (2012)
[3] Kyusun Lee et al., "The analysis method of the DRAM Cell pattern Hotspot" Proc SPIE9423-71(2015)

Breaking Through 1-D Layout Limitations and Regaining 2-D Design Freedom Part II: Stitching Yield Modeling and Optimization

Jun Zhou, Hongyi Liu, Ting Han and Yijian Chen*
School of Electronic and Computer Engineering
Shenzhen Graduate School, Peking University
Shenzhen 518055, Guangdong, China

ABSTRACT

In this paper, a stitch database is built from various identified stitching structures in an open-cell layout library. The corresponding stitching yield models are developed for the hybrid optical and self-aligned multiple patterning (hybrid SAMP). Based on the concept of probability-of-success (POS) function, we first develop a single-stitching yield model to quantify the effects of overlay errors and cut-hole CD variations. The overhang distance designed in a stitching process (or its mean value μ) is found to be critical to the stitching yield performance and can be optimized using this yield model. We also investigate the physical significance of several process parameters such as half pitch (HP), standard deviation (σ) of the random overhang distribution, and cut-hole CD (CL). Our study shows that certain types of stitching yield are sensitive to σ and HP, while in general high yield can be achieved for a large number of stitching types we examined. To improve the yield of certain challenging stitching structures, various layout modification strategies are proposed and discussed.

Keywords: stitching yield, overlay error, cut hole, overhang, probability of success (POS).

1. INTRODUCTION

Self-aligned multiple patterning (SAMP) is a promising technology to scale IC devices down to sub-10nm half pitch (HP) [1-2]. However, a major barrier preventing the SAMP processes from dominating the critical-layer patterning is that an efficient method for random 2-D layout decomposition and synthesis has not been found yet. In another paper [3], we propose a layout decomposition and synthesis algorithm that can decompose 2-D layouts in polynomial time for the hybrid optical and self-aligned multiple patterning (hereafter called "hybrid SAMP"). Similar to double patterning, the target layout is decomposed into two components. Each component contains one or multiple sets of unidirectional patterns (possibly in different directions) and can be patterned by a SAMP+cut/block process. Without using connecting vias, the final 2-D patterns are formed by directly stitching two components together. Therefore, yield performance of the stitching process is critical to the hybrid SAMP patterning technique.

In this paper, we study the processing/design issues related with stitching yield and develop analytic models to predict the yield performance. Based on the concept of "probability of success (POS)", a single/local-stitching yield model is developed first by incorporating overlay errors and cut-hole CD variations. POS function replaces the "probability-of-failure (POF)" function [4-5] and acts as a weight factor for the probability density function related with certain stitching condition. We assume that overlay errors and cut-hole CD variations obey a Gaussian distribution and POS is a linear function of the overlap area/distance of two stitching features. A database of

* Phone: (+86) 755-21536236, Email: chenyj@pkusz.edu.cn

Design-Process-Technology Co-optimization for Manufacturability IX, edited by John L. Sturtevant, Luigi Capodieci, Proc. of SPIE Vol. 9427, 942714 · © 2015 SPIE · CCC code: 0277-786X/15/$18 · doi: 10.1117/12.2085898

various stitching structures identified from a commonly seen layout library [3] is built, which includes three types of basic stitching structures, e.g., T shape, L shape and double-L shape, as well as their derivative structures. This stitch database covers most stitching structures generated in our layout decomposition and synthesis algorithm reported in [3]. The single-stitching yield model allows us to solve the optimal overhang for certain types of stitches. The information of these critical layout parameters can be incorporated into IC design, thus beneficial to efficiency improvement of layout decomposition and synthesis. Various layout modification techniques are proposed for certain types of challenging stitching features such as "double-L" structures.

2. SINGLE-STITCHING YIELD MODELING

2.1 Layout decomposition and synthesis

In an IC layout, features with different densities are often present in different regions and it is not necessary to decompose the whole layout using the SAMP+stitching algorithm [3]. To accommodate more CD flexibility, we propose incorporating multiple patterning into a hybrid SAMP process. The cut masks of two decomposed components can be used in a way similar to double patterning to generate random 2-D features (whose density is beyond the optical resolution of a scanner, but not high enough for implementing a SAMP+stitching process). Fig. 1 shows how double patterning can be incorporated into a hybrid SAMP process. We first divide the target layout into two parts: Part 1 (green) formed by a hybrid SAMP process, and part 2 (red) formed by double patterning. Two sets of decomposed patterns in part 2 will be assigned into two "cut" masks generated in a hybrid SAMP process. Both parts require a stitching process to form 2-D patterns (although with different densities), while the most challenging yield issues occur in Part 1 wherein two highest-density components (separately fabricated by SAMP processes) need to be directly stitched.

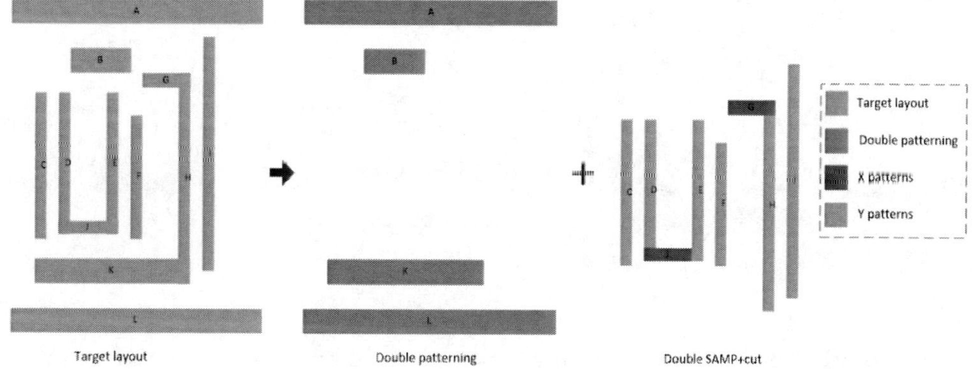

Fig. 1. A schematic demonstration of the hybrid SAMP layout decomposition incorporating double patterning [3]. Large-CD features, e.g., A, B, K, L, are collected for the double-patterning treatment. Other highest-density features are formed by the hybrid SAMP technique (e.g., double SAMP+cut processes).

2.2 Categorization of stitching structures

As shown in Fig. 2, a number of stitching types exist in layout synthesis. We build a stitch library based on (commonly seen) perpendicularly stitching structures such as T, L, double-L shapes, and their derivative structures. In addition, some types of stitches that may not be formed by perpendicularly overlapping, e.g., tip-to-tip and abreast stitches, are also covered. For instance, three basic patterns in Fig. 2 (labeled with A, B and C) correspond to the L, T and double-L shapes, while A1 and A2 are the derivative structures of the L-shape (A) patterns. The difference between a basic stitch and its derivative structures is that a neighboring line lies right next to the overhang extension. Apparently, such an additional line will lead to stitching yield loss since the overlay inaccuracy can cause the overhang part to touch the line. To avoid such a patterning failure, it is convenient to define a quantitative index, minimum safe distance (MSD), which is the "safe" distance between two neighboring lines, as shown in Fig. 2(b). In this paper, MSD is quantified as the sum of the Maximum Overhang (MO) in the original layout and $3\sigma_{\mathbf{overhang}}$. Here, overhang is considered as a random vector variable (x, y) whose value can change from die to die (and from wafer to wafer) due to the combinational effect of cut-hole CD variations and overlay errors. Its sample values follow certain statistical distribution and $\sigma_{\mathbf{overhang}}$ is the standard deviation (in a certain direction). The line-cut process is different from the conventional line-end printing (i.e., tip-to-tip printing in a single exposure) and hopefully will enable a patterning process with significantly higher accuracy of line-edge placement. When the distance between two neighboring lines (one with stitch) is larger than MSD, the probability of an unwanted line-touch event is negligible. It should be reminded that the line-cut process for a positive-tone SAMP process (spacer is line [1]) is used here only for the demonstration purpose, while the modeling/optimization strategy presented in this paper can be readily extended to the negative-tone SAMP process (spacer is trench) wherein blocking patterns may be required. The "Maximum Overhang" is a parameter related to specific circuits and layouts that are determined by IC/lithographic designers before a stitching process occurs. It will be unusual if the maximum overhang in a layout is close to or larger than the minimum half pitch (HP) that can be resolved by a scanner; therefore, we assume a MO of 0.3HP in certain calculations. However, this value is not necessarily optimal and MO in general is set to be a tunable parameter in our model such that we can adjust its value according to the layout input from various designs.

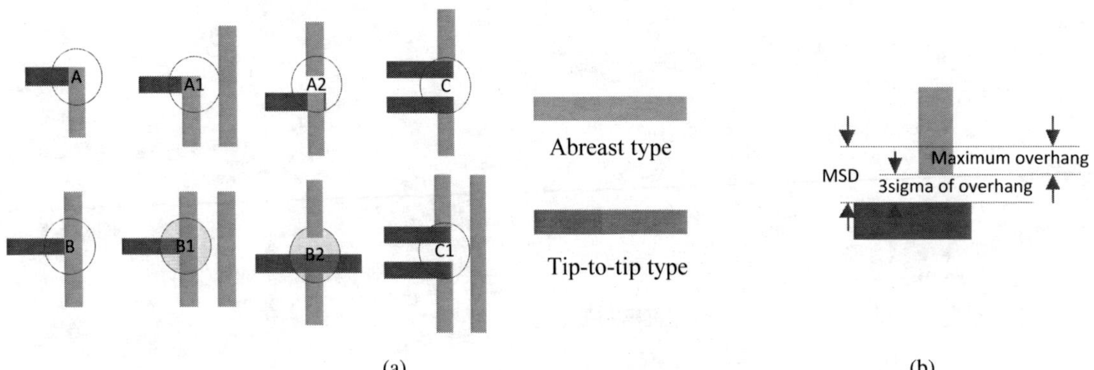

Fig. 2. (a) Categorization of the stitching types, (b) schematic description of the minimum safe distance (MSD). MSD is the maximum overhang (MO) plus $3\sigma_{\mathbf{overhang}}$.

2.3 POS functions in single-stitching yield

In order to facilitate the yield discussion, we define the stitching formed by two patterns perpendicular to each other as single orthogonal stitching (OS). The random variable of overhang distance is assumed to obey a statistical distribution, e.g., $f_X(x, \mu_X, \sigma_X)$ and $f_Y(y, \mu_Y, \sigma_Y)$ representing overhang's probability density functions in X and Y directions, respectively (μ_X: mean of the X-direction overhang, σ_X: standard deviation of the X-direction overhang; μ_Y: mean of the Y-direction overhang, σ_Y: standard deviation of the Y-direction overhang). The single OS yield is calculated by:

$$Y_{OS} = \iint POS_{OS}(x,y) f_X(x,\mu_X,\sigma_X) f_Y(y,\mu_Y,\sigma_Y)dxdy$$

Here, $POS_{OS}(x,y)$ is a modification/weight factor of the overhang probability density function, or simply called "probability of success (POS)" which is a function of random overhang variables x and y. It indicates the difference of various stitching conditions (e.g., with different overlap areas) in their probability to achieve a successful patterning process. For instance, a larger overlap area between two stitching features is expected to have a better stitching yield. Two types of POS functions can be identified according to whether there exists an overhang in a stitched pattern. If an overhang is present and long enough to touch the neighboring line in its extension direction, the POS value will be zero; otherwise, it will be one (see Fig. 3(a) and Fig. 3(b)). If an overhang does not exist and the two stitching features only partially overlap at the end, the POS value will be assumed to be the ratio of the (partially) overlapping area (e.g., "S" area shown in Fig. 3(c)) to the fully overlapping area (e.g., "A" area shown in Fig. 3(c)). This assumption results in a simple POS function linearly increasing with the overlap area, as shown in Fig. 3(d). The linear POS function is only a zeroth-order approximation to calculate the stitching yield; and the real fab data, once available, should be incorporated to improve the model accuracy.

Fig. 2(a) also demonstrates an abreast type of stitching (pattern overlapping between two parallel features). It is a different type of stitching for CD tuning [3] which requires a minimum overlap distance and may suffer from the yield loss due to LER and CD variation. Although this may not appear obvious, we can image if the overlap distance is close to or smaller than the LER or CD variations, process related fluctuations can lead to significant stitching failure or even zero yield. To characterize this type of yield loss, it is necessary to define a threshold value of overlap distance at which a step jump of stitching yield occurs. Another type of stitching, e.g., tip-to-tip stitching is similar to the single orthogonal stitching.

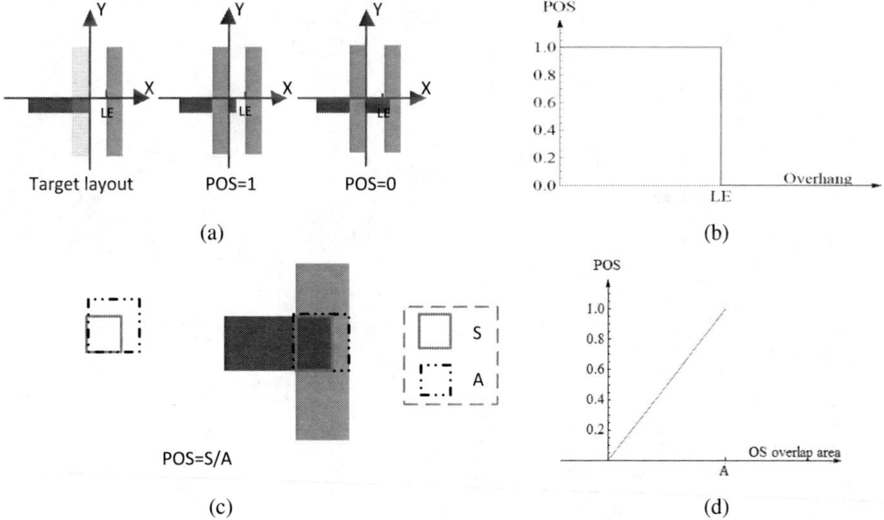

Fig. 3. The concept of probability-of-success (POS) function in a stitching process. (a) A schematic demonstration of various overhang conditions and the corresponding POS values. (b) If the right edge of overhang is located within the space between two vertical lines, the POS value will be one. If the overhang touches the neighboring line, the POS value will be zero. (c) A schematic demonstration of various overlapping conditions and the corresponding POS functions. (d) In general, if the overlap area of two stitching features is zero, the POS value will be zero; otherwise, the POS function will linearly increase with the overlap area.

2.4 Single-stitching yield modeling

In this sub-section, a single-stitching yield model will be developed. We simply assume the line and space CD are both half pitch (HP) unless specified otherwise; however, CD tuning can be accommodated by adjusting the related parameters of the model. As mentioned before, the variables x and y stand for the random overhang values in X and Y directions, respectively. To build the yield model for the stitching type A (see Fig. 2), we shall first discuss the POS functions in various integration domains.

- If $x < -HP$, POS will be equal to 0 regardless of the value of y (see Fig. 4(b)) because the overlap area of two stitching features is zero.
- If $-HP \leq x < 0$, the overlap area depends on the y value. When $y < -HP$, POS will be 0 due to the zero overlap area (see Fig. 4(c)). When $-HP \leq y < 0$, POS will be $(x + HP)(y + HP)/HP^2$ since the overlap area is $(x + HP)(y + HP)$ (see Fig. 4(d)). When $0 \leq y$, POS will be $(x + HP)/HP$ since the overlap area is $(x + HP)HP$ (see Fig. 4(e)).
- If $0 \leq x$, POS will be 0 when $y < -HP$ as the overlap area is 0 (see Fig. 4(f)). When $-HP \leq y < 0$, POS will be $(y + HP)/HP$ as the overlap area is $(y + HP)HP$ (see Fig. 4(g)). When $0 \leq y$, POS will be 1 as the overlap area is HP^2 (see Fig. 4(h)).

Namely, $POS_A(x,y)$ can be described as:

$$POS_A(x,y) = \begin{cases} \dfrac{(x+HP)}{HP}\dfrac{(y+HP)}{HP}, & -HP \leq x < 0 \; and \; -HP \leq y < 0; \\ \dfrac{(x+HP)}{HP}, & -HP \leq x < 0 \; and \; y \geq 0; \\ \dfrac{(y+HP)}{HP}, & x \geq 0 \; and \; -HP \leq y < 0; \\ 1, & x \geq 0 \; and \; y \geq 0; \\ 0, & Otherwise. \end{cases} \tag{1}$$

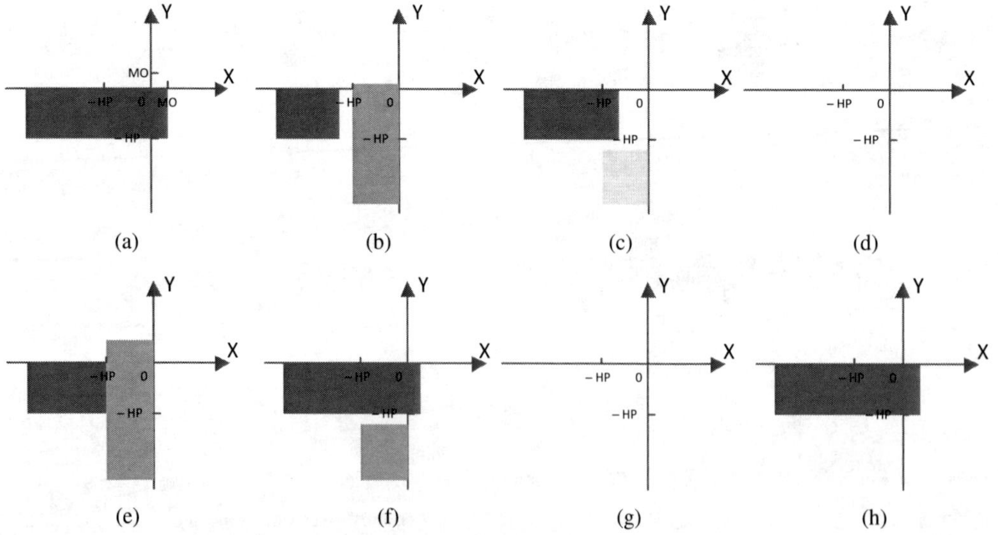

(a) (b) (c) (d)

(e) (f) (g) (h)

Fig. 4. Various overlapping conditions of stitching type A. (a) designed layout of stitching features, (b) $x < -HP$ and $-\infty \leq y \leq \infty$, (c) $-HP \leq x < 0$ and $y < -HP$, (d) $-HP \leq x < 0$ and $-HP \leq y < 0$, (e) $-HP \leq x < 0$ and $0 \leq y$, (f) $0 \leq x$ and $y < -HP$, (g) $0 \leq x$ and $-HP \leq y < 0$, (h) $0 \leq x$ and $0 \leq y$.

According to the discussion in sub-section 2.3, the yield of stitching type A can be integrated as:

$$Y_A(\mu_X,\sigma_X,\mu_Y,\sigma_Y) = \iint POS_A(x,y) f_X(x,\mu_X,\sigma_X) f_Y(y,\mu_Y,\sigma_Y) dxdy$$

To calculate the above integral, we set μ_X and μ_Y as MO (see Fig. 4(a)). We further assume a symmetric design/patterning in X and Y directions and introduce a new function G to replace the original distributions (assumed to be Gaussian) f_X and f_Y, wherein a parameter σ replaces σ_X and σ_Y:

$$G(x,MO,\sigma) = \frac{1}{\sqrt{2\pi}\sigma} e^{-\frac{(x-MO)^2}{2\sigma^2}}, \quad G(y,MO,\sigma) = \frac{1}{\sqrt{2\pi}\sigma} e^{-\frac{(y-MO)^2}{2\sigma^2}}.$$

The yield formula for stitching type A can be constructed as:

$$\begin{aligned} Y_A(\sigma) &= \int_{-\infty}^{+\infty}\int_{-\infty}^{+\infty} POS_A(x,y) G(x,MO,\sigma) G(y,MO,\sigma) dxdy \\ &= \left[\int_{-HP}^{0} G(x,MO,\sigma)\frac{(x+HP)}{HP}dx + \int_{0}^{+\infty} G(x,MO,\sigma)dx\right] \\ &\times \left[\int_{-HP}^{0} G(y,MO,\sigma)\frac{(y+HP)}{HP}dy + \int_{0}^{+\infty} G(y,MO,\sigma)dy\right] \end{aligned} \tag{2}$$

The yield models of other stitching types can be similarly developed. As shown in Fig. 5, the yield of stitching type B can be calculated as the sum of two parts.

- If $x \geq 0$, the POS value will be one (see Fig. 5(b)).
- If $-HP \leq x < 0$, the POS value will be $(x+HP)/HP$ (see Fig. 5(c)).

Namely,
$$Y_B(\sigma) = \int_{-HP}^{0} G(x,MO,\sigma)\frac{(x+HP)}{HP}dx + \int_{0}^{+\infty} G(x,MO,\sigma)dx \tag{3}$$

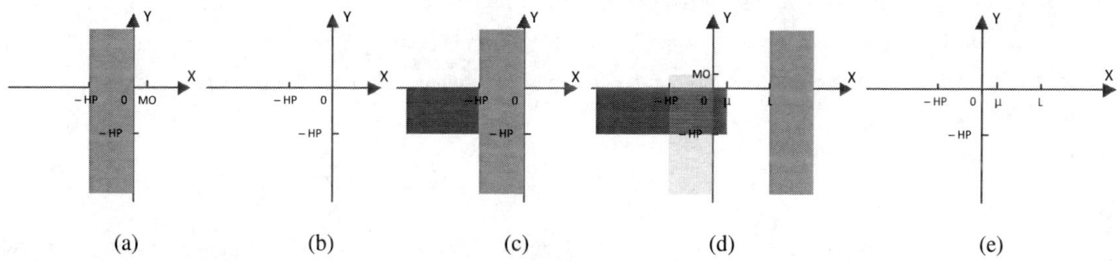

(a) (b) (c) (d) (e)

Fig. 5. Various overlapping conditions of stitching types B, A1, and B1. (a) designed layout of stitching type B, (b) $x \geq 0$, POS is 1, (c) $-HP \leq x < 0$, POS is $(x+HP)/HP$, (d) designed layout of stitching type A1, (e) designed layout of stitching type B1.

In order to develop the yield models of A1 and B1, as shown in Fig. 5(d) and (e), the distance between two vertical/parallel features is set to be L. The difference between stitching types A and A1 is that there is a neighboring line lies to the right of the stitching part. Mathematically, the differences between type A and A1's yield formulae are the integration interval and μ_X. Unlike setting μ_X and μ_Y as MO in stitching type A, μ_X is a parameter that needs to be optimized in stitching type A1. According to our discussion in section 2.3, if $x > L$, POS will be zero as the overhang in X direction will touch the neighboring line. The yield formula for stitching type A1 can be constructed as:

$$\begin{aligned} Y_{A1}(\mu_X,\sigma,L) &= \left[\int_{-HP}^{0} G(x,\mu_X,\sigma)\frac{(x+HP)}{HP}dx + \int_{0}^{L} G(x,\mu_X,\sigma)dx\right] \\ &\times \left[\int_{-HP}^{0} G(y,MO,\sigma)\frac{(y+HP)}{HP}dy + \int_{0}^{+\infty} G(y,MO,\sigma)dy\right] \end{aligned} \tag{4}$$

The yield formula for stitching type B1 is:

$$Y_{B1}(\mu_X,\sigma,L)=\int_{-HP}^{0}G(x,\mu_X,\sigma)\frac{(x+HP)}{HP}dx+\int_{0}^{L}G(x,\mu_X,\sigma)dx \tag{5}$$

To obtain the maximum yield, we need to calculate the derivatives of Y_{A1} and Y_{B1}:

$$Y'_{A1}(\mu_X,\sigma,L)_{\mu X}=\left[\int_{-HP}^{0}G(x,\mu_X,\sigma)\frac{(x-\mu_X)}{\sigma^2}\frac{(x+HP)}{HP}dx+\int_{0}^{L}G(x,\mu_X,\sigma)\frac{(x-\mu_X)}{\sigma^2}dx\right]$$
$$\times\left[\int_{-HP}^{0}G(y,MO,\sigma)\frac{(y+HP)}{HP}dy+\int_{0}^{+\infty}G(y,MO,\sigma)dy\right] \tag{6}$$

$$Y'_{B1}(\mu_X,\sigma,L)_{\mu X}=\int_{-HP}^{0}G(x,\mu_X,\sigma)\frac{(x-\mu_X)}{\sigma^2}\frac{(x+HP)}{HP}dx+\int_{0}^{L}G(x,\mu_X,\sigma)\frac{(x-\mu_X)}{\sigma^2}dx \tag{7}$$

For stitching type A1, μ_o, the optimal overhang, can be calculated from the equation: $Y'_{A1}(\mu_X,\sigma,L)_{\mu X}=0$. Similarly, for stitching type B1, an optimal overhang can be calculated from the equation: $Y'_{B1}(\mu_X,\sigma,L)_{\mu X}=0$.

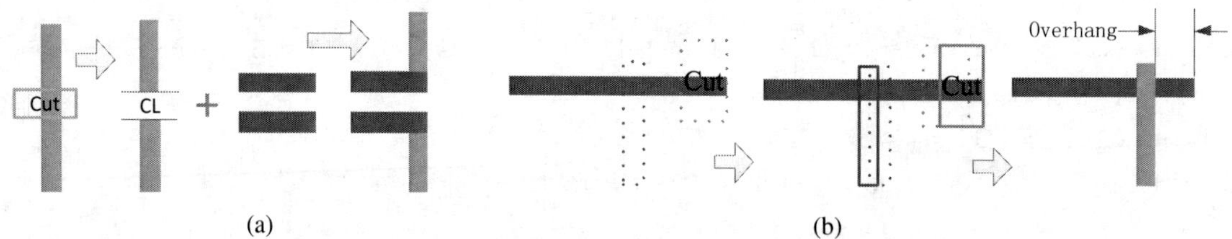

Fig. 6. (a) A schematic illumination of a "cut and stitch" process wherein a line in Y direction is first cut into two parts and then stitched by two horizontal lines to form the "double-L" patterns (CL is the cut-hole CD in Y direction). (b) The overhang generation process showing that the overhang value is impacted by both overlay inaccuracy and horizontal cut-hole CD variation. The dashed lines indicate the ideal line and cut-hole positions, while the solid lines indicate the actual (non-ideal) line and cut-hole positions such that two types of variations (overlay errors and horizontal cut-hole CD variation) can be accumulated.

The yield models of stitching types A2 and B2 are more complicated since they involve cut processes. We shall first develop the yield models for other stitching types C and C1, which also need cut processes as shown in Fig. 6(a). In general, an overhang is generated through a cut process and a stitching process, in which two overlay errors and one CD variation can be accumulated, as shown in Fig. 6(b). If the cut-hole CD variation is incorporated into this model, we shall have: $\sigma^2=2\sigma_{overlay}^2+\sigma_{cut}^2$, where $\sigma_{overlay}^2$ indicating the variance of overlay errors and σ_{cut}^2 indicating the variance of cut-hole CD deviation from the mean value. Note that the yield of stitching type C is jointly determined by two individual stitching structures. Naturally, the minimum yield of these two individual stitches (or the worst-case yield) will be the final yield of the stitching type C. Fig. 7 shows the geometrical information of several physical parameters used in the yield model of stitching type C with $CL<HP$ and $x>0$ (where CL is the cut-hole CD in Y direction). In Fig. 7, w is defined to be the middle-point position of the cut hole in Y direction, which is a random variable due to the overlay errors. We assume that w obeys the same Gaussian distribution as that of overhang since the stitching design is often symmetric and the mean of w is zero. The physical intuition behind this assumption is that w and overhang are both impacted by the same overlay errors.

- If $x>0$, when $-(HP+CL)\le w\le-(HP-CL)$, POS will be $(3HP-CL+2w)/(2HP)$ (see Fig. 7(c)). When $-(HP-CL)<w<(HP-CL)$, POS will be 1 (see Fig. 7(d)). When $(HP-CL)\le w\le$

$(HP + CL)$, POS will be $(3HP - CL - 2w)/(2HP)$ (see Fig. 7(e)). In other conditions, POS is 0.

- If $-HP \le x \le 0$, when $-(HP + CL) \le w \le -(HP - CL)$, POS will be $(x + HP)(3HP - CL + 2w)/(2HP^2)$. When $-(HP - CL) < w < (HP - CL)$, POS will be $(x + HP)/HP$. When $(HP - CL) \le w \le (HP + CL)$, POS will be $(x + HP)(3HP - CL - 2w)/(2HP^2)$. In other conditions, POS is 0.
- If $x < -HP$, POS will be 0 because of zero overlap area.

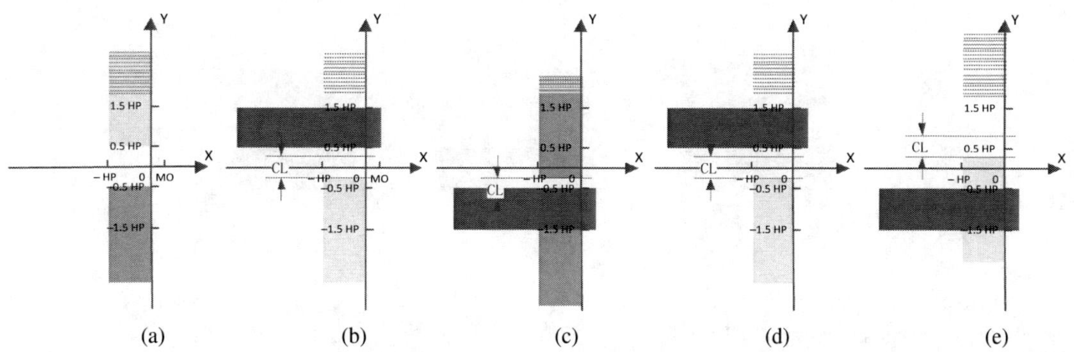

Fig. 7. Various overlapping and cutting conditions of stitching types C with $CL < HP$ and $x > 0$. (a) designed layout of stitching C, (b) actual cut process with $CL < HP$, (c) $-(HP + CL)/2 \le w \le (CL - HP)/2$, (d) $(CL - HP)/2 < w < (HP - CL)/2$, (e) $(HP - CL)/2 \le w \le (HP + CL)/2$.

The POS function of stitching type C is constructed as:

$$POS_C(x,w) = \begin{cases} \dfrac{(3HP - CL + 2w)}{2HP}, & 0 < x \text{ and } -(HP + CL) \le w \le -(HP - CL); \\ 1, & 0 < x \text{ and } -(HP - CL) < w < (HP - CL); \\ \dfrac{(3HP - CL - 2w)}{2HP}, & 0 < x \text{ and } (HP - CL) < w < (HP + CL); \\ \dfrac{(HP + x)(3HP - CL + 2w)}{2HP^2}, & -HP \le x \le 0 \text{ and } -(HP + CL) \le w \le -(HP - CL); \\ \dfrac{(HP + x)}{HP}, & -HP \le x \le 0 \text{ and } -(HP - CL) < w < (HP - CL); \\ \dfrac{(HP + x)(3HP - CL - 2w)}{2HP^2}, & -HP \le x \le 0 \text{ and } (HP - CL) < w < (HP + CL); \\ 0, & Otherwise. \end{cases}$$

The yield formula for stitching type C with $CL < HP$ can be written as an integral:

$$Y_C(\sigma, CL) = \iint POS_C(x,w) G(x, MO, \sigma) G(w, 0, \sigma) dx dw$$

Taking the symmetry of the stitching design into account, $Y_C(\sigma)$ can be simplified to:

$$Y_C(\sigma, CL) = 2\left[\int_{-HP}^{0} G(x, MO, \sigma)\frac{x + HP}{HP}dx + \int_{0}^{+\infty} G(x, MO, \sigma)dx\right] \times \left[\int_{0}^{(HP-CL)/2} G(w, 0, \sigma)dw + \int_{(HP-CL)/2}^{(HP+CL)/2} G(w, 0, \sigma)\frac{3HP - CL - 2w}{2HP}dw\right], CL < HP; \quad (8)$$

Fig. 8 schematically shows the yield formula for stitching type C when $HP \le CL \le 3HP$ and $x > 0$. In the region of $0 \le w < (3HP - CL)/2$, POS is $(3HP - CL - 2w)/2HP$. When $HP \le CL \le 3HP$ and $x \le 0$, the yield behavior is similar to the above case with $HP \le CL \le 3HP$ and $x > 0$. Thus, the yield formula for stitching type C with $HP \le CL \le 3HP$ can be expressed as:

$$Y_C(\sigma, CL) = 2\int_0^{+\infty} G(x, MO, \sigma)dx \int_0^{(3HP-CL)/2} G(w, 0, \sigma)\frac{3HP - CL - 2w}{2HP}dw$$

$$+2\int_{-HP}^{0} G(x, MO, \sigma)\frac{x + HP}{HP}dx \int_0^{(3HP-CL)/2} G(w, 0, \sigma)\frac{3HP - CL - 2w}{2HP}dw, 3HP \ge CL \ge HP. \quad (9)$$

If $CL > 3HP$, the yield of stitching type C is zero.

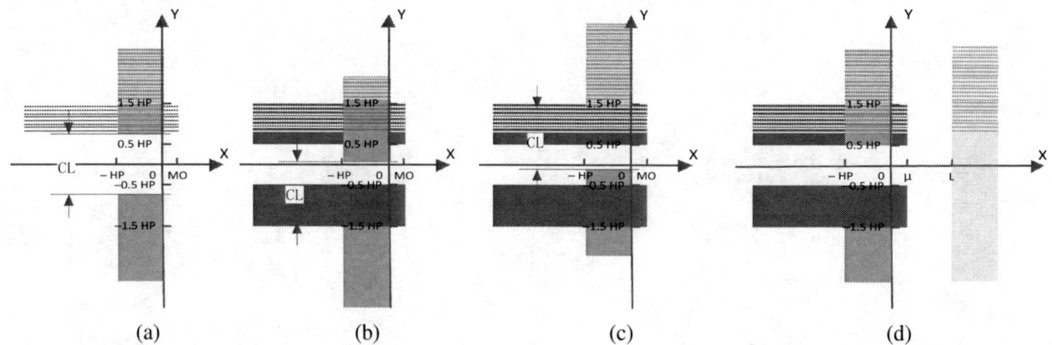

Fig. 8. (a) The actual cut process with $HP \le CL \le 3HP$, (b) $(CL - 3HP)/2 \le w < 0$, (c) $0 \le w < (3HP - CL)/2$, (d) designed layout of stitching C1.

As shown in Fig. 8(d), the yield formula for stitching type C1 can be expressed as:

$$Y_{C1}(\mu_X, \sigma, L, CL) = 2\left[\int_{-HP}^{0} G(x, \mu_X, \sigma)\frac{x + HP}{HP}dx + \int_0^L G(x, \mu_X, \sigma)dx\right]$$

$$\times\left[\int_0^{(HP-CL)/2} G(w, 0, \sigma)dy + \int_{(HP-CL)/2}^{(HP+CL)/2} G(w, 0, \sigma)\frac{3HP - CL - 2w}{2HP}dw\right], CL < HP; \quad (10)$$

$$Y_{C1}(\mu_X, \sigma, L, CL) = 2\int_0^L G(x, \mu_X, \sigma)dx \int_0^{(3HP-CL)/2} G(w, 0, \sigma)\frac{3HP - CL - 2w}{2HP}dw$$

$$+2\int_{-HP}^{0} G(x, \mu_X, \sigma)\frac{x + HP}{HP}dx \int_0^{(3HP-CL)/2} G(w, 0, \sigma)\frac{3HP - CL - 2w}{2HP}dw, 3HP \ge CL \ge HP. \quad (11)$$

If $CL > 3HP$, the yield of stitching type C1 is 0. We can derive the derivative of $Y'_{C1}(\mu_X, \sigma)_{\mu_X}$ and solve the optimal μ_o from the equation: $Y'_{C1}(\mu_X, \sigma)_{\mu_X} = 0$.

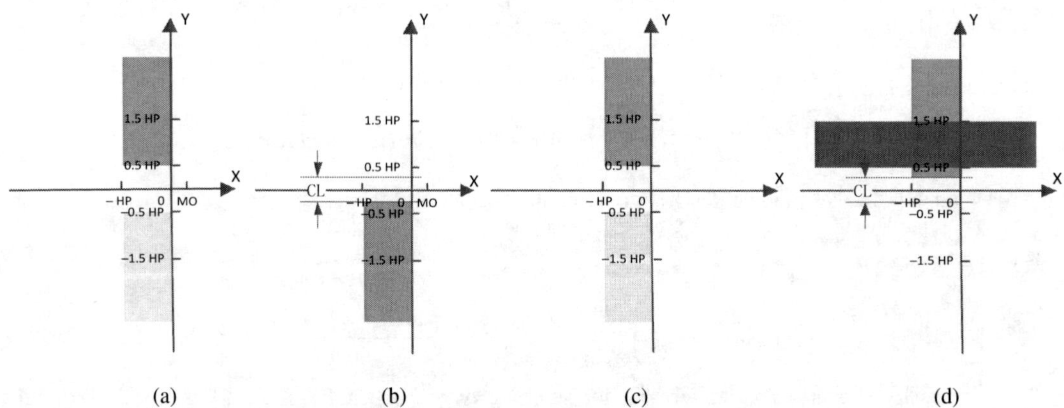

Fig. 9. Various overlapping and cutting conditions of stitching types A2 and B2. (a) designed layout of stitching A2, (b) actual cut process of stitching A2, (c) designed layout of stitching B2, (d) actual cut process of stitching B2.

The yield models of stitching types A2 and B2 (see Figs. 9(b) and 9(d)) are similar to that of the stitching type C, and their derivation will not be repeated and only the yield formulae will be listed below:

$$Y_{A2}(\sigma,CL)=\left[\int_{-HP}^{0}G(x,MO,\sigma)\frac{x+HP}{HP}dx+\int_{0}^{+\infty}G(x,MO,\sigma)dx\right]$$
$$\times\left[\int_{-\infty}^{(HP-CL)/2}G(w,0,\sigma)dw+\int_{(HP-CL)/2}^{(HP+CL)/2}G(w,0,\sigma)\frac{3HP-CL-2w}{2HP}dw\right],CL<HP; \quad (12)$$

$$Y_{A2}(\sigma,CL)=\left[\int_{-HP}^{0}G(x,MO,\sigma)\frac{x+HP}{HP}dx+\int_{0}^{+\infty}G(x,MO,\sigma)dx\right]\cdot$$
$$\times\left[\int_{-\infty}^{(HP-CL)/2}G(w,0,\sigma)dw+\int_{(HP-CL)/2}^{(3HP-CL)/2}G(w,0,\sigma)\frac{3HP-CL-2w}{2HP}dw\right],3HP\geq CL\geq HP. \quad (13)$$

$$Y_{B2}(\sigma,CL)=\int_{-\infty}^{(HP-CL)/2}G(w,0,\sigma)dw+\int_{(HP-CL)/2}^{(HP+CL)/2}G(w,0,\sigma)\frac{3HP-CL-2w}{2HP}dw,CL<HP; \quad (14)$$

$$Y_{B2}(\sigma,CL)=\int_{-\infty}^{(HP-CL)/2}G(w,0,\sigma)dw+\int_{(HP-CL)/2}^{(3HP-CL)/2}G(w,0,\sigma)\frac{3HP-CL-2w}{2HP}dw,3HP\geq CL\geq HP. \quad (15)$$

If $CL>3HP$, the yields of stitching A2 and B2 are both zero.

3. SINGLE-STITCHING YIELD RESULTS

Using equations (2) and (3), the yield results of stitching types A and B vs. σ for varying HP (5nm to 15nm) are calculated and shown in Figs. 10(a) and 10(b), respectively. For the same HP and σ, the stitching yield of type A is lower than that of type B because the yield of type A is affected by the overhang variations in two directions, while the yield of type B is affected by the overhang variation in one direction only. Stitching yields of both types drop with σ, especially at smaller HP.

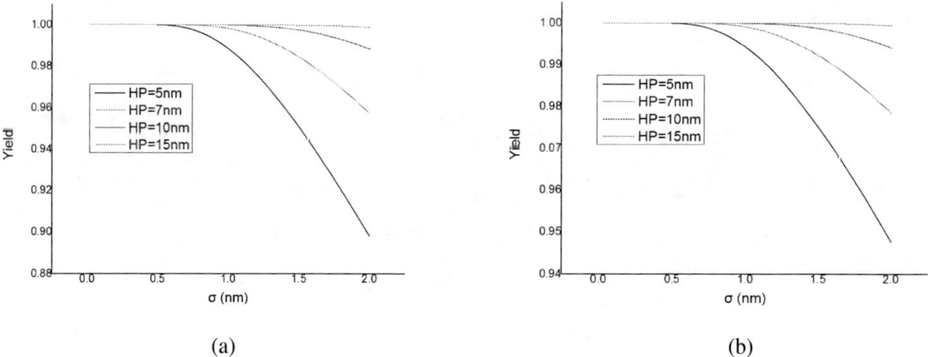

Fig. 10. (a) The stitching yield of type A vs. σ. (b) The stitching yield of type B vs. σ.

Y_{A1} and Y_{B1} are the function of μ_X,σ and L (see equations (4) and (5)). Given constant L and σ, in order to achieve the maximum yield, an optimal overhang μ_o must be solved from the equation: $Y'_{A1}(\mu_X,\sigma,L)_{\mu_X}=0$. The 3D graphs of $Y_{A1}(\mu_o,\sigma,L)$ vs. σ and L for varying HP are plotted in Fig. 11, which show that the gap distance L does not play a critical role in stitching yield performance at larger HP. Unlike L which only impacts the yield performance at smaller HP, σ is always a critical parameter for the stitching yield. The Y_{A1} yield curves (with an optimal μ_o) vs. σ are plotted in Fig. 12, in which two HP and three L values are examined to confirm the above tendency.

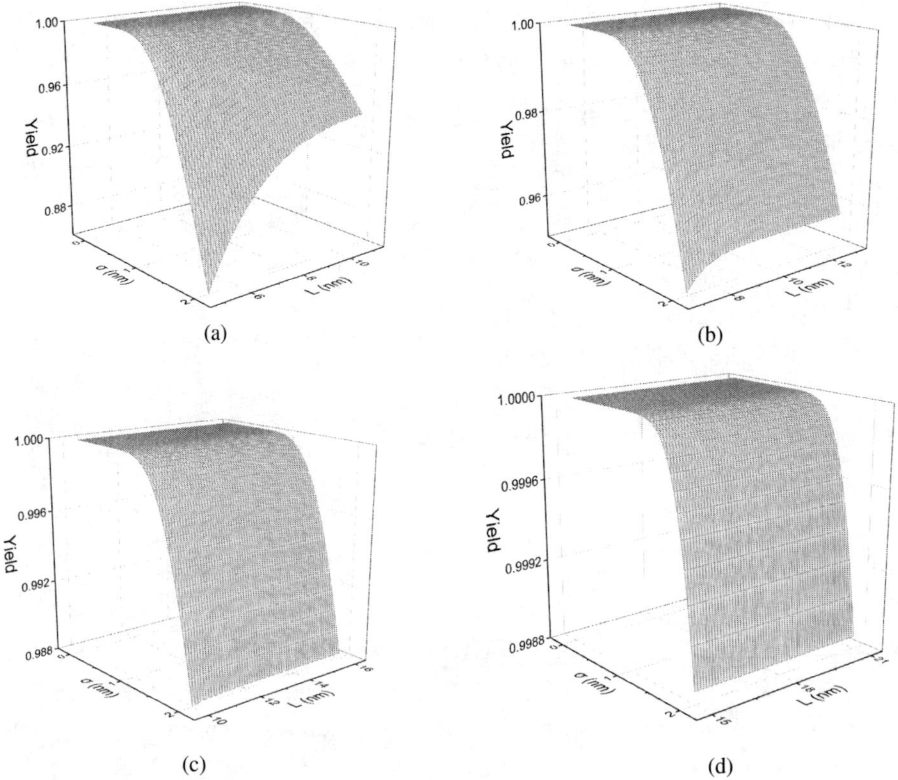

Fig. 11. The stitching yield of type A1 vs. σ and L. (a) HP=5nm. (b)HP=7nm. (c) HP=10nm. (d) HP=15nm.

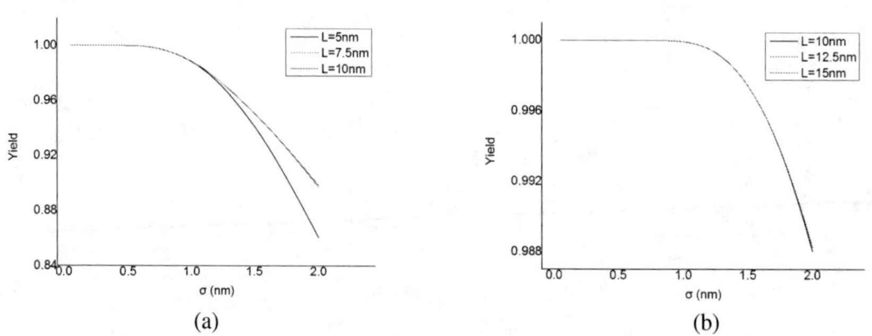

Fig. 12. The optimized stitching yield of type A1 vs. σ for varying L values. (a) HP=5nm. (b)HP=10nm.

We also investigate how the overhang design affects the stitching yield. In Fig. 13, we plot the stitching yield vs. overhang (normalized by half pitch, σ=1.5nm) and L with different μ and HP. Apparently, an optimal overhang is required to obtain the maximum yield (see Fig. 13(a)), reflecting a balance between the yield loss due to a large overhang touching the neighboring line and the yield loss due to a small overhang lowering the POS function in the presence of overlay inaccuracy. It is evident that the optimal edge position of overhang is not at the middle point of a gap, which is similar to the case of cut-hole landing on lines [4]. By identifying the optimal μ_o, the stitching yield can be improved significantly. For instance, about 10% and 6% yield improvement can be achieved at 5-nm and 10-nm HP, respectively (see Figs. 13(b) and 13(c)). Compared with the edge-placement yield of cut-hole landing (on lines) [4], the stitching yield in general is higher, especially at smalerl HP. This opens another door to continuous IC scaling through the "SAMP+stitching" approach once the 1D gridded design heavily relying on via/cut patterning meets prohibitive yield barriers [3, 4].

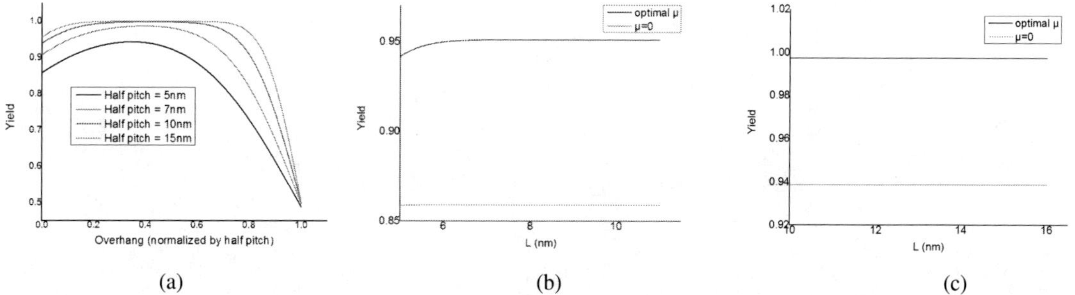

(a) (b) (c)

Fig. 13. (a)The stitching yield of type A1 vs. overhang (normalized by half pitch, σ=1.5nm), (b) the non-optimized and optimized stitching yields of type A1 vs. L for varying μ (σ=1.5nm, HP=5nm), (c) the non-optimized and optimized stitching yields of type A1 vs. L for varying μ (σ=1.5nm, HP=10nm).

From equations (12) and (13), we calculate and plot the stitching yield of type A2 vs. σ and CL for varying HP in Fig. 14. It shows that the stitching yield is sensitive to cut-hole CD (CL) variation, and Fig. 15 confirms that smaller CL helps to improve the stitching yield. As shown in Figs. 16-17, the stitching yield of type C is significantly worse than that of type A2 even their yield behavior is quite similar. Due to the processing challenges in fabricating sub-10-nm holes, some layout modification strategies to improve the double-L stitching yield will be discussed in the next section.

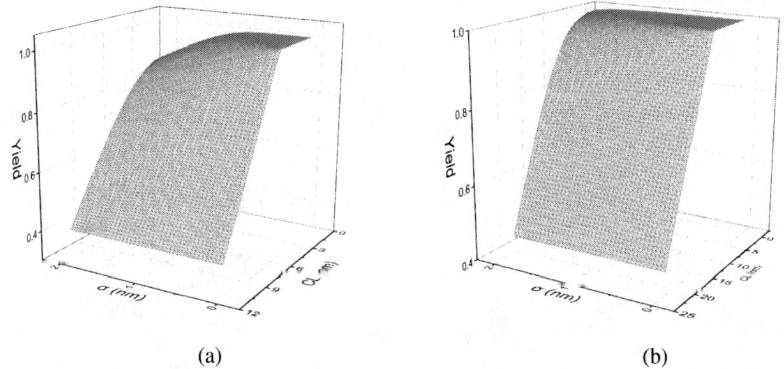

(a) (b)

Fig. 14. The stitching yield of type A2 vs. σ and CL. (a) HP=5nm. (b) HP=10nm.

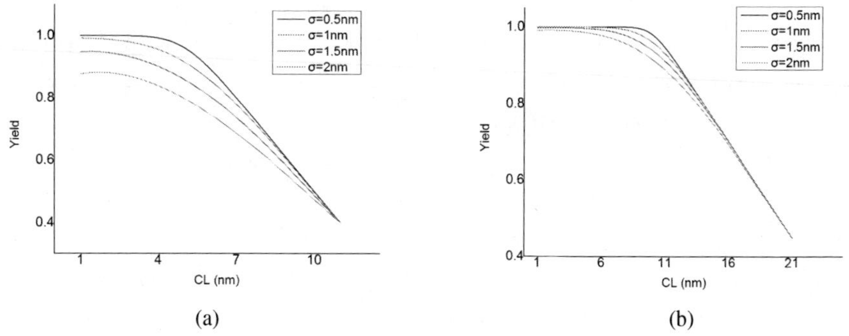

(a) (b)

Fig. 15. The stitching yield of type A2 vs. CL with varying σ. (a) HP=5nm. (b) HP=10nm.

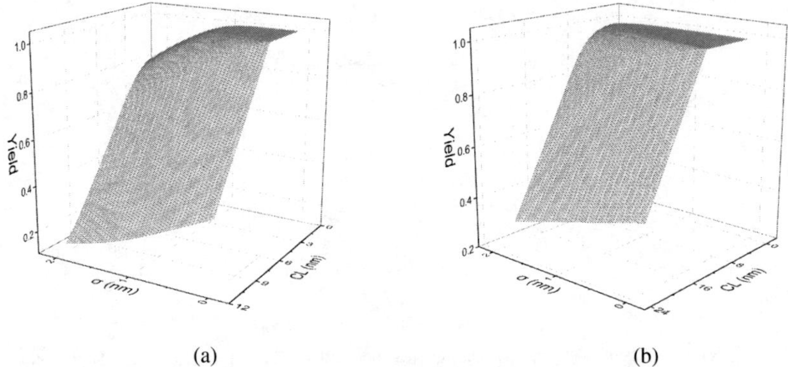

(a) (b)

Fig. 16. The stitching yield of type C vs. σ and CL. (a) HP=5nm. (b) HP=10nm.

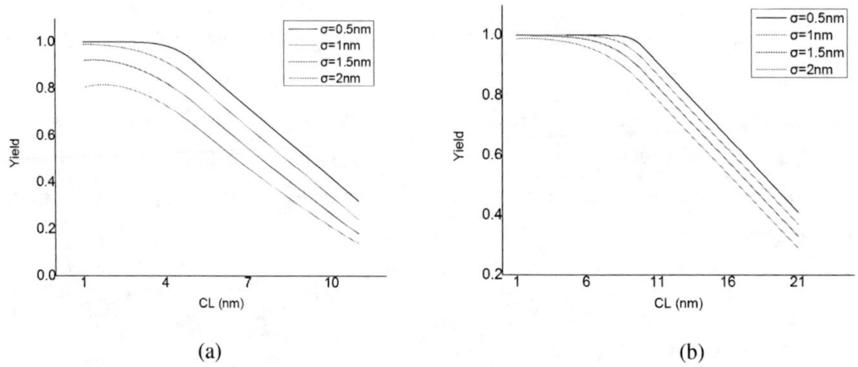

(a) (b)

Fig. 17. The stitching yield of type C vs. CL for varying σ. (a) HP=5nm. (b) HP=10nm.

4. LAYOUT MODIFICATION FOR YIELD IMPROVEMENT

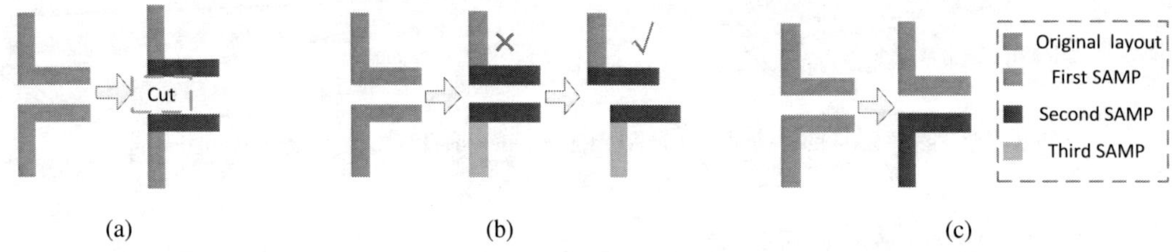

(a) (b) (c)

Fig. 18. Layout modification strategies. (a) Enlarging the cut-hole CD, (b) making a staggering double-L structure, which can be formed by introducing a spacer gap to accurately control the staggering distance or by simply adding another SAMP (third) process, (c) patterning each "L" feature with one single "SAMP+cut" process to avoid stitches, which requires the CD uniformity within each "L" structure.

There are some stitches (e.g., double-L type) with low stitching yield on a chip and they may be modified to improve the yield performance. Several layout modification strategies will be discussed in this section. As demonstrated in Fig. 18, several layout modification strategies can be considered to improve the stitching yield. The first method is to enlarge the tip-to-tip distance and consequently the cut-hole CD, as shown in Fig. 18(a). The second approach is to modify the layout design and use a staggering double-L structure, which can be formed by introducing a spacer gap to accurately control the staggering distance or by simply adding another SAMP (third SAMP) process. An alternative patterning method is to define each "L" feature with one single "SAMP+cut" process to avoid stitches, which requires the CD uniformity within each "L" structure and correspondingly certain modification of design rules and layout decomposition algorithm [3].

5. CONCLUSION

A single-stitching yield model is developed based on a stitch database built from open-cell layout library. By defining the probability-of-success (POS) function and incorporating overlay errors and cut-hole CD variations, a series of yield formulae for different types of stitches are constructed. The overhang value is shown to be the key design parameter which needs to be optimized to achieve the best yield performance. Our calculation results indicate that σ is a sensitive factor for all the examined HP values, while L only impacts the yield at smaller HP. For certain stitch structures, the stitching yield is expected to be significantly degraded due to the severe processing challenges to form extremely small cut holes. Several layout modification strategies are proposed to overcome these barriers and improve the stitching yield performance.

ACKNOWLEDGMENT

This research is supported by the Shenzhen City Fund for Fundamental Research and the start-up research fund from Shenzhen Graduate School of Peking University.

REFERENCES

[1] W. Kang, C. Feng, Y. Chen, "Mask strategy and layout decomposition for self-aligned quadruple patterning," *Proc. of SPIE*, Vol. 8684, 86840E, 2013.

[2] J. You, H. Liu, Y. Chen, "A layout decomposition algorithm for self-aligned multiple patterning," *Proc. of SPIE*, Vol. 8327, 83270N, 2014.

[3] H. Liu, J. Zhou, Y. Chen, "Breaking through 1-D layout limitations and regaining 2-D design freedom, part I: 2-D layout decomposition and stitching techniques for hybrid optical and self-aligned multiple patterning," *SPIE Advanced Lithography*, 2015.

[4] P. Zhang, C. Hong, Y. Chen. "A generalized edge-placement yield model for the cut-hole patterning process," *Proc. of SPIE*, Vol. 9052, 90521Q, 2014.

[5] C. H. Stapper, F. M. Armstrong, K. Saji, "Integrated circuit yield statistics," *Proceedings of the IEEE* 71(4), 453-470, 1983.

[6] Y. Ma, J. A. Torres, G. Fenger, Y. Granik, J. Ryckaert, G. Vanderberghe, J. Bekaert, J. Word, "Challenges and opportunities in applying grapho-epitaxy DSA lithography to metal cut and contact/via applications," *Proc. of SPIE*, Vol. 9231,92310T, 2014.

Automatic DFM methodology for Bit-Line Pattern Dummy

Mohamed Bahr
Mentor Consulting Division, Mentor Graphics, Egypt

ABSTRACT

This paper presents an automated DFM solution to generate Bit Line Pattern Dummy (BLPD) for memory chips. Dummy shapes are aligned with memory functional bits lines to ensure uniform and reliable memory device. This paper will present a smarter approach that uses an analysis based technique for adding the dummy fill shapes that have different types according to the space available. Experimental results based on layout of a memory test chip.

Keywords: DFM, bit-line, pattern dummy, physical verification

1. INTRODUCTION

Design for Manufacturability (DFM) has evolved from concept to mandatory practice in conjunction with Integrated Circuit (IC) dimension shrinkage. DFM is needed to achieve high manufacturing yield, overcoming problems that are related to the manufacturing process. In memory chips the functional bit-lines are characterized by complex interconnection patterns. The complex interconnection patterns are due to large number of functional bit-lines forming memory chip connections. This complexity produces width and density variations across memory bit-lines. Those variations in width and density increase the effect of process induced variations. Critical Dimension (CD) variation is a process induced variation which occurs due to different etching rates for memory bit lines. CD variation for interconnect metal levels neighboring bit lines in memory chips causes metal structure thinning and leaning. Dummy fill patterns are needed to compensate the density variations across the chips. The required patterns must be inserted with respect to memory functional bit-lines.

Dummy patterns can be defined as the additionally inserted patterns which do not play a part in the electrical operation of semiconductor device. Dummy patterns are usually floating shapes. Then main goals of these additional patterns is improving lithographic performance and controlling pattern density. Better line width roughness and process window area can be achieved by adopting dummy pattern insertion. Dummy patterns can play a role as to balance etch loads for manufactured feature.

This paper presents the problem affecting memory chips, it addresses the complexity of the interconnection patterns and the complexity and restrictions for the added dummy fill patterns. An automated DFM solution to generate Bit Line Pattern Dummy for memory chips is presented. Dummy shapes are aligned with memory functional bits to ensure uniform and reliable memory device. This paper will present a smarter approach that uses an analysis based technique for adding the dummy shapes that have different types according to the space available. This automated DFM solution uses Mentor Graphics® Standard Verification Rule Format and Calibre® YieldEnhancer(YE) toolset for its implementation. This flow was tested on various test chips with hundreds of different metal interconnects combinations and proven efficiency and reasonable run time. The solution is technology independent.

2. PROBLEM STATEMENT

Memory chips continue their drive toward ever increasing complexity. This is not only due to new design techniques, theorems and topologies; but also increasingly demanding requirements from IC fabricators. One of the most common yield detractors is interconnect metal lines collapsing due to uncontrolled CD variations. The complex interconnects patterns for memory chips introduce metal structures which are vulnerable to CD variations related failures such as metal thinning and metal leaning. Figure 1 shows an example for complex interconnection pattern. The metal structure thinning effect is a decrease in narrow metal structures width due to lack of supporting metal structures which causes different

Design-Process-Technology Co-optimization for Manufacturability IX, edited by John L. Sturtevant, Luigi Capodieci, Proc. of SPIE Vol. 9427, 942715 · © 2015 SPIE · CCC code: 0277-786X/15/$18 · doi: 10.1117/12.2185545

etching rates. The metal structure leaning effect is a displacement of metal structure due to wide space available on the metal structure side.

In traditional density constraints, dummy metal structures can be inserted with fixed dimensions to decrease density variations. An example for traditional density constraint solution is the box type dummy. In memory chips, parasitic capacitance and duty rules for Optical Proximity Correction (OPC) make the traditional solutions not applicable, even if it satisfies the density manufacturing constraint. The duty rules are the set of constraints between shapes depending on shapes width. The memory bit lines need a special type of dummy insertion that complies with density constraints and metal duty rules. The solution must add dummy shapes depending on design layout and interconnect metal shapes instead of fixed dummy shapes. The added dummy shapes added must have a bit line shapes, floating and extendable.

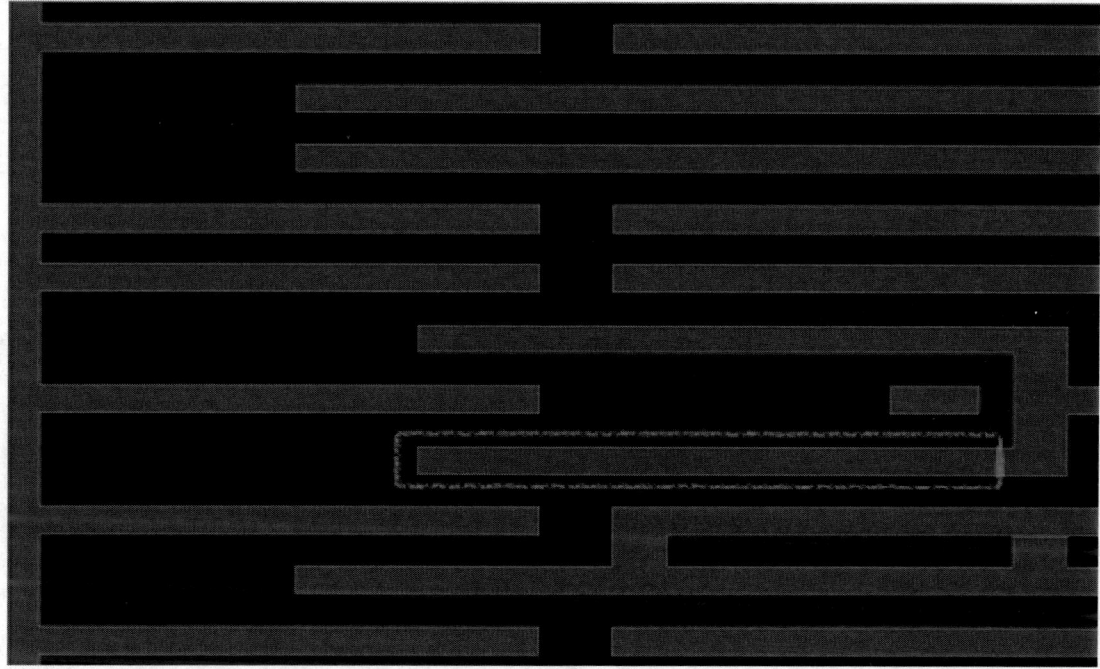

Figure 1: Memory bit lines- an example for a weak metal structure is highlighted

3. PROPOSED SOLUTION

A DFM solution for automatic pattern dummy insertion for memory bit lines was designed and implemented. The solution is based on Mentor Graphics Calibre® Standard Verification Rule Format (SVRF) language and YieldEnhancer (YE) toolset. The alignment of pattern dummy with main metal feature is the main anchor for the solution success. Figure 2 shows the basic solution steps.

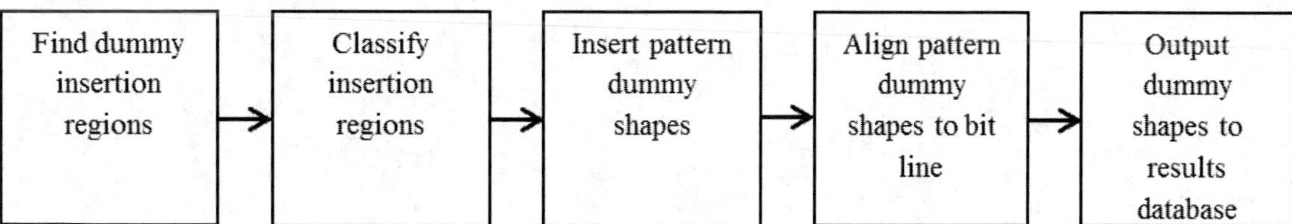

Figure 2: Proposed Solution

3.1 Find dummy insertion regions

The first step of the solution is to find the regions in the design layout that doesn't contain interconnect metal level that will undergo dummy insertion. A candidate region for insertion is a region with width more width cut-off(Wc) value shown in Equation 1 below. W_c is width cut off value, W_{min} is minimum metal width and S_{min} is minimum space between metal shapes.

$$W_c = W_{min} + 2 \times S_{min} \tag{1}$$

This steps requires Boolean operations using Calibre® Standard Verification Rule Format(SVRF) language. The output from this step is regions in the design layout that are candidates for pattern dummy insertion. Figures 3 and 4 shows a clip of design layout metal interconnect level and all available fill regions respectively.

3.2 Classify insertion regions

This step classifies the insertion regions from step one into two main types, narrow metal to narrow metal regions, narrow metal to wide metal regions.

Figure 3: A clip of design layout, only interconnect metal is shown

Figure 4: Available fill regions for the shown clip of design layout

3.3 Insert pattern dummy shapes

In this step pattern dummy shapes are added into the insertion regions according to the classification in step two. All dummy shapes are added with spacing equal to S_{min} previously defined in Equation 1. Figure 5 shows an example of added dummy fill patterns.

3.4 Align pattern dummy shapes to bit line

In this step pattern dummy shapes are aligned to the nearest bit line. Alignment is done on neighbor bit line tip width value. This step requires mathematical based shape modifications using Calibre® YieldEnhancer(YE) toolset. The alignment result is shown in Figure 6.

The added pattern dummy shapes don't violate metal duty rules as they are aligned with the original bit line; the added pattern dummy shares the same neighboring bit line with the bit line used for alignment, as the bit line used for alignment is following the metal duty rule therefore the pattern dummy added will not violate metal duty rule with neighboring bit lines. The different type pattern dummy shapes are added in ascending order, this guarantees that added dummy shapes in each type are aligned to the previous types added. Figure 7 shows an example of insertion order with alignment propagation to later types of pattern dummy.

3.5 Output dummy shapes to results database

The final step, pattern dummy shapes which were previously added and aligned are transferred into desired output database by the user.

4. EXPERIMENTAL RESULTS

4.1 Design Data

The developed solution successfully ran on several test chips for DRAM memory chips. Below is the result from running on a very dense large test chip. The solution is scalable on any number of CPU cores. The below results are based on a multi-threaded run on 8-CPU cores.

4.2 Results

Table 1 summarizes the results data for Bit Line Pattern Dummy solution. Comparison with manual work done before to generate pattern dummy shapes is included. In addition to improvement mentioned in the results below, manual work is always error prone.

5. CONCLUSION

The implemented solution is highly optimized in terms of run time with a very high improvement over manual methods. The solution is technology independent, providing higher yield and consistent results. The proposed solution can be extended to application on high density memory chips or logic devices.

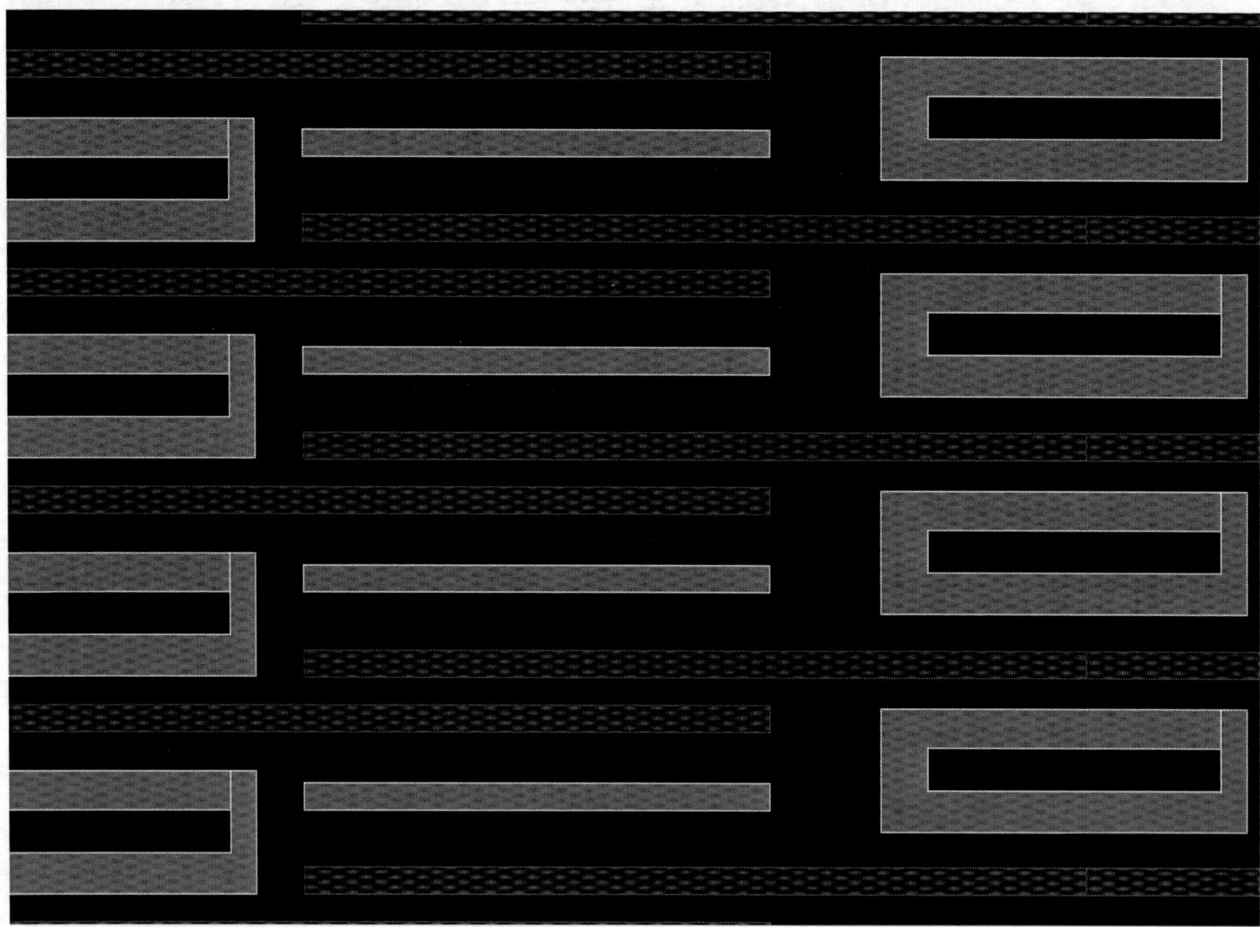

Figure 5: An example of added dummy fill patterns

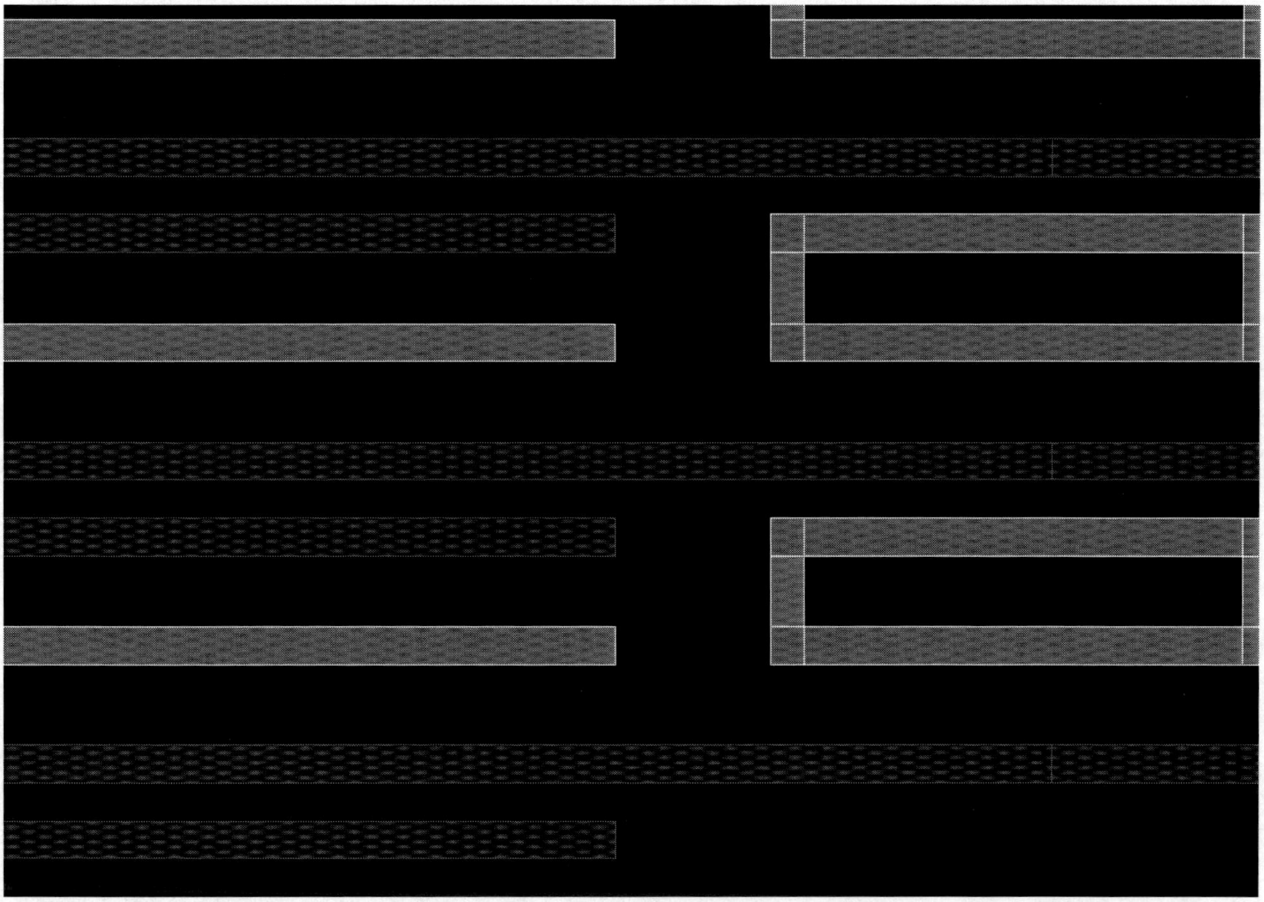

Figure 6: Inserted pattern dummy after alignment

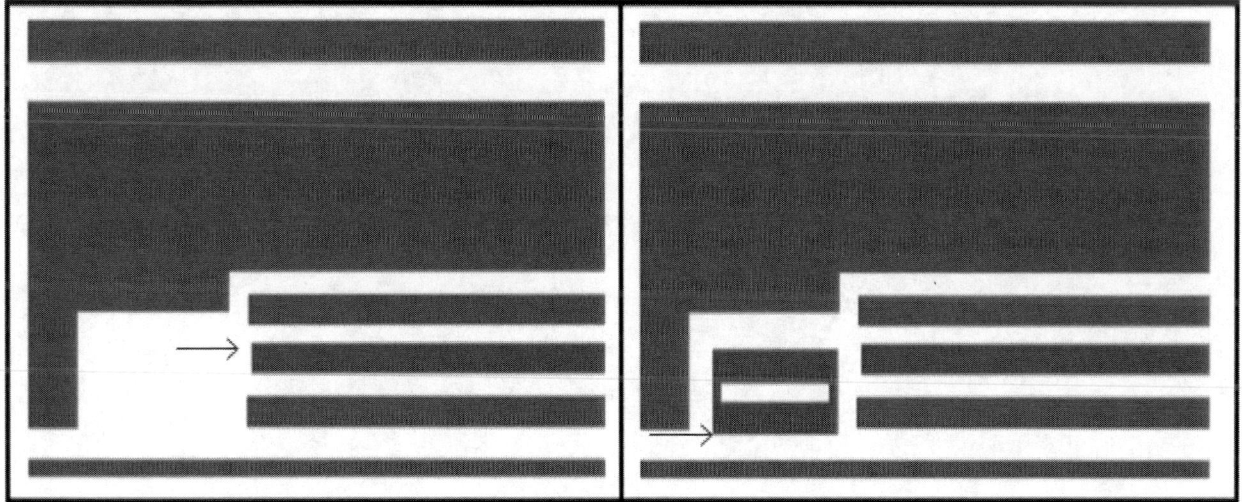

Figure 7: Left - First type of narrow to narrow metal pattern dummy is added, Right - Second type of narrow to narrow metal pattern dummy is added, aligned to bit line and 1s first pattern dummy already inserted

Table 1: Bit Line Pattern Dummy results data, comparison and improvement

Comparison Item	Manual	Automatic BLPD	Improvement
TAT	10 person * 7 days (33,600 minutes) estimated	1 person (6 minutes)	56,000x
Design Rule Check Errors	500 estimated	0	Error Free

REFERENCES

[1] Brian Lee et al., "Using smart dummy fill and selective reverse etch back for pattern density", Proc. CMP-MIC, pp. 255-258, Santa Clara, CA, March 2000.

[2] James Moon, "Self Assembled Dummy Patterns for Lithography Process Margin Enhancement", Proc. SPIE 6521, Design for Manufacturability through Design-Process Integration, 652118 (2007)

[3] Jongwon Jang, "Model Based Pattern Dummy Generation for Logic Devices", Proc. SPIE 9052, Optical Microlithography XXVII, 90521W (2014)

[4] Guoxiang Ning, "Reticle and Wafer CD Variation for Different Dummy Pattern", Proc. SPIE 8522, Photomask Technology 2012, 85222J (2012)

[5] Standard Verification Rule Format Manual, Mentor Graphics®

[6] Calibre® YieldAnalyzer and YieldEnhancer User's and Reference Manual, Mentor Graphics®